MOBILE COMMUNICATION SYSTEMS AND SECURITY

MOBILE COMMUNICATION SYSTEMS AND SECURITY

Man Young Rhee

Endowed Chair Professor
Kyung Hee University, Republic of Korea

IEEE PRESS

IEEE Communications Society, Sponsor

John Wiley & Sons (Asia) Pte Ltd

Other Wiley Editorial Offices

John Wiley & Sons, Ltd, The Atrium, Southern Gate, Chichester, West Sussex, PO19 8SQ, UK

John Wiley & Sons Inc., 111 River Street, Hoboken, NJ 07030, USA

Jossey-Bass, 989 Market Street, San Francisco, CA 94103-1741, USA

Wiley-VCH Verlag GmbH, Boschstr. 12, D-69469 Weinheim, Germany

John Wiley & Sons Australia Ltd, 42 McDougall Street, Milton, Queensland 4064, Australia

John Wiley & Sons Canada Ltd, 5353 Dundas Street West, Suite 400, Toronto, ONT, M9B 6H8, Canada

Wiley also publishes its books in a variety of electronic formats. Some content that appears in print may not be available in electronic books.

Library of Congress Cataloging-in-Publication Data

Rhee, Man Young.
  Mobile communication systems and security/Man Young Rhee.
    p. cm.
  Includes bibliographical references and index.
  ISBN 978-0-470-82336-1 (cloth)
  1. Mobile communication systems. 2. Wireless communication systems. 3. Mobile communication
systems–Security measures. 4. Wireless communication systems Security measures. I. Title.
  TK5103.2.R47 2009
  005.8–dc22   2008029106

ISBN 978-0-470-82336-1 (HB)

Typeset in 10/12pt Times by Thomson Digital, Noida, India.
Printed and bound in Singapore by Markono Print Media Pte Ltd, Singapore.
This book is printed on acid-free paper responsibly manufactured from sustainable forestry in which at least two trees are planted for each one used for paper production.

Contents

Preface

This book presents the evolution and migration of mobile radio technologies from first-generation (1G) to third-generation (3G) and beyond systems, with an emphasis on wireless security. 1G, or circuit-switched analog systems, consist of voice-only communication; 2G and beyond systems, comprising both voice and data communications, largely rely on packet-switched wireless mobile technologies.

The book covers the technological development of mobile radio communications in compliance with each iterative generation over the past decade. Currently, mobile data services have been rapidly transforming to facilitate and ultimately profit from the increased demand for non-voice services. Through aggressive 3G deployment plans, the world's major operators boast attractive and homogeneous portal offerings in all of their markets, notably in music and video multimedia services. Despite the improbability of any major changes in the next 4–5 years, rapid technological advances have already bolstered talks for 3.5G and even 4G systems. Orthogonal Frequency Division Multiplexing (OFDM) is a technique that allows high data rate transmission over extremely hostile channels at a comparably low complexity. For IMT 2000 3G systems and beyond, WCDMA/HSPA and CDMA2000 1x EV-DO/1x EV-DV based on MIMO-OFDM technologies will be key dynamics for high capacity broadband services. With wider radio channels greater than 10 MHz, OFDM-based technologies such as UMB, LTE, WiBro, and Mobile WiMax have emerged as viable options to deliver wider-bandwidth mobile broadband services. Thus IMT 2000 beyond systems, based on OFDM multiplexing technology, will allow for new entrants to compete with current incumbent mobile operators.

The book progresses in a systematic manner, covering mobile radio communications, with detailed presentation. The material in this book will present the theory and practice on wireless mobile technologies and its security problems through a rigorous and thorough in-depth presentation. The book can be used as a main text for graduate courses in the area of Wireless Networking and Security or Mobile Cellular Communications. It can also be used as suggested reading or a reference text for advanced undergraduates and postgraduates and is also suitable for practicing engineers in industry and research scientists in the laboratories. The book consists of 10 chapters; the following is a summary of the contents of each chapter.

Chapter 1 deals with the GSM technology, a Time Division Multiple Access (TDMA) radio access system, which is the most widely deployed digital cellular network. The 2G GSM technology is discussed by way of the following four parts: The essential entities that are required for call control and network management in GSM system architecture, the exchange of signaling messages pertaining to various functions of mobility, radio resource, and connection management and interface in GSM transmission network architectures, the call establishment, paging, call maintenance, and synchronization of the signaling channels on the air interface;

and three major security algorithms (A3, A5, and A8), developed by the Security Algorithm Group of Experts (SAGE) of GSM members, which describe the GSM security architecture.

Chapter 2 presents the CDMA air-interface standard as a 2G network. The 2G CDMA standard, based on the TIA/EIA IS-95, cdmaOne IS-95A provides an attractive technique for wireless access to broadband services based on spread spectrum communication. The 2G cdmaOne IS-95A system becomes the common platform on which cdmaOne IS-95B (2.5G), CDMA2000 1xEV-DO (3G), and 1xEV-DV (3.5G) are built. The topics covered provide a thorough analysis of both the Reverse CDMA Channel and Forward CDMA Channel of the cdmaOne system structure. The Reverse CDMA Channel is the reverse link from the mobile station to the base station; it is composed of access channels and reverse traffic channels. These code channels share the same CDMA frequency assignment using direct-sequence CDMA technologies. Each traffic channel is identified by the distinct user long code sequence, while each access channel is identified by a distinct access channel long code sequence. Thus, the long code uniquely identifies a mobile station on the reverse traffic channel and separates multiple access channel(s) on the Reverse CDMA Channel. All data transmitted on the Reverse CDMA Channel are added to the Frame Quality Indicator (CRC) and the encoder tail bits followed by convolutional encoding, block-interleaved, modulated by the 64-ary orthogonal modulation using Walsh index, and direct-sequence spread prior to transmission.

On the other hand, the Forward CDMA Channel is complied with one or more code channels that are transmitted on a CDMA frequency assignment using a particular pilot PN offset. The Forward CDMA Channel consists of the pilot channel, up to one sync channel, up to seven paging channels, and a number of forward traffic channels. Each of these code channels is orthogonally spread by the appropriate Walsh function to provide orthogonal channelization among all code channels on a given Forward CDMA Channel. Following the orthogonal spreading, each code channel is spread by a quadrature pair of pilot PN sequences at a fixed chiprate of 1.2288 Mcps. After spreading operation, the I and Q impulses are applied to the I and Q baseband filters. Following baseband filtering, the binary I and Q at the output of the quadrature spreading are mapped into specified phase transitions. The cdmaOne IS-95A will become the backbone to the CDMA2000 1x family of standards discussed in Chapters 7 and 8.

Chapter 3 discusses the General Packet Radio Services (GPRS), the packet-mode extension of GSM standardized by the ETSI (European Telecommunications Standards Institute); the first GPRS networks were launched in the spring of 2001 in Europe. GPRS (2.5G) is a new bearer service that is greatly improved and simplified for the wireless Internet access enabling user data packets to transfer between GSM mobile stations and external packet data networks. In fact, 2G data services based on circuit-switched radio transmission fail to fulfill the needs of users and providers because of slow data rates, long connection setup, and inefficient resource utilization for bursty traffic. GPRS packet transmission offers packet-switched bearer services for bursty Internet traffic, resulting in a more efficient utilization of traffic channels. This chapter will cover the GPRS system architecture that includes the data and signaling interfaces, the GSNs required to support GPRS functionality, the GPRS-PLMN backbone network, the mobile station registration with the network, the radio resource assignment, the mobility management for packet transfer and routing in GPRS system, and the layered protocol architecture providing for the data and control planes. The chapter also includes the ciphering algorithm for protecting the GPRS system.

Chapter 4 presents a clear concept of Third Generation Partnership Projects (3GPP and 3GPP2). 3GPP (3G Partnership Project) is a collaborative agreement and a crucial technical decision that was established by five standard organizations in December 1998. Those organizations decided to cooperate and combine their efforts for accelerating the joint project aimed for guaranteed global communications interoperability. 3GPP organizational partners, responsible for determining its general policy and strategy, include the Association of Radio Industries and Business (ARIB, Japan), TI Committee of the Telecommunications of the American National Standards Institute (ANSI TI, USA), European Telecommunications Standard Institute (ETSI, Europe), Telecommunications Technology Association (TTA, Korea), and Telecommunication Technology Committee (TTC, Japan). Soon after China Wireless Telecommunications Standard Group (CWTS, China) joined as the sixth partner. The original scope of 3GPP was to produce globally applicable Technical Specifications (TSs) and Technical Reports (TRs) for a 3G mobile system based on evolved GSM core networks and radio access technology on Wideband Code Division Multiple Access (WCDMA). Almost one year later ANSI decided to establish 3GPP2, a 3G partnership project to evolve ANSI/EIA-41 networks. 3GPP is responsible for developing 3G standards for GSM-based systems, while 3GPP2 is responsible for the development of IS-95-based CDMA systems. 3GPP was created from an IMT-2000 initiative of ITU. The prime purpose of 3GPP partners is to coordinate standards for technologies and applications that are interoperable at both the network and terminal levels. There are also 3GPP's market representation partnerships with organizations such as the GSM Association, UMTS Forum, Global Mobile Suppliers Association, IPv6 Forum, 3G Americas, and 3G.IP. Their role is to provide markets inputs and requirements for 3GPP standardization groups. Like 3GPP, 3GPP2 also has organizational partners and market representation partners. 3GPP2 is the international organization in charge of the standardization of CDMA 2000 1x systems, producing the fruitful results of combined research carried out by the TR 45.5 (North America) and TTA (Korea) standardization groups.

Chapter 5 introduces the pioneering work for UMTS standardization, which is applicable to 3GPP mobile systems. After the ETSI's initiatives, the 3GPP international organization virtually took over standardization work with the aim of joint adaptation of the WCDMA/FDD technology for worldwide coverage. UMTS standardization is based on the standardization of UTRA, a UMTS core network for easy migration from 2G to 3G for GSM operators, and the improvement of radio coverage for new 3G high-speed services. The 3GPP released several versions of the UMTS standard: Release 99, Release 4, Release 5, and Release 6. The UMTS system consists of a number of logical network elements (UE, UTRAN, and CN) that have their own functionalities. This chapter covers terrestrial interfaces based on the principle that protocol stacks and planes are logically independent of each other. Specifically, the UTRAN performs radio transmission and receptions; and the core network (CN) handles switching, routing, and service control. The I_u interface, which divides the system into UTRAN and CN, connects UTRAN and CN. There are two occasions: I_u CS protocol structure (I_u circuit-switched case connecting UTRAN to CS) and I_u PS protocol structure (I_u packet-switched case connecting UTRAN to PS). Finally, the chapter discusses the UMTS security related features. In Spring 2002, SAGE initiated the task of designing a new encryption algorithm for GSM, ECSD, GPRS, and EGPRS encryption. These new algorithms were intended to implement dual-mode handsets for operating with both GSM and UMTS modes. The 3GPP Task Force specified three encryption algorithms: the A5/3 algorithm for GSM and ECSD; the GEA3 algorithm for GPRS (including EGPRS); and the f8 algorithm for UMTS. The common aspect of all three encryption algorithms is KGCORE; the KGCORE function is based on the KASUMI block cipher specified for the GEA3 algorithm for GPRS and for the f8 algorithm for UMTS.

Chapter 6 describes HSDPA (High Speed Downlink Packet Access) features relating to Adaptive Modulation and Coding (AMC), Hybrid ARQ, Fast Cell Selection, Scheduling at the NodeB, and MIMO Antenna technologies. The operational techniques are primarily aimed to increase throughput, to reduce delay, and to achieve high peak rates. These functionalities rely on a new type of HS-DSCH transport channel to which the NodeB is terminated. The HSDPA functionality should be able to operate in an environment where certain cells are not updated with HSDPA functionality. This chapter also fully covers MAC architecture-UE side (details of MAC-d, MAC-c/sh, MAC-hs, and MAC-ehs) and MAC architecture-UTRAN side (MAC-c/sh, MAC-hs, MAC-ehs). OFDM technology applicable to the advanced mobile systems for the next generation is also included in this chapter.

Chapter 7 introduces the CDMA2000 1xEV-DO technology of 3GPP2. The layered interface model over the air interface consists of seven layers in which their respective protocols and functions are specified. Each layer in the reference model complies with one or more protocols that specify their functionality. Brief descriptions of each layer are included as shown below:

- The Application Layer provides multiple applications that provide the signaling and packet applications for transporting air-interface protocol messages.
- The Stream Layer provides the multiplexing of distinct application streams.
- The Session Layer provides address management, protocol negotiation, protocol configuration, and state maintenance services.
- The Connection Layer provides air link connection establishment and maintenance services.
- The Security Layer provides authentication and encryption Services.
- The MAC Layer defines the procedures used for receiving and transmitting over the physical layer.
- The Physical Layer provides structure, frequency, power output, modulation, and encoding specifications for the forward and reverse links.

More detailed and elaborated presentation arefound in this chapter with regard to CDMA2000 1x EV-DO technology. The CDMA2000 1x standard evolved from cdmaOne IS-95A/B networks as the first 3GPP2 technology to be commercially deployed. Compared to IS-95 networks, it provides double voice capacity and offers packet data speed of 153 Kbps (Release 0) and 307 Kbps (Release 1) in a single CDMA channel with a bandwidth of 1.25 MHz. 1x EV-DO is bi-mode (1x for voice and EV-DO for data). The first 1x EV-DO networks were launched by SK Telecom in January 2002 and KTF in May 2002 in South Korea. The 1x EV-DO system in 2 GHz band was brought into operation (by KDDI) in October 2004 in Japan.

Chapter 8 covers the CDMA2000 1x EV-DV technology. This 1x EV-DV standard provides integrated voice with simultaneous high speed packet data services such as video conferencing and other multimedia services at speeds up to 3.09 Mbps. The 1x EV-DV standard was approved by the 3GPP2 in June 2002 and submitted to ITU for approval, but at the time of writing wireless operators have not launched it for commercial purposes. The scope of 1x EV-DV extends from the mobile station CDMA operation to the base station CDMA operation. Particular emphasis is placed on the complete analysis of Reverse CDMA Channel (consists of the set of 13 code channels) and the Forward CDMA Channel (consists of the set of 20 code channels). This chapter also describes all component entities and interfaces (through the set of code channels) within the layers in the CDMA2000 1x EV-DV structure. Applications and upper layer protocols corresponding to OSI layers 3 to 7, use the services provided by

CDMA2000 Link Access Control (LAC) services. Those services may include signaling services, voice services, packet data applications, and circuit data applications. It seems that UMTS (WCDMA) and CDMA2000 are continuing to diverge both in IP core network and radio access. 3GPP has been developing HSDPA/HSUPA, while 3GPP2 has developed 1x EV-DV (Revision D), an evolved version of 1x RTT for voice and data. It has however found many similarities between HSDPA and 1x EV-DV and there will be an opportunity for harmonization after complete standardization. The ultimate goal should be to achieve minimal difference and a single access in near future. Harmonization should solve some of the issues that arose from the unsuccessful attempt to create a single common technology. Both HSDPA and 1x EV-DV should produce a single technology to enhance downlink packet data performances and to improve spectral efficiency for data services by using the following items: SDPDCH, high-order modulation (16-QAM, 64-QAM), AMC, HARQ retransmission schemes, fast scheduling for packet data, FCS (Fast Call Setup), and shorter frame size.

Chapter 9 presents cryptographic algorithms for providing the data security problem in the air-interface security layer of mobile communications. The topics in this chapter cover the Advanced Encryption Standard (AES), Public-key Cryptoalgorithms (PKCs), and Elliptic Curve Crypto-systems (ECCs). In 2001, the AES became an FIPS-approved symmetric block cipher (developed by Daemen and Rijman in 1999) capable of data blocks 128 bits long using the key size of 128, 192, and 256 bits. All commercial public key cryptosystems rely on the difficulty of discrete log problems and prime factoring techniques, which are difficult to compute when the modulus is very large. The Elliptic Curve discrete algorithm appears to be substantially more difficult and somewhat harder than the existing discrete logarithm problem. Comparing two systems (PKC and ECC), implementations show that ECC provides us with both increased speed and decreased key size for a given equal level of security. For each case, the addition or doubling of two points on the elliptic curve, the computation for the EC solutions over the prime field Z_p, as well as the finite Galios Field $GF(2^m)$, are shown through various examples. The Euler's criterion can be used to find a quadratic residue by applying Fermat's theorem. Two points (x, y_1) and (x, y_2) on the elliptic curve can be computed for the value of x. All practical public-key systems, such as Diffie-Hellman, RSA, ElGamal, Schnorr, DSA, and many other public key algorithms, can be implemented with elliptic curves over large finite fields. Many examples converting from traditional public-key protocols to the elliptic-curve protocols are demonstrated in this chapter.

Chapter 10 introduces several algorithms in order to compute the message digest by employing several hash functions. The hash functions to be dealt with include MD5 (1992), SHA-1 (1995), and HMAC (1996). The HMAC is a secret key hashed message authentication algorithm, providing both data integrity and data origin authentication for packets sent between two parties. The data expansion function, P-hash (secret, data), uses a single hash function to expand a secret and seed in an arbitrary quantity of output. The Pseudorandom Function (PRF) can be used to expand data in blocks for the purposes of key generation and validation. The PRF takes relatively small values such as a secret, a seed, and an identifying label as input and generates an output of arbitrary longer blocks of data. Using the techniques mentioned above, ample examples are given for the reader to better understand the theory and applications.

Some important and useful supplements are available on the book's companion website at the following URL: www.wiley.com/go/rhee.

Man Young Rhee
Seoul, Korea

Acknowledgement

The 3GPPs (3GPP and 3GPP2) and their Organization Partners have made every effort to develop wireless mobile technologies that provide globally applicable technical protocols and specifications. They have also issued numerous volumes of documents relating to crucial technical standards. The 3GPPs truly deserve to get the credit for their tremendous contributions in this specific area. Although some technical documents may still be subject to further approval processes and require subsequent revisions, we should pay a tribute of admiration for their tremendous efforts. The accomplishments of the 3GPPs have not only had a great influence on me, but also provided a great deal of inspiration for the writing of this book.

Man Young Rhee

About the Author

Man Young Rhee is an Endowed Chair Professor at Kyung Hee University and has over 45 years of research and teaching experience in the field of communication technologies, coding theory, cryptography, and information security. His career in academia includes professorships at Hanyang University (he also held the position of Vice President at this university), Virginia Tech, Seoul National University, and the University of Tokyo. Dr Rhee has held a number of high level positions in both government and corporate sectors: President of Samsung Semiconductor Communications; President of Korea Telecommunications Company; Chairman of the Korea Information Security Agency at the Ministry of Information and Communication; President of the Korea Institute of Information Security and Cryptology; and Vice President of the Agency for Defense Development at the Ministry of National Defense. He is a Member of the National Academy of Sciences, Senior Fellow at the Korea Academy of Science and Technology, and an Honorary Member of the National Academy of Engineering of Korea. His awards include the "Dongbaek" Order of National Service Merit and the "Mugunghwa" Order of National Service Merit, the highest grade honor for a scientist in Korea, NAS Prize, the National Academy of Sciences, NAEK Grand Prize, the National Academy of Engineering of Korea, and Information Security Grand Prize, KIISC. He has also published the following four books: *Coding, Cryptography, and Mobile Communications: Internet Security*, John Wiley, 2003; *CDMA Cellular Mobile Communications and Network Security*, Prentice Hall, 1998; *Cryptography and Secure Communications*, McGraw-Hill, 1994; and *Error Correcting Coding Theory*, McGraw-Hill, 1989. Dr Rhee has a B.S. in Electrical Engineering from Seoul National University, and an M.S. in Electrical Engineering and a Ph.D. from the University of Colorado.

Abbreviations

1G	First Generation
16QAM	16 Quadrature Amplitude Modulation
2G	2nd Generation Wireless Technologies
2.5G	2.5 Generation Wireless Technologies
3DES	Triple Data Encryption Standard
3G	3rd Generation Wireless Technologies
3GPP	3rd Generation Partnership Project
3GPP2	3G Partnership Project 2
4G	4th Generation
A-MIMO	Adaptive Multiple Input Multiple Output (Antenna)
AAA	Authentication, Authorization, and Accounting
AAL2	ATM Adaptation Layer Type 2
AAL5	ATM Adaptation Layer Type 5
AAS	Adaptive Antenna System (also Advanced Antenna System)
ACK	Acknowledgement
ACL	Access Control List
ACM	Adaptive Coding and Modulation
ACSS	Access Control SubSystem
ADC	Analog-to-Digital Conversion
ADSL	Asymmetric Digital Subscriber Line
AES	Advanced Encryption Standard
AGCH	Access Grant Channel (GSM)
AH	Authentication Header
AI	Air Interface
AKA	Authentication and Key Agreement
ALCAP	Access Link Control Application Part
AMC	Adaptive Modulation and Coding
AM	Acknowledged Mode
AMPS	Advanced Mobile Phone System
AMR	Adaptive Multiple Rate codec
AMS	Adaptive MIMO Switching
ANSI	American National Standards Institute
AP	Application Provider
AP	Access Point
AP	Access Preamble

AP-AICH	Access Preamble Acquisition Indication Channel
API	Access Preamble Indicator
APSK	Amplitude and Phase Shift Keying
ARIB	Association of Radio Industries and Business
ARPU	Average Revenue Per User
ARQ	Automatic Repeat Request
AS	Application Server
ASC	Access Service Class
ASIC	Application Specific Integrated Circuit
ASK	Amplitude-Shift Keying
ASN	Access Service Network
ASP	Application Service Provider
ATI	Access Terminal Identifier
ATM	Asynchronous Transfer Mode
AuC	(or AC) Authentication Center
BAIC	Barring of All Incoming Calls
BAN	BRAIN Access Network
BAOC	Barring of All Outgoing Calls
BAR	BRAIN Access Router
BC	Bearer Control
BCCH	Broadcast Control Channel
BCH	Broadcast Channel
BCMP	BRAIN Candidate Mobility Protocol
BCSM	Basic Call State Model
BE	Best Effort
BER	Bit Error Rate
BG	Border Gateway
BGAN	Broadband Global Area Network (Inmarsat)
BGCF	Breakout Gateway Control Function
BGP	Border Gateway Protocol
BGPv4	Border Gateway Protocol version 4
BIC	Barring of Incoming Calls
BIFS	Binary Format for Scenes
BLER	Block Error Rate
BMAC	Broadcast/Multicast Medium Access Control
BMC	Broadcast/Multicast Control
BMG	BRAIN Mobility Gateway
BM-IWF	Broadcast Multicast Interworking Function
BoD	Bandwidth on Demand
BOIC	Barring of Outgoing International Calls
BOIC-exHC	BOIC except those directed toward the home PLMN country
BPF	Bandpass Filter
BPSK	Binary Phase Shift Keying
BQB	Bluetooth Qualification Body
BRAIN	Broadband Radio Access For IP-based Networks
BRAN	Broadband Radio Access Network
BRENTA	BRAIN End Terminal Architecture

BREW	(Qualcomms) Binary Run-time Environment for Wireless
BS	Base Station
BS	Bearer Service
BSC	Base Station Controller
BSS	Base Station System
BSS	Base Station Subsystem
BSS	Base Station Subsystem (GSM)
BSSAP	BSS Application Part
BSSGP	Base Station System GPRS Protocol
BTS	Base Transceiver Station
BTSM	Base Transceiver System Management
BWA	Broadband Wireless Access
CA	Channel Assignment
CA	Collision Avoidance
CABAC	Context-based Adaptive Binary Arithmetic Coding
CAC	Connection Admission Control
CAI	Channel Assignment Indicator
CAMEL	Customized Application Mobile Enhanced Logic
CAP	Camel Application Part
CAR	Committed Access Rate
CAVLC	Context-adaptive Variable Length Coding
CBC	Cell Broadcast Center
CBR	Continuous Bit Rate
CBS	Cell Broadcast Service
CC	Call Control
CCC	CPCH Control Command
CCCH	Common Control Channel (GSM)
CCH	Control Channel
CCI	Co-Channel Interference
CCITT	International Telegraph and Telephone Consultative Committee
CCK	Complementary Code Keying
CCM	Counter with Cipher-block chaining Message authentication code
CCPCH	Common Control Physical Channel
CCS	Content Creation Subsystem
CCTrCH	Coded Composite Transport Channel
CD	Collision Detection
CD/CA-ICH	Collision Detection/Channel Assignment Indication Channel
CDI	Collision Detection Indicator
CDK	Complementary Code Keying
CDMA	Code Division Multiple Access
CDP	Content Delivery Platform
CDPD	Cellular Digital Packet Data
CDR	Call Detail Record
CEPT	Conference of European Post and Telecommunications
CERT	Computer Emergency Response Team
CF	(B/NRy/NRc/U) Call Forwarding (Busy/No Reply/Not Reachable/ Unconditional)

CFN	Connection Frame Number
CGALIES	Coordination Group on Access to Location Information by Emergency Services
CGF	Charging Gateway Function
CGI	Cell Global Identity
CGI RTT	Cell Global Identity Round Trip Time
CGI TA	Cell Global Identity Timing Advance
CGSN	Combined GPRS Support Node (SGSN GGSN)
CINR	Carrier to Interference Noise Ratio
CLIP	Calling Line Identification Presentation
CLIR	Calling Line Identification Restriction
CM	Connection Management
CMAC	block Cipher-based Message Authentication Code
CMOS	Complementary Metal Oxide Semiconductor
CN	Core Network
COS	Class of Service
CP	Control Plane
CPCH	Common Packet Channel
CPCId	Common Physical Channel Identifier
CPE	Customer Premise Equipment
CPICH	Common Pilot Channel
CPL	Call Processing Language
CQI	Channel Quality Indicator
CRC	Cyclic Redundancy Check
CRCI	CRC Indicator
CRNC	Controlling Radio Network Controller
CS	Circuit Switched
CS/CCA	Carrier Sense/Clear Channel Assessment
CSCF	Call Session Control Function
CSD	Circuit Switched Data
CSICH	Common Packet Channel Status Indication Channel
CSMA	Carrier Sense Multiple Access
CSMA/CA	Carrier Sense Multiple Access with Collision Avoidance
CSN	Connectivity Service Network
CSTD	Cyclic Shift Transmit Diversity
CTC	Convolutional Turbo Code
CTCId	Common Transport Channel Identifier
CTP	Card Telephony Protocol
CTP	Common Transport Protocol
CTS	Clear To Send
CUG	Closed User Group
CW	Call Waiting
CWTS	China Wireless Telecommunication Standard Group (China)
D8PSK	Differential 8 Phase-Shift Keying
DAB	Digital Audio Broadcasting
DAC	Digital to Analog Converter
DAMA	Demand Assigned Multiple Access

D-AMPS	Digital Advanced Mobile Phone System
DARPA	(US) Defense Advanced Research Project Agency
DCA	Dynamic Channel Allocation
DCCH	Dedicated Control Channel (GSM)
DCF	Distributed Coordination Function
DCH	Dedicated Transport Channel or Dedicated Channel
DCS	Digital Cellular System
DCS	Dynamic Channel Selection
DDoS	Distributed Denial of Service
DECT	Digital Enhanced Cordless Telecommunications
DES	Data Encryption Standard
DFS	Dynamic Frequency Selection
DGNA	Dynamic Group Number Assignment (TETRA)
DHCP	Dynamic Host Configuration Protocol
DL	Downlink (Forward Link)
DLCI	Data Link Connection Identifier
DMAP	DECT Multimedia Access Profile
DMO	Direct Mode Operation (TETRA)
DN	Directory Number
DoD	(US) Department Of Defense
DoS	Denial Of Service
DPCCH	Dedicated Physical Control Channel
DPCH	Dedicated Physical Channel
DPDCH	Dedicated Physical Data Channel
DPRS	DECT Packet Radio Service
DQPSK	Differential Quadrature Phase-Shift Keying
DRA	Dynamic Rate and Coding Adaptation
DRM	Digital Rights Management
DRNC	Drift Radio Network Controller
DRNS	Drift Radio Network Subsystem
DS-CDMA	Direct-Sequence Code Division Multiple Access
DSCH	Downlink Shared Channel
DSDV	Destination-Sequenced Distance-vector
DSL	Digital Subscriber Line
DSLAM	Digital Subscriber Loop Access Multiplexer
DSMA-CD	Digital Sense Multiple Access (Collision Detection
DSM-CC	Digital Storage Media Command and Control
DSSS	Direct Sequence Spread Spectrum
DTAP	Direct Transfer Application Process
DTX	Discontinuous Transmission
DVB	Digital Video Broadcasting
DwPCH	Downlink Pilot Channel
DwPTS	Downlink Pilot Time Slot
E2ENP	End-to-end Negotiation Protocol
EAP	Extensible Authentication Protocol
ECC	Electronic Communications Committee
ECC	Elliptic Curve Cryptography (or Cryptosystem)

ECDSA	Elliptive Curve Digital Signature Algorithm
EDCF	Enhanced Distributed Coordination Function
EDGE	Enhanced Data GSM Environment or Enhanced Data rates for Global Evolution
EHF	Extremely High Frequency
EHS	Enhanced Hotline Service
EIR	Equipment Identity Register
EIR	Equipment Identity Register (GSM)
EIRP	Effective Isotropic Radiated Power
EMS	Enhanced Message Service
EP	Elementary Procedure
ESN	Electronic Serial Number
ESP	Encapsulating Security Protocol
ETSI	European Telecommunications Standard Institute
EU	European Union
EWC	Enhanced Wireless Consortium
F-ACKCH	Forward Acknowledgement Channel
F-APICH	Forward Dedicated Auxiliary Pilot Channel
F-ATDPICH	Forward Auxiliary Transmit Diversity Pilot Channel
F-BCCH	Forward Broadcast Control Channel
f-btch	Forward Broadcast Traffic Channel
F-CACH	Forward Common Assignment Channel
F-CCCH	Forward Common Control Channel
F-CPCCH	Forward Common Power Control Channel
F-DCCH	Forward Dedicated Control Channel
F-DPCH	Fractional Dedicated Physical Channel
F-FCH	Forward Fundamental Channel
F-GCH	Forward Grant Channel
F-PCH	Forward Paging Channel
F-PDCCH	Forward Packet Data Control Channel
F-PDCH	Forward Packet Data Channel
F-PICH	Forward Pilot Channel
F-QPCH	Forward Quick Paging Channel
F-RCCH	Forward Rate Control Channel
F-SCCH	Forward Supplemental Code Channel
F-SCH	Forward Supplemental Channel
F-TDPICH	Forward Transmit Diversity Pilot Channel
FACCH	Fast Associated Control Channel (GSM)
FACH	Forward Access Channel
FAP	Fair Access Policy
FBSS	Fast Base Station Switching
FCC	Federal Communications Commission
FCCH	Frequency Correction Channel (downlink)
FCH	Frame Control Header
FCP	Flow Control Protocol
FCS	Frame Check Sequence

FDD	Frequency Division Duplex (UMTS)
FDM	Frequency Division Multiplexing
FDMA	Frequency Division Multiple Access
FE-ID	Functional Equipment Identifier
FEC	Forward Error Correction
FEMA	Federal Emergency Management Agency
FER	Frame Error Rate
FFS	For Further Study
FFT	Fast Fourier Transform
FH	Frequency Hopping
FHSS	Frequency Hopping Spread Spectrum
FIPS	Federal Information Processing Standard
FLSS	Forward Link SubSystem
F-SYNCH	Forward Sync Channel
F-SYNG	Forward Sync Channel
FT	Frame Type
FTC	Forward Traffic Channel
FTP	File Transfer Protocol
FTTH	Fiber To The Home
Ga	GGSN-CGF interface
GAP	Generic Access Profile
GEA3	GPRS Encryption Algorithm 3
GERAN	GSM EDGE Radio Access Network
GFSK	Gaussian Frequency Shift Keying
GGSN	Gateway GPRS Support Node
GHz	Gigahertz (thousands of MHz)
Gi	GGSN-PDN (public) interface
GIS	Geographical Information System
GLC	Gateway Location Center
GloMo	Global Mobile Information Systems
GMLC	Gateway Mobile Location Center
GMLC/GMPC	Gateway Mobile Location/Positioning Center (LBS)
GMM	GPRS Mobility Management
GMSC	Gateway Mobile Switching Center
GMSS	Gateway Management SubSystem
Gn	GGSN-SGSN (private) interface
GPRS	General Packet Radio Service
GPS	Global Positioning System
GSIM	GSM Service Identity Module
GSM	Global System for Mobile Communications
GT	Global Title
GTP	GPRS Tunneling Protocol
GTT	Global Title Translation
GUI	Graphical User Interface
GUP	Generic User Profile
HARQ	Hybrid Automatic Repeat Request

HHO	Hard Hand-Off
HIA	Host Interface for Administration
HiSWAN	High Speed Wireless Access Network
HLR	Home Location Register (GSM)
HLR-PS	HLR-Provisioning Server
HMAC	Hash Message Authentication Code
HO	Hand-Off or Hand Over
HPA	High Power Amplifier
HPLMN	Home Public Land Mobile Network
HS-DPCCH	High Speed Dedicated Physical Control Channel (uplink)
HS-DSCH	High Speed Downlink Shared Channel
HS-PDSCH	High Speed Physical Downlink Shared Channel
HS-SCCH	High Speed Shared Control Channel for HS-DSCH
HS-SICH	HSDPA Shared Information Channel
HSCSD	High Speed Circuit Switched Data
HSDPA	High Speed Downlink Packet Access
HSPA	High Speed Packet Access
HSS	Home Subscriber Server
I&A	Identification and Authentication
ICH	Indicator Channel
ICI	Inter Carrier Interface
IDC	International Data Corporation
IDEA	International Data Encryption Algorithm
IDL	Interface Definition Language
IEC	International Electrotechnical Commission
IEEE	Institute of Electrical and Electronics Engineers
IETF	Internet Engineering Task Force
IF	Intermediate Frequency
IFFT	Inverse Fast Fourier Transform
IFL	Inter-Facility Link
IFS	Inter-frame Space
IKE	Internet Key Exchange
IMEI	International Mobile Equipment Identity
IMSI	International Mobile Subscriber Identity
IMT-2000	International Mobile Telecommunications 2000
IN	Intelligent Network
INA	Internet Name and Address Management
INAP	IN Application Part
InP	Indium Phosphide
IOT	Interoperability Testing
IP	Internet Protocol
IP2W	IP To Wireless (interface)
IP-CAN	IP-Connectivity Access Network
IPCP	IP Configuration Protocol
IPDR	IP Detail Record
IPsec	Internet Protocol Security
IPv4	Internet Protocol version 4

IPv6	Internet Protocol version 6
IPX	Internet Packet Exchange
IR	Incremental Redundancy
IRDB	International Roaming Database
ISDN	Integrated Services Digital Network
ISDR	International Strategy for Disaster Reduction
ISI	Intersymbol Interference
ISO	International Standards Organization
ITU	International Telecommunications Union
IV	Initialization Vector
IVR	Interactive Voice Response
IWF	Interworking Function
Kbps	Kilobits per second
KG	Key Generator
KHz	Kilohertz
KSG	Key Stream Generator
L1	Layer 1 (physical Layer)
L2	Layer 2 (data Link Layer)
L2CAP	Logical Link Control and Adaptation Protocol
L3	Layer 3 (network Layer)
LAC	Link Access Control
LAC	Local Area Code
LAN	Local Area Network
LAPDm	Modified Link Access Protocol for the D channel
LBC	Low Bit-rate Coding
LBS	Location-Based Service
LDPC	Low-Density Parity-Check (coding)
LFSR	Linear Feedback Shift Register
LLC	Logical Link Control
LMU	Location Measurement Unit
LNB	Low Noise Block
LNS	L2TP Network Server
LOCUS	Location of Cellular Users for Emergency Services
LoS	Line-of-sight
LPD	Link Protocol Discriminator
LTE	Long Term Evolution
LTOA	Latest Time Of Arrival
LUP	Location Update Protocol
M3UA	SS7 MTP3-User Adaptation Layer
MAC	Medium Access Control layer
MAC-hs	Medium Access Control (high speed RL Radio Link
MAI	Multiple Access Interference
MAP	Media Access Protocol
MAP	Mobile Application Part
MBMS	Multimedia Broadcast/Multicast
Mbps	Megabits per second
MCC	Mobile Country Code

MCD	Multi-Carrier Demodulator
Mcps	Mega chip per second
MCS	Modulation and Coding Scheme
MDDA	Mobile Directory Dial Assistance
MDHO	Macro Diversity Hand Over
MDS	Multimedia Distribution Subsystem
MExE	Mobile Execution Environment
MHz	Megahertz (million Hertz)
MIA	Mobile Internet Access
MIMO	Multiple Input Multiple Output
MIP	Mobile IP
MLC	Mobile Location Center (LBS)
MM	Mobility Management
MMAC	Multimedia Mobile Access Communications
MMM	MultiMedia Messaging
MMR	MIND Mobile Router
MMS	Mobile Multimedia Messaging Service; Mobility Management Sublayer (GSM)
MMS	MOGIS Management System; Mobility Management Sublayer (GSM)
MMSC	Multimedia Messaging Service Center
MN	Mobile Node
MNC	Mobile Network Code
MNP	Mobile Number Portability
ModCod	Modulation and Coding
MPC	Mobile Positioning Center
MPCC	Multiparty Call Control
MPLS	Multi-Protocol Label Switching
MPS	Mobile Positioning Service (LBS)
MPTY	Multiparty Calling
MS	Mobile Station
MSC	Mobile services Switching Center (GSM/UMTS)
MSIB	Mobile Status Indicator Bit
MSISDN	Mobile Subscriber ISDN Number
MSNS	Multimedia Streaming Notification Service
MSO	Multi-Services Operator
MSP	Mobile Subscriber Provisioning
MSU	Message Storage Unit
MT	Mobile Terminal
MTBF	Mean Time Between Failure
MTI	Moving Target Indicator
MTP	Message Transport Part.
NACK	Not Acknowledge
NAI	Network Access Identifier
NAP	Network Access Provider
NAS	Network Access Server
NAT	Network Address Translation
NBAP	Node B Application Part

NFS	Number Field Sieve
NIC	Network Interface Controller
NIST	National Institute of Standards and Technology
NLoS	Non-Line-of-Sight
NMT	Nordic Mobile Telephone
nrtPS	Non-Real-Time Polling Service
NSA	National Security Agency
NSGS	Network Side GSM Server
NSS	Network and Switching Subsystem
NTIA	National Telecommunications and Information Administration
O&M	Operation And Maintenance
OAM	Operations, Administration & Maintenance
OFDM	Orthogonal Frequency Division Multiplexing
OFDMA	Orthogonal Frequency Division Multiple Access
OFM	Output Feedback Mode
OGC	Open Geospatial Consortium
OMA	Open Mobile Alliance
OMC	Network Operation and Maintenance Center
OSI	Open Systems Interconnection
OSS	Operational Support Systems (or Operating SubSystems)
OTA	Over The Air
OTAP	Over-The-Air Protocol
OTP	One-Time Password
OVSP	Orthogonal Variable Spreading Factor (Codes)
P2P	Peer-to-Peer
PAM	Pulse Amplitude Modulation
PAMR	Public Access Mobile Radio
PC	Power Control
PCCPCH	Primary Common Control Physical Channel
PCF	Point Coordination Function
PCH	Paging Channel (downlink)
PCM	Personnel Call Manager
PCM	Pulse Code Modulation
PCPCH	Physical Common Packet Channel
PCPICH	Primary Common Pilot Channel
PCS	Personal Communications Services
PCU	Packet Control Unit
PDCH	Packet Data Channels
PDCP	Packet Data Coverage Protocol
PDN	Packet Data Network
PDP	Packet Data Protocol for example, IP
PDSCH	Physical Dedicated Shared Channel
PDU	Protocol Data Unit
PER	Packet Error Rate
PhCH	Physical Channel
PHY	Physical Layer
PICH	Page Indication Channel

PIN	Personal Identification Number
PKI	Public Key Infrastructure
PKM	Public Key Management
PLL	Physical Link Layer
PLMN	Public Land Mobile Network
PMM	Packet Mobility Management
PMR	Professional Mobile Radio
PN	Pseudo-random Noise
PPM	Pulse Position Modulation
PPTP	Point-to-Point Tunneling Protocol
PRACH	Physical Random Access Channel
PRNET	Packet Radio Network
PS	Packet Switched
PSAP	Public Safety Access Point
PSI	Public Service Identity
PSK	Phase Shift Keying
PSPDN	Packet Switched Public Data Network
PSTN	Public Switched Telephone Network
PWA	Personal Wireless Assistant P-WLAN Public WLAN
QAM	Quadrature Amplitude Modulation
QOF	Quasi-Orthogonal Function
QoS	Quality of Service
QPSK	Quadrature Phase Shift Keying
R-ACH	Reverse Access Channel
R-ACKCH	Reverse Acknowledgement Channel
R-CCCH	Reverse Common Control Channel
R-CQICH	Reverse Channel Quality Indicator Channel
R-DCCH	Reverse Dedicated Control Channel
R-EACH	Reverse Enhanced Access Channel
R-FCH	Reverse Fundamental Channel
R-PDCCH	Reverse Packet Data Control Channel
R-PDCH	Reverse Packet Data Channel
R-PICH	Reverse Pilot Channel
R-REQCH	Reverse Request Channel
R-SCCH	Reverse Supplemental Code Channel
R-SCH	Reverse Supplemental Channel
RAB	Radio Access Bearer
RACH	Random Access Channel (uplink)
RAN	Radio Access Network
RANAP	Radio Access Network Application Part
RATI	Random Access Terminal Identifier
RAU	Routing Area Update (GPRS)
RF	Radio Frequency
RFID	Radio Frequency Identification
RFS	Radio Frequency Subsystem
RG	Relative Grant
RL	Radio Link

RLC	Radio Link Control
RLP	Radio Link Protocol
RNC	Radio Network Controller (UMTS)
RNS	Radio Network Subsystem
RNSAP	Radio Network Subsystem Application Part
RNTI	Radio Network Temporary Identity
RR	Radio Resource
RRC	Radio Resource Control
RRI	Reverse Rate Indicator
RRM	Radio Resource Management
RS	Reed Solomon (coding)
RSA	Rivest-Shamir-Adelman
RSC	Recursive Systematic Convolutional Coder
RSCom	Radio Spectrum Committee
RSCP	Received Signal Code Power
RSN	Robust Security Networks
RSPG	Radio Spectrum Policy Group
RTC	Reverse Traffic Channel
RTS	Request To Send
RTSP	Real Time Streaming Protocol
RTT	Round-trip Time
RX	Receive
SAAL-NNI	Signaling ATM Adaptation Layer for Network-to-Network Interface
SACCH	Slow Associated Control Channel (GSM)
SAGE	Security Algorithms Group of Experts
SAPI	Service Access Point Identifier
SAS	Stand-Alone SMLC
SB3G	Systems Beyond 3G
SCCP	Signaling Connection Control Part
SCCPCH	Secondary Common Control Physical Channel
SCF	Service Control Function
SCH	Synchronization Channel
SCI	Synchrorized Capsule Indicator
SCP	Service Control Point
SCPC	Single Channel Per Carrier
SCPICH	Secondary Common Pilot Channel
SCS	Service Capability Server
SCTP	Simple Control Transmission Protocol
SDCCH	Stand-alone Dedicated Control Channel (GSM)
SDPDCH	Stand-alone Dedicated Packet Data Channel
SDU	Service Data Unit
SeNTRE	Security Network for Technological Research in Europe
SF	Spreading Factor
SFN	Single Frequency Network
SFN	System Frame Number
SGF	Signaling Gateway Function
SGSN	Serving GPRS Support Node

SHA	Secure Hash Algorithm
SI	Status Indicator
SIG	SS7 Interworking Gateway
SIM	Subscriber Identity Module
SIMO	Single Input Multiple Output
SINR	Signal to Interference Noise Ratio
SIP	Session Initiation Protocol
SIR	Signal to Interference Ratio
SLF	Subscription Locator Function
SLP	Signaling Link Protocol
SM	Spatial Multiplexing
SME	Small and Medium Enterprise
SMLC	Serving Mobile Location Center
SMS	Short Message Services
SN	Subscriber Number
SNDCP	Subnetwork Dependent Convergence Protocol
SNIR	Signal to Noise Interference Ratio
SNP	Signaling Network Protocol
SNR	Signal to Noise Ratio
SOFDMA	Scalable Orthogonal Frequency Division Multiple Access
SOM	Start–Of–Message
SPC	Signaling Point Code
SRBF	Signaling Radio Burst Control
SRES	Signed Response
SRNC	Serving Radio Network Controller
SRNS	Serving Radio Network Subsystem
SRP	Service Request Processor
SS	Subscriber Station
SS7	Signaling System 7
SSC	Secondary Synchronization Code
SSN	Sub-System Number
ST	Satellite Terminal
STB	Set-Top Box
STC	Signaling Transport Converter
SSCF	Service Specific Coordination Function
SSCOP	Service Specific Connection Oriented Protocol
STC	Space Time Coding
SURAN	Survivable Adaptive Radio Network
TCH	Traffic Channel (GSM)
TCP	Transport Control Protocol
TD	Transmit Diversity
TDD	Time Division Duplex (UMTS)
TDM	Time Division Multiplexing
TDMA	Time Division Multiple Access
TE	Terminal Equipment
TEK	Traffic Encryption Key
TF	Transport Format

TFC	Transport Format Combination
TFCI	Transport Format Combination Indicator
TFI	Temporary Flow Identity
TFI	Transport Format Indicator
TFRI	Transport Format and Resource Indicator
TFRC	Transport Format Resource Combination
TFSS	Time and Frequency SubSystem
TFT	Traffic Flow Template
TFTP	Trivial File Transfer Protocol
TGI	Task Group I
TI	Transaction Identifier
TID	Tunnel Identifier
TINA-C	Telecommunication Information Networking Architecture Consortium
TISS	Terrestrial Interface SubSystem
TKIP	Temporal Key Integrity Protocol
TLLI	Temporary Link Layer Identifier
TLS	Transport Layer Security protocol
TMSI	Temporary Mobile Subscriber Identifier
TNL	Transport Network Layer
ToA	Time Of Arrival
ToAWE	Time Of Arrival Window Endpoint
ToAWS	Time Of Arrival Window Startpoint
ToS	Type of Service
TPC	Transmission Power Control
Tr	Transport segment (requirements)
TrCH	Transport Channel
TrGW	Translation Gateway
TSGS	Terminal Side GSM Server
TSTD	Time Switched Transmit Diversity
TTA	Telecommunications Technology Association (Korea)
TTC	Telecommunications Technology Committee (Korea)
TTG	Transmit/receive Transition Gap
TTI	Transmission Time Interval
TX	Transmit
UAM	User Authorization Manager program
UAProf	User Agent Profile
UATI	Unicast Access Terminal Identifier
UC-ID	UTRAN Cell Identifier
UDDI	Universal Description, Discovery and Integration
UDP	User Datagram Protocol
UE	User Equipment
UGS	Unsolicited Grant Service
UL	Uplink (Reverse Link)
UM	Unacknowledged Mode
UMB	Ultra Mobile Broadband
UMS	Unified Messaging System
UMTS	Universal Mobile Telecommunications System

UP	User Plane
UpPCH	Uplink Pilot Channel
UpPTS	Uplink Pilot Time Slot
URA	UTRAN Registration Area
URL	Uniform Resource Locator
USCH	Uplink Shared Channel
USIM	Universal Service [Subscriber] Identity Module
UT	User Terminal (BGAN)
UTC	Coordinated Universal Time
UTRA	UMTS Terrestrial Radio Access
UTRAN	Universal Terrestrial Radio Access Network
UWC	Universal Wireless Communications
VBR	Variable Bit Rate
VHE	Virtual Home Environment
VIE	Visual Information Engineering
VLR	Visitors Location Register
VMS	Voice Messaging System
VoD	Video on Demand
VoIP	Voice over IP
VPLMN	Visited Public Land Mobile Network
VPN	Virtual Private Network
VRL	Virtual Radio Link
VSAT	Very Small Aperture Terminal
VSF	Variable Spreading Factor
VSM	Vertical Spatial Multiplexing
W3C	World Wide Web Consortium
WAP	Wireless Application Protocol
WCDMA	Wide-band Code Division Multiple Access
WDM	Wavelength Division Multiplexing
WEP	Wired Equivalent Privacy
WEP2	Wired Equivalent Privacy 2
WG	Wireless Gateway
WG-1000	Wireless Gateway 1000
WiBro	Wireless Broadband (Service)
WiFi	Wireless Fidelity (IEEE 802.11b wireless networking)
WIG	Wireless Interworking Group
WiMAX	Worldwide Interoperability for Microwave Access
WPA	Wireless Protected Access
WRAN	Wireless Radio Access Network
WRC	World Radio Conference
WS	Wireless Stream
WSB	Wireless Services Broker
WSDL	Web Service Description Language
WTP	Wireless Transaction Protocol
WWAN	Wireless Wide Area Network
WWW	World Wide Web
ZRP	Zone Routing Protocol

1

Global System for Mobile Communications

The GSM standard (Global System for Mobile Communications) for mobile telephony was introduced in the mid-1980s and is the European initiative for creating a new cellular radio interface. The GSM system uses a TDMA radio access system employed in 135 countries, operating in 200 KHz channels with eight users per channel. It is the most widely deployed digital network in the world today, used by 10.5 million people in more than 200 countries.

1.1 GSM Bandwidth Allocation

GSM can operate four distinct frequency bands:

- *GSM 450:* GSM 450 supports very large cells in the 450 MHz band. It was designed for countries with a low user density such as in Africa. It may also replace the original 1981 NMT 450 (Nordic Mobile Telephone) analog networks used in the 450 MHz band. NMT is a first-generation wireless technology.
- *GSM 900:* When speaking of GSM, the original GSM system was called GSM 900 because the original frequency band was represented by 900 MHz. To provide additional capacity and to enable higher subscriber densities, two other systems were added afterward:
- *GSM 1800:* GSM 1800 (or DCS 1800) is an adapted version of GSM 900 operating in the 1800 MHz frequency range. Any GSM system operating in a higher frequency band requires a large number of base stations than for an original GSM system. The availability of a wider band of spectrum and a reduction in cell size will enable GSM 1800 to handle more subscribers than GSM 900. The smaller cells, in fact, give improved indoor coverage and low power requirements.
- *GSM 1900 (or PCS 1900):* PCS 1900 (Personal Communications System) is a GSM 1800 variation designed for use on the North American Continent, which uses the 1900 MHz band. Since 1993, phase 2 of the specifications has included both the GSM 900 and DCS 1800 (Digital Cellular System) in common documents. The GSM 1900 system has been added to

Mobile Communication Systems and Security   Man Young Rhee
© 2009 John Wiley & Sons (Asia) Pte Ltd

the IS-136 D-AMPS (Digital Advanced Mobile Phone System) and IS-95 Code Division Multiple Access (CDMA) system, both operated at the 1900 MHz band.

The ITU (International Telecommunication Union) has allocated the GSM radio spectrum with the following bands:

- GSM 900: Uplink: 890–915 MHz
            Downlink: 935–960 MHz

- GSM 1800: Uplink: 1710–1785 MHz
             Downlink: 1805–1880 MHz

- GSM 1900: Uplink: 1850–1910 MHz
             Downlink: 1930–1990 MHz

In the above, uplink designates connection from the mobile station to the base station and downlink denotes connection from the base station to the mobile station.

1.2 GSM System Architecture

A cell containing a Mobile Station (MS) is formed by the radio coverage area of a Base Transceiver Station (BTS). Several BTSs together are controlled by one Base Station Controller (BSC). The BTS and BSC form the Base Station Subsystem (BSS). The combined traffic of the MSs in their respective cells is routed through the Mobile Switching Center (MSC). Several databases are required for call control and network management: the Home Location Register (HLR), the Visitor Location Register (VLR), the Authentication Center (AuC), and the Equipment Identity Register (EIR). The GSM system architecture comprised with a set of essential components is illustrated in Figure 1.1.

The GSM system network can be divided into three subgroups that are interconnected using standardized interfaces: Mobile Station (MS), Base Station Subsystem (BSS), and Network SubSystem (NSS). These subgroups are further comprised of the components in the following sections.

1.2.1 Mobile Station (SIM + ME)

The Mobile Station (MS) can refer to a handset or mobile equipment (ME). The Subscriber Identity Module (SIM) card in a GSM handset is a microprocessor smart card that securely stores various critical information such as the subscriber's identity as well as the authentication and encryption algorithms responsible for providing legitimate access to the GSM network. Each SIM card has a unique identification number called the International Mobile Subscriber Identity (IMSI). In addition, each MS is assigned to a unique hardware identification called theInternational Mobile Equipment Identity (IMEI). An MS can also be a terminal (M-ES) that acts as a GSM interface, that is, for a laptop computer.

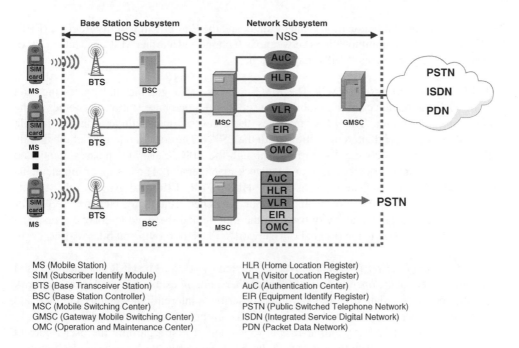

MS (Mobile Station)   HLR (Home Location Register)
SIM (Subscriber Identify Module)   VLR (Visitor Location Register)
BTS (Base Transceiver Station)   AuC (Authentication Center)
BSC (Base Station Controller)   EIR (Equipment Identify Register)
MSC (Mobile Switching Center)   PSTN (Public Switched Telephone Network)
GMSC (Gateway Mobile Switching Center)   ISDN (Integrated Service Digital Network)
OMC (Operation and Maintenance Center)   PDN (Packet Data Network)

Figure 1.1   GSM system architecture

1.2.2 Base Station Subsystem (BSS)

The Base Station Subsystem (BSS) consists of the Base Transceiver Station (BTS) and the Base Station Controller (BSC). The BSS ensures transmission and management of radio resources.

- *Base Transceiver Station (BTS)*: The BTS is responsible for providing the wireless connection between the handset and the wireless network. The GSM uses a series of radio transceivers called BTSs that provide the points of entry to the GSM network. A BTS is comprised of a set of radio transmitters and receivers, and antennas to connect the mobile to a cellular network for pursuing the required call handling tasks. The BTS takes in the calls within its coverage zone and ensures their proper handling.
- *Base Station Controller (BSC)*: The primary function of the BSC is call maintenance. As shown in Figure 1.1, the BSC manages the routing of communications from one or more base stations. A BSC controls a cluster of cell towers. It is responsible for setting up a voice or data call with the mobile terminal and managing handoff when the phone moves from one cell tower boundary to another, without disruption of service. In other words, the BSC manages radio resources and ensures the handover; that is, the passing of a subscriber from one cell to another with no degradation of the quality of the communication. The BSC also serves as the switch for concentration towards the Gateway Mobile Switching Center (GMSC).

1.2.3 Network SubSystem (NSS)

The Network SubSystem (NSS) is made up of the two essential elements, MSC and GMSC, along with its supporting elements: the Home Location Register (HLR), the Visitor Location

Register (VLR), the Authentication Center (AuC), and the Equipment Identity Register (EIR). The NSS establishes communications between a cell phone and another MSC, and takes care of the Short Message Services (SMS) transmission.

- *Mobile Switching Center (MSC)*: The MSC controls call signaling and processing, and coordinates the handover of the mobile connection from one base station to another as the mobile roams around. The MSC manages the roles of inter-cellular transfer, mobile subscriber visitors, and interconnections with the PSTN. The combined traffic of the mobile stations in their respective cells is routed through the MSC. Several databases mentioned above are available for call control and network management. Those supporting elements include the location registers consisting of HLR, VLR, EIR, and AuC. Each MSC is connected through GMSC to the local Public Switched Telephony Network (PSTN or ISDN) to provide the connectivity between the mobile and the fixed telephone users. The MSC may also connect to the Packet Data Networks (PDN) to provide mobiles with access to data services.
- *Home Location Register (HLR)*: The NSS is assisted by HLRs. The HLR is a database used for management of the operator's mobile subscribers. For all users registered with a network operator, permanent data (the user's profile, subscriber's international identity number, and telephone number) and temporary data (the user's current location) are stored in the HLR. In the case of a call to a user, the HLR is always queried first regarding the user's current location. The main information, stored in the HLR, concerns the location of each mobile station in order to route calls to the mobile subscribers managed by each HLR.
- *Visitor Location Register (VLR)*: The VLR is responsible for a group of location areas, and stores the data of those users who are currently in its area of responsibility. This may include the permanent user data that have been transmitted from the HLR to VLR for faster access. But the VLR may also assign and store local data such as a temporary identification. Concerning subscriber mobility, the VLR comes into play by verifying the characteristics of the subscriber and ensuring the transfer of location information. The VLR contains the current location of the MS and selected administrative information from the HLR. It is necessary for call control and provision of the services for each mobile currently located in the zones controlled by the VLR. A VLR is connected to one MSC and normally integrated into the MSC's hardware.
- *Authentication Center (AuC)*: The AuC holds a copy of the 128-bit secret key that is stored in each subscriber's SIM card. These security-related keys are used for authentication and encryption over the radio channel. Figure 1.5 on page 11 illustrates the GSM authentication scheme.
- *Equipment Identification Register (EIR)*: The GSM distinguishes explicitly between the user and the equipment, and deals with them separately. The EIR registers equipment data rather than subscriber data. It is a database that contains a list of all valid mobile station equipments within the GSM network, where each mobile station is identified by its International Mobile Equipment Identity (IMEI). Thus, the IMEI uniquely identifies a mobile station internationally. The IMEI (a kind of serial number) is allocated by the equipment manufacturer and registered by the network operator who stores it in the EIR. The International Mobile Subscriber Identity (IMSI) identifies uniquely each registered user and is stored in the SIM. A mobile station can only be operated if a SIM with a valid IMSI is inserted into equipment with a valid IMEI.

1.2.4 Operating SubSystem (OSS)

The Operating SubSystem (OSS) constitutes the network Operation and Maintenance Center (OMC) as the operator's network management tool.

- *NetworkOperation and Maintenance Center (OMC)*: The OMC is a management system, which oversees the GSM functional blocks. The OMC assists the network operator in maintaining satisfactory operation of the GSM network. The OMC is responsible for controlling and maintaining the MSC, BSC, and BTS.

1.3 GSM Transmission Network Architecture

The GSM transmission network architecture is depicted in Figure 1.2. It is also called the GSM protocol architecture used for the exchange of signaling messages pertaining to various functions of mobility, radio resource, and connection management and interface. The protocol layering consists of the physical layer (Layer 1), the data link layer (Layer 2), and the message management layer (Layer 3). Brief explanations for these items are described in the following sections.

1.3.1 Message Management Layer (Layer 3)

The GSM Layer 3 protocols are used for the communication of radio resource, mobility, code format, and call-related connection management between the various network entities

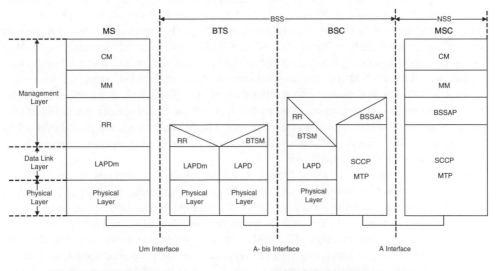

CM : Connection management
MM : Mobility management
RR : Radio resource
LAPD : Link access protocol for the D channel

BTSM : BTS management
SCCP : Signaling connection control part
MTP : Message transfer part
BSSAP : BSS application part

Figure 1.2 GSM transmission network architecture

involved. The Layer 3 protocol is made up of three sublayers called the radio resource (RR) (implemented over the link between the MS and the BSS), the mobility management (MM) (connecting between the MS and MSC), and the connection management (CM) (exchanging information with the peer) for providing the communications between the MS and MSC. Layer 3 also implements the message transport part (MTP) and the signaling connection control part (SCCP) of the CCITT SS7, on the link between the BSC and MSC (the A interface), to provide the transport and addressing functions for signaling messages belonging to the various calls routed through the MSC.

- *Radio Resource (RR) Management Sublayer:* The RR management sublayer terminates at the BSS and performs the functions of establishing physical connections over the radio for the purpose of transmitting call-related signaling information such as the establishment of signaling and traffic channels between a specific mobile user and the BSS. The RR management functions are basically implemented in the BSS. The roll of the RR management sublayer is to establish and release stable connection between MSs and an MSC for the duration of a call, and to maintain it despite user movements. RR messages are mapped to the BSS Application Part (BSSAP) in the BSC. The BTS Management (BTSM) is used to transfer all OAM-related information to the BTS. The Message Transfer Part (MTP) and the Signal Connection Control Part (SCCP) are used to support the transfer of signaling messages between the MSC and the BSS. The SCCP is used to provide a referencing mechanism to identify a particular transaction relating to a particular call. The SCCP can also be used to enhance message routing, operation, and maintenance information. The MTP is used between the BSS and the MSC. The MTP provides a mechanism for reliable transfer of signaling messages. The BSSAP provides the channel switching and aerial function, and performs the RR management and the interworking functions between the data link protocols used on the radio and the BSS-MSC side for transporting signaling-related messages.
- *Mobility Management (MM) Sublayer:* The MM sublayer is terminated at the MSC and the relayed messages from or to the mobile station (MS) are relayed transparently in the BSS using the direct transfer application process (DTAP). These are procedures used to establish, maintain, and release a MM connection between the MS and the MSC, over which an entity of the CM sublayer can change information with its peer. The mobility management (MM) handles the control functions relating to the registration of subscriber location, paging, authentication, handover, and channel allocation. The MM messages are not interpreted by the BTS or the BSC. They are transferred over the *A-bis* interface as transparent messages and over the *A interface* using the DTAP.
- *Connection Management (CM) Sublayer:* The CM sublayer terminates at the MSC and contains entities that consist of the call control (CC), the short message services (SMS), and the call-independent supplementary services (SS). An MM connection is initiated by a CM service request message, which identifies the requesting CM entity, and the type of service required of the MM connection. The MM connections provide services to the different entities of the upper CM sublayer. Once a MM connection has been established, the CM can use it for information transfer, using the DTAP process. The DTAP process is used for the transparent transfer of MM/CM signaling messages between the MS and the MSC. The DTAP function provides the transport level protocol inter-working function for transferring Layer 3 signaling messages from and to the MS and to and from the MSC without any analysis of the message contents.

Bit	8	7	6	5	4	3	2	1
	Unused	LPD			SAPI		C/R	EA

LPD   : Link Protocol Discriminator   C/R : Command or Response Frame
SAPI : Service Access Point Identifier   EA   : Extended Address

Figure 1.3   LAPDm format for representing the address field

1.3.2 Data Link Layer (Layer 2)

The data link layer over the radio link, connecting the MS to the BSS, is based on a LAPDm or a modified LAPD (Link Access Protocol for the D channel). LAPDm uses no flags for frame delimitation; instead frame delimitation is done by the physical layer that defines the transmission frame boundaries.

LAPDm uses a length indicator field to distinguish the information carrying field used to fill the transmission frame. LAPDm uses an address field in its frame format to carry the service access point identifier (SAPI, 3 bits). When using command/response frames, the SAPI identifies the user for which a command frame is intended, and the user transmitting a response frame. Figure 1.3 shows the format for the address field. The link protocol discriminator (LPD, 2 bits) is used to specify the use of LAPDm. The C/R (1 bit) specifies a command or response frame as used in LAPD. An extended address (EA, 1 bit) is used to extend the address field to more than one octet. The 1 bit unused field is reserved for future uses.

LAPDm uses a control field to carry the sequence number, and to specify the type of frame as used in LAPD. LAPDm specifies three types of frames for supervisory functions, unnumbered information transfer, and control functions as defined in LAPD. LAPDm uses no cyclic redundancy check bits for error detection. Error correction and detection schemes are provided by a combination of block and convolutional coding used in conjunction with bit interleaving in the physical layer. A frame format of LAPDm is shown in Figure 1.4. Signaling transport between the data link layer (Layer 2) on the radio side and the SS7 on BSS-MSC link is provided by a distribution data unit within the information field of SCCP. There are also parameters known as the discrimination parameter and the data link connection identifier (DLCI) parameter. The discrimination parameter (size: 1 octet) uses a single bit to address a message either to the DTAP or the BSSAP processes. The DLCI parameter (sized one octet) is made up of two subparameters that identify the radio channel type and the SAPI value in the LAPDm protocol used for the message on the radio link.

1.3.3 Physical Layer (Layer 1)

The physical layer on the radio link is based on a TDMA structure that is implemented on multiple access schemes. The multiple access scheme used in the GSM is a combination of

Address Field	Control Field	Length Indicator	Information Field	EA

Figure 1.4   Frame format of LAPDm

FDMA and TDMA. The Conference of European Post and Telecommunications (CEPT) has made available for use two frequency bands for the GSM system: 890–915 MHz for the uplink from the mobile station to the base station, and 935–960 MHz for the downlink from the base station to the mobile terminal.

Two types of channels are considered as traffic channels and control channels. The traffic channels are intended to carry encoded voice or user data, whereas the control channels are intended to carry signaling and synchronization data between the base station and the mobile station. The CCITT SS7 MTP and SCCP protocols are used to implement both the data link and the Layer 3 transport functions for carrying the call control and mobility management (MM) signaling messages on the BSS-MSC link. The MM and CM sublayer signaling information from the mobile station is routed over signaling channels (such as the broadcast control channel (BCCH), the slow associated control channel (SACCH), and the fast associated control channel (FACCH)) to the BSS from where they are transparently relayed through the DTAP process to an SCCP on the BSS-MSC link for transmission to the peer call control (CC) entity in the MSC for processing. Alternatively, any call signaling information initiated by the MSC on the SCCP connection is relayed through the DTAP process in the BSS to the assigned signaling channel, using the LAPDm data link protocol, for delivery to the mobile station.

1.4 Signaling Channels on the Air Interface

The signaling channels are used for call establishment, paging, call maintenance, and synchronization.

1.4.1 Broadcast Channels (BCHs)

The broadcast channels (downlink only) are mainly responsible for synchronization, frequency correction, and broadcast control.

- *Synchronization Channel (SCH)*: The SCH for downlink only will provide the MS with all the information (frame synchronization of the mobiles and identification of the BS) needed to synchronize with a BTS.
- *Frequency Correction Channel (FCCH)*: The FCCH for downlink only provides correction of MS frequencies and transmission of the frequency standard to the MS. It is also used for synchronization of an acquisition by providing the boundaries between timeslots and the position of the first time slot of a TDMA fame.
- *Broadcast Control Channel (BCCH)*: The BCCH broadcasts to all mobiles general information regarding its own cell as well as the neighboring cells, that is, local area code (LAC), network operator, access parameters, and so on. The MS receives signals via the BCCH from many BTSs within the same network and/or different networks.

1.4.2 Common Control Channels (CCCHs)

The common control channels are used in both downlinks and uplinks between the MS and the BTS. These channels are used to convey information from the network to MSs and provide access to the network.

The CCCHs include the following channels:

- *Access Grant Channel (AGCH)*: The AGCH is used on downlinks only for assignment of a dedicated channel (DCH) after a successful random access. The BTS allocates a traffic channel (TCH) or the stand-alone dedicated control channel (SDCCH) to the MS, thus allowing the MS access to the network.
- *Random Access Channel (RACH)*: The RACH is used in uplink only for random access attempts by the mobiles. It allows the MS to request an SDCCH in response to a page or due to a call.
- *Paging Channel (PCH)*: The PCH is used in downlink only for paging to mobiles. The MS is informed by the BTS for incoming calls via the PCH. Specifically, the paging message for mobiles is sent via the BSSAP to the BSS as a connectionless message through the SCCP/MTP. A single paging message transmitted to the BSS may contain a list of cells in which the page is to be broadcast. The paging messages received from the MSC are stored in the BTS and corresponding paging messages are transmitted over the radio interface at the appropriate time. Each paging message relates to one mobile station only and the BSS has to pack the pages into the relevant paging message. When a paging message is broadcast over the radio channel, if a response message is received from the mobile, the relevant signaling connection is set up towards the MSC and the page response message is passed to the MSC.

1.4.3 Dedicated Control Channel (DCCH)

The dedicated control channels are used on both downlinks and uplinks. The DCCHs are responsible for roaming, handovers, encryption, and so on. The DCCHs include the following channels:

- *Standalone Dedicated Control Channel (SDCCH)*: The SDCCH is a communication channel between the mobile (MS) and the BTS. This channel is used for the transfer of call control signaling to and from the mobile during call setup.
  - *Slow Associated Control Channel (SACCH)*: The 26 multiframe is used to define traffic channels (TCH) and their slow and fast associated control channels (SACCH and FACCH) that carry link control information between the mobile and the base stations. The TCH has been defined to provide the following different forms of services: the full-rate speech or data channels supporting effective bit-rates of 13 Kbps (for speech), 2.4, 4.8, and 9.6 Kbps (for data); and the half-rate channels with effective bit-rates of 6.5 Kbps (for speech), 1.2, 2.4, and 4.8 Kbps for data. Like the TCHs, the SDCCH has its own slow associated control channel (SACCH) and is released once call setup is complete.
  - *Fast Associated Control Channel (FACCH)*: The FACCH is obtained on demand by stealing frames from TCH and is used when a very fast exchange of information is needed such as a handover. The FACCH is used by either end for signaling the transfer characteristics of the physical path, or for other purposes such as connection handover control messages. The stealing of a TCH slot for FACCH signaling is indicated through a flag within the TCH slot.

1.5 GSM Security Architecture

GSM security architecture can be implied as a symmetric key cipher system. GSM uses three major security algorithms (A3, A5, and A8) developed by the GSM Memorandum of Understanding (MoU) member countries. Unfortunately, GSM members did not follow an open review process. The public academic review of the security algorithms is a crucial consideration for the long-term viability of a cryptographic system. In fact, the Security Algorithm Group of Experts (SAGE) of GSM members developed security algorithms privately. Although SAGE, comprised of leading cryptographers, developed the GSM crypto-graphic algorithms (A3, A5, and A8) successfully, the GSM members did not benefit from global review or analysis of the security architecture.

The three major algorithms (developed by SAGE) are as follows:

- *A3 Algorithm:* The first important step in the GSM security architecture is authentication. It must definitely provide that a user and handset are authorized to use the GSM network. The A3 algorithm is used to authenticate a handset to a GSM network.
- *A5 Algorithm:* The A5 algorithms are used to encrypt voice and data after a successful authentication. A5/0 is the dummy cipher definition that the communication is unprotected. The A5/1 is the original algorithm primarily used in Western Europe, whereas the A5/2 is a weaker version of the A5/1 algorithm used in other parts of the world. A5/3 is the newer algorithm introduced for 3GPP mobile communications which is based on KGCORE function.
- *A8 Algorithm:* The A8 algorithm is used to generate symmetric encryption keys to use in A5/1 or A5/2.

The initial security architecture was developed in the early 1990s. At that time, the keys of 64-bit size seemed reasonable. Unfortunately, rapid improvements in computing power and cryptographic analysis now make 64-bit keys increasingly more susceptible to brute force attacks. In 1998, researchers at UC Berkeley uncovered the specific weakness in the A3/A8 algorithms. Although the A3 authentication algorithm claimed to be 64-bit keys, the last 10 bits of the keys have been left blank, resulting in an actual key size of only 54 bits. Because 56-bit keys such as the Data Encryption Standard (DES) have been proven susceptible, the 54-bit A3 keys could be successfully attacked with brute force attacks, or differential cryptanalysis against DES (S-boxes), which was more efficient than brute force (introduced by Biham and Shamir in 1990).

1.5.1 GSM Authentication

The VLR is responsible for call control and provides the services for each mobile located in the zones controlled by the VLR. The VLR tracks the user and allocates the calls so that the network knows where to route the call when a call is placed to a roaming user. When a call is placed from a mobile phone, the GSM network's VLR authenticates the individual subscriber's phone. The VLR immediately communicates with the HLR, which in turn retrieves the subscriber's information from the AuC.

Because the authentication key (128 bits in length), K_i, is the most essential component in the authentication process, it must not be transmitted over the air without protection, to avoid being

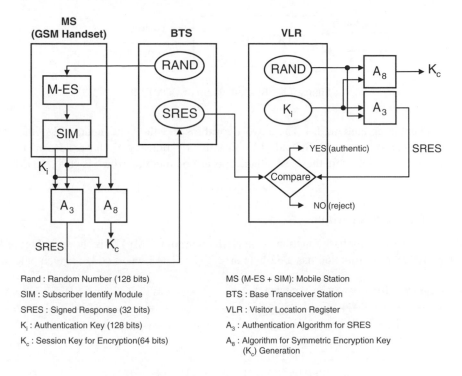

Rand : Random Number (128 bits)
SIM : Subscriber Identify Module
SRES : Signed Response (32 bits)
K_i : Authentication Key (128 bits)
K_c : Session Key for Encryption(64 bits)

MS (M-ES + SIM): Mobile Station
BTS : Base Transceiver Station
VLR : Visitor Location Register
A_3 : Authentication Algorithm for SRES
A_8 : Algorithm for Symmetric Encryption Key
        (K_c) Generation

Figure 1.5   GSM authentication scheme

susceptible to interception. Therefore, the K_i is recommended to be placed only in the tamper-proved SIM card in the MS and in the AuC, VLR, and HLR databases in the MSC. Figure 1.5 shows GSM authentication scheme (A3) for signed response (SRES) and key generation algorithm (A8) for generation of the symmetric encryption key.

The subscriber's information retrieved from the AuC is forwarded to the VLR and the following process commences:

- The BTS generates a 128-bit random challenge value (called RAND) and transmits it to the GSM handset phone.
- The handset encrypts a RAND using the A3 algorithm with the authentication key K_i. This encryption is resulted in a 32-bit signed response (an SRES).
- At the same time, the VLR can easily calculate the SRES because the VLR possesses a RAND, the K_i, and a copy of the A3 algorithm.
- The handset transmits the SRES to a BTS and it is forwarded to a VLR.
- The VLR checks the SRES value from the handset against the SRES calculated by the VLR itself.
- If both SRES values are matched, the authentication is accepted and the subscriber can use the GSM network (see Figure 1.5).
- If the SRES values are not matched, the connection is terminated and the failure is reported to the handset.

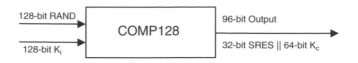

Figure 1.6   Block diagram of COMP128

A3 and A8 are integrated in SIM. These two algorithms take the same input: a 128-bit random challenge (RAND) sent from MSC and a 128-bit authentication key, K_i, that resides in SIM. The service providers implement these two algorithms as one module, which is called COMP128.

1.5.1.1 COMP128 Algorithm

COMP128 is a one-way hash function. A block diagram of COMP128 is shown in Figure 1.6. The COMP128 hash function has a 32-byte array X[ ]. The pseudo code is given below.

FUNCTION COMP128 (RAND, Ki, OUTPUT)

1	X[16-31] := RAND
2	FOR i := 0 to 7 DO
	BEGIN X[0-15] := Ki
2.1	CALL Compression (5 rounds)
2.2	CALL BytesToBit
2.3	IF i < 7 THEN
	BEGIN Permute
	END
2.4	END
	OUTPUT := Compress the 16-byte result to 12-byte
	RETURN
3	
4	

COMP128 compression function is a Butterfly-structure as shown in Figure 1.7. It uses five levels of compression. In each compression level a resulting byte depends on two input bytes. The two input bytes are used to determine the index of the lookup table, which the lookup table entry will use to update the resulting byte. The lookup table used by level i is referred to as Table T_i and it contains 2^{9-i} entries of $(8-i)$-bit values. For example:

Level	Table name	Number of entries	Value
0	T_0	512	8-bit
1	T_1	256	7-bit
2	T_2	128	6-bit
3	T_3	64	5-bit
4	T_4	32	4-bit

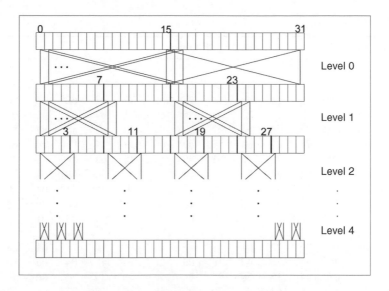

Figure 1.7   COMP128 compression function

Table entries are below:

Table	Entries															
T_0	66	b1	ba	a2	02	9c	70	4b	37	19	08	0c	fb	c1	f6	bc
	6d	d5	97	35	2a	4f	bf	73	e9	f2	a4	df	d1	94	6c	a1
	fc	25	f4	2f	40	d3	06	ed	b9	a0	8b	71	4c	8a	3b	46
	43	1a	0d	9d	3f	b3	dd	1e	d6	24	a6	45	98	7c	cf	74
	f7	c2	29	54	47	01	31	0e	5f	23	a9	15	60	4e	d7	e1
	b6	F3	1c	5c	c9	76	04	4a	f8	80	11	0b	92	84	f5	30
	95	5a	78	27	57	e6	6a	e8	af	13	7e	be	ca	8d	89	b0
	fa	1b	65	28	db	e3	3a	14	33	b2	62	d8	8c	16	20	79
	3d	67	cb	48	1d	6e	55	d4	b4	cc	96	b7	0f	42	ac	c4
	38	c5	9e	00	64	2d	99	07	90	de	a3	a7	3c	87	d2	e7
	ae	a5	26	f9	e0	22	dc	e5	d9	d0	f1	44	ce	bd	7d	ff
	ef	36	a8	59	7b	7a	49	91	75	ea	8f	63	81	c8	c0	52
	68	aa	88	eb	5d	51	cd	ad	ec	5e	69	34	2e	e4	c6	05
	39	fe	61	9b	8e	85	c7	ab	bb	32	41	b5	7f	6b	93	e2
	b8	da	83	21	4d	56	1f	2c	58	3e	ee	12	18	2b	9a	17
	50	9f	86	6f	09	72	03	5b	10	82	53	0a	c3	f0	fd	77
	b1	66	a2	ba	9c	02	4b	70	19	37	0c	08	c1	fb	bc	f6
	d5	6d	35	97	4f	2a	73	bf	f2	e9	df	a4	94	d1	a1	6c
	25	fc	2f	f4	d3	40	ed	06	a0	b9	71	8b	8a	4c	46	3b
	1a	43	9d	0d	b3	3f	1e	dd	24	d6	45	a6	7c	98	74	cf
	c2	F7	54	29	01	47	0e	31	23	5f	15	a9	4e	60	e1	d7
	f3	b6	5c	1c	76	c9	4a	04	80	f8	0b	11	84	92	30	f5

(continued)

Table (*continued*)

Table	Entries															
	5a	95	27	78	e6	57	e8	6a	13	af	be	7e	8d	ca	b0	89
	1b	fa	28	65	e3	db	14	3a	b2	33	d8	62	16	8c	79	20
	67	3d	48	cb	6e	1d	d4	55	cc	b4	b7	96	42	0f	c4	ac
	c5	38	00	9e	2d	64	07	99	de	90	a7	a3	87	3c	e7	d2
	a5	ae	f9	26	22	e0	e5	dc	d0	d9	44	f1	bd	ce	ff	7d
	36	ef	59	a8	7a	7b	91	49	ea	75	63	8f	c8	81	52	c0
	aa	68	eb	88	51	5d	ad	cd	5e	ec	34	69	e4	2e	05	c6
	fe	39	9b	61	85	8e	ab	c7	32	bb	b5	41	6b	7f	e2	93
	da	b8	21	83	56	4d	2c	1f	3e	58	12	ee	2b	18	17	9a
	9f	50	6f	86	72	09	5b	03	82	10	0a	53	f0	c3	77	fd
T_1	13	0b	50	72	2b	01	45	5e	27	12	7f	75	61	03	55	2b
	1b	7c	46	53	2f	47	3f	0a	2f	59	4f	04	0e	3b	0b	05
	23	6b	67	44	15	56	24	5b	55	7e	20	32	6d	5e	78	06
	35	4f	1c	2d	63	5f	29	22	58	44	5d	37	6e	7d	69	14
	5a	50	4c	60	17	3c	59	40	79	38	0e	4a	65	08	13	4e
	4c	42	68	2e	6f	32	20	03	27	00	3a	19	5c	16	12	33
	39	41	77	74	16	6d	07	56	3b	5d	3e	6e	4e	63	4d	43
	0c	71	57	62	66	05	58	21	26	38	17	08	4b	2d	0d	4b
	5f	3f	1c	31	7b	78	14	70	2c	1e	0f	62	6a	02	67	1d
	52	6b	2a	7c	18	1e	29	10	6c	64	75	28	49	28	07	72
	52	73	24	70	0c	66	64	54	5c	30	48	61	09	36	37	4a
	71	7b	11	1a	35	3a	04	09	45	7a	15	76	2a	3c	1b	49
	76	7d	22	0f	41	73	54	40	3e	51	46	01	18	6f	79	53
	68	51	31	7f	30	69	1f	0a	06	5b	57	25	10	36	74	7e
	1f	26	0d	00	48	6a	4d	3d	1a	43	2e	1d	60	25	3d	34
	65	11	2c	6c	47	34	42	39	21	33	19	5a	02	77	7a	23
T_2	34	32	2c	06	15	31	29	3b	27	33	19	20	33	2f	34	2b
	25	04	28	22	3d	0c	1c	04	3a	17	08	0f	0c	16	09	12
	37	0a	21	23	32	01	2b	03	39	0d	3e	0e	07	2a	2c	3b
	3e	39	1b	06	08	1f	1a	36	29	16	2d	14	27	03	10	38
	30	02	15	1c	24	2a	3c	21	22	12	00	0b	18	0a	11	3d
	1d	0e	2d	1a	37	2e	0b	11	36	2e	09	18	1e	3c	20	00
	14	26	02	1e	3a	23	01	10	38	28	17	30	0d	13	13	1b
	1f	35	2f	26	3f	0f	31	05	25	35	19	24	3f	1d	05	07
T_3	01	05	1d	06	19	01	12	17	11	13	00	09	18	19	06	1f
	1c	14	18	1e	04	1b	03	0d	0f	10	0e	12	04	03	08	09
	14	00	0c	1a	15	08	1c	02	1d	02	0f	07	0b	16	0e	0a
	11	15	0c	1e	1a	1b	10	1f	0b	07	0d	17	0a	05	16	13
T_4	0f	0c	0a	04	01	0e	0b	07	05	00	0e	07	01	02	0d	08
	0a	03	04	09	06	00	03	02	05	06	08	09	0b	0d	0f	0c

Example 1.1. Authentication and session key generation using COMP128 is shown below.

```
Given,
128-bit authentication key = 46 5b 5c e8 b1 99 b4 9f aa 5f 0a 2e e2 38 a6
bc
128-bit RAND = 23 55 3c be 96 37 a8 9d 21 8a e6 4d ae 47 bf 35
Step 1:
X[16-31] = 46 5b 5c e8 b1 99 b4 9f aa 5f 0a 2e e2 38 a6 bc
Step 2:
Round 1:
        X[1-15] = 23 55 3c be 96 37 a8 9d 21 8a e6 4d ae 47 bf 35
        X[16-31] = 46 5b 5c e8 b1 99 b4 9f aa 5f 0a 2e e2 38 a6 bc
Compression:
        Round 1:
              X[1-15]=ff 08 09 ac 82 af 6d 41 db 23 53 d0 65 91 0c ce
              X[16-31]=0f 02 8e 57 7f 70 02 bb 18 65 ab ec cf cd 40 ed
        Round 2:
              X[1-15] = 3a 13 4a 65 65 51 78 36 5b 2d 6d 55 5d 34 4d 4c
              X[16-31] = 14 18 48 06 3b 7f 1c 1e 29 5d 40 75 6f 36 17 74
        Round 3:
              X[1-15] = 15 1f 2d 0e 2e 05 33 34 0c 0c 3b 13 22 34 03 31
              X[16-31] = 19 1c 34 15 1e 3b 07 3e 3b 12 13 3c 02 1f 04 20
        Round 4:
              X[1-15] = 0a 17 0d 18 04 16 1f 16 1d 0c 1e 07 1d 03 17 10
              X[16-31] = 05 12 1c 19 0b 1f 06 1a 00 00 13 14 00 09 11 08
        Round 5:
              X[1-15] = 05 07 0d 04 0a 0f 07 06 00 0b 01 04 04 0d 02 0f
              X[16-31] = 00 0b 0d 03 00 00 08 0b 0f 0f 09 08 04 00 0c 0e
        Bytes to Bit:
           0101 0111 1101 0100 1010 1111 0111 0110 0000 1011 0001 0100
0100 1101 0010 1111
           0000 1011 1101 0011 0000 0000 1000 1011 1111 1111 1001 1000
0100 0000 1100 1110
Permutation:
        X[1-15] = 05 07 0d 04 0a 0f 07 06 00 0b 01 04 04 0d 02 0f
        X[16-31] = 0a c0 85 3a 4b 0c 17 5e 27 85 59 97 ce d7 69 fa
Round 2:
        X[1-15] = 23 55 3c be 96 37 a8 9d 21 8a e6 4d ae 47 bf 35
        X[16-31]= 0a c0 85 3a 4b 0c 17 5e 27 85 59 97 ce d7 69 fa
Compression:
        Round 1:
          X[1-15] = 1e 8e 0e 9d 8a e1 c7 f8 b0 2d de 62 a9 09 38 a0
          X[16-31] =b6 be f0 91 3a 62 6a de 13 90 d3 43 8b 57 1f e6
```

```
Round 2:
  X[1-15] = 0d 1a 46 41 10 6c 22 58 60 38 19 49 3c 01 54 52
  X[16-31] = 10 74 29 0a 4c 1b 5c 48 4b 61 1a 6d 23 04 6c 24
Round 3:
  X[1-15] = 2a 2f 19 35 3e 37 2c 09 36 2d 02 13 3f 35 29 3a
  X[16-31] = 39 3e 26 08 0d 06 2c 1e 04 28 2f 1f 16 3c 37 05
Round 4:
  X[1-15] = 04 10 16 1e 03 13 1d 1f 0d 1e 0e 16 14 02 02 15
  X[16-31] = 01 06 0f 19 08 1d 12 0f 0c 1c 1f 0a 19 12 1a 05
Round 5:
  X[1-15] = 01 05 04 0e 00 06 09 06 00 05 08 04 05 0e 01 06
  X[16-31] = 02 05 0c 02 0a 02 0a 09 01 06 09 05 0d 01 01 06
Bytes to Bit:
0001 0101 0100 1110 0000 0110 1001 0110 0000 0101 1000 0100 0101 1110
0001 0110
        0010 0101 1100 0010 1010 0010 1010 1001 0001 0110 1001 0101
1101 0001 0001 0110
Permutation:
  X[1-15] = 01 05 04 0e 00 06 09 06 00 05 08 04 05 0e 01 06
  X[16-31] = 13 10 05 8e 60 86 37 76 14 9c 97 86 44 97 8e 41
Similarly performing up to round 8 the value will be:
X[1-15] = 0f 05 03 05 02 01 02 04 01 08 05 06 0b 03 01 09
X[16-31] = 04 02 05 03 06 01 08 08 08 09 0b 05 0a 07 02 0e
Step 3:
  Output = f5 35 21 24 4d 86 22 26 d6 9c b8 00
Therefore,
  Signed response, SRES = f5 35 21 24 and Session key, Kc = 4d 86 22
26 d6 9c b8 00                                                        □
```

On 13 April, 1998, Marc Briceno and colleagues described an attack on COMP128 with which it was possible to find out the secret key K_c. Using this key one can clone a SIM-card and misuse it, because of the weakness in the compression method of COMP128. The GSM reacted to that attack and worked on new versions of COMP128. The original version was renamed as COMP128-1 and two other versions (COMP128-2 and COMP128-3) were developed, although these remain unpublished.

Following this description of the GSM authentication problem, the next section considers GSM confidentiality.

1.5.2 GSM Confidentiality

After a successful authentication, the session key for encryption must be generated. This 64-bit session key K_c is used to encrypt the message data between the handset and the network. Using data and K_c, the GSM network and handset will complete an encryption process for establishing encrypted data link. The encryption steps are outlined as follows:

Figure 1.8   Block diagram of A5/1

- The SIM card in the handset stores the received 128-bit RAND from the BTS.
- Because the GSM network knows the RAND and has a copy of the authentication key K_i, a 64-bit session key K_c is generated using the key-generation algorithm A8.
- Using the encryption algorithm A5, the message data with the encryption key K_c can be created into the encrypted data block.

1.5.2.1 GSM A5/1 Algorithm

The stream cipher A5/1 is initialized with the 64-bit session key K_C generated by A8 and the 22-bit number of the frame being encrypted or decrypted (see Figure 1.8). The same K_C is used throughout the call, but the frame number changes during the call, thus generating a unique sequence of 114-bit keystream, which is XORed with the 114-bit plaintext.

A5/1 is built from three short linear feedback shift registers (LFSR) of lengths 19, 22, and 23 bits, which are denoted by R1, R2, and R3 respectively. The outputs of the three registers are XORed together and the XOR represents one keystream bit. The rightmost bit in each register is labeled as bit zero. The taps of R1 are at bit positions 13, 16, 17, and 18; the taps of R2 are at bit positions 20 and 21; and the taps of R3 are at bit positions 7, 20, 21, and 22 (see Figure 1.9).

The registers are clocked in a stop/go fashion using a majority rule. Each register has an associated clocking bit (bit 8 for R1, bit 10 for R2, and bit 10 for R3). At each cycle, the clocking bit of all three registers is examined and the majority bit is determined. A register is clocked if the clocking bit agrees with the majority bit. For example, if the clock bits of the three registers are 1, 1, and 0, the first two registers are clocked, or if the clock bits are 0, 1 and 0, the first and third register are clocked. Hence at each step two or three registers are clocked.

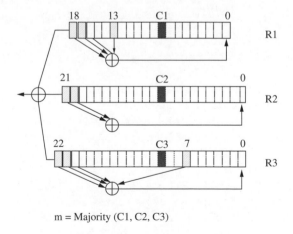

m = Majority (C1, C2, C3)

Figure 1.9   A5/1 stream cipher

When a register is clocked, its taps are XORed together and the result is stored in the rightmost bit of the left-shifted register. At each round the outputs of each register are XORed together to generate one keystream bit.

The process of generating keystream bits from the key K_C and frame counter F_n is carried out in four steps:

- The three registers are zeroed, and then clocked for 64 cycles (ignoring the stop/go clock control). In this step each bit of K_C (from least significant bit (lsb) to most significant bit (msb)) is XORed in parallel into the lsb's three registers.
- The three registers are clocked for 22 additional cycles (ignoring the stop/go clock control). In this step the successive bits of F_n (from lsb to msb) are again XORed in parallel into the lsb's three registers.
- The three registers are then clocked for 100 additional cycles with the stop/go clock control and the generated keystream bits are discarded. This is done in order to mix the frame number and keying material together.
- The three registers are then clocked for 228 additional cycles with the stop/go clock control in order to produce two 114-bit sequences of output keystream, one for each direction. At each clock cycle, one output keystream bit is produced as the XOR of the msb's three registers.

Example 1.2. Keystream generation using the A5/1 algorithm is shown below.

```
Given,
Session Key, KC = 4a 86 22 26 d6 9c b8 00
Number of frames, Fₙ = 134;
The 228-bit keystream will be calculated.
At the first step the Kᴄ is loaded into the registers:
R1 = 101 1011 1001 1110 0000
R2 = 10 1100 0001 0001 1011 0001
R3 = 000 0000 1101 1111 1010 1101
Next Fₙ is loaded into the registers:
R1 = 011 1010 1101 1111 1010
R2 = 11 1111 0001 0010 1101 0010
R3 = 101 0001 0100 1110 0101 1110
After continuing 100 additional cycles with stop/go clock control
the values of the registers will be:
R1 = 0 0001 1011 1010 1110
R2 = 01 0111 1010 1010 1011 1000
R3 = 000 1010 0011 0011 1010 1100
The three registers are then clocked for 228 additional cycles with
the stop/go clock control in order to produce two 114-bit sequences
of output keystream, one for MS to BTS and another for BTS to MS
direction.
```

```
114 bit keystream from MS to BTS:
29 ef 7a fb 23 f0 20 84 cf bf 40 e5 10 ba 80
114 bit keystream from BTS to MS:
a6 28 66 c2 43 5e 77 2f e1 3b 51 02 5a f8 40
```
□

1.5.2.2 GSM A5/3 Algorithm

The GSM A5/3 algorithm produces two 114-bit keystream strings, one of which is used for uplink encryption/decryption and the other for downlink encryption/decryption. This algorithm will be described in terms of the core function KGCORE.
GSM A5/3 inputs are specified as:

Frame dependent input: COUNT (COUNT[0]...COUNT[21])
Cipher key: K_c (K_c[0]...K_c[KLEN-1])

where KLEN is the range 64...128 bits inclusive, but the specification of the A5/3 algorithm only allows KLEN to be of value 64.
GSM A5/3 outputs are specified as:

BLOCK1 (BLOCK1[0]...BLOCK1[113])
BLOCK2 (BLOCK2[0]...BLOCK2[113])

Mapping the GSM A5/3 inputs onto the inputs of the KGCORE function is as follows:

```
CA[0]...CA[7] = 00001111
CB[0]...CB[4] = 00000
CC[0]...CC[9] = 000000000
CC[10]...CC[31] = COUNT[0]...COUNT[21] (Frame dependent input)
CD[0] = 0
CE[0]...CE[15] = 000000000000000
CK[0]...CK[KLEN-1] = K_c[0]...K_c[KLEN-1] (if KLEN<128 then K_c is
cyclically repeated to fill the 128 bits of CK)
CL = 228
```

This input mapping is depicted in Figure 1.10. Using these inputs, the KGCORE function is applied to derive the output CO[0]...CO[227]. Note that the DIRECTION bit for GSM is not applicable and set to zero. An exclusive detail of KGCORE function will be covered in Chapter 5.
Mapping the output of KGCORE onto the output of GSM A5/3 is as follows:

BLOCK1[0]...BLOCK1[113] = CO[0]...CO[113]
BLOCK2[0]...BLOCK2[113] = CO[114]...CO[227]

Instead of the A5 algorithm, an alternative algorithm is to think of triple encryption with three independent keys. The possible vulnerability of DES to a brute-force attack brings us to find an

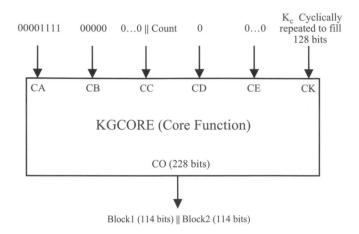

Figure 1.10   GSM A5/3 keystream generator function

alternative algorithm. Triple DES is a widely accepted approach, which uses multiple encryption using three independent keys.

1.5.3 Multiple Encryption

Multiple encryption is a widely accepted approach that encrypts a message block with multiple keys.

1.5.3.1 Triple Encryption

To improve the security of a block algorithm, it needs sometimes to encrypt a message block three times with three different keys, even if the key length becomes longer. Triple encryption with three independent keys (K1, K2, K3) is more secure than triple encryption with two keys (K1, K2, K3 = K1) proposed by Tuchman at IBM in 1979. In fact, triple DES with three different keys results in a great increase in cryptographic strength.

Denote the message block as M and its ciphertext as C. The encryption process by Encryption-Decryption-Encryption (EDE) mode is written as:

$$C = E_{K3}[D_{K2}[E_{K1}(M)]]$$

The sender encrypts the data block M with the first key K1, decrypts its result with the second key K2, and finally encrypts with the third key K3, as shown in the above equation. Decryption by the DED mode is required so that three keys are applied in the reverse order:

$$M = DK_1[E_{K2}[D_{K3}(C)]]$$

The receiver decrypts the ciphertext C with the third key K3, encrypts with the second key K2, and finally decrypts with the first key K1 in order to recover the message block M.

Example 1.3. Consider the triple DES computation as shown below:

```
Given three keys:
       K1 = 0x 260b152f31b51c68
       K2 = 0x 321f0d61a773b558
       K3 = 0x 519b7331bf104ce3
   and the message block
M = 0x403da8a295d3fed9
```

The 16-round DES keys corresponding to each given key K1, K2, or K3 are computed as shown in the following table:

Round	K1	K2	K3
1	000ced9158c9	5a1ec4b60e98	03e4ee7c63c8
2	588490792e94	710c318334c6	8486dd46ac65
3	54882eb9409b	c5a8b4ec83a5	575a226a8ddc
4	a2a006077207	96a696124ecf	aab9e009d59b
5	280e26b621e4	7e16225e9191	98664f4f5421
6	e03038a08bc7	ea906c836569	615718ca496c
7	84867056a693	88c25e6abb00	4499e580db9c
8	c65a127f0549	245b3af0453e	93e853d116b1
9	2443236696a6	76d38087dd44	cc4a1fa9f254
10	a311155c0deb	1a915708a7f0	27b30c31c6a6
11	0d02d10ed859	2d405ff9cc05	0a1ce39c0c87
12	1750b843f570	2741ac4a469a	f968788e62d5
13	9e01c0a98d28	9a09b19d710d	84e78833e3c1
14	1a4a0dc85e16	9d2a39a252e0	521f17b28503
15	09310c5d42bc	87368cd0ab27	6db841ce2706
16	53248c80ee34	30258f25c11d	c9313c0591e3

Encryption

Compute the ciphertext C through the EDE mode operation on M.

Each stage in Triple DES-EDE sequences is computed as:

```
First stage: E_{K1} (M) = 0x7a39786f7ba32349
Second stage: D_{K2} [E_{K1}(M)] = 0x9c60f85369113aea
Third stage: E_{K3} [D_{K2}[E_{K1}(M)]] = 0xe22ae33494beb930
= C (ciphertext)
```

Decryption

Using the ciphertext C obtained above, the message block M is recovered as follows:

```
Fourth stage: D_{K3} (C) = 0x9c60f85369113aea
Fifth stage: E_{K2} [D_{K3} (C)]= 0x7a39786f7ba32349
Final stage: D_{K1} [E_{K2} [D_{K3} (C)]]= 0x403da8a295d3fed9
```
= M (message block)   □

1.5.3.2 DES-CBC Mode Operation

The successive use of the DES algorithm in Cipher Block Chaining (CBC) mode offers a strong confidentiality mechanism. DES-CBC mode requires an explicit Initialization Vector (IV) of 64 bits that is the same size as the message block M. The IV must be a random value that prevents the generation of identical ciphertext. There are two possible triple-encryption modes: inner CBC and outer CBC as shown in Figure 1.11. In inner CBC, the $(IV)_1$ is XORed with the first message block M1 before it is encrypted. Encryption with K1 produces first ciphertext S_0, which is XORed with $(IV)_2$ for use of the next decryption stage and is also sent to the second message block M2. For successive triple DES-EDE operations making reference to Figure 1.11, the mode operations can be expressed as below.

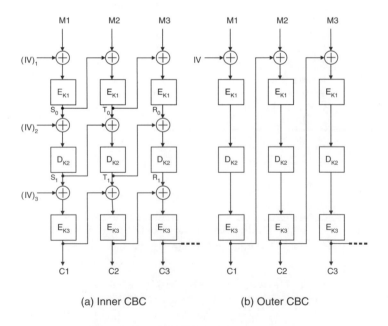

(a) Inner CBC   (b) Outer CBC

Figure 1.11   Triple DES-EDE in CBC mode

Inner CBC
This mode operation requires three different IVs.

$$S_0 = E_{K1}(M_1 \oplus (IV)_1), T_0 = E_{K1}(M_2 \oplus S_0), R_0 = E_{K1}(M_3 \oplus T_0)$$
$$S_1 = D_{K2}(S_0 \oplus (IV)_2), T_1 = D_{K2}(T_0 \oplus S_1), R_1 = D_{K2}(R_0 \oplus T_1)$$
$$C_1 = E_{K3}(S_1 \oplus (IV)_3), C_2 = E_{K3}(T_1 \oplus C_1), C_3 = E_{K3}(R_1 \oplus C_2)$$

Outer CBC
This mode operation requires only one IV.

$$C_1 = E_{K3}(D_{K2}(E_{K1}(M_1 \oplus IV)))$$
$$C_2 = E_{K3}(D_{K2}(E_{K1}(M_2 \oplus C_1)))$$
$$C_3 = E_{K3}(D_{K2}(E_{K1}(M_3 \oplus C_2)))$$

Example 1.4. Consider the triple DES-EDE operation with outer CBC mode (See Figure 1.11(b)).

Suppose three message blocks M_1, M_2, and M_3, and IV are given as:

$$M_1 = 0 \times 317f2147a6d50c38$$
$$M_2 = 0 \times c6115733248f702e$$
$$M_3 = 0 \times 1370f341da552d79$$
$$\text{and } IV = 0 \times 714289e53306f2e1$$

Assume that three keys K1, K2, and K3 used in this example are exactly the same keys as given in Example 1.3.

Computation of ciphertext blocks (C_1, C_2, C_3) at each EDE stage is as follows:

1. C_1 computation with first EDE operation:

$M_1 \oplus IV = 0x403da8a295d3fed9$
$E_{K1}(M_1 \oplus IV) = 0x7a39786f7ba32349$
$D_{K2}(E_{K1}(M_1 \oplus IV)) = 0x9c60f85369113aea$
$C_1 = E_{K3}(D_{K2}(E_{K1}(M_1 \oplus IV))) = 0xe22ae33494beb930$

2. C_2 computation with second EDE operation:

$M_2 \oplus C_1 = 0x243bb407b031c91e$
$E_{K1}(M_2 \oplus C_1) = 0xfeb7c33e747abf74$
$D_{K2}(E_{K1}(M_2 \oplus C_1)) = 0x497f548f78af6e6f$
$C_2 = E_{K3}(D_{K2}(E_{K1}(M_2 \oplus C_1))) = 0xe4976149de15ca176$

3. C_3 computation with third EDE operation:

$$M_3 \oplus C_2 = 0x5a06e7dc3b098c0f$$
$$E_{K1}(M_3 \oplus C_2) = 0x0eb878e2680e7f78$$
$$D_{K2}(E_{K1}(M_3 \oplus C_2)) = 0xc6c8441ee3b5dd1c$$
$$C_3 = E_{K3}(D_{K2}(E_{K1}(M_3 \oplus C_2))) = 0xf980690fc2db462d$$

Thus, all three ciphertext blocks (C_1, C_2, C_3) are completely obtained using outer CBC mechanism.   □

1.5.4 Encryption by AES Rijndael Algorithm

The Advanced Encryption Standard (AES) is a FIFS-approved cryptographic algorithm developed by Daemen and Rijmen in 1999. The AES is a symmetric block cipher that encrypts a 128-bit data block using encryption keys of 128, 192, and 256 bits. Referring to Figure 1.5, the A8 algorithm generates the 64-bit session key for encryption. In order to use the AES algorithm for encryption, it is suggested to store the first session key (64 bits) in register A and the second session key (64 bits) in register B. These two session keys are concatenated to produce the 128-bit session key. Thus, the AES algorithm for encryption can be achieved with the 128-bit data block and the 128-bit concatenated key. Chapter 9 will contain a detailed discussion of the AES key expansion routine, encryption by cipher, and decryption by inverse cipher.

The GSM system was primarily designed for voice communications, leaving very limited data transfer capabilities with this system. High Speed Circuit Switched Data (HSCSD) is the first stage in GSM evolution. However, the GPRS is a genuine evolution of the GSM and is usually called a 2.5G cellular system. The GPRS introduced packet transmission for data service, replacing GSM's circuit mode. However, the drawback of HSCSD compared to GPRS is that it uses several time slots in circuit-switched mode, whereas with GPRS it is not only possible to use several time slots in packet-switched mode, but also possible to multiplex several users on the same resource.

Enhanced Data rate for GSM Evolution (EDGE) is the final stage in the evolution of GSM technology and basically consists of optimizing GPRS. EDGE uses 8PSK modulation. Its new modulation mode makes it possible to increase actual throughput on the GSM radio interface. EDGE is an upgraded version of GPRS.

3G networks in Europe use the Frequency Division Duplex (FDD) mode of Universal Mobile Telecommunications System (UMTS) or Wide-band Code Division Multiple Access (WCDMA). UMTS (WCDMA/FDD), favored by the Europeans, is based on wireless WCDMA access and its core network is based on the GSM/GPRS network, allowing easy migration from 2G to 3G for GSM operators. UMTS standardization is based on the WCDMA radio interface, known as UMTS Terrestrial Radio Access (UTRA). UMTS meets the requirements for improving radio coverage, fixed-mobile convergences, and new 3G high-speed services.

High Speed Downlink Packet Access (HSDPA) presents a change in WCDMA systems, and is considered as a 3.5G system. With Multiple Input and Multiple Output (MIMO) function, speeds of 20 Mbps can be reached. The operator is looking to get ahead of its competitors in the mobile market that use CDMA2000 1xEV-DV. The GSM 1x interprets GSM to CDMA2000 migration. The CDMA2000 standards family, including CDMA2000 1x,

CDMA 1x EV-DO, and CDMA 1x EV-DV, is a 3G solution for GSM operators. The CDMA2000 1x standard evolved from cdmaOne IS-95A/B networks in late 1999. With GSM 1x, GSM operators can seamlessly exercise a great influence on their existing GSM core networks and services while enhancing the data capabilities and spectral efficiencies of their radio access with CDMA2000 infrastructure. All these related technologies will be treated separately in Chapters 5–8.

2

cdmaOne IS-95A Technology

The Code Division Multiple Access (CDMA) air-interface standard is used in both 2G and 3G networks. 2G CDMA standards, based on the TIA/EIA IS-95 in 1996, are called cdmaOne and include IS-95A and IS-95B. cdmaOne provides a family of related services including cellular, PCS, and fixed wireless access systems. CDMA consistently provides better capacity for voice and data communications than other commercial mobile technologies. This CDMA system will become the common platform on which 2G and 3G technologies are built.

Cellular CDMA is an attractive technique for wireless access to broadband services based on spread spectrum communication. The CDMA system, pioneered by Qualcomm, is based on a multiple access system developed in cooperation with a number of participating operators and equipment manufacturers. cdmaOne describes a complete wireless system based on the TIA/EIA IS-95 standard, including 2G IS-95A and 2.5G IS-95B revisions.

The IS-95 was first published in July 1993. The IS-95A revision was published in May 1995 and is the basis for many of the commercial 2G CDMA systems around the world. cdmaOne is the fastest growing 2G wireless technology reaching 100 million subscribers after only six years of commercial deployment. The cdmaOne system is used for voice and data communications. Roaming is a key functionality for wireless systems and cdmaOne offers many advantages to enable roaming.

South Korea was the first country to commercialize CDMA technology for mobile telephony. This early adoption contributes to South Korea's position in CDMA markets. The CDMA system is used in several countries in Asia (particularly in South Korea), in addition to North America, South America, and a few African countries.

In this chapter, we primarily deal with the subject on the TIA/EIA IS-95A + STB74 standard published in 27 February, 1996. The American standardization body (the UWCC) promoted the TDMA-136 (30 KHz band) for use of operators on the American continent, but it has limited data transfer capabilities. The TDMA-136 standard (IS-54) introduced in 1994 had its limitations in data transfers due to the utilization of a relatively narrowband, and these drawbacks forced its demise in December 2001.

Mobile Communication Systems and Security   Man Young Rhee
© 2009 John Wiley & Sons (Asia) Pte Ltd

2.1 Reverse CDMA Channel

The Reverse CDMA Channel is the reverse link from the mobile station to the base station. The Reverse CDMA Channel is composed of access channel (AC) and reverse traffic channels. These channels share the same CDMA frequency assignment using direct-sequence CDMA technologies. Each traffic channel is identified by a distinct user long code sequence, whereas each AC is identified by a distinct access channel long code sequence. Thus, the long code uniquely identifies a mobile station on the reverse traffic channel and separates the multiple ACs on the Reverse CDMA Channel. Multiple Reverse CDMA Channels can be used by a base station in a frequency division multiplexed manner. Data transmitted on the Reverse CDMA Channel is grouped into 20 ms frames. All data transmitted on the Reverse CDMA Channel, composed of access and reverse traffic channels, is convolutionally encoded, block interleaved, modulated by the 64-ary orthogonal modulation, and direct-sequence spread prior to transmission. Typical structures of the Reverse CDMA Channel are shown in Figures 2.1 and 2.2.

2.1.1 Reverse Traffic Channel

After adding the Frame Quality Indicator and encoder tail bits, the data frames are transmitted on the reverse traffic channel at data rates of 9.6, 4.8, 2.4, or 1.2 Kbps for Rate Set 1 or at rates of 14.4, 7.2, 3.6, or 1.8 Kbps for Rate Set 2. The reverse traffic channel (RTC) may use any of these four data rates in its rate set for transmission.

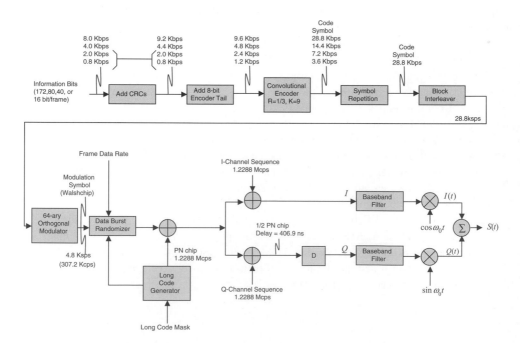

Figure 2.1 Reverse traffic channel structure (Reproduced under written permission from Telecommunications Industry Association)

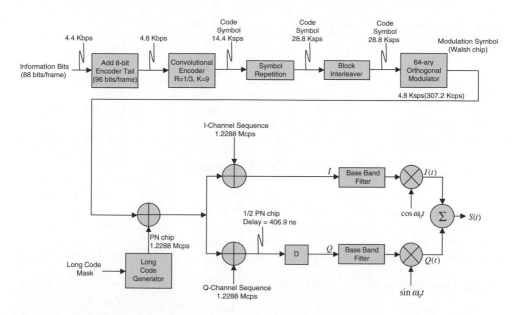

Figure 2.2   Access channel structure (Reproduced under written permission from Telecommunications Industry Association)

2.1.1.1 Modulation Parameters

The modulation parameters for the RTC are shown in Tables 2.1 and 2.2. The reverse traffic channels may use any of four data rates classified in Rate Sets 1 or 2. However, the Transmission

Table 2.1   Reverse traffic channel modulation parameters for Rate Set 1

Parameter	Data rate (bps)				Units
	9600	4800	2400	1200	
PN chip rate	1.2288	1.2288	1.2288	1.2288	Mcps
Code rate	1/3	1/3	1/3	1/3	bits/code symbol
Transmission duty cycle	100	50	25	12.5	%
Code symbol repetition	1	2	4	8	repeated code symbols/ code symbol
Repeated code symbol rate	28 800	28 800	28 800	28 800	sps
Modulation	6	6	6	6	repeated code symbols/ modulation symbol
Modulation symbol rate	4800	4800	4800	4800	sps
Walsh chip rate	307.20	307.20	307.20	307.20	Kcps
Modulation symbol duration	208.33	208.33	208.33	208.33	μs
PN chips/Repeated code symbol	42.67	42.67	42.67	42.67	PN chips/repeated code symbol
PN chips/Modulation symbol	256	256	256	256	PN chips/modulation symbol
PN chips/Walsh chip	4	4	4	4	PN chips/Walsh chip

Table 2.2 Reverse traffic channel modulation parameters for Rate Set 2

Parameter	Data rate (bps)				Units
	14400	7200	3600	1800	
PN chip rate	1.2288	1.2288	1.2288	1.2288	Mcps
Code rate	1/2	1/2	1/2	1/2	bits/code symbol
Transmission duty cycle	100	50	25	12.5	%
Code symbol repetition	1	2	4	8	repeated code symbols/ code symbol
Repeated code symbol rate	28 800	28 800	28 800	28 800	sps
Modulation	6	6	6	6	repeated code symbols/ modulation symbol
Modulation symbol rate	4800	4800	4800	4800	sps
Walsh chip rate	307.20	307.20	307.20	307.20	Kcps
Modulation symbol duration	208.33	208.33	208.33	208.33	μs
PN chips/Repeated code symbol	42.67	42.67	42.67	42.67	PN chips/repeated code symbol
PN chips/Modulation symbol	256	256	256	256	PN chips/modulation symbol
PN chips/Walsh chip	4	4	4	4	PN chips/Walsh chip

Duty Cycle (TDC) on the RTC varies with the transmission data rate. Specifically, the TDC for 14.4 and 9.6 Kbps frames is 100%, the TDC for 7.2 and 4.8 Kbps frames is 50%, the TDC for 3.6 and 2.4 Kbps frames is 25%, and the TDC for 1.8 and 1.2 Kbps frames is 12.5%. The duty cycle for transmission varies proportionately with the data rate, and therefore the actual burst transmission rate is fixed at 28 800 code symbols per second. Because six code symbols are modulated as one of 64 modulation symbols for transmission, the modulation symbol transmission rate is fixed at 4800 (=28 800/6) modulation symbols per second, which results in a fixed Walsh chip rate of 307.2 (=4.8 × 64) Kcps. As the rate of the spreading PN sequence is fixed at 1.2288 Mcps, each Walsh chip is spread by four PN chips. Tables 2.1 and 2.2 define the data rates and their relationship for the various modulation parameters on the reverse traffic channel.

The mobile station should support Rate Set 1, but may also support Rate Set 2 on the reverse traffic channel. The mobile station supports variable data rate operation with all four elements (9600, 4800, 2400, and 1200 bps) of Rate Set 1 and four elements (14 400, 7200, 3600, and 1800 bps) of Rate Set 2.

2.1.1.2 RTC Frame Quality Indicator

Each frame with Rate Set 2 and the 9600 and 4800 bps frames of Rate Set 1 will include a Frame Quality Indicator. The Frame Quality Indicator is a CRC. No Frame Quality Indicator is used for the 2400 and 1200 bps transmission rates of Rate Set 1. The Frame Quality Indicator (CRC) should be calculated on all bits within the frame, except the Frame Quality Indicator itself and the encoder tail bits. The 9600 bps transmission with Rate Set 1 and 14 400 bps transmission with Rate Set 2 will use a 12-bit Frame Quality Indicator. The 7200 bps transmission with Rate Set 2 will use a 10-bit Frame Quality Indicator. The 4800 bps transmissions with Rate Set 1 and the 36 000 bps transmissions with Rate Set 2 will use an 8-bit Frame Quality Indicator. The 1800 bps transmissions with Rate Set 2 will use a 6-bit Frame Quality Indicator.

The generator polynomials for the Frame Quality Indicator are:

$$g(x) = x^{12} + x^{11} + x^{10} + x^9 + x^8 + x^4 + x + 1$$

for the 12-bit Frame Quality Indicator;

$$g(x) = x^{10} + x^9 + x^8 + x^7 + x^6 + x^4 + x^2 + 1$$

for the 10-bit Frame Quality Indicator;

$$g(x) = x^8 + x^7 + x^4 + x^3 + x + 1$$

for the 8-bit Frame Quality Indicator;
and

$$g(x) = x^6 + x^2 + x + 1$$

and for the 6-bit Frame Quality Indicator.

The Frame Quality Indicators (CRCs) mentioned above are computed according to the following procedure using the respective logic:

- All shift register stages are initially set to logical one and the switch is set in the down position as well as Gates 1 and 2.
- The register is clocked a number of times equal to the number of Erasure Indicators and information bits in the frame with those bits as input. For Rate Set 1, where the Frame Quality Indicator is used, the number of information bits per frame is 172 and 80 for the 9600 and 4800 bps transmission rates, respectively. For Rate Set 2, the number of Erasure Indicator and information bits per frame is 268, 126, 56, and 22 for the 14 400, 7200, 3600 and 1800 bps transmission rates, respectively.
- The switches, including Gates 1 and 2, are set in the up position so that the output is a modulo-2 addition with "0."
- The register is clocked an additional number of times equal to the number of bits in the Frame Quality Indicator, that is, 12, 10, 8, or 6, respectively. These additional bits will be the Frame Quality Indicator bits.
- The bits are transmitted in the order calculated.

The RTC Frame Quality Indicators can be computed using the logic shown in Figures 6.1.3.3.2.1-1 through 6.1.3.3.2.1-4 in pp. 6-44–42 of IS-95A + TSB74 published on 27 February, 1996.

As an example, Figure 2.3 shows the 8-bit Frame Quality Indicator (CRC) computation circuit for the 4800 bps data rate of Rate Set 1.

Example 2.1. Referring to Figure 2.1, data rates of 9.6, 4.8, 2.4, and 1.2 Kbps in Rate Set 1 are calculated by taking the following steps:

- Data rates of 9.2 and 4.4 Kbps are each computed from information rates of 8.6 and 4.0 Kbps by adding the Frame Quality Indicator (i.e. add the 12-bit CRC for the 9.2 Kbps rate and the

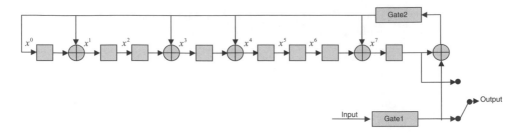

Figure 2.3   CRC computation circuit at the 4800 bps data rate

8-bit CRC for the 4.4 Kbps rate):

$$172 + 12 = 184 \text{ bits}/20 \text{ ms} = 9200 \text{ bps}$$
$$80 + 8 = 88 \text{ bits}/20 \text{ ms} = 4400 \text{ bps}$$

- Adding further the additional 8-bit encoder tail, the transmission rates of 9.6 and 4.8 Kbps are computed from the 9.2 and 4.4 Kbps rates as follows:

$$184 + 8 = 192 \text{ bits}/20 \text{ ms} = 9600 \text{ bps}$$
$$88 + 8 = 96 \text{ bits}/20 \text{ ms} = 4800 \text{ bps}$$

- Next, the transmission rate 2.4 Kbps or 1.2 Kbps is calculated from the information rate 2.0 Kbps or 0.8 Kbps as follows:

$$40 + 8 = 48 \text{ bits}/20 \text{ ms} = 2400 \text{ bps}$$
$$16 + 8 = 24 \text{ bits}/20 \text{ ms} = 1200 \text{ bps}$$ □

Example 2.2.   Consider the 80-bit information sequence for the 4800 bps transmission rate of Rate Set 1. Using Figure 2.3, compute the Frame Quality Indicator bits with the 80-bit information input of

```
1010110011    1001101111    0010010100  0100110011
0011001000    1110010110    1001100110  0110011110
```

The register is clocked 80 times for the 96-bit frame (20 ms). The CRC bits can be calculated as shown in Table 2.3.

Table 2.3 Frame quality indicator (CRC) computation for the 480 bps frame by setting register initial contents as (11111111)

Shift no.	Information bits	Feedback bits	Register contents
1	1	$1 \oplus 1 = 0$	01111111
2	0	$0 \oplus 1 = 1$	11100110
3	1	$1 \oplus 0 = 1$	10101010
35	1	$1 \oplus 1 = 0$	00000000
36	1	$1 \oplus 0 = 1$	11011001
37	0	$0 \oplus 1 = 1$	10110101
78	1	$1 \oplus 1 = 0$	00110000
79	1	$1 \oplus 0 = 1$	11000001
80	0	$0 \oplus 1 = 1$	10111001

As shown from Table 2.3, the register content for $n = 80$ is 10011101. When the register is now clocked an additional 80 times, the register output will be the CRC bits. Finally, the RTC frame sequence for the 4800 bps rate will be computed by simply adding the encoder tail bits (all-zero 8 bits), as shown below:

```
1010110011  1001101111  0010010100  0100110011  0011001000
1110010110  1001100110  0110011110  1001110100  000000
```

This frame sequence will be the input to the (3, 1, 8) convolutional encoder.   □

2.1.1.3 RTC Frame Structure

The RTC bit allocations are classified in the following manner (see Figure 2.1).

- *Rate Set 1*: RTC frames sent with Rate Set 1 at four different transmission rates are composed of either Information + CRC + Tail bits or Information + Tail bits depending on transmission rates.
    - at the 9600 bps Transmission Rate:
      172 information bits followed by 12 Frame Quality Indicator (CRC) bits and 8 encoder tail bits: as a result, frame bits = 192 bits.
    - at the 4800 bps Transmission Rate:
      80 information bits followed by 8 CRC bits and 8 encoder tail bits: as a result, frame bits = 96 bits.
    - at the 2400 bps Transmission Rate:
      40 information bits followed by 8 encoder tail bits: frame bits = 48 bits.
    - at the 1200 bps Transmission Rate:
      16 information bits followed by 8 encoder tail bits: frame bits = 24 bits.
- *Rate Set 2*: RTC frames sent with Rate Set 2 at four different transmission rates are composed of:

— at the 14.4 Kbps Transmission Rate:
　One erasure indicator bit followed by 267 information bits, 12 Frame Quality Indicator
　　(CRC) bits, and 8 encoder rail bits: frame bits = 288 bits.
— at the 7.2 Kbps Transmission Rate:
　One erasure indicator bit followed by 125 information bits, 10 CRC bits, and 8 encoder
　　rail bits: frame bits = 144 bits.
— at the 3.6 Kbps Transmission Rate:
　One erasure indicator bit followed by 55 information bits, 8 CRC bits, and 8 encoder rail
　　bits: frame bits = 72 bits.
— at the 1.8 Kbps Transmission Rate:
　One erasure indicator bit followed by 21 information bits, 6 CRC bits, and 8 encoder rail
　　bits: frame bits = 36 bits.

The summary of the RTC frame structure is shown in Table 2.4.

2.1.1.4 RTC Convolutional Encoding

The mobile station should convolutionally encode the data transmitted on the RTC and the AC
prior to interleaving. An (n, k, m) convolutional code is implemented with a k-input, n-output
linear sequential circuit with the memory order m. In general n and k are small integers with
$k<n$, but m is relatively large. The advance of convolutional coding spawned a number of
practical applications to digital transmission over wire and radio (wireless) communication
channels.

An (n, k, m) convolutional code designates the code rate $R=k/n$ with encoder stages of
$m=K-1$, where K is called the constraint length of the code.

The RTC convolutional encoder (n, k, m) has three generators g_0, g_1, and g_2 and three output
symbols (c_0, c_1, c_2) corresponding to the encoder input bits. The initial state of the convolutional
encoder is assumed to be all-zero.

Table 2.4 RTC frame structure summary

Rate Set	Transmission rate (bps)	Number of bits per frame				
		Total	Erasure indicator	Information	Frame quality indicator	Encoder tail
1	9600	192	0	172	12	8
	4800	96	0	80	8	8
	2400	48	0	40	0	8
	1200	24	0	16	0	8
2	14400	288	1	267	12	8
	7200	144	1	125	10	8
	3600	72	1	55	8	8
	1800	36	1	21	6	8

The last eight bits of each reverse traffic channel, called the encoder tail bits, should be set to logical zeros (00000000).

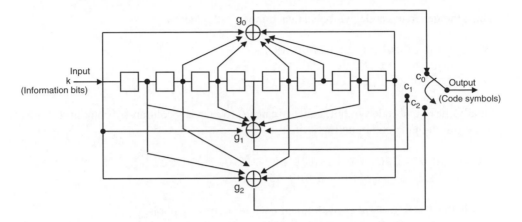

Figure 2.4 $k = 9$, rate 1/3 convolutional encoder

Convolutional encoding involves the modulo-2 addition of selected taps of a serially time delayed data sequence. The length of the data sequence delay is equal to $m = K - 1 = 8$, where K denotes the code constant length. Figure 2.1 illustrates the convolutional encoder for the RTC.

For the AC and Rate Set 1 of the RTC, the convolutional code rate is $R = 1/3$. For Rate Set 2 of the RTC, the convolutional code rate is $R = 1/2$.

Example 2.3. Referring to the RTC structure depicted in Figure 2.1, the 1200 bps transmission rate of Rate Set 1 is shown as a simple demonstration. The RTC frame at 1200 bps transmission rate consists of 24 bits that is composed of 16 information bits followed by 8 encoder tail bits. Suppose the 16 information bits are 1110001011010011 and the 8 encoder tail bits are 00000000. Then the input frame sequence to the convolutional encoder becomes 111000101101001100000000. Since the code rate is $R = 1/3$, the encoder generates three code symbols (c_0, c_1, c_2) for each input bit. Using Figure 2.4, three code symbols can be computed as shown in Table 2.5.

Table 2.5   Generation of three code symbols (c_0, c_1, c_2)

Shift no.	Input bit	Register contents	Output code symbols (c_0, c_1, c_2)
1	1	00000000	111
2	1	10000000	100
3	1	11000000	001
4	0	11100000	000
14	0	01011010	100
15	1	00101101	111
16	1	10010110	100
23	0	00000011	001
24	0	00000001	111

Thus, the generated code symbols at each modulo-2 adder are

```
c₀ = (11000011 10000111 00000001)
c₁ = (10000011 11101010 10100101)
c₂ = (10101100 11111010 00001011)
```

Finally, the output code symbol sequence can be produced by column-by-column shifting of c_0, c_1, c_2 through the commutating switch:

```
c= (111 100 001 000 001 001 100 100 111 011 011 001
011 100 111 100 010 000 010 000 001 010 001 111)
```

This is the $K = 9$, $R = 1/3$ convolutional encoder output symbols.

The computation of the output symbol sequence of the $(3, 1, 8)$ convolutional encoder at the RTC 1200 bps transmission rate of Rate Set 1 is again considered with several alternative methods as in the following.

Let d be the information frame including encoder tail bits as

```
d = (1110 0010 1101 0011 0000 0000)
```

The generator functions for this code are

$$g_0 = 557(\text{octal}) = (101101111)(\text{binary})$$
$$= D^8 + D^6 + D^5 + D^3 + D^2 + D + 1(\text{polynomial})$$
$$g_1 = 663(\text{octal}) = (110110\,011) = D^8 + D^7 + D^5 + D^4 + D + 1$$
$$g_2 = 711(\text{octal}) = (111001001) = D^8 + D^7 + D^6 + D^3 + 1 \qquad \square$$

Example 2.4. Solution by Discrete Convolution:

```
c₀ = d * g₀ = (111000101101001100000000) * (101101111)
           = (1100 0011 1000 01111 0000 0001)
c₁ = d * g₁ = (111000101101001100000000) * (110110011)
           = (1000 0000 1110 1010 1010 0101)
c₂ = d * g₂ = (111000101101001100000000) * (111001001)
           = (1010 1100 1111 1010 0000 1011)
```

The code symbol sequence coming out of the encoder is then computed by concatenating unit symbols (c_0, c_1, c_2) (consisting of three bits) at a time as

```
C = (111 100 001 000 001 001 100 100 111 011 011 001
011 100 111 100 010 000 010 000 001 010 001 111) (I)
```

                                                                                        $\square$

Example 2.5. Solution by Scalar Matrix Product:

$$C = d \cdot G$$

where d denotes the input information data and G is a semi-infinite generator matrix:

$$G = \begin{bmatrix} G_0 & G_1 & \cdots & G_n & & & \\ & G_0 & G_1 & \cdots & G_n & & \\ & & G_0 & G_1 & \cdots & G_n & \\ & & \cdots & \cdots & \cdots & \cdots & \end{bmatrix}$$

Component matrices G_i, $0 \le i \le 8$, are easily obtained from g_0, g_1, and g_2 as follows:

$$G_0 = 111, \ G_1 = 011, \ G_2 = 101, \ldots, G_7 = 110, G_8 = 111,$$

Thus, the output code symbol C is computed as

C = d·G
 = (111 000 101 101 001 100 000 000) ·

$$\begin{bmatrix} 111 & 011 & 101 & 110 & 010 & 101 & 100 & 110 & 111 & & & & & & & & & \\ & 111 & 011 & 101 & 110 & 010 & 101 & 100 & 110 & 111 & & & & & & & \\ & & \cdots & & & & & & & & \cdots & & & & & & \\ & & & \cdots & & & & & & & & \cdots & & & & & \\ & & & & 111 & 011 & 101 & 110 & 010 & 101 & 100 & 110 & 111 \end{bmatrix}$$

   = (111 100 001 000 001 001 100 100 111 011 011 001 011
      100 111 100 010 000 010 000 001 010 001 111)

This shows the code symbols coming out of the encoder.   □

Example 2.6. Solution by Polynomial Matrix:
The information data are expressed in the polynomial form as

$$d(D) = D^{23} + D^{22} + D^{21} + D^{17} + D^{15} + D^{14} + D^{12} + D^9 + D^8$$

The generator matrix in the polynomial form represents

$$\begin{aligned} G(D) &= [g_0(D), g_1(D), g_2(D)] \\ &= [D^8 + D^6 + D^5 + D^3 + D^2 + D + 1, D^8 + D^7 + D^5 + D^4 + D + 1, D^8 + D^7 \\ &\quad + D^6 + D^3 + 1] \end{aligned}$$

The code symbol polynomial matrix at the modulo-2 adders is obtained as

$$
\begin{aligned}
C(D) &= d(D)G(D) = [C_0(D), C_1(D), C_2(D)] \\
&= [D^{31} + D^{30} + D^{25} + D^{24} + D^{23} + D^{18} + D^{17} + D^{16} + D^{8}, \\
&\quad D^{31} + D^{23} + D^{22} + D^{21} + D^{19} + D^{17} + D^{15} + D^{13} + D^{10} + D^{8}, \\
&\quad D^{31} + D^{29} + D^{27} + D^{26} + D^{23} + D^{22} + D^{21} + D^{20} + D^{19} + D^{17} + D^{11} + D^{9} + D^{8}]
\end{aligned}
$$

Finally, the delay operation by the communicating switch will produce the encoder output symbols as shown below:

$$
\begin{aligned}
C(D) &= D^2 C_0(D^3) + D C_1(D^3) + C_2(D^3) \\
&= D^2(D^{93} + D^{90} + D^{75} + D^{72} + D^{69} + D^{54} + D^{51} + D^{48} + D^{24}) \\
&\quad + D(D^{93} + D^{69} + D^{66} + D^{63} + D^{57} + D^{51} + D^{45} + D^{39} + D^{30} + D^{24}) \\
&\quad + (D^{93} + D^{87} + D^{81} + D^{78} + D^{69} + D^{66} + D^{63} + D^{60} + D^{57} + D^{51} + D^{33} + D^{27} + D^{24}) \\
&= D^{95} + D^{94} + D^{93} + D^{92} + D^{87} + D^{81} + D^{78} + D^{77} + D^{74} + D^{71} + D^{70} + D^{69} \\
&\quad + D^{67} + D^{66} + D^{64} + D^{63} + D^{60} + D^{58} + D^{57} + D^{56} + D^{53} + D^{52} + D^{51} + D^{50} \\
&\quad + D^{46} + D^{40} + D^{33} + D^{31} + D^{27} + D^{26} + D^{25} + D^{24} \\
&= (111\,100\,001\,000\,001\,001\,100\,100\,111\,011\,011\,001 \\
&\quad 011\,100\,111\,100\,010\,000\,010\,000\,001\,010\,001\,111) \qquad \square
\end{aligned}
$$

Thus, it is proved that alternative solutions through three different approaches are identical to the output symbol sequence obtained in Example 2.3.

For Rate Set 2 of the RTC, the convolutional code rate will be $R = 1/2$. The generator functions for this code will be $g_0 = 753(\text{octal}) = (0111\,0101\,0011)$ and $g_1 = 561(\text{octal}) = (0101\,0110\,0001)$. This convolutional code generates two code symbols for each data bit input to the encoder. These code symbols are output so that the code symbol c_0 encoded with g_0 will be the first output and the code symbol c_1 encoded with g_1 will be the last output. The mobile station also supports variable data rate operation with all four elements of Rate Set 2. It is no problem at all to solve for the $(2, 1, 8)$ convolutional encoder by taking exactly the same steps as done for the $(3, 1, 8)$ convolutional encoder.

2.1.1.5 Code Symbol Repetition

Code symbols output from the convolutional encoder are repeated before being interleaved when the data rate is lower than 9600 bps for Rate Set 1 and 14 400 bps for Rate Set 2. The code symbol repetition rate on the RTC varies with data rate. Code symbols should not be repeated for 14 400 bps and 9600 bps data rates. Each code symbol at the 7200 and 4800 bps data rates will be repeated one time (i.e. each symbol occurs two consecutive times). Each code symbol at the 3600 and 2400 bps data rates should be repeated three times (each symbol occurs four consecutive times). Each code symbol at the 1800 and 1200 bps data rates should be repeated seven times (each symbol occurs eight consecutive times). For all the data rates for Rate Sets 1 and 2, each code symbol should be repeated as shown in Table 2.6.

For all the data rates, this results in a constant repeated code symbol rate of 28 800 cps (code symbol per second). On the RTC these repeated code symbols must not be transmitted multiple

Table 2.6 Code symbol repetition

Rate Set	Data rate (bps)	Repetition times/symbol	Consecutive occurring times/symbol
1	9600	0	1
	4800	1	2
	2400	3	4
	1200	7	8
2	14400	0	1
	7200	1	2
	3600	3	4
	1800	7	8

times. Rather, the repeated code symbols are input to the block interleaver function, and all but one of the code symbol repetitions should be deleted prior to actual transmission due to the variable transmission duty cycle.

2.1.1.6 Block Interleaving

The mobile station interleaves all repeated code symbols on the RTC channel prior to orthogonal modulation and transmission. A block interleaver spanning 20 ms must be used. The interleaver forms a 576-cell array with 32 rows and 18 columns. The RTC Interleaver Memory (Write Operation) for various specified data rates belonging to Rate Sets 1 and 2 is shown in Tables 2.7 through 2.10.

RTC repeated code symbols will be output from the interleaver by rows. For Rate Sets 1 and 2, the interleaver rows will be output in the following order depending on the data rate:

- At 9600 bps (Rate Set 1) and 14 400 bps (Rate Set 2):

1	2	3	4	5	6	7	8	9	10	11	12	13	14	15	16
17	18	19	20	21	22	23	24	25	26	27	28	29	30	31	32

- At 4800 bps (Rate Set 1) and 7200 bps (Rate Set 2):

$$(1 + 4i, 3 + 4i, 2 + 4i, 4 + 4i, 0 \le i \le 7)$$

1	3	2	4	5	7	6	8	9	11	10	12	13	15	14	16
17	19	18	20	21	23	22	24	25	27	26	28	29	31	30	32

- At 2400 bps (Rate Set 1) and 3600 bps (Rate Set 2):

$$(1 + 8i, 5 + 8i, 2 + 8i, 6 + 8i, 3 + 8i, 7 + 8i, 4 + 8i, 8 + 8i, 0 \le i \le 3)$$

1	5	2	6	3	7	4	8	9	13	10	14	11	15	12	16
17	21	18	22	19	23	20	24	25	29	26	30	27	31	28	32

Table 2.7   RTC interleaver memory (Write Operation) for 9600 and 14400 bps

1	33	65	97	129	161	193	225	257	289	321	353	385	417	449	481	513	545
2	34	66	98	130	162	194	226	258	290	322	354	386	418	450	482	514	546
3	35	67	99	131	163	195	227	259	291	323	355	387	419	451	483	515	547
4	36	68	100	132	164	196	228	260	292	324	356	388	420	452	484	516	548
5	37	69	101	133	165	197	229	261	293	325	357	389	421	453	485	517	549
6	38	70	102	134	166	198	230	262	294	326	358	390	422	454	486	518	550
7	39	71	103	135	167	199	231	263	295	327	359	391	423	455	487	519	551
8	40	72	104	136	168	200	232	264	296	328	360	392	424	456	488	520	552
9	41	73	105	137	169	201	233	265	297	329	361	393	425	457	489	521	553
10	42	74	106	138	170	202	234	266	298	330	362	394	426	458	490	522	554
11	43	75	107	139	171	203	235	267	299	331	363	395	427	459	491	523	555
12	44	76	108	140	172	204	236	268	300	332	364	396	428	460	492	524	556
13	45	77	109	141	173	205	237	269	301	333	365	397	429	461	493	525	557
14	46	78	110	142	174	206	238	270	302	334	366	398	430	462	494	526	558
15	47	79	111	143	175	207	239	271	303	335	367	399	431	463	495	527	559
16	48	80	112	144	176	208	240	272	304	336	368	400	432	464	496	528	560
17	49	81	113	145	177	209	241	273	305	337	369	401	433	465	497	529	561
18	50	82	114	146	178	210	242	274	306	338	370	402	434	466	498	530	562
19	51	83	115	147	179	211	243	275	307	339	371	403	435	467	499	531	563
20	52	84	116	148	180	212	244	276	308	340	372	404	436	468	500	532	564
21	53	85	117	149	181	213	245	277	309	341	373	405	437	469	501	533	565
22	54	86	118	150	182	214	246	278	310	342	374	406	438	470	502	534	566
23	55	87	119	151	183	215	247	279	311	343	375	407	439	471	503	535	567
24	56	88	120	152	184	216	248	280	312	344	376	408	440	472	504	536	568
25	57	89	121	153	185	217	249	281	313	345	377	409	441	473	505	537	569
26	58	90	122	154	186	218	250	282	314	346	378	410	442	474	506	538	570
27	59	91	123	155	187	219	251	283	315	347	379	411	443	475	507	539	571
28	60	92	124	156	188	220	252	284	316	348	380	412	444	476	508	540	572
29	61	93	125	157	189	221	253	285	317	349	381	413	445	477	509	541	573
30	62	94	126	158	190	222	254	286	318	350	382	414	446	478	510	542	574
31	63	95	127	159	191	223	255	287	319	351	383	415	447	479	511	543	575
32	64	96	128	160	192	224	256	288	320	352	384	416	448	480	512	544	576

- At 1200 bps (Rate Set 1) and 1800 bps (Rate Set 2):

   $(1 + 16i, 9 + 16i, 2 + 16i, 10 + 16i, 3 + 16i, 11 + 16i, 4 + 16i, 12 + 16i, 5 + 16i,$
   $13 + 16i, 6 + 16i, 14 + 16i, 7 + 16i, 15 + 16i, 8 + 16i, 16 + 16i, i = 0, 1)$

1	9	2	10	3	11	4	12	5	13	6	14	7	15	8	16
17	25	18	26	19	27	20	28	21	29	22	30	23	31	24	32

Example 2.7. Referring to Table 2.10 at 1200 and 1800 bps, the interleaver output symbols are calculated by means of the encoder output symbols (I) obtained in Example 2.4. The interleaver output will be obtained through the steps shown in Table 2.11.

Table 2.8 RTC or AC interleaver memory (Write Operation) for 4800 and 7200 bps

1	17	33	49	65	81	97	113	129	145	161	177	193	209	225	241	257	273
1	17	33	49	65	81	97	113	129	145	161	177	193	209	225	241	257	273
2	18	34	50	66	82	98	114	130	146	162	178	194	210	226	242	258	274
2	18	34	50	66	82	98	114	130	146	162	178	194	210	226	242	258	274
3	19	35	51	67	83	99	115	131	147	163	179	195	211	227	243	259	275
3	19	35	51	67	83	99	115	131	147	163	179	195	211	227	243	259	275
4	20	36	52	68	84	100	116	132	148	164	180	196	212	228	244	260	276
4	20	36	52	68	84	100	116	132	148	164	180	196	212	228	244	260	276
5	21	37	53	69	85	101	117	133	149	165	181	197	213	229	245	261	277
5	21	37	53	69	85	101	117	133	149	165	181	197	213	229	245	261	277
6	22	38	54	70	86	102	118	134	150	166	182	198	214	230	246	262	278
6	22	38	54	70	86	102	118	134	150	166	182	198	214	230	246	262	278
7	23	39	55	71	87	103	119	135	151	167	183	199	215	231	247	263	279
7	23	39	55	71	87	103	119	135	151	167	183	199	215	231	247	263	279
8	24	40	56	72	88	104	120	136	152	168	184	200	216	232	248	264	280
8	24	40	56	72	88	104	120	136	152	168	184	200	216	232	248	264	280
9	25	41	57	73	89	105	121	137	153	169	185	201	217	233	249	265	281
9	25	41	57	73	89	105	121	137	153	169	185	201	217	233	249	265	281
10	26	42	58	74	90	106	122	138	154	170	186	202	218	234	250	266	282
10	26	42	58	74	90	106	122	138	154	170	186	202	218	234	250	266	282
11	27	43	59	75	91	107	123	139	155	171	187	203	219	235	251	267	283
11	27	43	59	75	91	107	123	139	155	171	187	203	219	235	251	267	283
12	28	44	60	76	92	108	124	140	156	172	188	204	220	236	252	268	284
12	28	44	60	76	92	108	124	140	156	172	188	204	220	236	252	268	284
13	29	45	61	77	93	109	125	141	157	173	189	205	221	237	253	269	285
13	29	45	61	77	93	109	125	141	157	173	189	205	221	237	253	269	285
14	30	46	62	78	94	110	126	142	158	174	190	206	222	238	254	270	286
14	30	46	62	78	94	110	126	142	158	174	190	206	222	238	254	270	286
15	31	47	63	79	95	111	127	143	159	175	191	207	223	239	255	271	287
15	31	47	63	79	95	111	127	143	159	175	191	207	223	239	255	271	287
16	32	48	64	80	96	112	128	144	160	176	192	208	224	240	256	272	288
16	32	48	64	80	96	112	128	144	160	176	192	208	224	240	256	272	288

□

2.1.1.7 Orthogonal Modulation

Modulation for the Reverse CDMA Channel is 64-ary orthogonal modulation. The modulation symbol is one of 64 mutually orthogonal waveforms generated using the Walsh function. One of 64 possible modulation symbols are given in Table 2.12 and are numbered 0 through 63. These modulation symbols are selected according to the following modulation symbol index (MSI):

$$MSI = C_0 + 2C_1 + 4C_2 + 8C_3 + 16\,C_4 + 32C_5$$

where C_5 represents the most recent and C_0 the oldest binary valued repeated code symbol of each group of six repeated code symbols that form a modulation symbol index.

Table 2.9 RTC interleaver memory (Write Operation) for 2400 and 3600 bps

1	9	17	25	33	41	49	57	65	73	81	89	97	105	113	121	129	137
1	9	17	25	33	41	49	57	65	73	81	89	97	105	113	121	129	137
1	9	17	25	33	41	49	57	65	73	81	89	97	105	113	121	129	137
1	9	17	25	33	41	49	57	65	73	81	89	97	105	113	121	129	137
1	9	17	25	33	41	49	57	65	73	81	89	97	105	113	121	129	137
2	10	18	26	34	42	50	58	66	74	82	90	98	106	114	122	130	138
2	10	18	26	34	42	50	58	66	74	82	90	98	106	114	122	130	138
2	10	18	26	34	42	50	58	66	74	82	90	98	106	114	122	130	138
2	10	18	26	34	42	50	58	66	74	82	90	98	106	114	122	130	138
3	11	19	27	35	43	51	59	67	75	83	91	99	107	115	123	131	139
3	11	19	27	35	43	51	59	67	75	83	91	99	107	115	123	131	139
3	11	19	27	35	43	51	59	67	75	83	91	99	107	115	123	131	139
3	11	19	27	35	43	51	59	67	75	83	91	99	107	115	123	131	139
4	12	20	28	36	44	52	60	68	76	84	92	100	108	116	124	132	140
4	12	20	28	36	44	52	60	68	76	84	92	100	108	116	124	132	140
4	12	20	28	36	44	52	60	68	76	84	92	100	108	116	124	132	140
4	12	20	28	36	44	52	60	68	76	84	92	100	108	116	124	132	140
5	13	21	29	37	45	53	61	69	77	85	93	101	109	117	125	133	141
5	13	21	29	37	45	53	61	69	77	85	93	101	109	117	125	133	141
5	13	21	29	37	45	53	61	69	77	85	93	101	109	117	125	133	141
5	13	21	29	37	45	53	61	69	77	85	93	101	109	117	125	133	141
6	14	22	30	38	46	54	62	70	78	86	94	102	110	118	126	134	142
6	14	22	30	38	46	54	62	70	78	86	94	102	110	118	126	134	142
6	14	22	30	38	46	54	62	70	78	86	94	102	110	118	126	134	142
6	14	22	30	38	46	54	62	70	78	86	94	102	110	118	126	134	142
7	15	23	31	39	47	55	63	71	79	87	95	103	111	119	127	135	143
7	15	23	31	39	47	55	63	71	79	87	95	103	111	119	127	135	143
7	15	23	31	39	47	55	63	71	79	87	95	103	111	119	127	135	143
7	15	23	31	39	47	55	63	71	79	87	95	103	111	119	127	135	143
8	16	24	32	40	48	56	64	72	80	88	96	104	112	120	128	136	144
8	16	24	32	40	48	56	64	72	80	88	96	104	112	120	128	136	144
8	16	24	32	40	48	56	64	72	80	88	96	104	112	120	128	136	144
8	16	24	32	40	48	56	64	72	80	88	96	104	112	120	128	136	144

Using a Hadamard matrix H_N, where $N = 2^m$ (a power of 2), Walsh functions are constructed as follows: A Hadamard matrix is an orthogonal $N \times N$ matrix of the entries $+1$ and -1 such that any row differs from any other row in exactly $N/2$ positions. One row of the matrix contains all $+1$s, while the other rows contain evenly $+1$s and -1s of $N/2$ each. Furthermore, all the entries in the first row and the first column of H_N have all $+1$s. Changing the $+1$s to 0s and the -1s to 1s, the Hadamard matrix for $N = 2$ is expressed as

$$H_2 = \begin{bmatrix} +1 & +1 \\ +1 & -1 \end{bmatrix} = \begin{bmatrix} 0 & 0 \\ 0 & 1 \end{bmatrix}$$

Table 2.10 RTC interleaver memory (Write Operation) for 1200 and 1800 bps

1	5	9	13	17	21	25	29	33	37	41	45	49	53	57	61	65	69
1	5	9	13	17	21	25	29	33	37	41	45	49	53	57	61	65	69
1	5	9	13	17	21	25	29	33	37	41	45	49	53	57	61	65	69
1	5	9	13	17	21	25	29	33	37	41	45	49	53	57	61	65	69
1	5	9	13	17	21	25	29	33	37	41	45	49	53	57	61	65	69
1	5	9	13	17	21	25	29	33	37	41	45	49	53	57	61	65	69
1	5	9	13	17	21	25	29	33	37	41	45	49	53	57	61	65	69
1	5	9	13	17	21	25	29	33	37	41	45	49	53	57	61	65	69
2	6	10	14	18	22	26	30	34	38	42	46	50	54	58	62	66	70
2	6	10	14	18	22	26	30	34	38	42	46	50	54	58	62	66	70
2	6	10	14	18	22	26	30	34	38	42	46	50	54	58	62	66	70
2	6	10	14	18	22	26	30	34	38	42	46	50	54	58	62	66	70
2	6	10	14	18	22	26	30	34	38	42	46	50	54	58	62	66	70
2	6	10	14	18	22	26	30	34	38	42	46	50	54	58	62	66	70
2	6	10	14	18	22	26	30	34	38	42	46	50	54	58	62	66	70
2	6	10	14	18	22	26	30	34	38	42	46	50	54	58	62	66	70
3	7	11	15	19	23	27	31	35	39	43	47	51	55	59	63	67	71
3	7	11	15	19	23	27	31	35	39	43	47	51	55	59	63	67	71
3	7	11	15	19	23	27	31	35	39	43	47	51	55	59	63	67	71
3	7	11	15	19	23	27	31	35	39	43	47	51	55	59	63	67	71
3	7	11	15	19	23	27	31	35	39	43	47	51	55	59	63	67	71
3	7	11	15	19	23	27	31	35	39	43	47	51	55	59	63	67	71
3	7	11	15	19	23	27	31	35	39	43	47	51	55	59	63	67	71
3	7	11	15	19	23	27	31	35	39	43	47	51	55	59	63	67	71
4	8	12	16	20	24	28	32	36	40	44	48	52	56	60	64	68	72
4	8	12	16	20	24	28	32	36	40	44	48	52	56	60	64	68	72
4	8	12	16	20	24	28	32	36	40	44	48	52	56	60	64	68	72
4	8	12	16	20	24	28	32	36	40	44	48	52	56	60	64	68	72
4	8	12	16	20	24	28	32	36	40	44	48	52	56	60	64	68	72
4	8	12	16	20	24	28	32	36	40	44	48	52	56	60	64	68	72
4	8	12	16	20	24	28	32	36	40	44	48	52	56	60	64	68	72
4	8	12	16	20	24	28	32	36	40	44	48	52	56	60	64	68	72

The Hadamard matrix for $N = 2^6$ is the 64×64 orthogonal Walsh function shown in Table 2.12. This 64×64 matrix can be generated by means of the following recursive procedure:

$$H_1 = 0 \quad \text{for } m = 0$$

$$H_2 = \begin{bmatrix} H_1 & H_1 \\ H_1 & \overline{H_1} \end{bmatrix} = \begin{bmatrix} 0 & 0 \\ 0 & 1 \end{bmatrix} \quad \text{for } m = 1$$

$$H_4 = \begin{bmatrix} H_2 & H_2 \\ H_2 & \overline{H_2} \end{bmatrix} = \begin{bmatrix} 0 & 0 & 0 & 0 \\ 0 & 1 & 0 & 1 \\ 0 & 0 & 1 & 1 \\ 0 & 1 & 1 & 0 \end{bmatrix}$$

Table 2.11 Computation of interleaver output symbols for 1200 bps data rate

Row number	Interleaver output	Row number	Interleaver output
1	101000111001000011	17	100110100110000101
9	100011110101100001	25	100000011110010001
2	101000111001000011	18	100110100110000101
10	100011110101100001	26	100000011110010001
3	101000111001000011	19	100110100110000101
11	100011110101100001	27	100000011110010001
4	101000111001000011	20	100110100110000101
12	100011110101100001	28	100000011110010001
5	101000111001000011	21	100110100110000101
13	100011110101100001	29	100000011110010001
6	101000111001000011	22	100110100110000101
14	100011110101100001	30	100000011110010001
7	101000111001000011	23	100110100110000101
15	100011110101100001	31	100000011110010001
8	101000111001000011	24	100110100110000101
16	100011110101100001	32	100000011110010001

$$H_8 \;=\; \begin{bmatrix} H_4 & H_4 \\ H_4 & \overline{H_4} \end{bmatrix} \;=\; \begin{bmatrix} 0 & 0 & 0 & 0 & 0 & 0 & 0 & 0 \\ 0 & 1 & 0 & 1 & 0 & 1 & 0 & 1 \\ 0 & 0 & 1 & 1 & 0 & 0 & 1 & 1 \\ 0 & 1 & 1 & 0 & 0 & 1 & 1 & 0 \\ 0 & 0 & 0 & 0 & 1 & 1 & 1 & 1 \\ 0 & 1 & 0 & 1 & 1 & 0 & 1 & 0 \\ 0 & 0 & 1 & 1 & 1 & 1 & 0 & 0 \\ 0 & 1 & 1 & 0 & 1 & 0 & 0 & 1 \end{bmatrix}$$

$$\vdots$$

$$H_{2N} \;=\; \begin{bmatrix} H_N & H_N \\ H_N & \overline{H_N} \end{bmatrix}$$

where N is a power of 2 and $\overline{H_N}$ denotes the binary complement of H_N.

Since one of 64 modulation symbols is transmitted for each six repeated code symbols, the modulation symbol rate will become $28\,800/6 = 4800$ sps (symbols per second). Therefore, the period of time required to transmit a single modulation symbol should be equal to 1/4800 second (208.333 μs). One-sixty-fourth (1/64) of a modulation symbol is defined as a Walsh chip. The period of time associated with a Walsh chip will become $1/4800 \times 64 = 1/307\,200 = 3.255$ μs. With a modulation symbol, Walsh chips should be transmitted in the order of $0, 1, 2, \ldots, 64$.

Example 2.8. At the 1200 bps transmission rate (Rate Set 1), the interleaver output symbols were computed as shown in Table 2.11.

```
Row No.1: 101000  111001 000011
Row No.2: 100011  110101 100001
. . . . . . . . . . . . . . . . .
Row No.32: 100000 011110 010001
```
☐

Table 2.12   64-ary orthogonal symbol set

Walsh Chip within Symbol

(Rows = Modulation Symbol Index; column groups = Walsh chip positions 0–63)

Mod. Symbol Index	0–3	4–7	8–11	12–15	16–19	20–23	24–27	28–31	32–35	36–39	40–43	44–47	48–51	52–55	56–59	60–63
0	0000	0000	0000	0000	0000	0000	0000	0000	0000	0000	0000	0000	0000	0000	0000	0000
1	0101	0101	0101	0101	0101	0101	0101	0101	0101	0101	0101	0101	0101	0101	0101	0101
2	0011	0011	0011	0011	0011	0011	0011	0011	0011	0011	0011	0011	0011	0011	0011	0011
3	0110	0110	0110	0110	0110	0110	0110	0110	0110	0110	0110	0110	0110	0110	0110	0110
4	0000	1111	0000	1111	0000	1111	0000	1111	0000	1111	0000	1111	0000	1111	0000	1111
5	0101	1010	0101	1010	0101	1010	0101	1010	0101	1010	0101	1010	0101	1010	0101	1010
6	0011	1100	0011	1100	0011	1100	0011	1100	0011	1100	0011	1100	0011	1100	0011	1100
7	0110	1001	0110	1001	0110	1001	0110	1001	0110	1001	0110	1001	0110	1001	0110	1001
8	0000	0000	1111	1111	0000	0000	1111	1111	0000	0000	1111	1111	0000	0000	1111	1111
9	0101	0101	1010	1010	0101	0101	1010	1010	0101	0101	1010	1010	0101	0101	1010	1010
10	0011	0011	1100	1100	0011	0011	1100	1100	0011	0011	1100	1100	0011	0011	1100	1100
11	0110	0110	1001	1001	0110	0110	1001	1001	0110	0110	1001	1001	0110	0110	1001	1001
12	0000	1111	1111	0000	0000	1111	1111	0000	0000	1111	1111	0000	0000	1111	1111	0000
13	0101	1010	1010	0101	0101	1010	1010	0101	0101	1010	1010	0101	0101	1010	1010	0101
14	0011	1100	1100	0011	0011	1100	1100	0011	0011	1100	1100	0011	0011	1100	1100	0011
15	0110	1001	1001	0110	0110	1001	1001	0110	0110	1001	1001	0110	0110	1001	1001	0110
16	0000	0000	0000	0000	1111	1111	1111	1111	0000	0000	0000	0000	1111	1111	1111	1111
17	0101	0101	0101	0101	1010	1010	1010	1010	0101	0101	0101	0101	1010	1010	1010	1010
18	0011	0011	0011	0011	1100	1100	1100	1100	0011	0011	0011	0011	1100	1100	1100	1100
19	0110	0110	0110	0110	1001	1001	1001	1001	0110	0110	0110	0110	1001	1001	1001	1001
20	0000	1111	0000	1111	1111	0000	1111	0000	0000	1111	0000	1111	1111	0000	1111	0000
21	0101	1010	0101	1010	1010	0101	1010	0101	0101	1010	0101	1010	1010	0101	1010	0101
22	0011	1100	0011	1100	1100	0011	1100	0011	0011	1100	0011	1100	1100	0011	1100	0011
23	0110	1001	0110	1001	1001	0110	1001	0110	0110	1001	0110	1001	1001	0110	1001	0110
24	0000	0000	1111	1111	1111	1111	0000	0000	0000	0000	1111	1111	1111	1111	0000	0000
25	0101	0101	1010	1010	1010	1010	0101	0101	0101	0101	1010	1010	1010	1010	0101	0101
26	0011	0011	1100	1100	1100	1100	0011	0011	0011	0011	1100	1100	1100	1100	0011	0011
27	0110	0110	1001	1001	1001	1001	0110	0110	0110	0110	1001	1001	1001	1001	0110	0110
28	0000	1111	1111	0000	1111	0000	0000	1111	0000	1111	1111	0000	1111	0000	0000	1111
29	0101	1010	1010	0101	1010	0101	0101	1010	0101	1010	1010	0101	1010	0101	0101	1010
30	0011	1100	1100	0011	1100	0011	0011	1100	0011	1100	1100	0011	1100	0011	0011	1100
31	0110	1001	1001	0110	1001	0110	0110	1001	0110	1001	1001	0110	1001	0110	0110	1001
32	0000	0000	0000	0000	0000	0000	0000	0000	1111	1111	1111	1111	1111	1111	1111	1111
33	0101	0101	0101	0101	0101	0101	0101	0101	1010	1010	1010	1010	1010	1010	1010	1010
34	0011	0011	0011	0011	0011	0011	0011	0011	1100	1100	1100	1100	1100	1100	1100	1100
35	0110	0110	0110	0110	0110	0110	0110	0110	1001	1001	1001	1001	1001	1001	1001	1001
36	0000	1111	0000	1111	0000	1111	0000	1111	1111	0000	1111	0000	1111	0000	1111	0000
37	0101	1010	0101	1010	0101	1010	0101	1010	1010	0101	1010	0101	1010	0101	1010	0101
38	0011	1100	0011	1100	0011	1100	0011	1100	1100	0011	1100	0011	1100	0011	1100	0011
39	0110	1001	0110	1001	0110	1001	0110	1001	1001	0110	1001	0110	1001	0110	1001	0110
40	0000	0000	1111	1111	0000	0000	1111	1111	1111	1111	0000	0000	1111	1111	0000	0000
41	0101	0101	1010	1010	0101	0101	1010	1010	1010	1010	0101	0101	1010	1010	0101	0101
42	0011	0011	1100	1100	0011	0011	1100	1100	1100	1100	0011	0011	1100	1100	0011	0011
43	0110	0110	1001	1001	0110	0110	1001	1001	1001	1001	0110	0110	1001	1001	0110	0110
44	0000	1111	1111	0000	0000	1111	1111	0000	1111	0000	0000	1111	1111	0000	0000	1111
45	0101	1010	1010	0101	0101	1010	1010	0101	1010	0101	0101	1010	1010	0101	0101	1010
46	0011	1100	1100	0011	0011	1100	1100	0011	1100	0011	0011	1100	1100	0011	0011	1100
47	0110	1001	1001	0110	0110	1001	1001	0110	1001	0110	0110	1001	1001	0110	0110	1001
48	0000	0000	0000	0000	1111	1111	1111	1111	1111	1111	1111	1111	0000	0000	0000	0000
49	0101	0101	0101	0101	1010	1010	1010	1010	1010	1010	1010	1010	0101	0101	0101	0101
50	0011	0011	0011	0011	1100	1100	1100	1100	1100	1100	1100	1100	0011	0011	0011	0011
51	0110	0110	0110	0110	1001	1001	1001	1001	1001	1001	1001	1001	0110	0110	0110	0110
52	0000	1111	0000	1111	1111	0000	1111	0000	1111	0000	1111	0000	0000	1111	0000	1111
53	0101	1010	0101	1010	1010	0101	1010	0101	1010	0101	1010	0101	0101	1010	0101	1010
54	0011	1100	0011	1100	1100	0011	1100	0011	1100	0011	1100	0011	0011	1100	0011	1100
55	0110	1001	0110	1001	1001	0110	1001	0110	1001	0110	1001	0110	0110	1001	0110	1001
56	0000	0000	1111	1111	1111	1111	0000	0000	1111	1111	0000	0000	0000	0000	1111	1111
57	0101	0101	1010	1010	1010	1010	0101	0101	1010	1010	0101	0101	0101	0101	1010	1010
58	0011	0011	1100	1100	1100	1100	0011	0011	1100	1100	0011	0011	0011	0011	1100	1100
59	0110	0110	1001	1001	1001	1001	0110	0110	1001	1001	0110	0110	0110	0110	1001	1001
60	0000	1111	1111	0000	1111	0000	0000	1111	1111	0000	0000	1111	0000	1111	1111	0000
61	0101	1010	1010	0101	1010	0101	0101	1010	1010	0101	0101	1010	0101	1010	1010	0101
62	0011	1100	1100	0011	1100	0011	0011	1100	1100	0011	0011	1100	0011	1100	1100	0011
63	0110	1001	1001	0110	1001	0110	0110	1001	1001	0110	0110	1001	0110	1001	1001	0110

Modulation Symbol Computation with Row No.1

Note that there is one of 64 modulation symbols per each 6-code symbol.

- For the first 6-symbol input 101000 ($C_0 = 1, C_1 = 0, C_2 = 1, C_3 = C_4 = C_5 = 0$), MSI $= C_0 + 4C_2 = 1 + 4 = 5$. Using Table 2.12, the corresponding modulation symbol is obtained as

```
010110 100101 101001 011010 010110 100101
101001 011010 010110 100101 1010
```

- For the second 6-symbol input 111001($C_0 = 1$, $C_1 = 1$, $C_2 = 1$, $C_3 = 0$, $C_4 = 0$, $C_5 = 1$), MSI $= C_0 + 2C_1 + 4C_2 + 32C_5 = 39$. The corresponding modulation symbol is

```
011010 010110 100101 101001 011010 011001
011010 010110 100101 101001 0110
```

- For the third 6-symbol input 000011 ($C_0 = C_1 = C_2 = C_3 = 0$, $C_4 = 1$, $C_5 = 1$), MSI $= 16C_4 + 32C_5 = 48$. The corresponding modulation symbol is

```
000000 000000 000011 111111 111111 111111
111111 111111 000000 000000 0000
```

Modulation Symbol Computation with Row No.9

- Input symbol = 100011, MSI = 49:

```
Modulation symbol = 010101 010101 010110 101010 101010 101010
101010 101010 010101 010101 0101
```

- Input symbol = 110101, MSI = 43:

```
Modulation symbol = 011001 101001 100101 100110 100110 011001
100101 100110 100110 010110 0110
```

- Input symbol = 100001, MSI = 33:

```
Modulation symbol = 010101 010101 010101 010101 010101 011010
101010 101010 101010 101010 1010
```

Thus, modulation symbols corresponding only to rows 1 and 9 in Table 2.11 are evaluated by concatenation of all six component modulation symbols. This represents only 36 cells out of 576 cells filling the complete 32 × 18 matrix.

2.1.1.8 Data Burst Randomizer

The RTC interleaver output stream is time gated to allow transmission of certain interleaver output symbols and the deletion of others, as illustrated in Figure 2.5. For the 9600 or 14 400 bps transmission rate, the transmission gate allows 100% of all interleaver output symbols to be transmitted. For the 4800 or 7200 bps transmission rate, the transmission gate allows one-half (50%) of the interleaver output symbols to be transmitted. With the 2400 or 3600 bps transmission rate, the gate permits a quarter (25%) of the interleaver output symbols to be transmitted. Finally, for the 1200 or 1800 bps transmission rate, the gate permits 12.5% of the interleaver output symbols to be transmitted. The data burst randomizer ensures that every code symbol input to the repetition process is transmitted exactly once.

The data burst randomizer generates a masking pattern of 0s and 1s that randomly masks out the redundant data generated by the code repetition. The masking pattern is determined by the data rate of the frame and by a block of 14 bits taken from the long code. These 14 bits consist of the least 14 bits of the long code used for spreading in the previous power control group (PCG 14) to the last power control group (PCG 15) of the previous frame. Let these 14 bits be denoted as $\alpha_0, \alpha_1, \alpha_2, \alpha_3, \alpha_4, \alpha_5, \alpha_6, \alpha_7, \alpha_8, \alpha_9, \alpha_{11}, \alpha_{12}, \alpha_{13}$, where α_0 represents the oldest bit and α_{13} represents the latest bit. These 14 bits occur exactly one power control group (1.25 ms) before each RTC frame boundary.

Transmission will occur on PCGs numbered from 0 to 15 according to the data rate selected. The data burst randomizer algorithm will be as follows:

Data Burst Randomizer Algorithm.

Data rate selected (bps)	PCGs numbered
9600 or 14 400	0, 1, 2, 3, 4, 5, 6, 7, 8, 9, 10, 11, 12, 13, 14, 15
4800 or 7200	$\alpha_0, 2 + \alpha_1, 4 + \alpha_2, 6 + \alpha_3, 8 + \alpha_4, 10 + \alpha_5, 12 + \alpha_6, 14 + \alpha_7$
2400 or 3600	α_0 if $\alpha_8 = 0$, or $2 + \alpha_1$ if $\alpha_8 = 1$ $4 + \alpha_2$ if $\alpha_9 = 0$, or $6 + \alpha_3$ if $\alpha_9 = 1$ $8 + \alpha_4$ if $\alpha_{10} = 0$, or $10 + \alpha_5$ if $\alpha_{10} = 1$ $12 + \alpha_6$ if $\alpha_{11} = 0$, or $14 + \alpha_7$ if $\alpha_{11} = 1$
1200 or 1800	α_0 if $\alpha_8 = 0$ and $\alpha_{12} = 0$, or $2 + \alpha_1$ if $\alpha_8 = 1$ and $\alpha_{12} = 0$ $4 + \alpha_2$ if $\alpha_9 = 0$ and $\alpha_{12} = 1$, or $6 + \alpha_3$ if $\alpha_9 = 1$ and $\alpha_{12} = 1$ $8 + \alpha_4$ if $\alpha_{10} = 0$ and $\alpha_{13} = 0$, or $10 + \alpha_5$ if $\alpha_{10} = 1$ and $\alpha_{13} = 0$ $12 + \alpha_6$ if $\alpha_{11} = 0$ and $\alpha_{13} = 1$, or $14 + \alpha_7$ if $\alpha_{11} = 1$ and $\alpha_{13} = 1$

Example 2.9. For the 1200 bps transmission rate, the information bits will be equal to 24 bits (20 ms) per frame, that is the input to the convolutional encoder. Since the encoder code rate is $R = 1/3$, the encoder symbol output is 72 bits. The repetition process will produce 576 ($= 72 \times 8$) code symbols. These code symbols are interleaved prior to modulation

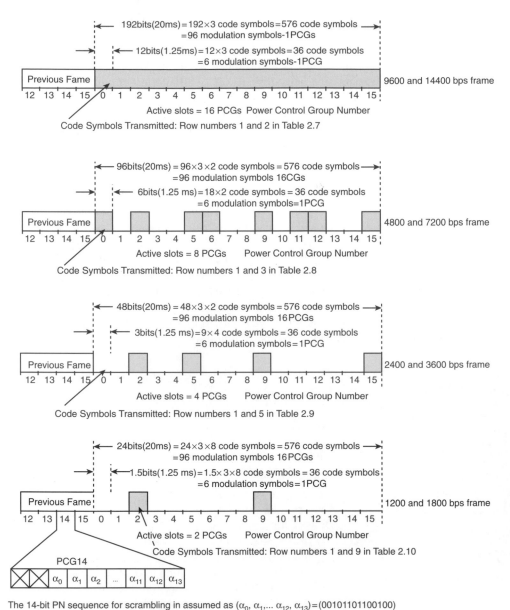

Figure 2.5 The data burst randomizer pulse trains varying with the transmission rate (Reproduced under written permission from Telecommunications Industry Association)

and transmission. Thus, the modulation symbols are 96 (=576/6). One frame (20 ms) consists of 16 PCGs. For a single PCG, Time duration = 1.25 (=20/16) ms, Information bits = 1.5 (=24/16) bits, Encoder code symbol = 4.5(=1.5 × 3) bits, Repetition symbols = 36 (=4.5 × 8) bits, and Walsh modulation = 6 (36/6) modulation symbols.

Suppose scrambling PN sequence is $(\alpha_0, \alpha_1, \ldots, \alpha_{13}) = (00101101100100)$ from whence $\alpha_0 = 0, \alpha_1 = 0, \alpha_2 = 1, \alpha_3 = 0, \alpha_4 = 1, \alpha_5 = 1, \alpha_6 = 0, \alpha_7 = 1, \alpha_8 = 1, \alpha_9 = 0, \alpha_{10} = 0, \alpha_{11} = 1, \alpha_{12} = 0$ and $\alpha_{13} = 0$.

The power control group numbers for the 1200 bps data rate are determined using the data burst randomizer algorithm as follows:

- Since $\alpha_8 = 1$ and $\alpha_{12} = 0$, $2 + \alpha_1 = 2 + 0 = 2$, which corresponds to PCG No.2.
- Since $\alpha_{10} = 0$ and $\alpha_{13} = 0$, $8 + \alpha_4 = 8 + 1 = 9$, which corresponds to PCG No.9.

Thus, it is clear, as shown in Figure 2.5, that the 1200 bps transmission gate allows only 12.5% of the interleaver output symbols, that is, PCG numbers 2 and 9. □

Example 2.10. Using the same masking streams (00101101100100) as used in Example 2.9, the gate-on power control groups with the frame are calculated by means of the data burst randomizer algorithm in the following:

- For 2400 or 3600 bps data rate:

$$2 + \alpha_1 \text{ if } \alpha_8 = 1, 2 + \alpha_1 = 2 \text{ indicates } (\rightarrow) \text{ PCG No. 2}$$
$$4 + \alpha_2 \text{ if } \alpha_9 = 0, 4 + \alpha_2 = 5 \rightarrow \text{PCG No. 5}$$
$$8 + \alpha_4 \text{ if } \alpha_{10} = 0, 8 + \alpha_4 = 9 \rightarrow \text{PCG No. 9}$$
$$14 + \alpha_7 \text{ if } \alpha_{11} = 1, 14 + \alpha_7 = 15 \rightarrow \text{PCG No. 15}$$

Transmission should occur on power control groups numbered: 2, 5, 9, 15.

- For 4800 or 7200 bps data rate:

$$\alpha_0 = 0, \text{ PCG No. 0} \qquad\qquad 8 + \alpha_4 = 8 + 1 = 9, \text{ PCG No. 9}$$
$$2 + \alpha_1 = 2 + 0 = 2, \text{ PCG No. 2} \quad 10 + \alpha_5 = 10 + 1 = 11, \text{ PCG No. 11}$$
$$4 + \alpha_2 = 4 + 1 = 5, \text{ PCG No. 5} \quad 12 + \alpha_6 = 12 + 0 = 12, \text{ PCG No. 12}$$
$$6 + \alpha_3 = 6 + 0 = 6, \text{ PCG No. 6} \quad 14 + \alpha_7 = 14 + 1 = 15, \text{ PCG No. 15}$$

Transmission should occur on power control groups numbered: 0, 2, 5, 6, 9, 11, 12, and 15.

- For 9600 or 14 400 bps data rate:

Transmission should occur on power control groups numbered: 0, 1, 2, 3, 4, 5, 6, 7, 8, 9, 10, 11,12,13,14, and 15. □

For these data rates, we see that the transmission gate allows 100% of all interleaver output symbols to be transmitted.

Thus, through Examples 2.9 and 2.10, our number computations of the gate-on power control groups proved to be correct as illustrated in Figure 2.5.

2.1.1.9 Direct Sequence Spreading by the Long Code

Direct sequence spreading using the long code should be applied to the Reverse CDMA Channel, which consists of the RTC and the AC. For the RTC, this spreading operation involves modulo-2 addition of the data burst randomizer output and the long code. This long code is periodic with period $2^{42} - 1$ chips and should satisfy the linear recursion specified by the Linear Feedback Shift Register (LFSR) characteristic polynomial $p(x)$ of the sequence generator:

$$p(x) = x^{42} + x^{35} + x^{33} + x^{31} + x^{27} + x^{26} + x^{25} + x^{22} + x^{21} + x^{19}$$
$$+ x^{18} + x^{17} + x^{16} + x^{10} + x^7 + x^6 + x^5 + x^3 + x^2 + x + 1$$

Each PN chip of the long code is generated by the modulo-2 inner product of a 42-bit mask and the 42-bit state vector of the LFSR sequence generator, as shown in Figure 2.6. The mask used for the long code varies depending on the channel type on which the mobile station is transmitting. The long code provides limited privacy. When transmitting on the RTC, the mobile station should use one of two long code masks unique to that mobile station: a public long code mask unique to the mobile station's ESN or a private long code mask. The public long code mask is created as follows:

$M_{41} - M_{32}$: Set to 1100011000.
$M_{31} - M_0$: Set to a permutation of the mobile station's ESN bits.

Defining the mobile station's ESN as

$ESN = (E_{31}, E_{30}, E_{29}, E_{28}, \ldots, E_3, E_2, E_1, E_0),$

a permutation of the ESN bits should be specified so that this permutation prevents high correlation between long codes as shown below:

$$
\begin{aligned}
\text{Permuted ESN} \quad = \quad & (E_0, E_{31}, E_{22}, E_{13}, E_4, E_{26}, E_{17}, E_8, \\
& E_{30}, E_{21}, E_{12}, E_3, E_{25}, E_{16}, E_7, E_{29}, \\
& E_{20}, E_{11}, E_2, E_{24}, E_{15}, E_6, E_{28}, E_{19}, \\
& E_{10}, E_1, E_{23}, E_{14}, E_5, E_{27}, E_{18}, E_9)
\end{aligned}
$$

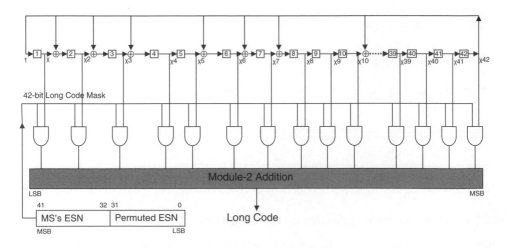

Figure 2.6   Long code generator

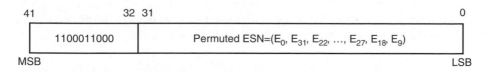

41   32 31   0

| 1100011000 | Permuted ESN=(E_0, E_{31}, E_{22}, ..., E_{27}, E_{18}, E_9) |

MSB   LSB

Figure 2.7   Public long code mask format

The public long code mask format is illustrated in Figure 2.7. The private long code mask is not available at this moment because distribution for its foreign use is controlled by TIA.

Example 2.11. For the 1200 bps transmission rate, compute the long code PN chips. If the mobile station's ESN bits are chosen as

```
ESN = (0100 1100 1011 0011 1010 0111 0110 0111),
then the permuted ESN becomes
```

```
Permuted ESN = (1001 0111 1100 0100 1010 1100 1110 1101)
```

Thus, the public long code mask format (see Figure 2.7) is represented as

Long Code mask = (1011 0111 0011 0101 0010 0011 1110 1001 0001 1000 11)

       ↑   ↑

       LSB   MSB

Referring to Figure 2.6, this 42-bit long code mask and the 42-state LFSR output are subjected to input to the AND gates. The resulted AND gate outputs are to be modulo-2 added. The individual long code bit will be the binary value "0" if the sum from the modulo-2 adder is even and will be the binary value "1" if the summed number is odd.

The vector form representing the characteristic polynomial $p(x)$ of the sequence generator is (1111 0111 0010 0000 1111 0110 0111 0001 0101 0000 00). Assume that the initial contents of LFSR are (1000 0000 0000 0000 0000 0000 0000 0000 0000 0000 00).

Applying these two sequences and the 42-bit long code mask to Figure 2.6, the long code PN chips can be calculated as shown below:

```
1011 0111 0011 0101 0010 0011 1110 1001 0001 1000
1111 1011 1010 1000 0010 0011 1000 0001 1000 1100
0000 0001 0100 0100 0011 ........
```

The sequence represents the partial long code PN chips obtained by only the first 100 shifts out of $2^{42} - 1$ shifts.

Finally, consider *direct sequence spreading* (DSS) by XORing the orthogonal modulation symbols and the long code PN chips.

For the 1200 bps data rate, the modulation symbols (or the data burst randomizer output) are obtained as

```
0101 1010 0101 1010 0101 1010 0101 1010
0101 1010 0101 1010 0101 1010 0101 1010,
```

which is computed from MSI = 5 corresponding to the 6-symbol input (101000). Recall that the PN chips/Walsh chip is 4 so that computation would be as follows:

```
Modulation symbols: 0000 1111 0000 1111 1111 0000 1111 0000 0000 1111
Long code chips: 1011 0111 0011 0101 0010 0011 1110 1001 0001 1000
EX-OR (DSS) ((⊕): 1011 1000 0011 1010 1101 0011 0001 1001 0001 0111
```

Thus, DSS by the long code PN chips is produced as

1011 1000 0011 1010 1101 0011 0001 1001 0001 0111 ...   □

Suppose $d_w(t)$ denotes the data sequence modulated by Walsh chips and T_b is the 4-bit data time interval. The Walsh modulated data sequence is modulo-2 added by the spreading PN chips of the long code $c(t)$. Each pulse of $c(t)$ is called a chip and T_c denotes the chip time interval such that $T_b = 4T_c$. The rate of the spreading PN sequence is fixed at 1.2288 Mcps. Since six code symbols are modulated by one of 64 time-orthogonal Walsh functions, the modulated symbol transmission rate is fixed at 28.8/6 = 4.8 Ksps. Therefore, each Walsh chip is spread by four PN chips, that is, $1.2288 \times 10^6/307.3 \times 10^3 = 4$. The direct sequence spreading $D_s(t)$ of $d_w(t)$ by the long code PN chips of 1.2288 Mcps is illustrated in Figure 2.8.

2.1.1.10 Quadrature Spreading

Following the DSS, the RTC and AC are spread in quadrature as shown in Figures 2.1 and 2.2. The sequences used for this spreading are the zero-offset I and Q pilot PN sequences. These PN

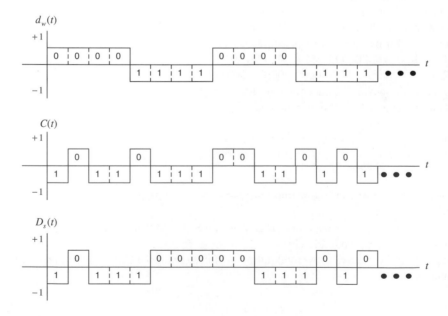

Figure 2.8   Direct sequence spreading of $d_w(t)$ by the long code PN chips $C(t)$

sequences are periodic with period 2^{15} chips and should be based on the following characteristic polynomials, respectively:

For the in-phase (I) sequence, it is given as

$$P_I(x) = x^{15} + x^{13} + x^9 + x^8 + x^7 + x^5 + 1$$

and for the quadrature-phase (Q) sequence, it gives

$$P_Q(x) = x^{15} + x^{12} + x^{11} + x^{10} + x^6 + x^5 + x^4 + x^3 + 1$$

The reciprocal polynomials of $P_I(x)$ and $P_Q(x)$ of period 2^{15} are generated as

$$i(x) = x^{15} P_I(x^{-1}) = x^{15} + x^{10} + x^8 + x^7 + x^6 + x^2 + 1$$

and

$$q(x) = x^{15} P_Q(x^{-1}) = x^{15} + x^{12} + x^{11} + x^{10} + x^9 + x^5 + x^4 + x^3 + 1$$

The maximum length LFSR sequences $\{i(n)\}$ and $\{q(n)\}$, based on these reciprocal polynomials of period 2^{15}, can be generated by using the following linear recursions:

$$i(n) = i(n-15) \oplus i(n-10) \oplus i(n-8) \oplus i(n-7) \oplus i(n-6) \oplus i(n-2)$$

and

$$q(n) = q(n-15) \oplus q(n-12) \oplus q(n-11) \oplus q(n-10) \oplus q(n-9)$$
$$\oplus q(n-5) \oplus q(n-4) \oplus q(n-3)$$

where $i(n)$ and $q(n)$ are binary-valued ("0" and "1") and the additions are modulo-2.

In order to obtain the I and Q pilot PN sequence (of period 2^{15}), a "0" is inserted in $\{i(n)\}$ and $\{q(n)\}$ after 14 consecutive "0" output. The mobile station should align the I and Q pilot PN sequences in such a way that the first chip is the "1" after the 15 consecutive "0"s. The I and Q pilot PN sequences, $\bar{P}_I$ and $\bar{P}_Q$, are then generated from $i(n)$ and $q(n)$ under the initial state content of LFSR with IC $= (1000000000000000)$.

The pilot PN sequences repeat every 26.666 ms $(=2^{15}/1\,228\,800$ seconds). There are exactly 75 repetitions in every 2 seconds. The data spread by the Q pilot PN sequence should be delayed by half a PN chip time (406.901 ns) with respect to the data spread by the zero-offset I pilot PN sequence.

Example 2.12. Compute $\bar{P}_I$ and $\bar{P}_Q$ using $i(n)$ and $q(n)$ of period $2^{15} = 32\,768$. Giving $i(n) = 0$ for $1 \leq n \leq 15$ and $i(16) = 1$, the in-phase pilot PN sequence $\bar{P}_I$ and the quadrature-phase pilot PN sequence P_Q are calculated by the following linear recursions:

Computation of $\bar{P}_I$ using $\displaystyle\sum_{k=15,10,8,7,6,2} \oplus\, i(n-k)$, $1 \leq n \leq 32\,768$:

$$i(17) = i(2) \oplus i(7) \oplus i(9) \oplus i(10) \oplus i(11) \oplus i(15) = 0$$

......

$$i(26) = i(11) \oplus i(16) \oplus i(18) \oplus i(19) \oplus i(20) \oplus i(24) = 1$$

......

$$i(35) = i(20) \oplus i(25) \oplus i(27) \oplus i(28) \oplus i(29) \oplus i(33) = 1$$

......

Continuing these recursive computations, $\bar{P}_I$ can be obtained as

$\bar{P}_I$ = 0000 0000 0000 0001 0101 0010 0111 0100 0110 1111...

Computation of $\bar{P}_Q$ using $\displaystyle\sum_{k=15,12,11,10,9,5,4,3} \oplus\, q(n-k)$, $1 \leq n \leq 32\,768$:

$$q(17) = q(2) \oplus q(5) \oplus q(6) \oplus q(7) \oplus q(8) \oplus q(12) \oplus q(13) \oplus q(14) = 0$$

......

$$q(26) = q(11) \oplus q(14) \oplus q(15) \oplus q(16) \oplus q(17) \oplus q(21) \oplus q(22) \oplus q(23) = 1$$

......

$$q(35) = q(19) \oplus q(23) \oplus q(24) \oplus q(25) \oplus q(26) \oplus q(30) \oplus q(31) \oplus q(32) = 1$$

......

Thus, the quadrature-phase pilot PN sequence $\bar{P}_Q$ is calculated as:

$\bar{P}_Q$ = 0000 0000 0000 0001 0011 1101 0111 0110 1010 0111   □

The I-channel spreading I and the Q-channel spreading Q are readily computed from the results of XORing the I and Q pilot chips, $\bar{P}_I$ and $\bar{P}_Q$, and the DSS input stream.

Example 2.13. For the 1200 bps transmission rate, the in-phase channel sequence I is computed by XORing the DSS (obtained in Example 2.11) and $\bar{P}_I$ as shown below:

```
DSS = 10111000 00111010 11010011 00011001 00010111
⊕
P̄_I = 00000000 00000001 01010010 01110100 0110111
```

```
In-phase I = 10111000 00111011 10000001 01101101 01111000
```

and the quadrature-phase channel sequence Q can be computed as follows:

DSS = 10111000 00111010 11010011 00011001 00010111
$\oplus$
$\bar{P}_Q$ = 00000000 00000001 00111101 01110110 10100111

Quadrature-phase Q = 10111000 00111011 11101110 01101111 10110000

$\square$

Following the quadrature spreading operation, the I and Q impulses are applied to the inputs of the I and Q baseband filters as shown in Figure 2.1.

2.1.1.11 Baseband Filtering

The baseband filters will have a frequency response S(f) that satisfies the limits as given in Figure 2.9. Specifically, the normalized frequency response of the filter is contained within $\pm\delta_1$ in the passband $0 \leq f \leq f_p$ and should be less than or equal to $-\delta_2$ in the stopband of $f \geq f_s$. The numerical values for the parameters are $\delta_1 = 1.5\,\text{dB}$, $\delta_2 = 40\,\text{dB}$, $f_p = 590\text{KHz}$, and $f_s = 740\text{KHz}$.

Let $s(t)$ be the impulse response of the baseband filter. Then $s(t)$ should satisfy the following equation:

$$\text{Mean Squared Error} = \sum_{k=0}^{\infty} [\alpha s(kT_S-\tau)-h(k)]^2 \leq 0.03$$

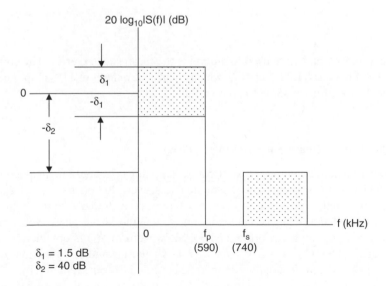

Figure 2.9   Frequency response limits of baseband filters (after IS-95A)

Table 2.13 Values of coefficients $h(k)$

k	$h(k)$
0, 47	−0.025288315
1, 46	−0.034167931
2, 45	−0.035752323
3, 44	−0.016733702
4, 43	0.021602514
5, 42	0.064938487
6, 41	0.091002137
7, 40	0.081894874
8, 39	0.037071157
9, 38	−0.021998074
10, 37	−0.060716277
11, 36	−0.051178658
12, 35	0.007874526
13, 34	0.084368728
14, 33	0.126869306
15, 32	0.094528345
16, 31	−0.012839661
17, 30	−0.143477028
18, 29	−0.211829088
19, 28	−0.140513128
20, 27	0.094601918
21, 26	0.441387140
22, 25	0.785875640
23, 24	1.0

where the constants α and τ are used to minimize the mean squared error. The constant T_s is equal to 203.451 ns ($=1/4(1.2288 \times 10^6)$), which equals one quarter of a PN chip. The values of the coefficients $h(k)$, for $k < 48$, are given in Table 2.13.

2.1.1.12 Quadrature Phase Shift Keying (QPSK)

After baseband filters, two output pulse streams, $I(t)$ and $Q(t)$, should be mapped into phase transitions. The pilot PN sequences in quadrature repeat every 26.666 ms ($=2^{15}/1.2288 \times 10^6$). Hence, there are exactly 75 repetitions in every 2 seconds (i.e. $2/26.666 \times 10^{-3}$). In non-offset QPSK, the two baseband streams coincide in time so that the carrier phase would be changed only once every T_b seconds. An orthogonal QPSK waveform $S(t)$ is obtained by amplitude modulation of $I(t)$ and $Q(t)$ ("0"s or "1"s) each onto the cosine and sine functions of a carrier wave. The in-phase stream $I(t)$ amplitude-modulates with the cosine function by an amplitude of $+1$("0") or $−1$("1"), which produces a BPSK waveform, whereas the quadrature-phase stream $Q(t)$ modulates with the sine function, resulting in a BPSK waveform orthogonal to the cosine function, resulting in a PSK waveform orthogonal to the cosine function. Thus, the sum of these two orthogonal waveforms yields the QPSK waveform.

A spectrally efficient modulation technique for CDMA channels is to maximize bandwidth efficiency, which requires simultaneous transmission on two carriers in phase quadrature. Quadrature modulation is of primary important in a spread-spectrum system and is less sensitive to some types of jamming.

Let $S(t)$ be the QPSK waveform expressed as

$$S(t) = I(t)\cos\omega_0 t + Q(t)\sin\omega_0 t$$
$$= \sqrt{2}\cos(\omega_0 t - \theta(t))$$

where $I(t) = \sqrt{2}\cos\theta(t), Q(t) = \sqrt{2}\sin\theta(t)$, and $\theta(t) = \tan^{-1}\frac{Q(t)}{I(t)}$.

The phase $\theta(t)$ is an offset from the reference phase of the periodic sequence. Thus, the QPSK stream $S(t)$ corresponding to the specific values of $I(t)$ and $Q(t)$ can be determined according to the values of $\theta(t)$ as follows:

1. For $\theta(t) = \pi/4$, $I(t)$ and $Q(t)$ become:

$$I(t) = \sqrt{2}\cos(\pi/4) = 1,$$

$$Q(t) = \sqrt{2}\sin(\pi/4) = 1, \text{ and } S(t) = \sqrt{2}\cos(\omega_0 t - \pi/4).$$

2. For $\theta(t) = 3\pi/4$, it gives $I(t) = \sqrt{2}\cos(3\pi/4) = -1,$

$$Q(t) = \sqrt{2}\sin(3\pi/4) = 1, \text{ and } S(t) = \sqrt{2}\cos(\omega_0 t - 3\pi/4).$$

3. For $\theta(t) = -3\pi/4$, we have $I(t) = \sqrt{2}\cos(-3\pi/4) = -1,$

$$Q(t) = \sqrt{2}\sin(-3\pi/4) = -1, \text{ and } S(t) = \sqrt{2}\cos(\omega_0 t + 3\pi/4).$$

4. For $\theta(t) = -\pi/4$, we have $I(t) = \sqrt{2}\cos(-\pi/4) = 1,$

$$Q(t) = \sqrt{2}\sin(-\pi/4) = -1, \text{ and } S(t) = \sqrt{2}\cos(\omega_0 t + \pi/4).$$

Based on the above analysis, $I(t)$ and $Q(t)$ mapping into phase for the Reverse CDMA Channel (either RTC or AC) can be shown as in Table 2.14.

Using Table 2.10, the signal constellation mapping and phase transition are shown in Figure 2.10. Data for the QPSK waveform plot is tabulated in Table 2.15.

The data stream Q spread by the Q-channel pilot PN chips was delayed by half a chip time. The I and Q data streams obtained in Section 2.1.1.10 are read to the I and Q data baseband filters. $I(t)$ and $Q(t)$ are two output streams coming out of the baseband filters. The timing of these two pulse streams $I(t)$ and $Q(t)$ is offset by $T_b/2$ seconds due to delaying Q by half a chip time.

The offset quadrature phase shift keying (OQPSK) waveform $S(t)$ caused by delaying Q by $T_b/2$ is illustrated in the following example.

Table 2.14 $I(t)$ and $Q(t)$ mapping for Reverse CDMA Channel

$\theta\,(t)$	$I(t)$	$Q(t)$		$I(t)$	$Q(t)$
$\pi/4$	1	1		0	0
$3\pi/4$	-1	1	Or	1	0
$-3\pi/4$	-1	-1		1	1
$-\pi/4$	1	-1		0	1
	(NRZ Value)			(Binary Value)	

Example 2.14. Refer to the two I-phase and Q-phase channel sequences, I and Q, computed in Example 2.13. They are

```
I = (10111000 00111011 10000001 01101101 01111000...),
Q = (10111000 00111011 11101110 01101100 10011000...)
```

For illustrative purposes, consider the pair of third elements $I_3 = (1000\,0001)$ and $Q_3 = (1110\,1110)$ from the I and Q streams, respectively. The offset QPSK pulse trains caused by Q_3 by half a chip time, $T_b/2$, can be illustrated as shown in Figure 2.11.   □

2.1.2 Access Channel

The access channel (AC) is used by the mobile station to initiate communication with the base station and to respond to paging channel messages. Each AC is associated with a single paging channel. Access channels are uniquely identified by their long code, that is, each AC is identified by a distinct access channel long code sequence.

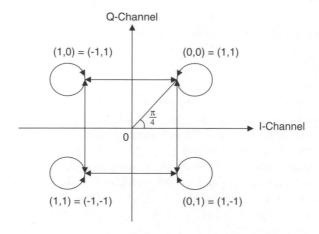

Figure 2.10   Signal constellation and phase transition applicable to Reverse CDMA Channel

Table 2.15 Data for the QPSK waveform plot

$\omega_0 t$	$S(t) = \sqrt{2}\cos(\omega_0 t - \theta(t))$			
	$(1, 1)$, $\theta\,(t) = \pi/4$	$(-1, 1)$, $\theta\,(t) = 3\pi/4$	$(-1, -1)$, $\theta\,(t) = -3\pi/4$	$(1, -1)$, $\theta\,(t) = -\pi/4$
0	1	-1	-1	1
$\pi/4$	$\sqrt{2}$	0	$-\sqrt{2}$	0
$\pi/2$	1	1	-1	-1
$3\pi/4$	0	$\sqrt{2}$	0	$-\sqrt{2}$
π	-1	1	1	-1
$-3\pi/4$	$-\sqrt{2}$	0	$\sqrt{2}$	0
$-\pi/2$	-1	-1	1	1
$-\pi/4$	0	$-\sqrt{2}$	0	$\sqrt{2}$
2π	1	-1	1	-1

The mobile station transmits information on the AC at a fixed data rate of 4800 bps. An AC frame contains 96 bits (20 ms in duration) and consists of 88 information bits and eight Encoder Tail Bits, as shown in Figure 2.12(a).

All data transmitted on the AC is convolutionally encoded, block interleaved, modulated by the 64-ary orthogonal modulation, and direct-sequence spread prior to transmission, exactly the

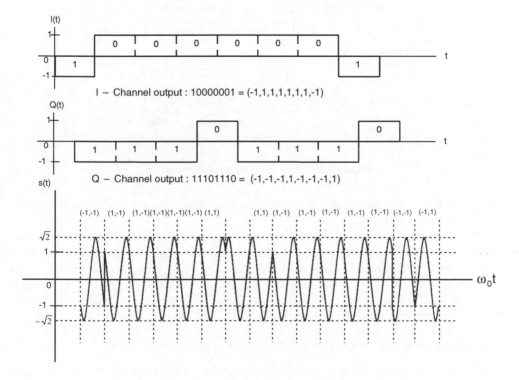

Figure 2.11 The offset QPSK waveform for 13 and Q3 sequences at the 1200 bps transmission rate

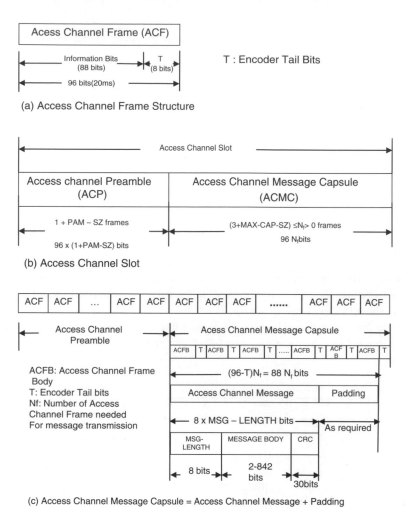

Figure 2.12   Access channel structural layout

same as with the RTC. But, note that the data burst randomizer is not used when the mobile station transmits on the AC.

The AC's numerology is quite similar to the RTC except that the transmission rate is fixed at 4800 bps after adding eight Encoder Tail Bits. Therefore, we will not go through the AC's structural analyses in detail because the same processes are repeated as with the RTC. The modulation parameters for the AC are identical to those for the RTC, except that the transmit duty cycle is 100% for the AC data rate of 4800 bps. The main properties of the AC are described briefly in the following sections.

2.1.2.1 Access Channel Preamble and Message Capsule

The Reverse CDMA Channel may contain up to 32 access channels numbered 0 through 31 per supported paging channel. An AC transmission consists of the AC preamble and the AC message capsule. The AC preamble is transmitted to aid the base station in acquiring an AC transmission. An AC message capsule consists of an AC message and padding as shown in Figure 2.12(b) and (c). The mobile station should transmit the AC message immediately following the preamble. Each AC message includes a length field (8 bits), a message body (2 to 842 bits), and a CRC (30 bits) in that order, not inclusive of the preamble or padding.

The length of the AC message capsule is an integer number of AC frames which is given by

$$
\text{Capsule Size} \; = \; \left\lceil \frac{\text{Message Length} + \text{Message Body Length} + \text{CRC}}{\text{Message Frame without Encoder Tail}} \right\rceil
$$

$$
= \; \left\lceil \frac{8 + (2 \text{ to } 842) + 30}{88} \right\rceil
$$

The message body size is selected such that the capsule size does not exceed $3 + \text{MAX} - \text{CAP} - \text{SZ}$ (the maximum number of AC frames in an access channel message capsule).

An AC message contains the 30-bit CRC sequence. This CRC sequence will be generated using the LFSR represented by the following characteristic polynomial:

$$
g(x) = x^{30} + x^{29} + x^{21} + x^{20} + x^{15} + x^{13} + x^{12} + x^{11} + x^{8} + x^{7} + x^{6} + x^{2} + x + 1
$$

The CRC field is computed by the following procedure and the corresponding CRC encoder illustrated in Figure 2.13:

1. Initial contents of the shift register are set to binary one in order to make the CRC field to nonzero values even for all-zero data.

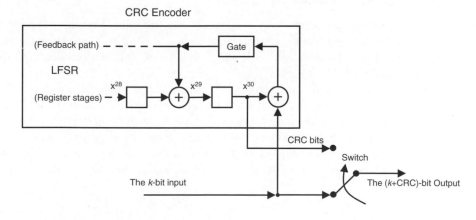

Figure 2.13   Encoder block diagram for generation of CRC field

2. Set the switch in the down position and close the gate in the feedback path of the CRC encoder.
3. Then the k-input bits start to transmit into the encoder output as well as the feedback path of the shift register. The k-bit input represents 8 + message body length in bits.
4. The shift register shall be clocked k times with k-bit input.
5. Set the switch to the up position and open the gate in the feedback path.
6. Clocking the register by an additional 30 times, the CRC field will be produced.
7. Thus, the bits will be transmitted in the order of the message length field, the message body length, and the CRC bits at the CRC encoder output.

The mobile station transmits padding consisting of zero or more "0" bits immediately following the AC message. The length of padding is confined itself as:

$$8 + \text{Message Body Length} + 30 + \text{Padding Length} = 88 \times \text{Capsule Size}$$

Example 2.15. Determine the capsule size and padding length when the message body length is assumed to be 460 bits.

$$\text{Capsule Sige} = \left\lceil \frac{8 + 460 + 30}{88} \right\rceil = 6$$

Padding Length $= (88 \times 6) - 498 = 528 - 498 = 30$ bits (all zeros). Since the capsule size cannot exceed $3 + N_{mf}$, the maximum number of AC frames N_{mf} should be at least due to $b \leq 3 + N_{mf}$. □

2.1.2.2 Access Channel Convolutional Encoding

The mobile station should convolutionally encode the data transmitted on the AC prior to interleaving. A constraint length of the convolutional code is 9. For the AC and the RTC with Rate Set 1, the convolutional code rate is 1/3. When generating AC data, the encoder should be initiated at the end of each 20 ms frame.

The generator functions for this code are $g_0 = 557$(octal), $g_1 = 663$(octal), and $g_2 = 711$ (octal). This code will generate three code symbols for each data bit input to the encoder. These code symbols are outputs so that the code symbol c_0 encoded with the generator function g_0 is the first code symbol, the code symbol c_1 encoded with g_1 is the second symbol, and the code symbol c_2 encoded with g_2 is the output last. The state of the convolutional encoder, upon initialization, should be the all-zero state. This convolutional encoder is illustrated in Figure 2.4.

2.1.2.3 Access Channel Code Symbol Repetition

For the AC, which is a fixed data rate of 4800 bps, each code symbol output from the convolutional encoder is repeated one time (i.e. each code symbol occurs two consecutive times). On the AC both repeated code symbols must be transmitted.

2.1.2.4 Access Channel Interleaving

The mobile station will interleave all repeated code symbols on the AC prior to modulation and transmission. AC repeated code symbols must be output from the interleaver by rows. The interleaver rows will be output in the following order:

1	17	9	25	5	21	13	29	3	19	11	27	7	23	15	31
2	18	10	26	6	22	14	30	4	20	12	28	8	24	16	32

The purpose of using block interleaving is not only to correct burst errors while sending the data through a multipath fading environment, but also to achieve excess redundancy for improving performance. A block interleaver framing 20 ms in duration is used. The 4800 bps block interleaver memory applicable to the AC (a 576-cell array having 32 rows and 18 columns) is shown in Table 2.16.

Table 2.16 AC interleaver memory (Write Operation) for 4800 bps

1	17	33	49	65	81	97	113	129	145	161	177	193	209	225	241	257	273
1	17	33	49	65	81	97	113	129	145	161	177	193	209	225	241	257	273
2	18	34	50	66	82	98	114	130	146	162	178	194	210	226	242	258	274
2	18	34	50	66	82	98	114	130	146	162	178	194	210	226	242	258	274
3	19	35	51	67	83	99	115	131	147	163	179	195	211	227	243	259	275
3	19	35	51	67	83	99	115	131	147	163	179	195	211	227	243	259	275
4	20	36	52	68	84	100	116	132	148	164	180	196	212	228	244	260	276
4	20	36	52	68	84	100	116	132	148	164	180	196	212	228	244	260	276
5	21	37	53	69	85	101	117	133	149	165	181	197	213	229	245	261	277
5	21	37	53	69	85	101	117	133	149	165	181	197	213	229	245	261	277
6	22	38	54	70	86	102	118	134	150	166	182	198	214	230	246	262	278
6	22	38	54	70	86	102	118	134	150	166	182	198	214	230	246	262	278
7	23	39	55	71	87	103	119	135	151	167	183	199	215	231	247	263	279
7	23	39	55	71	87	103	119	135	151	167	183	199	215	231	247	263	279
8	24	40	56	72	88	104	120	136	152	168	184	200	216	232	248	264	280
8	24	40	56	72	88	104	120	136	152	168	184	200	216	232	248	264	280
9	25	41	57	73	89	105	121	137	153	169	185	201	217	233	249	265	281
9	25	41	57	73	89	105	121	137	153	169	185	201	217	233	249	265	281
10	26	42	58	74	90	106	122	138	154	170	186	202	218	234	250	266	282
10	26	42	58	74	90	106	122	138	154	170	186	202	218	234	250	266	282
11	27	43	59	75	91	107	123	139	155	171	187	203	219	235	251	267	283
11	27	43	59	75	91	107	123	139	155	171	187	203	219	235	251	267	283
12	28	44	60	76	92	108	124	140	156	172	188	204	220	236	252	268	284
12	28	44	60	76	92	108	124	140	156	172	188	204	220	236	252	268	284
13	29	45	61	77	93	109	125	141	157	173	189	205	221	237	253	269	285
13	29	45	61	77	93	109	125	141	157	173	189	205	221	237	253	269	285
14	30	46	62	78	94	110	126	142	158	174	190	206	222	238	254	270	286
14	30	46	62	78	94	110	126	142	158	174	190	206	222	238	254	270	286
15	31	47	63	79	95	111	127	143	159	175	191	207	223	239	255	271	287
15	31	47	63	79	95	111	127	143	159	175	191	207	223	239	255	271	287
16	32	48	64	80	96	112	128	144	160	176	192	208	224	240	256	272	288
16	32	48	64	80	96	112	128	144	160	176	192	208	224	240	256	272	288

Table 2.16 can also be applicable to the reverse traffic channel for 4800 bps.

2.1.2.5 Access Channel Modulation

The AC data will be modulated as specified in Section 2.1.1.7. Modulation for the AC is 64-ary orthogonal modulation. The modulation symbol is one of 64 mutually orthogonal waveforms generated using Walsh functions. The data burst randomizer is *not* used when the mobile station transmits on the AC. The mobile station transmits on the AC. The mobile station should not gate off any power control group while transmitting on the AC.

2.1.2.6 Access Channel Direct Sequence Spreading

Direct sequence spreading using the long code is also applied to the AC. The AC should be spread by the long code as specified in Section 2.1.1.9. For the AC, direct sequence spreading operation involves modulo-2 addition of the 64-ary orthogonal modulator output and the long code. Figure 2.14 illustrates the AC long code mask format.

2.1.2.7 Access Channel Quadrature Spreading

Following the direct sequence spreading, the AC is spread in quadrature, as shown in Figure 2.2. The sequences used for this spreading are the zero-offset I and Q pilot PN sequences as specified in Section 2.1.1.10. These PN sequences are periodic with period 2^{15} chips.

2.1.2.8 Access Channel Baseband Filtering

The access channel should be filtered as specified in Section 2.1.1.11. Following the quadrature spreading operation, that is, $I = DSS \oplus \overline{P_I}$ and $Q = DSS \oplus \overline{P_Q}$, the I and Q data are applied to the inputs of the I and Q baseband filters.

Since the data spread by the Q pilot PN sequence is delayed by half a chip time, offset quadrature phase shift keying (OQPSK) is used for the spreading modulation, which is advantageous for transmitting simultaneously on two carriers in phase quadrature.

2.1.3 Multiplex Option 1 Information

Multiplex Option 1 applies to Rate Set 1, providing for the transmission of primary traffic and signaling or secondary traffic. Signaling traffic may be transmitted via blank-and-burst, with the signaling traffic using all of the frame, or via dim-and-burst with the primary traffic and signaling traffic sharing the frame. Multiplex Option 1 also supports the transmission of

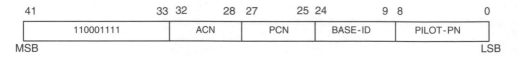

41		33	32	28	27	25	24	9	8	0
	110001111			ACN		PCN		BASE-ID		PILOT-PN

MSB   LSB

ACN: Access Channel Number
PCN: Paging Channel Number
BASE-ID: Base Station Identification
PILOT-PN: PN offset for Forward CDMA Channel

Figure 2.14   Access channel long code mask format

Table 2.17   RTC information bits for Multiplex Option 1

Transmit rate (bits/sec)	Format bits			Primary traffic bits/frame	Signaling traffic bits/frame	Secondary traffic bits/frame
	Mixed Mode (MM)	Traffic Type (TT)	Traffic Mode (TM)			
	"0"	—	—	171	0	0
	"1"	"0"	"00"	80	88	0
	"1"	"0"	"01"	40	128	0
	"1"	"0"	"10"	16	152	0
9600	"1"	"0"	"11"	0	168	0
*	"1"	"1"	"00"	80	0	88
*	"1"	"1"	"01"	40	0	128
*	"1"	"1"	"10"	16	0	152
*	"1"	"1"	"11"	0	0	168
4800	—	—	—	80	0	0
2400	—	—	—	40	0	0
1200	—	—	—	16	0	0

Note that secondary traffic structures, marked with *, are optional.

secondary traffic. When primary traffic is available, secondary traffic is transmitted via dim-and-burst with the primary traffic and secondary traffic sharing the frame. The information bit structures for primary, signaling, and secondary traffic sharing the same frame are specified in Table 2.17 and Figure 2.15. Information bits for secondary traffic, as shown in Figure 2.15, were superimposed as indicated by bits inside the square frame.

2.1.4 Multiplex Option 2 Information

Multiplex Option 2 applies to Rate Set 2, providing for the transmission of primary traffic, secondary traffic, and signaling traffic. The mobile station may support Multiplex Option 2. If so, it should support the transmission of primary traffic and signaling traffic using the information bit structures as shown in Table 2.18. When the mobile station supports secondary traffic, the mobile station should also use the information bit structure specified in Table 2.18.

Signaling traffic may be transmitted via blank-and-burst with the signaling traffic using all the frames, via dim-and-burst with the primary traffic and signaling traffic sharing the frame, or via dim-and-burst with the primary traffic, secondary traffic, and signaling traffic sharing the same frame.

2.2 Forward CDMA Channel

The Forward CDMA Channel is complied with one or more code channels that are transmitted on a CDMA frequency assignment using a particular pilot PN offset. The Forward CDMA Channel consists of the pilot channel, up to one sync channel, up to seven paging channels, and a number of forward traffic channels, Each of these code channels is orthogonally spread by the appropriate Walsh function to provide orthogonal channelization among all code channels on a

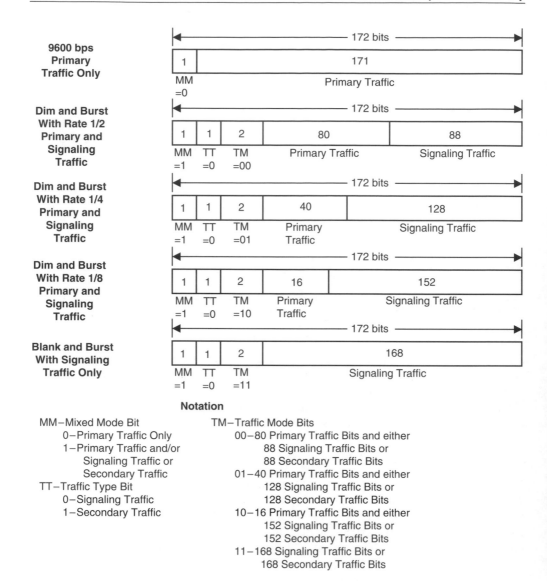

Figure 2.15   Information bits for primary traffic and signaling traffic, including secondary traffic

given Forward CDMA Channel. Following the orthogonal spreading, each code channel is spread by a quadrature pair of pilot PN sequences at a fixed chip rate of 1.2288 Mcps. After the spreading operation, the I and Q impulses are applied to the I and Q baseband filters. Following baseband filtering, the binary I and Q at the output of the quadrature spreading is mapped into specified phase transitions. Multiple Forward CDMA Channels may be used within a base station in a frequency division multiplexed manner.

A typical example of the code channels transmitted by a base station is shown in Figure 2.16. Out of the 64 code channels available for use, this example depicts the pilot channel (always

Table 2.18   RTC information bits for Multiplex Option 2

Transmit rate (bits/sec)	Format bits		Primary traffic (bits/frame)	Signaling traffic (bits/frame)	Secondary traffic (bits/frame)
	Mixed Mode (MM)	Frame Mode (FM)			
14400	"0"	—	266	0	0
	"1"	"0000"	124	138	0
	"1"	"0001"	54	208	0
	"1"	"0010"	20	242	0
	"1"	"0011"	0	262	0
*	"1"	"0100"	124	0	138
*	"1"	"0101"	54	0	208
*	"1"	"0110"	20	0	242
*	"1"	"0111"	0	0	262
*	"1"	"1000"	20	222	20
7200	"0"	—	124	0	0
	"1"	"000"	54	67	0
	"1"	"001"	20	101	0
	"1"	"010"	0	121	0
*	"1"	"011"	54	0	67
*	"1"	"100"	20	0	101
*	"1"	"101"	0	0	121
*	"1"	"110"	20	81	20
3600	"0"	—	54	0	0
	"1"	"00"	20	32	0
	"1"	"01"	0	52	0
*	"1"	"10"	20	0	32
*	"1"	"11"	0	0	52
1800	"0"	—	20	0	0
*	"1"	—	0	0	20

Note that the mobile station support of the secondary traffic structures, marked with *, are optional.

required), one sync channel, seven paging channels (the maximum number allowed), and 55 traffic channels. Code channel number zero (W_0) is always assigned to the pilot channel. If the sync channel is present, it will be assigned to code channel number 32 (W_{32}). If paging channels are present, they should be assigned to code channel numbers one through seven (W_1 to W_7 inclusive) in sequence. The remaining code channels are available to the forward traffic channels (W_8, W_9, ..., W_{31}, W_{32}, ..., W_{63}). Another possible configuration could replace all paging channels and one sync channel with traffic channels something like a maximum of one pilot channel, zero paging channels, zero sync channels, and 63 traffic channels.

The sync channel operates at a fixed rate of 1200 bps. The paging channel supports fixed data rate operation at 9600 or 4800 bps. The forward traffic channels are grouped in sets called Rate Sets. Rate Set 1 contains four elements, that is, 14 400, 7200, 3600, and 1800 bps. The base

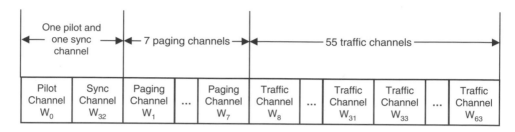

Pilot Channel W_0	Sync Channel W_{32}	Paging Channel W_1	...	Paging Channel W_7	Traffic Channel W_8	...	Traffic Channel W_{31}	Traffic Channel W_{33}	...	Traffic Channel W_{63}

Figure 2.16   A Forward CDMA Channel transmitted by a base station

station should support Rate Set 1 on the forward traffic channel. The base station may support Rate Set 2 on the forward traffic channel. The base station shall support variable data rate operation with all four elements of each supported Rate Set.

The overall structures of the Forward CDMA Channel are shown in Figures 2.17, 2.18, 2.22, and 2.28. Requirements that are specific to CDMA base station operation will be discussed in the following sections.

2.2.1 Pilot Channel

A pilot channel is transmitted at all times by the base station on each active Forward CDMA Channel. The pilot channel is an unmodulated spread spectrum signal that is used for synchronization by a mobile station operating within the coverage area of the base station. The mobile station monitors the pilot channel for every CDMA channel supported by the base station. The mobile station monitors the pilot channel at all times except when not receiving in the slotted mode. The pilot channel is a reference channel that allows a mobile station to acquire the timing of the Forward CDMA Channel and thus provides a phase reference for coherent demodulation. The pilot channel is illustrated in Figure 2.17.

In the pilot channel acquisition substate, the mobile station acquires the pilot channel of the selected CDMA system. Upon entering this substate, the mobile station tunes to the CDMA channel number equal to CDMACHs, sets its code channel for the pilot channel, and searches

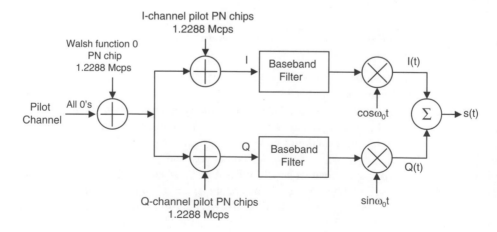

Figure 2.17   Pilot channel structure

for the pilot channel. If the mobile station acquires the pilot channel within $T_{20m} = 15$ seconds, the mobile station enters the sync channel acquisition substate. If the mobile station does not acquire the pilot channel within T_{20m} seconds, the mobile station enters the system determination substate. T_{20m} denotes the maximum time to remain in the pilot channel acquisition substate of the mobile station initialization state.

In the CDMA cellular telephone system, each cell site transmits a pilot carrier signal at each like frequency. This pilot carrier is used by the mobile station to obtain initial system synchronization and to provide robust time, frequency, and phase tracking of the signals from the cell site. This signal is tracked continuously by each mobile station. Variations in the transmitted power level of the pilot signal control the coverage area of the cell. The pilot carriers are transmitted by each cell site using the same code but with different spread spectrum code phase offsets, allowing them to be distinguished. The fact that the pilots all use the same code allows the mobile station to find system timing synchronization by a single search through all code phases. The strongest signal found corresponds to the code phase of the cell site.

2.2.1.1 Pilot PN Sequence Offset

Each base station uses a time base reference from which the pilot PN sequence will be derived. The time base reference should be time-aligned to CDMA System Time in order to synchronize each base station's time base reference to CDMA System Time.

Each base station should use a time offset of the pilot PN sequence to identify a Forward CDMA Channel. Time offsets may be reused within a CDMA cellular system. Distinct pilot channels are identified by an offset index (0 through 511 inclusive), specifying the offset value from the zero offset pilot PN sequence. The zero offset pilot PN sequence designating the start of the sequence will be output at the beginning of every even second with reference to the base station transmission time. The start of the zero offset pilot PN sequence for the I or Q sequence should be defined as the state of the sequence for which the previous 15 outputs were zeros ("0"s).

512 values are possible for the pilot PN sequence offset. The offset in chips for a given pilot PN sequence from the zero shift pilot PN sequence equals the index value multiplied by 64. As an example, if the pilot PN sequence offset index is 15, then the pilot PN sequence will be $15 \times 64 = 960$ PN chips. In this case the pilot PN sequence will start 781.25 μs after the start of every *even* second of time, referenced to the base station transmission time.

2.2.1.2 Pilot Channels Orthogonal Spreading

Each code channel transmitted on the Forward CDMA Channel is spread with a Walsh function at a fixed chip rate of 1.2288 Mcps to provide orthogonal channelization among all code channels on a given Forward CDMA Channel. One of 64 time-orthogonal Walsh functions, as defined in Table 2.12, should be used. A code channel that is spread using Walsh function n shall be assigned to code channel number n ($n = 0-63$). Code channel number $n = 0$ is always assigned to the pilot channel. This 64-zero Walsh chip is input to the quadrature spreading.

2.2.1.3 Pilot Channel Quadrature Spreading

Following the orthogonal spreading, the pilot channel is spread in quadrature, as discussed fully in Section 2.1.1.10. The quadrature spreading sequence of length 2^{15} chips is based on the following characteristics polynomials:

$P_I(x) = x^{15} + x^{13} + x^9 + x^8 + x^7 + x^5 + 1$ (for the in-phase sequence)
$P_Q(x) = x^{15} + x^{12} + x^{11} + x^{10} + x^6 + x^5 + x^4 + x^3 + 1$ (for the quadrature-phase sequence)

The maximum length LFSR $\{i(x)\}$ and $\{q(x)\}$ based on the above polynomials are of length 2^{15}-1 and can be generated by the following linear recursion:

$$i(n) = i(n-15) \oplus i(n-10) \oplus i(n-8) \oplus i(n-7) \oplus i(n-6) \oplus i(n-2)$$

and

$$q(n) = q(n-15) \oplus q(n-12) \oplus q(n-11) \oplus q(n-10) \oplus q(n-9) \oplus q(n-5) \oplus q(n-4) \oplus q(n-3)$$

where $i(n)$ and $q(n)$ are binary values ("0" and "1") and $\oplus$ denotes the modulo-2 addition. In order to obtain the I and Q pilot PN sequences of period 2^{15}, a "0" must be inserted in $\{i(n)\}$ and $\{q(x)\}$ for producing one run of 15 consecutive "0" outputs. The chip rate for the pilot PN sequence is 1.2288 Mcps. The pilot PN sequence period is $32\,768/1\,228\,800 = 26.666$ ms and exactly 75 pilot PN sequence repetitions occur every 2 seconds. The pilot PN sequence offset was specified in Section 2.1.1.10.

2.2.1.4 Pilot Channel Filtering

Following the quadrature operation, the I and Q impulses are applied to the inputs of the I and Q baseband filters as specified in Section 2.1.1.11. The base station should provide phase equalization for the transmit signal path. The equalization filter is designed to provide the equivalent baseband transfer function:

$$H(\omega) = K \frac{\omega^2 + j\alpha\omega\omega_0 - \omega_0^2}{\omega^2 - j\alpha\omega\omega_0 - \omega_0^2}$$

where K is an arbitrary gain, ω is the radiation frequency, $j = \sqrt{-1}$, $\alpha = 1.36$ and $\omega_0 = 2\pi \times 3.15 \times 10^5$. The equalizing filter implementation is equivalent to applying baseband filters with this transfer function individually to the baseband I and Q waveforms.

After baseband filtering, the binary I and Q pulse trains are mapped into phase transitions according to signal constellation.

2.2.2 Sync Channel

The sync channel is assigned to the code channel 32 (W32) in the Forward CDMA Channel, which transports the synchronization message to the mobile station. The sync channel will operate at a fixed rate of 1200 bps as shown in Figure 2.18. The sync channel is convolutionally encoded before transmission and each encoded symbol is repeated one time, that is, occurring two consecutive times prior to block interleaving. All symbols after repetition on the sync channel are block interleaved. The sync channel uses a block interleaver spanning 26.667 ($=128/4800$) ms that is equivalent to 128 modulation symbols at the symbol rate of 4800 sps. A sync channel frame is 26.667 ms in duration. The sync channel transmitted on the Forward CDMA Channel is spread with a Wash function at a fixed chip rate of 1.2288 Mcps.

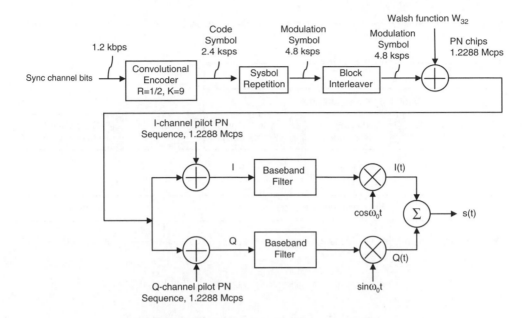

Figure 2.18 Sync channel structure (1200 bps)

In summary, the sync channel is an encoded, interleaved, spread and modulated spread spectrum signal that is used by mobile stations to acquire initial time synchronization within the coverage area of the base station.

The I and Q channel pilot sequences for the sync channel use the same pilot PN sequence offset as the pilot channel for a given base station. The sync channel is used during the system acquisition stage. Once the mobile station acquires the system, it will not normally reuse the sync channel until it powers on again. Once the mobile station achieves pilot PN sequence synchronization by acquiring the pilot channel, the synchronization for the sync channel becomes immediately known. This is because the sync channel is spread with the same pilot PN sequence, and because the frame and interleaver timing on the sync channel are aligned with the pilot PN sequence.

A sync channel superframe is formed by three sync channel frames ($26.667 \times 3 = 80$ ms). Message transmission on the sync channel begins only at the start of a sync channel superframe, as shown in Figure 2.19. Finally, the modulation parameters of a sync channel are listed in Table 2.19.

2.2.2.1 Sync Channel Encoding

The base station should convolutionally encode the data transmitted on the sync channel. In general, an (n,k,m) convolutional code designates the code rate $R = k/n$ with encoder stages of $m = K - 1$, where K is the constraint length of the code. The sync channel of Rate Set 1 is encoded by the (2,1,8) convolutional encoder with the code rate $R = 1/2$ and a constraint length of $K = 9$, as illustrated in Figure 2.20. The generator function for the rate 1/2 code will be $g_0 = 753$(octal) $= (111101011)$(binary) and $g_1 = 561$(octal) $= (101110001)$(binary), respectively. This code thus generates two code symbols for each data bit input to the encoder.

On the assumption that the initial content of encoder stages is all zeros, the code symbol c_0 encoded with $g_0 = 753$ is the first output c_0 and the code symbol c_1 encoded with $g_1 = 561$ is the

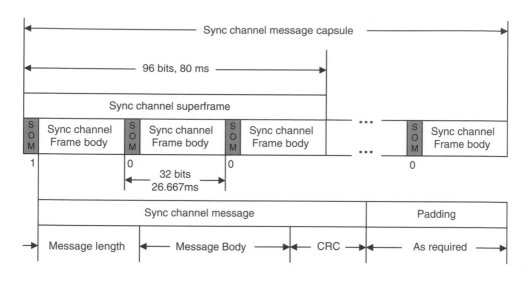

SOM: Start of Message bit

Figure 2.19 Sync channel superframe structure

second symbol. Multiplexing and pairwise concatenating, $(c_0 \| c_1)$, will result in the encoder output symbols.

Example 2.16. Consider the (2,1,8) convolutional encoder with two generator functions $g_0 = (111101011)$ and $g_1 = (101110001)$ for applying to the sync channel. The 24-bit message input is assumed as

m = (100111010011100010100101)

When the message sequence reads in the encoder bit by bit, the 32-bit encoder output symbols c_0 and c_1 are computed as

c_0 = (1110001000 0000110100 0110101001 11)
c_1 = (1010000010 0111110010 1010000101 01)

Table 2.19 Sync channel modulation parameters

Parameters	Data rate 1200 bps	Units
PN chip rate	1.2880	Mcps
Code rate	1/2	bits/code symbol
Code repetition	2	mod symbol/code symbol*
Modulation symbol rate	4800	sps
PN chips/Modulation symbol	256	PN chips/mod symbol
PN chips/bit	1024	PN chips/bit

* Each repetition of a code symbol is a modulation symbol.

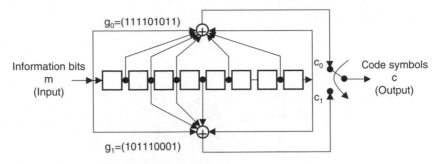

$g_0=(111101011)$

Information bits
m
(Input)

c_0

Code symbols
c
(Output)

c_1

$g_1=(101110001)$

(a) Encoder based on the generator vectors

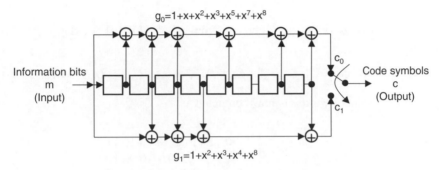

$g_0=1+x+x^2+x^3+x^5+x^7+x^8$

Information bits
m
(Input)

c_0

Code symbols
c
(Output)

c_1

$g_1=1+x^2+x^3+x^4+x^8$

(b) Encoder based on the generator polynomials

Figure 2.20   The (2,1,8) convolutional encoder: (a) based on g_0 and g_1; (b) based on $g_0(x)$ and $g_1(x)$

Pairwise concatenation such as $(c_0\|c_1)$ will result in the 64-bit output symbol sequence as

c = (1110110000 0010000100 0001010111
1100100100 0110110010 0010010011 1011)   □

Example 2.17. In this example, alternative methods to produce the identical solution are presented.
   Discrete Convolution Method:

$$c_0 = m * g_0 = (1001110100111100010100101) \otimes (111101011)$$
$$= (1110\ 0010\ 0000\ 0011\ 0100\ 0110\ 1010\ 0111)$$
$$c_1 = m * g_1 = (1001\ 1101\ 0011\ 1000\ 1010\ 0101) \otimes (101110001)$$
$$= (1010\ 0000\ 1001\ 1111\ 0010\ 1010\ 0001\ 0101)$$

The code symbol word c can be obtained by pairwise concatenation of $(c_0\|c_1)$ as shown below:

c = (11 10 11 00 00 00 10 00 01 00 00 01 01 01 11 11
00 10 01 00 01 10 11 00 10 00 10 01 00 11 10 11)

Scalar Matrix Product Method:

$$C = m \cdot G$$

where G is a semi-infinite generator matrix.

$$G = \begin{bmatrix} G_0 & G_1 & G_2 & \cdots & G_k & & & \\ & G_0 & G_1 & G_2 & \cdots & G_k & & \\ & & G_0 & G_1 & G_2 & \cdots & G_k & \\ & & & \ddots & & & & \ddots \\ & & & & \ddots & & & \\ & & & & & \ddots & & \\ & & & & & & \ddots & \end{bmatrix}$$

Component matrix G_k, $0 \le k \le 8$, is obtained from $g_0 = (111101011)$ and $g_1 = (101110001)$ as follows:

$$G_0 = 11, G_1 = 10, G_2 = 11, G_3 = 11, G_4 = 01, G_5 = 10, G_6 = 00, G_7 = 10, G_8 = 11.$$

Thus, the code symbol word c is now computed as

$C = m \cdot G$

$$= (1001110100111000 10100101) \cdot \begin{bmatrix} 11\ 10\ 11\ 11\ 01\ 10\ 00\ 10\ 11 & & \\ & 11\ 10\ 11\ 11\ 01\ 10\ 00\ 10\ 11 & \\ & & 11\ 10\ 11\ 11\ 01\ 10\ 00\ 10\ 11 \\ & \ddots & & \ddots \\ & & \ddots & & \ddots \\ & & & 11\ 10\ 11\ 11\ 01\ 10\ 00\ 10\ 11 \end{bmatrix}_{24 \times 64}$$

$$= (11101100000010000100000101011111001001000110)$$

This is the code symbol word c as expected.

Polynomial Matrix Method:
The message polynomial is represented by

$$m(D) = D^{23} + D^{20} + D^{19} + D^{18} + D^{16} + D^{13} + D^{12} + D^{11} + D^7 + D^5 + D^2 + 1$$

The generator polynomial matrix is expressed as

$$\begin{aligned} G(D) &= [g_0(D), g_1(D)] \\ &= [D^8 + D^7 + D^6 + D^5 + D^3 + D + 1, D^8 + D^6 + D^5 + D^4 + 1] \end{aligned}$$

The code symbol polynomial matrix at the modular-2 adders is obtained as

$$\begin{aligned} [C_0(D), C_1(D)] = [&(D^{31} + D^{30} + D^{29} + D^{25} + D^{17} + D^{16} + D^{14} + D^{10} + D^9 + D^7 + D^5 + D^2 + D + 1), \\ &(D^{31} + D^{29} + D^{23} + D^{20} + D^{19} + D^{18} + D^{17} + D^{16} + D^{13} + D^{11} + D^9 + D^4 + D^2 + 1)] \end{aligned}$$

Next, the delay operation by the commutation switch will produce the encoder output symbol sequence as follows:

$$
\begin{aligned}
C(D) &= DC_0(D^2) + C_1(D^2) \\
&= D(D^{62} + D^{60} + D^{58} + D^{50} + D^{34} + D^{32} + D^{28} + D^{20} + D^{18} + D^{14} + D^{10} \\
&\quad + D^4 + D^2 + 1) + (D^{62} + D^{58} + D^{46} + D^{40} + D^{38} + D^{36} + D^{34} + D^{32} + D^{26} \\
&\quad + D^{22} + D^{18} + D^8 + D^4 + 1) \\
&= D^{63} + D^{62} + D^{61} + D^{59} + D^{58} + D^{51} + D^{46} + D^{40} + D^{38} + D^{36} + D^{35} + D^{34} \\
&\quad + D^{33} + D^{32} + D^{29} + D^{26} + D^{22} + D^{21} + D^{19} + D^{18} + D^{15} + D^{11} + D^8 + D^5 \\
&\quad + D^4 + D^3 + D + 1
\end{aligned}
$$

$$
\begin{aligned}
c = \ &(11\ 10\ 11\ 00\ 00\ 00\ 10\ 00\ 01\ 00\ 00\ 01\ 01\ 01\ 11\ 11 \\
&\ \ 00\ 10\ 01\ 00\ 01\ 10\ 11\ 00\ 10\ 00\ 10\ 01\ 00\ 11\ 10\ 11)
\end{aligned}
$$

☐

Thus, we have demonstrated how to obtain the identical encoder output symbols through several alternative methods. Readers are recommended to understand each of these alternative approaches.

Notice that the state of the sync channel convolutional encoder should not be reset between sync channel frames.

2.2.2.2 Sync Channel Code Symbol Repetition

The transmission rate supporting the sync channel is fixed at 1200 bps. For the sync channel, each convolutionally encoded symbol will be repeated one time (i.e. each symbol occurs 2 consecutive times) prior to block interleaving.

2.2.2.3 Sync Channel Interleaving

For the sync channel with Rate Set 1, all the symbols after symbol repetition should be interleaved. The purpose of block interleaving is to avoid burst errors and to provide the access redundancy for achieving performance improvements.

The sync channel is divided into 80 ms superframes. Since the sync channel frame consists of 32 bits and 26.667 ($=32/1200$) ms, each superframe is divided into three 26.667 ms frames. Thus, the sync channel uses a block interleaver spanning 26.667 ms, which is equivalent to 128 ($=16 \times 8$) modulation symbols at the symbol rate of 4800 sps. Notice that each repetition of a code symbol indicates a modulation symbol.

The input (array write) symbol sequence to the sync channel interleaver is given in Table 2.20. The table is read down by columns from the left to the right. That is, the first input symbol (1) is at the top left, the second input symbol (1) is just below the first input symbol, the seventeenth input symbol (9) is just to the right of the first input symbol, and the eighteenth input symbol (9) is just to the right of the second input symbol (1). In this input symbol table, symbols with the same number denote repeated code symbols. The output (array read) symbol sequence is given in Table 2.21. That is, the first output symbol (1) is at the top left, the second output symbol (33) is just below the first output symbol, and the seventeenth output symbol (3) is just to the right of the first output symbol.

Table 2.20   Sync channel interleaver input (Array Write Operation)

1	9	17	25	33	41	49	57
1	9	17	25	33	41	49	57
2	10	18	26	34	42	50	58
2	10	18	26	34	42	50	58
3	11	19	27	35	43	51	59
3	11	19	27	35	43	51	59
4	12	20	28	36	44	52	60
4	12	20	28	36	44	52	60
5	13	21	29	37	45	53	61
5	13	21	29	37	45	53	61
6	14	22	30	38	46	54	62
6	14	22	30	38	46	54	62
7	15	23	31	39	47	55	63
7	15	23	31	39	47	55	63
8	16	24	32	40	48	56	64
8	16	24	32	40	48	56	64

2.2.2.4 Sync Channel Orthogonal Spreading

The sync channel transmitted on the Forward CDMA Channel is spread with a Walsh function at a fixed chip rate of 1.2288 Mcps.

The sync channel that is spread using the Walsh function 32 (which is one of 64 time-orthogonal Walsh functions) will be assigned to code channel number 32 (W_{32}). The Walsh function index (modulation symbol index) n for the sync channel is designated by $n = 32$ either in the row heading or in the column heading shown in Table 2.12. The Walsh function spreading sequence repeats with a period of 52.083 µs ($=64/1.2288$ Mcps).

The Walsh chips within a Walsh index 32 are 00000000 00000000 00000000 00000000 11111111 11111111 11111111 11111111. Since the interleaver output consists of 128 symbols,

Table 2.21   Sync channel interleaver output (Array Read Operation)

1	3	2	4	1	3	2	4
33	35	34	36	33	35	34	36
17	19	18	20	17	19	18	20
49	51	50	52	49	51	50	52
9	11	10	12	9	11	10	12
41	43	42	44	41	43	42	44
25	27	26	28	25	27	26	28
57	59	58	60	57	59	58	60
5	7	6	8	5	7	6	8
37	39	38	40	37	39	38	40
21	23	22	24	21	23	22	24
53	55	54	56	53	55	54	56
13	15	14	16	13	15	14	16
45	47	46	48	45	47	46	48
29	31	30	32	29	31	30	32
61	63	62	64	61	63	62	64

the Walsh chips within a Walsh function index 32 will be used twice. Due to the fact that the modulation symbol rate is 4800 sps and PN chip rate is 1.2288 Mcps, the ratio of these two rates is 256, which corresponds to 256 Walsh chips with respect to 1 interleaver output symbol.

2.2.2.5 Sync Channel Quadrature Spreading

Following the orthogonal spreading, the sync channel should be PN spread as specified in Section 2.1.1.10. The spreading sequence is a quadrature sequence of length 2^{15} (i.e. 32 768 PN chips in length). This spreading sequence is called the pilot PN sequence and based on $P_I(x)$ and $P_Q(x)$, as indicated in Section 2.1.1.10.

The maximum length LFSR sequences $\{i(n)\}$ and $\{q(n)\}$ based on $P_I(x)$ and $P_Q(x)$ are of length $2^{15} - 1$. $\{i(n)\}$ and $\{q(n)\}$ can be determined by the linear recursions $i(n)$ and $q(n)$ that are binary-valued ("0" and "1"). In order to obtain the I and Q pilot PN sequences of period 2^{15}, they should have one run of 15 consecutive "0" outputs.

Since the chip rate for the pilot PN sequence is 1.2288 Mcps, its period is 26.667 ($=$32 768/ 1 228 800) ms, and exactly 75 pilot PN sequence repetitions occur every 2 seconds.

2.2.2.6 Sync Channel Filtering

Following the quadrature spreading operation, the I and Q impulses are applied to the inputs of the I and Q baseband filters. Filtering for the sync channel is as specified in Section 2.1.1.11.

2.2.2.7 Sync Channel Message Capsule and Message Structure

This section specifies requirements for the signaling message format transmitted on the sync channel. The sync channel is divided into 80 ms, 96 bits superframes. Each superframe is divided into three 26.667 (32/1200) ms frames because the sync channel operates at a fixed rate of 1200 bps. The first bit of each frame is a Start-of-Message (SOM) bit and the remaining bits in the frame comprise the sync channel frame body.

A sync channel message capsule is composed of a sync channel message and padding. A sync channel message consists of an 8-bit message length field, a message body, and 30-bit CRC field. Padding bits are set to zero ("0") and appended to the end of the sync channel message.

A typical example of sync channel superframe structure is depicted in Figure 2.19. Sync channel message capsules begin with the first bit of the first sync channel frame body of a sync channel superframe. The base station will set the SOM bit immediately preceding the beginning of a sync channel message capsule to "1" and set all other SOM bits to "0."

The base station transmits the sync channel message in consecutive sync channel frame bodies. The base station should include sufficient padding bits in each sync channel message capsule to extend it through the bit preceding the SOM bit at the beginning of the next sync channel superframe. The base station will begin a new sync channel message capsule in the first sync channel frame of that superframe. The base station limits the maximum sync channel message length to 148 octets, that is $148 * 8 = 1184$ bits.

A 30-bit CRC is computed for the sync channel signaling message, which includes the 8-bit message length field, the message body field (2–2002 bits), and the 30-bit CRC. The CRC field can be generated by the maximum length LFSR characteristics polynomial of

$$g(x) = x^{30} + x^{29} + x^{21} + x^{20} + x^{15} + x^{13} + x^{12} + x^{11} + x^8 + x^7 + x^6 + x^2 + x + 1$$

The CRC computation procedure for the sync channel is identical to the procedure for the access channel CRC computation, described in Section 2.1.2.1.

2.2.3 Paging Channel

The paging channel is an encoded, interleaved, spread, and modulated spread spectrum signal that is used by mobile stations operating within the coverage area of the base station. The base station uses the paging channel to transmit system overhead information and mobile station specific messages.

- *Modulation Rates and Time Alignment:* The paging channel transmits information at a fixed data rate of 9600 or 4800 bps and its frame is 20 ms in duration. All paging channels in a given system should transmit information at the same data rate.

  The I and Q channel pilot PN sequences for the paging channel use the same pilot PN sequence offset as the pilot channel for a given base station.

  The first paging channel frame will occur at the start of base station transmission time. The paging channel is divided into paging channel slots that are each 80 ms in duration. For the non-slotted mode of operation, paging and control messages for a mobile station can be received in any of the paging channel slots. For the slotted mode of operation, a mobile station monitors the paging channel only during certain assigned slots. Figure 2.21 shows an example of the mobile station's slotted length of 1.28 seconds, in which the computed value of a paging slot is equal to 6. Consequently, the mobile station's slot cycle begins when the slot number equals 6. The mobile station will begin monitoring the paging channel at the start of the slot number 6. The minimum length slot cycle consists of 16 slots of 80 ms each, as shown in Figure 2.21, 1.28 ($=16 \times 80$ ms) seconds.

  The paging channel slot number is determined as

  $$\text{Slot No} = \lfloor t/4 \rfloor \bmod 2048$$

  where t is the system time in frames and mod 2048 denotes the maximum length slot cycle (2048 slots).

  The start of the interleaver block and frame of the paging channel should align with the start of the zero-offset pilot PN sequence at every even second time mark.
- *Paging Channel Modulation Parameters:* The modulation parameters for a paging channel are as shown in Table 2.22.

The paging channel structure for the data rate of either 9600 bps or 4800 bps is shown in Figure 2.22.

2.2.3.1 Paging Channel Convolutional Encoding

The paging channel data is convolutionally encoded in the same manner as the sync channel encoding. For the paging channel with Rate Set 1, the code rate is 1/2 and a constraint length is 9. The generator functions for the rate 1/2 code are $g_0 = 753$ (octal) and $g_1 = 561$ (octal). This code generates two code symbols for each data bit input to the encoder. Convolutional encoding involves the modulo-2 addition of selected taps of a serially time-delayed data sequence by means of pairwise concatenation switching.

For the paging channel as well as the sync channel and forward traffic channel Rate Set 1, the (2,1,8) convolutional encoder will be employed and its encoder scheme is illustrated as shown in Figure 2.20.

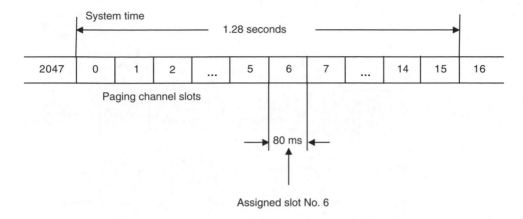

Figure 2.21   Mobile station's slotted mode operation

2.2.3.2 Paging Channel Code Symbol Repetition

For the paging channel, each convolutionally encoded symbol should be repeated prior to block interleaving whenever the information rate is lower than 9600 bps. Each code symbol at the 4800 bps rate is repeated one time (i.e. each symbol occurs two consecutive times).

2.2.3.3 Paging Channel Interleaving

For the paging channels with Rate Set 1, all code symbols after symbol repetition should be block interleaved. The paging channels use the interleaver spanning 20 ms, which is equivalent to 384 modulation symbols at the modulation symbol rate of 19 200 sps.

The interleaver block should align with the paging channel frame. The alignment shall be such that the first bit of the frame influences the first 18 (for 9600 bps) or 36 (for 4800 bps) modulation symbols input to the interleaver. Since the paging channel is not convolutionally encoded by blocks, the last 8 bits of a paging channel frame influence symbols in the successive interleaver block.

Table 2.22   Paging channel modulation parameters

Parameters	Data rate (bps)		Units
	9600	4800	
PN chip rate	1.2288	1.2288	Mcps
Code rate	1/2	1/2	bits/code symbol
Code symbol repetition	1	2	mod symbols/code symbol*
Modulation symbol rate	19 200	19 200	sps
PN chips/Modulation symbol	64	64	PN chips/mod symbol
PN chips/bit	128	256	PN chips/bit

* Each repetition of a code symbol is a modulation symbol.

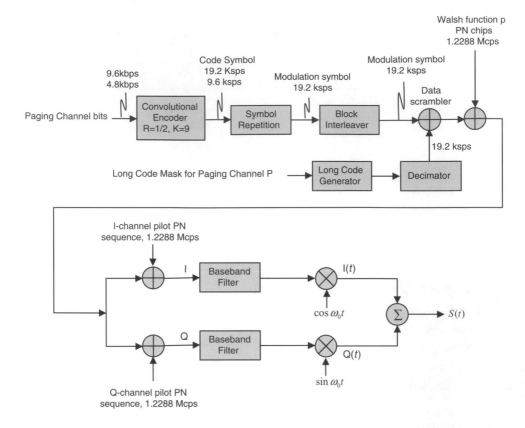

Figure 2.22   Paging channel structure for the data rate of 9600 bps or 4800 bps

2.2.3.4 Paging Channel Data Scrambling

Data scrambling is applied to the paging channel (as well as the forward traffic channel) by using the paging channel long code mask. This long code mask consists of a 42-bit binary sequence as shown in Figure 2.23. The data scrambling is accomplished by performing modulo-2 addition of the interleaver output symbol with the decimated binary value of the long code PN chip that is valid at the start of the transmission period for that symbol, as illustrated in Figure 2.24.

Example 2.18. Consider the data scrambling problem based on Figures 2.23 and 2.24. Suppose the paging channel number (PCN) is 001 and the pilot PN sequence offset index is 100100011. Then the long code mask sequence will become

$$110001100110100000010000000000000100100011.$$

The long code is periodic with period $2^{42} - 1$ chips specified by the LFSR characteristic polynomial $p(x)$ of code generator:

$$
\begin{aligned}
p(x) = {} & x^{42} + x^{35} + x^{33} + x^{31} + x^{27} + x^{26} + x^{25} + x^{22} + x^{21} + x^{19} + x^{18} + x^{17} \\
& + x^{16} + x^{10} + x^{7} + x^{6} + x^{5} + x^{3} + x^{2} + x + 1 \text{ (Polynomial Form)}
\end{aligned}
$$

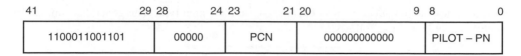

41		29 28	24 23	21 20		9 8	0
1100011001101		00000	PCN	000000000000		PILOT – PN	

PCN: Paging Channel Number
PILOT- PN: Pilot sequence offset index for the Forward CDMA Channel

Figure 2.23 Paging channel long code mask

or

$$P = (111101110010000011110110011100010101000000) \text{ (Vector Form)}$$

Each PN chip of the long code is generated by the modulo-2 inner product of a 42-bit mask and the 42-bit state vector of the sequence generator, as shown in Figure 2.6. Notice here that the

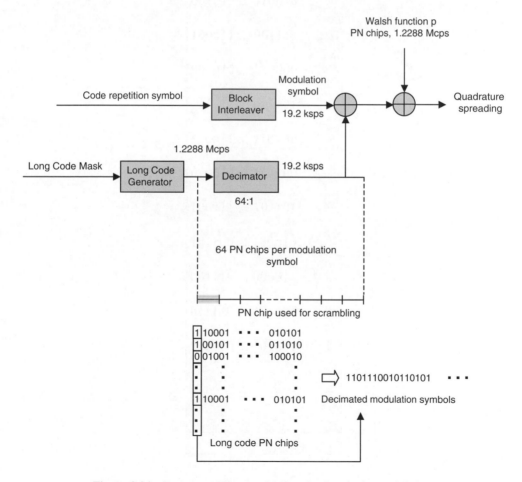

Figure 2.24 Data scrambling mechanism for functioning and timing

mask used for the long code varies depending on the channel type. The generated long code sequence is computed as follows:

	← 42 bits →
1	100010 ... 010101
1	001011 ... 011010
0	010011 ... 100010
1	011101 ... 010010
1	111001 ... 111011
1	010010 ... 010100
0	111100 ... 110001
0	000001 ... 010100
1	101100 ... 111111
0	001011 ... 110011
1	001111 ... 110010
1	101101 ... 111001
0	010001 ... 001101
1	110000 ... 011100
0	101011 ... 011110
1	100101 ... 000010
.	.
.	.

Decimated Code

Only the first chip (in the left most column) of every 64 bits will be used for the data scrambling. This decimated value from the decimator at a 19.2 Ksps rate will be XORing with the block interleaver output symbols as shown below:

Interleaver output (assumed):	1010110111010001 . . .
$\oplus$	
Decimated Code:	1101110010110101 . . .
Scrambled data:	0111000101100100 . . .

This scrambled data will be the input being orthogonal-spread with a Walsh function at a fixed chip rate of 1.2288 Mcps.   □

2.2.3.5 Paging Channel Orthogonal Spreading

Paging channels are assigned to code channel number 1 through 7 (inclusive) in sequence. For orthogonal spreading in an paging channel, one of 64 time-orthogonal Walsh functions $n(1 \leq n \leq 7)$ is used. The Walsh function spreading sequence repeats with a period of 52.083 μs (=64/1.2288 Mcps), which is equal to the duration of one paging channel modulation symbol.

2.2.3.6 Paging Channel Quadrature Spreading

Following the orthogonal spreading, each paging channel is spread in quadrature. Each paging channel is quadrature spread by the I and Q pilot PN sequences of period 2^{15} (i.e. 32 768 PN chips in length). The 15-stage PN chip generator is used for quadrature spreading based on $P_I(x)$ and $P_Q(x)$ as defined in Section 2.1.1.10.

2.2.3.7 Paging Channel Baseband Filtering

Following the quadrature spreading separation, the I and Q impulses are applied to the inputs of the I and Q baseband fillers as specified in Section 2.1.1.11.

2.2.3.8 Paging Channel Message Capsule Structure

The paging channel is divided into 10 ms long paging channel half frames, as shown in Figure 2.25. The first bit in any paging channel half frame is an Synchronized Capsule Indicator (SCI) bit.

- A paging channel message capsule is composed of a paging channel message and padding. A paging channel message consists of a length field, a message body, and a CRC field. Padding consists of zero or more bits.
- The base station may transmit synchronized or unsynchronized paging channel message capsules. A synchronized message capsule starts on the second bit of a paging channel half

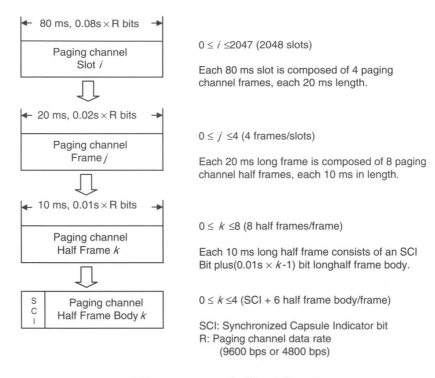

Figure 2.25   Paging channel slot and frame structure

frame. An unsynchronized message capsule begins immediately after the previous message capsule. After the end of a paging channel message, the base station acts as follows:
— If 8 bits or more remain before the next SCI bit, the base station may transmit an unsynchronized message capsule immediately following that message. The base station should not include any padding bits in a paging channel message capsule that is followed by an unsynchronized paging channel message capsule.
— If fewer than 8 bits remain before the next SCI bit, or if no unsynchronized message capsule is transmitted following a paging channel message capsule, the base station should include sufficient padding bits in that message capsule to extend it through the bit preceding the next SCI bit and transmit a synchronized message capsule immediately following that SCI bit. This implies that all bits transmitted on the paging channel are either SCI bits or are part of a message capsule. The base station sets all padding bits to "0."
— When a message capsule immediately follows an SCI bit, the base station sets the SCI bit to "1." The base station sets all other SCI bits to "0."

• The slots are grouped into cycles of 2040 slots (20 ms × 2040 = 16 384 seconds), which are referred to as the maximum slot cycles. The slots of each maximum slot cycle are numbered from 0 to 2047. Each maximum slot cycle begins at the start of the frame when the system time 80 ms per unit (modulo 2048) is zero. The base station transmits the first message that begins in each slot in a synchronized message capsule. This permits mobile stations operating

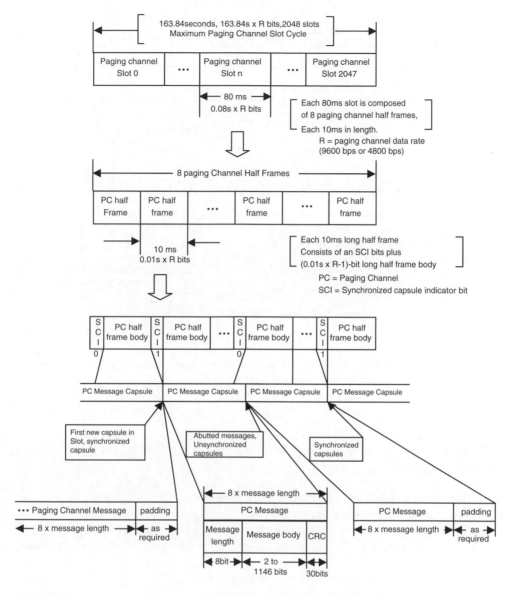

Figure 2.26 An example of paging channel message structure (Reproduced under written permission from Telecommunications Industry Association)

in the slotted mode to obtain synchronization immediately after becoming active. The overall structure of a paging channel message capsule is shown in Figure 2.26.

2.2.3.9 Paging Channel Message CRC

A 30-bit CRC is computed for each paging channel message. A paging channel consists of the message length field, the message body field, and the CRC field. The message length field is

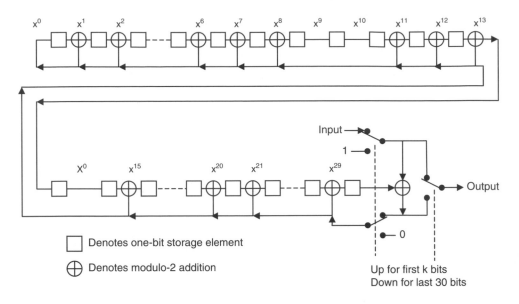

Figure 2.27   Paging channel CRC calculation

8 bits long. The base station limits the maximum paging channel message length to 148 octets, or 1184 bits. The generator polynomial for the CRC is given as

$$g(x) = x^{30} + x^{29} + x^{21} + x^{20} + x^{15} + x^{13} + x^{12} + x^{11} + x^8 + x^7 + x^6 + x^2 + x + 1$$

The CRC should be the values computed by the following procedure and its logic diagram (shown in Figure 2.27).

- All shift register should be initialized to logical one.
- The switches should be set in the up position.
- The information bit count K is defined as $8 +$ message body length in bits.
- The register is then clocked K times, with the length and message body fields of the message as the K input bits.
- The switches should be set in the down position.
- The register is then clocked an additional 30 times.
- The 30 additional outputs will be the CRC field.
- The bits should be transmitted in the order in which they appear at the output of the CRC encoder.

2.2.4 Forward Traffic Channel

The forward traffic channel (FTC) is used for the transmission of user and signaling information to a specific mobile station during a call. The FTC maximum number is equal to 63 minus the number of paging channels and sync channels operating on the same Forward

CDMA Channel. FTC modulation rates, time alignment, and frame structure are implemented as follows:

- The base station transmits information on the FTC at variable data rates:

 9600, 4800, 2400, and 1200 bps for Rate Set 1;
 14 400, 7200, 3600, and 1800 bps for Rate Set 2.

- The FTC frame is 20 ms in duration. Although the data rate may vary on a frame-by-frame basis (i.e. 20 ms), the modulation symbol rate is kept constant by code repetition at 19 200 sps.
- The I and Q channel pilot PN sequence for the FTC use the same pilot PN sequence offset as the pilot channel for a given base station.
- The modulation symbols transmitting at the lower data rates are transmitted using lower energy. For a few examples, the energy per modulation symbol, E_s, for supported data rates should be as:

 $E_s = E_b/2$ for 9600 bps rate, $E_s = E_b/16$ for 1200 bps rate with Rate Set 1;
 $E_s = 3E_b/4$ for 14 400 bps rate, $E_s = 3E_b/32$ for 1800 bps transmission rate with Rate Set 2, and so on.

- The amount of time offset implementing FTC frames is specified by the frame-offset parameter. A zero-offset FTC frame is such that every 100th frame should align with the even-second time mark referenced to the base station transmission time. An offset frame should begin 1.25 frame-offset ms later than the zero-offset traffic channel frame. The FTC block interleaver should always be aligned with the FTC frame.
- FTC frames sent with Rate Set 1 at the 9600 bps transmission rate consist of 192 bits. These 192 bits are composed of 172 information bits followed by 12 Frame Quality Indicator (CRC) bits and eight Encoder Tail Bits. FTC frames sent with Rate Set 2 at the 14 400 bps transmission rate consist of 288 bits. These 288 bits are composed of one reserved bit followed by 267 information bits, 12 Frame Quality Indicator (CRC) bits, and eight Encoder Tail Bits. The entire FTC frame bit allocations sent with Rate Sets 1 and 2 are summarized in Table 2.23.

Table 2.23   FTC frame structure summary

Rate Set	Transmission rate (bps)	Number of bits per frame				
		Total	Reserved	Information	Frame quality indicator	Encoder tail
1	9600	192	0	172	12	8
	4800	96	0	80	8	8
	2400	48	0	40	0	8
	1200	24	0	16	0	8
2	14400	288	1	267	12	8
	7200	144	1	125	10	8
	3600	72	1	55	8	8
	1800	36	1	21	6	8

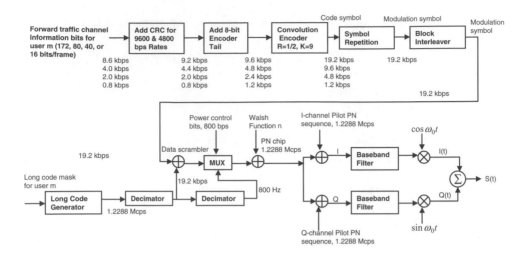

Figure 2.28   Forward traffic channel structure

- Data scrambling applies to the FTC. The data scrambling is accomplished by performing the module-2 addition of the interleaver output symbol with the decimated binary value ("0" or "1") of the long code PN chips.
- A power control subchannel is continuously transmitted on the FTC. The subchannel transmits at a rate of one bit every 1.25 ms (i.e. 800 bps), as discussed in this section.

The overall structure of the FTC is shown in Figure 2.28. The FTC modulation parameters for Rate Sets 1 and 2 are tabulated in Tables 2.24 and 2.25, respectively.

- For FTC Rate Set 2, an effective code rate of 3/4 is achieved by puncturing two of every six symbols after the symbol repetition. The effective code rate is the rate of the convolutional code (1/2) divided by the puncturing rate (4/6). The puncturing pattern is "110101," where a "0" means the symbol is deleted and a "1" means the symbol is passed. This means that the

Table 2.24   FTC modulation parameters for Rate Set 1

Parameter	Data rate (bps)				Units
	9600	4800	2400	1200	
PN chip rate	1.2288	1.2288	1.2288	1.2288	Mcps
Code rate	1/2	1/2	1/2	1/2	bits/code symbol
Code symbol repetition	1	2	4	8	modulation symbols/ code symbol*
Modulation symbol rate	19 200	19 200	19 200	19 200	sps
PN Chips/Modulation Symbol	64	64	64	64	PN chips/Modulation symbol
PN chips/bit	128	256	512	1024	PN chips/bit

*Each repetition of a code symbol is a modulation symbol.

Table 2.25   FTC modulation parameters for Rate Set 2

Parameter	Data rate (bps)				Units
	14400	7200	3600	1800	
PN chip rate	1.2288	1.2288	1.2288	1.2288	Mcps
Code rate	1/2	1/2	1/2	1/2	bits/code symbol
Code symbol repetition	1	2	4	8	modulation symbols/code symbol*
Puncturing rate	4/6	4/6	4/6	4/6	modulation symbols/repeated symbols
Effective code rate*	3/4	3/4	3/4	3/4	code rate/puncturing rate
Modulation symbol rate	19 200	19 200	19 200	19 200	sps
PN Chips/Modulation symbol	64	64	64	64	PN chips/modulation symbol
PN chips/bit	85.33	170.67	341.33	682.67	PN chips/bit

*Each repetition of a code symbol is a modulation symbol.

first, second, fourth, and sixth symbols are passed, while the third and fifth symbols are removed (refer to Table 2.25).

2.2.4.1 FTC Frame Quality Indicator

Each frame with Rate Set 2 and the 9600 and 4800 bps frames of Rate Set 1 should include a Frame Quality Indicator. This Frame Quality Indicator is a CRC, which is to determine whether the frame is in error or to assist in the determination of the data rate of the received frame. The CRC should be calculated on all the bits within the frame, except the Frame Quality Indicator itself and the encoder tail bits.

The 9600 bps transmission with Rate Set 1 and the 14 400 bps transmission with Rate Set 2 use the generator polynomial

$$g(x) = x^{12} + x^{11} + x^{10} + x^9 + x^8 + x^4 + x + 1 \qquad \text{for the 12-bit CRC.}$$

The 7200 bps transmission with Rate Set 2 uses the generator polynomial

$$g(x) = x^{10} + x^9 + x^8 + x^7 + x^6 + x^4 + x^3 + 1 \qquad \text{for the 10-bit CRC.}$$

The 4800 bps transmission with Rate Set 1 and the 3600 bps transmission with Rate Set 2 use the generator polynomial

$$g(x) = x^8 + x^7 + x^4 + x^3 + x + 1 \qquad \text{for the 8-bit CRC.}$$

The 1800 bps transmission with Rate Set 2 uses the generator polynomial

$$g(x) = x^6 + x^2 + x + 1 \qquad \text{for the 6-bitCRC.}$$

Example 2.19. The 4800 bps transmission rate with Rate Set 1 uses an 8-bit Frame Quality Indicator (CRC). Consider the CRC bits computation based on all bits within the 80 information bits/frame at the 4800 bps rate.

The CRC generator polynomial for FTC is

$$g(x) = x^8 + x^7 + x^4 + x^3 + x + 1 \quad \text{or} \quad g = (11011001)$$

Assume that the 80-bit information sequence is

$$d = (11000111000010100100$$
$$11001101110101011001$$
$$11001100101001110101$$
$$10011111000110011110)$$

All LFSR stages are initially set to logical ones:

$$\text{I.C.} = (11111111)$$

At 4800 bps frame, the frame length is 96 bits (20 ms):

Frame length(96 bits) = Information length(80 bits) + CRC(8 bits) + Encoder tail(8 bits).

Clocking the register 80 times with the 80-bit information input, the 8 additional output bits will be generated as the CRC bits. Clocking the register 8 times again, the CRC bits are produced as (01110011). □

2.2.4.2 FTC Convolutional Encoding

The forward traffic channel data is convolutionally encoded. For FTC Rate Set 1, the convolutional code rate is 1/2, whereas for FTC Rate Set 2 an effective code rate of 3/4 is achieved by puncturing two of every six symbols after the symbol repetition. When generating FTC data, the encoder should be initialized to the all-zero state at the end of each 20 ms frame. The generator functions for the rate 1/2 code will be =753 (octal) and =561 (octal).

2.2.4.3 FTC Code Symbol Repetition

The code symbol repetition rate on the FTCs with Rate Set 1 varies with the data rate. Code symbols should not be repeated for 14 400 and 9600 bps data rate. Each code symbol at the 7200 and 4800 bps data rates should be repeated one time (i.e. each symbol occurs two consecutive times). Each code symbol at the 3600 and 2400 bps data rate should be repeated three times (i.e. each symbol occurs four consecutive times). Each code symbol at the 1800 and 1200 bps data rate should be repeated seven times (i.e. each symbol occurs eight consecutive times).

For the FTCs with Rate Set 2, the code symbols resulting from the symbol repetition should be punctured. An effective code rate of 3/4 should be achieved by puncturing two of every six symbols after the symbol repetition. The effective code rate is the rate of the convolutional code (1/2) divided by the puncturing rate (4/6).

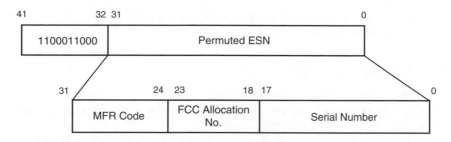

Figure 2.29 FTC public long code mask including the bit allocation of ESN

2.2.4.4 FTC Interleaving

For the FTCs with Rate Set 1, all the symbols after symbol repetition should be block interleaved. For the FTCs with Rate Set 2, all symbols after symbol repetition and subsequent puncturing should be block interleaved. The FTC uses the identical block interleaver spanning 20 ms, which is equivalent to 384 modulation symbols at the modulation symbol rate of 19 200 sps.

2.2.4.5 FTC Data Scrambling

The FTC data is scrambled by using the FTC long code mask illustrated in Figure 2.29. Data scrambling is accomplished by performing the modulo-2 addition of the block interleaver output with the decimated value ("0" or "1") of the long code PN chips. This PN sequence should be the equivalent of the long code operating at 1.2288 MHz clock rate where only the first output of every 64 chips is used for the data scrambling at a 19 200 sps rate.

The 42-bit FTC long code mask consists of the 10-bit mask pattern and the 32-bit permutated Electronic Serial Number (ESN). The 32-bit ESN (E31, E30, . . . , E1, E0) is assigned by the manufacturer for uniquely identifying the mobile station to any cellular system. The bit allocation of ESN (E0–E31) consists of MFR code (E24–E31), FCC allocation number (E18–E23), and serial number (E0–E17). The permutation of ESN is designed to prevent high correlation between long codes. The permuted ESN is given as

$$\text{Permuted ESN} = \{E_0, E_{31}, E_{22}, E_{13}, E_4, E_{26}, E_{17}, E_8, E_{30}, E_{21}, E_{12}, E_3,$$
$$E_{25}, E_{16}, E_7, E_{29}, E_{20}, E_{11}, E_2, E_{24}, E_{15}, E_6, E_{28}, E_{19},$$
$$E_{10}, E_1, E_{23}, E_{14}, E_5, E_{27}, E_{18}, E_9\}$$

The long code of the period $2^{42} - 1$ bits is divided into 64 bits length each and is arranged in the row-by-row array. The left most column, which is composed of all the first bit of each row, must represent the decimated binary values.

Voice privacy is provided in the CDMA system by means of the private long code mask used for the PN spreading. Voice privacy is provided on the traffic channels only. All calls are initiated using the public long code mask for PN spreading. The transmission to private long code mask is not performed if authentication is not performed.

To initiate a transition to the private or public long code, the base station or the mobile station sends a long code transition request order on the traffic channel. The mobile station or the base station will take action in response to receipt of this order. The base station may cause a transition to the public long code mask by sending the handoff direction message with the `PRIVATE_LCN` bit set appropriately.

2.2.4.6 FTC Power Control Subchannel

A power control subchannel transmits continuously on the FTC. The base station should insert on every FTC a power control subchannel. The subchannel should transmit at a rate of one bit ("0" or "1") every 1.25 ms at the 800 bps rate.

A "0" bit indicates to the mobile station to increase the mean output power level and a "1" bit indicates to the mobile station to decrease the mean output power level. The base station RTC receiver estimates the received signal strength of the particular mobile station to which it is assigned over a 1.25 ms period, equivalent to 6 ($=96 \times 1.25/20$) modulation symbols. The base station receiver uses the estimate to determine the value of the power control bit ("0" or "1").

The base station shall transmit the power control bit on the corresponding forward traffic channel using the puncturing technique described below. The transmission of the power control bit should occur on the FTC in the second power control group following the corresponding RTC power control group in which the signal strength was estimated. For example, if the signal is received on the RTC in power control group number 5, then the corresponding power control bit is transmitted on the FTC during power control group number 7 ($=5 + 2$). The power control bits should be inserted into the FTC data stream after the data scrambling, as observed in Figure 2.28.

For Rate Set 1, the length of one power control bit will correspond exactly to two modulation symbols of the FTC (i.e. $104.166 \mu s = (1.25 \times 10^{-3}/24) \times 2$). Each power control bit replaces two consecutive FTC modulation symbols by the technique of symbol puncturing and is transmitted with energy not less than E_b, namely the energy per information bit of the FTC, as shown in Figure 2.30.

For Rate Set 2, the length of the power control bit corresponds exactly to one modulation symbol of the FTC (i.e. $52.0833\ldots\mu s$). Each power control bit will replace one FTC modulation symbol and is transmitted with energy not less than $3E_b/4$, namely 3/4 of the energy per information bit of the FTC.

There are 16 possible starting positions for the power control bit as shown in Figure 2.31. Each position corresponds to one of the first 16 modulation symbols (numbered 0 through 15) of a 1.25 ms period. In each 1.25 ms period, a total of 24 bits from the long code are used for scrambling. These bits are numbered 0 through 23, where bit 0 is the first to be used and bit 23 the last in each 1.25 ms period.

The 4-bit binary number with values 0 through 15 formed by scrambling bits 23, 22, 21, and 20 will be used to determine the position of the power control bit as shown in Figure 2.31. Bit 20 is the least significant bit, and bit 23 is the most significant bit. In the example of Figure 2.31, the value of bits 23, 22, 21, and 20 are 0111 (14 decimal), and the power control bit starting position is the 14th. The relationship between the scrambled modulation symbols at the 19.2 Ksps rate and the punctured power control subchannel at the 800 bps rate is illustrated in Figure 2.28.

2.2.4.7 FTC Orthogonal Spreading

The FTC is spread with a Walsh function at a fixed chip rate of 1.2288 Mcps. One of 64 time-orthogonal Walsh functions, as defined in Table 2.12, is used. A code channel that is spread using Walsh function n should be assigned to code channel number n ($0 \le n \le 63$).

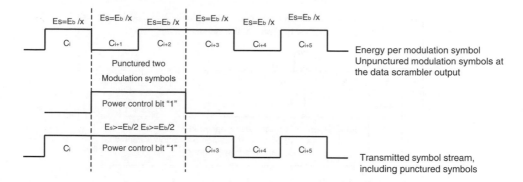

(a) Power control bit "1" (A decrease in the mean output power level)

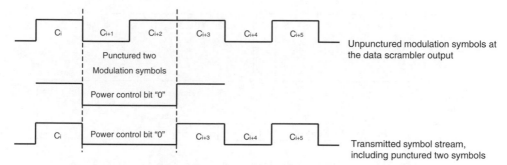

(b) Power control bit "0" (A increase in the mean output power level)

Transmission rate	Values of x
9.6 kbps	2
4.8 kbps	4
2.4 kbps	8
1.2 kbps	16

Figure 2.30   Power control bit and punctured modulation symbols

Each base station uses a time base reference (time-aligned to CDMA System Time) from which all time critical CDMA transmission components, including pilot PN sequences, frames, and Walsh functions, should be derived. Reliable, external means should be provided at each base station in order to synchronize each base station's time base reference to CDMA System Time.

Walsh function time alignment must be such that the first Walsh chip, designated by 0 in the column headings of Table 2.12, begins at an even second time mark referenced to base station transmission time. The Walsh function spreading sequence is repeated with a period of $52.083\ldots\mu s$ ($=64/1.2288$ Mcps), which is equal to the duration of one FTC modulation symbol.

2.2.4.8 FTC Quadrature Spreading

The FTC is PN spread as specified in Section 2.1.1.10. The spreading sequence (called the pilot PN sequence) is a quadrature sequence of length $2^{15} = 32\,768$ chips. The chip rate for the pilot

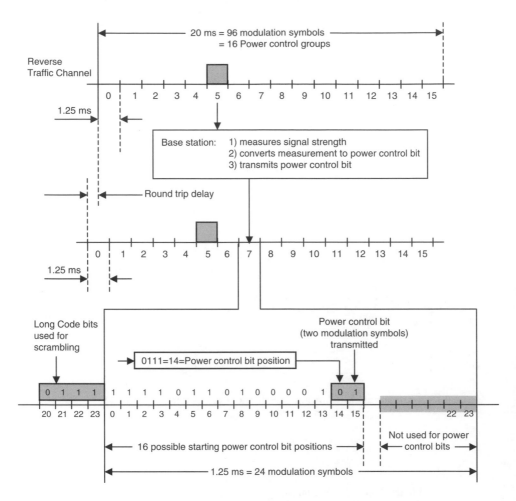

Figure 2.31 Randomization of power control bit position

PN sequence is 1.2288 Mcps. The pilot PN sequence period is $32\,768/1\,228\,800 = 26.666\ldots$ms, and exactly 75 pilot PN sequence repetitions occur every 2 seconds.

In order to obtain the I and Q pilot sequence of period 2^{15}, a "0" is inserted in the linear recursions, $\{i(n)\}$ and $\{q(n)\}$, after 14 consecutive "0" outputs. Hence, the pilot PN sequences have one run of 15 consecutive "0" outputs instead of 14 as explained in Section 2.1.1.10.

2.2.4.9 FTC Filtering

Filtering for the FTC is as specified in Section 2.1.1.11.

In this chapter we have presented the complete analyses with regard to the channel structures and their functionalities of Reverse/Forward CDMA Channels based on 2G cdmaOne IS-95A. The evolution and migration of this 2G cdmaOne IS-95A mobile technology to the 3GPP2 CDMA 1× family standards will be covered in Chapters 7 and 8.

3

General Packet Radio Service (GPRS)

All 2G mobile technologies including GSM have been designed and optimized for circuit-switched voice communications, leaving very limited data transfer capabilities in their systems. The General Packet Radio Service (GPRS) is the packet-mode extension of GSM, standardized by European Telecommunications Standards Institute (ETSI). GPRS (GSM Phase $2+$) is a new bearer service developed for GSM that is greatly improved and simplified for wireless Internet access, enabling user data packets to transfer between GSM mobile stations and external packet data networks. It is advantageous for users of GPRS to benefit from shorter access times and higher transfer rates. 2G wireless data services based on circuit-switched radio transmission do not fulfill the needs for users and providers, due to slow data rates, long connection setup, and inefficient resource utilization for bursty traffic. Therefore, it is obvious that packet-switched bearer services for bursty Internet traffic will result in a much better utilization of the traffic channels in an efficient way.

In addition, GPRS packet transmission offers volume-based billing rather than time-based billing due to the amount of transmitted data by circuit switched services.

This chapter covers the GPRS system architectures, the fundamental functionality, the mobile station registration with the network, radio resource management, mobility management for keeping track of its location, and routing in GPRS systems.

3.1 GPRS System Architecture

The first GPRS networks were launched in the spring of 2001 in Europe. GPRS introduced packet transmission for data services, replacing GMS's circuit mode. The existing GSM system architecture is integrated into the GPRS system by overlaying with a new class of network nodes (routers), called the Serving GPRS Support Node (SGSN) and the Gateway GPRS Support Node (GGSN).

GSNs (SGSN and GGSN) are responsible for the delivery and routing of packet data between the mobile stations and the external packet data networks (PDNs).

Mobile Communication Systems and Security   Man Young Rhee
© 2009 John Wiley & Sons (Asia) Pte Ltd

3.1.1 GPRS Network Support Nodes

This section deals with the GSNs required to support GPRS functionality. The SGSN is responsible for the delivery of data packets from and to the mobile stations within its service area. SGSNs are routers that bridge over the GPRS core network. They are linked to several Base Station Controllers (BSCs) and their functions are to manage packet delivery to terminals in a given area. A Packet Control Unit (PCU) is installed at BSC level to manage resources and the exchange protocol with the SGSN. The SGSN may route its packets over different GGSNs to reach different Packet Data Networks (PDNs). The SGSN also establishes a management context containing information pertaining to mobility and security for the mobile stations (MSs). The SGSN supports GPRS for A/G_p mode and/or I_u mode. In general, the SGSN's tasks include packet transfer and routing, logical link and mobility management, authentication, and charging functions.

The GGSN acts as an interface between the GPRS backbone network and the external packet data networks. The GGSN is the first point of PDN interconnection with a PLMN supporting GPRS via the G_i reference point. The GGSN communicates with multiple SGSNs and serves as the gateway (interface) to external PDNs such as the Internet.

The GGSN handles the IP address allocation to mobile stations. The GGSN converts the GPRS packets coming from the SGSN into an appropriate packet data protocol (PDP) format and sends them out on the corresponding PDN. To the MS direction, PDP addresses of incoming data packets are converted to the GSM address of the destination user. The re-addressed packets are sent to the responsible SGSN. The GGSN also contains routing information for PS-attached users. The GGSN may request location information from the HLR via the G_c interface.

Thus, GSNs are responsible for delivery and routing of data packets between the mobile stations and the external packet data networks with a method known as encapsulation and tunneling. The SGSN and GGSN nodes can be combined into a single entity called GPRS Support Node (GSN), but that is not a common development scenario.

3.1.2 Reference Points and Data Transfer Interfaces

The GPRS core network functionality is logically implemented on two network nodes (SGSN and GGSN), as described earlier. Figure 3.1 shows the interfaces between the new GPRS nodes and the GSM network. A GSN has functionality required to support GSM EDGE Radio Access Network (GERAN) and/or UMTS Terrestrial Radio Access Network) (UTRAN) access networks.

The SGSN is the node that is serving the mobile station and hence is responsible for the delivery of data packets from and to the mobile stations. The SGSN also keeps track of the location of an individual MS and performs security functions and access control. The GGSN is the node that provides intervention between SGSN and PDNs, and it is connected with SGSNs via an Intra-PLMN backbone network or an IP-based PDN.

The G_b interface connects the BSC with the SGSN. Via the G_n and G_p interfaces, user data and signaling data are transmitted between the GSNs. The G_n interface will be used if SGSN and GGSN are located in the same PLMN, whereas the G_p interface will be used if they are in different PLMNs. Network interworking takes place through the G_i reference point and the G_p interface. The G_p interface is defined between two SGSNs. This allows the SGSNs to exchange user profiles when an MS moves from one SGSN area to another.

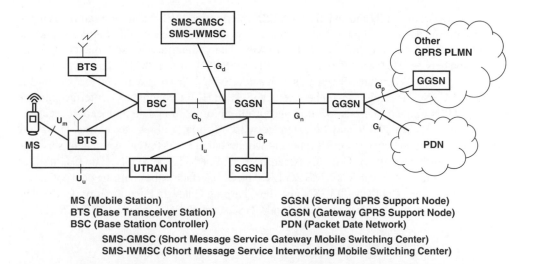

Figure 3.1   GPRS system architecture based on reference points and data transfer interface

In order to exchange the SMS message via GPRS, the G_d interface should be provided for interconnection of the SMS Gateway MSC (SMS-GMSC) with SGSN.

3.1.3 Signaling Transfer Interfaces

Figure 3.2 illustrates the signaling interfaces to support several service levels in an efficient manner.

The SGSN may send location information to the MSC/VLR via the optional G_s interface. The SGSN may also receive paging requests from the MSC/VLR via the G_s interface. The GGSN may request location information from HLR via the optional G_c interface. Across the G_f

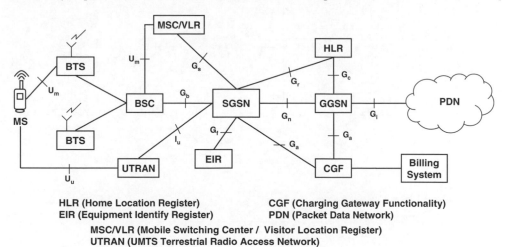

Figure 3.2   GPRS system architecture based on signaling transfer interface

interface between an SGSN and an EIR, the SGSN queries the IMEI of a mobile station trying to register with the network. The HLR stores the user profile, the current SGSN address, and the PDP address(es) for each GPRS user in the PLMN. The G_r interface is used to exchange that information between HLR and SGSN. Specifically, the SGSN informs the HLR about the current location of the MS. When the MS registers with a new SGSN, the HLR will send the user profile to the new SGSN. The signaling path between GGSN and HLR through the G_c interface may be used by GGSN to query a user's location and profile in order to update its location register.

In addition, the MSC/VLR may be extended with functions and register entries that allow efficient coordination between GPRS (packet-switched) and GSM (circuit-switched) services. Moreover, paging requests of GSM calls can be performed via the SGSN. The G_s interface connects the databases of SGSN and MSC/VLR. The G_a interface is the charging data collection interface between an SGSN or a GGSN and the Charging Gateway Functionality (CGF). The CGF collects charging records from SGSNs and GGSNs, which is described in 3GPP TS 32.215.

3.1.4 GPRS-PLMN Backbone Networks

Figure 3.3 shows two intra-PLMN backbone networks of different PLMNs connected with an inter-PLMN backbone network. The Border Gateways (BGs) are connected between the PLMNs and the external inter-PLMN backbone network. As seen from Figure 3.3, the G_b interface connects the BSC with the SGSN. Using the G_n and G_p interfaces, user and signaling data are transmitted between the GSNs. The G_n interface is used if the SGSN and the GGSN are located in the same PLMN I, whereas the G_p interface will be used if they are in a different PLMN II. All GSNs are connected via an IP-based GPRS backbone network. Within this

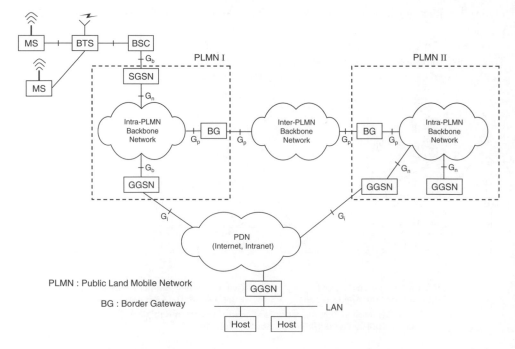

Figure 3.3   A routing example of GPRS system through Inter/Intra-PLMN backbone networks

backbone network, the GSNs encapsulate the PDN packets and transmit them using the GPRS Tunneling Protocol (GTP).

There are two kinds of backbone networks: Intra-PLMN backbone networks and Inter-PLMN backbone networks. GPRS supports interworking with networks based on the IP.

- Intra-PLMN backbone network is the IP-based network interconnecting GSNs within the same PLMN.
- Inter-PLMN backbone network is the IP-based network interconnecting GSNs and intra-PLMN backbone networks in different PLMNs.

Every intra-PLMN backbone network is a private IP network intended for GPRS packet domain data and signaling only. Figure 3.3 shows two intra-PLMN backbone networks of different PLMNs connected with an inter-PLMN backbone network. Two intra-PLMN backbone networks are connected via the G_p interface using BG and an inter-PLMN backbone network. The inter-PLMN backbone network is selected by a roaming agreement that includes the BG security functionality to protect the private intra-PLMN backbones against unauthorized users and attacks.

3.2 GPRS Logical Functions

Fundamental logical functions related to GPRS operations are briefly defined in the following sections.

3.2.1 Network Access Control

The access control defines a set of procedures by which a user is connected to the mobile side or the fixed side of the PDNs. The access protocol should enable the user to employ the services and facilities of the network.

- Registration is a link to a user's Mobile ID, which is associated with the user's packet data protocol and address within the PLMN, and with the user's access point to the packet data network.
- Admission control aims to calculate, determine, and reserve which network resources are required to provide the quality of services requested.
- The packet terminal should adapt data packets received/transmitted from/to the terminal equipment to form suitable for transmission of GPRS across PLMN.
- The authentication function performs the identification and authentication of the service requester to ensure that the user is authorized to use the particular network services.
- A screening function (for example, by use of Internet firewalls) is concerned with filtering out unauthorized messages through packet filtering functions.
- Data collection is necessary to support subscription and to charge traffic fees.

3.2.2 Packet Transfer and Routing

The transfer of messages within and between the PLMN(s) requires routing through the ordered nodes from the originating node, relaying nodes, and the destination node. The route for transmission of a message will follow a set of rules as stated below.

The routing function selects the transmission path for the next hop in the node and determines the network node to which a message should be forwarded by following the destination address of the message. The relay function should find a node to forward data from one node to the next node in the route.

Address translation converts one address to another address in order to make routing of packets within and between the PLMNs. Address mapping is used to map one network address to another network address for routing packets within and between PLMN(s).

Encapsulation is the addition of address and control information to a data unit for routing a packet within and between the PLMN(s). *Decapsulation* is the removal of address and control information from a packet to reveal the original data unit. *Tunneling* is the transfer of encapsulated data units within and between PLMNs from the point of encapsulation to the point of decapsulation.

The compression function optimizes use of the radio path capacity by transferring the Service Data Unit (SDU) as little as possible while preserving the information contained within it.

The ciphering function preserves the confidentiality and protects the PLMN from intruders. Security function provides user identity confidentiality, and user data and signaling confidentiality. Security-related network functions (confidentiality and authentication) will be described in Section 3.4.

To provide a practical summary, we now consider how the packets are routed in GPRS. A mobile station located in PLMN I sends IP packets to the Internet connected to a web server. The SGSN, registered with the MS, encapsulates the packets coming from the mobile station and examines the Packets Data Protocol (PDP) context, and routes them through the intra–PLMN backbone network to the appropriate GGSN (see Figure 3.3). The GGSN decapsulates the packets and sends them out on the IP network for transferring the packets to the access router of the destination network, and delivers the IP packets to the host.

On the other hand, assume that an IP address has been assigned to the MS by the GGSN of PLMN II. Then, the MS's IP address has the same network prefix as the IP address of GGSN in the PLMN II. The corresponding host is now sending IP packets to the MS. The packets are sent out of the IP-based PDN and are routed to the GGSN of PLMN II. This GGSN of PLMN II queries the HLR and obtains the information that the mobile station is currently located in PLMN I. It encapsulates the incoming IP packets and tunnels them through the inter-PLMN backbone network to the appropriate SGSN in PLMN I. The SGSN decapsulates the packets and delivers them to the MS.

3.2.3 Mobility Management

The GPRS mobility management (MM) provides functions that are used to keep track of the current location of a MS within the same PLMN or within another PLMN. The MM activities related to a subscriber are characterized by the three modes of the operation (for A/G_b or I_u) as described below.

- A/G_b mode of MS operation:
  — *Class A mode of operation:* The MS is attached to both PS domain (GPRS service) and conventional CS domain (GSM voice and SMS data services).

— *Class B mode of operation:* The MS is able to register with both PS and CS domain simultaneously, but the MS can only operate one of the two services (i.e. PS or CS service) at a time.

— *Class C mode of operation:* The MS is exclusively attached to the PS domain. Simultaneous registration at a time for GPRS and GSM services is not possible.

* I_u *mode of MS operation:* An I_u mode MS operates in one of three operation modes that are different from the ones of an A/G_b mode MS due to the capabilities of an I_u mode Radio Access Network (RAN) to multiplex CS and PS connections, or due to paging coordination for PS/CS services offered by the core network (CN).

— *CS/PS mode of operation:* The MS is attached to both the PS domain and CS domain, and the MS is capable of simultaneous signaling with the PS/CS core network domains. This mode of operation is comparable to the class-A mode of operation defined for A/G_b mode. The ability to operate PS/CS services simultaneously depends on the MS capabilities.

— *PS mode of operation:* The MS is attached to the PS domain only and may operate services of the PS domain only. This mode of operation is equivalent to the A/G_b mode GPRS class-C mode of operation.

— *CS mode of operation:* The MS is attached to the CS domain only and operates services of the CS domain only. However, this mode of operation does not prevent PS-like service being offered over the CS domain.

Any combination of A/G_b and I_u mode MSs can be allowed for GERAN and UTRAN multisystem terminals.

3.2.4 State Models for Location Management

The location management related to a GPRS subscriber is characterized by one of three different MM states. Each state describes a certain level of functionality and information allocated. The MS frequently sends location update messages to its current SGSN. The information sets held at the MS and the SGSN are called the MM context.

The state model has been defined for location management in GPRS. A MS can be in one of three states depending on its state. In A/G_b mode, the MM states for a GPRS subscriber are IDLE, STANBY, and READY. In I_u mode, the MM states for a GPRS subscriber are PMM-DETACHED, PMM-IDLE, and PMM-CONNECTED.

* *IDLE State:* The MS is not reachable and the subscriber is not attached to GPRS mobility management. The MS and SGSN contexts hold no valid location or routing information for the subscriber. Hence, the subscriber-related MM procedures are not performed. Data transmission to and from the mobile subscriber as well as the paging of the subscriber is not possible in this case.

* *STANDBY State:* The MS and SGSN have established MM contexts when the subscriber was attached to GPRS mobility management. To find out the current cell of an MS in STANDBY state, paging of the MS within a certain routing area must be performed. Pages for data or

signaling information transfers may be received in this state, but data reception and transmission are not possible.

The MS executes MM procedures to inform the SGSN when it has entered a new Routing Area (RA). Whenever an MS moves to a new RA, it sends a routing area update request to its assigned SGSN. The message contains the Routing Area Identity (RAI) of its old RA. The base station subsystem (BSS) adds the Cell Identifier (CI) of the new cell from which the SGSN can derive the new RAI.

The MS may initiate activation or deactivation of PDP contexts while in STANDBY state. A PDP context should be activated before data can be transmitted or received for this PDP context. The SGSN may have to send data or signaling information to an MS in STANDBY state. The SGSN then sends a paging request in the routing area where the MS is located if the Paging Proceed Flag (PPF) is set.

- *READY State:* The SGSN MM context corresponds to the STANDBY MM context extended by local information on the cell level for the subscriber. Performing a GPRS attach, the MS gets into READY state. With a GPRS detach, the MS may disconnect from the network. The MS may activate or deactivate a PDP context while in READY state. The MS may send and receive PDP PDUs in this state. The network initiates no GPRS page in READY state.

An MS in READY state informs its SGSN of every movement to a new cell. The SGSN will only be informed when an MS moves to a new RA. A timer supervises the READY state. An MM context moves from READY state to STANDBY state when the READY timer expires. In order to move from READY state to IDLE state, the MS initiates the GPRS Detach procedure.

3.2.5 State Transitions of a Mobile Station

The GPRS state model has been defined for state transitions and for location management. Figure 3.4 illustrates state transitions of a MS. The movement from one state to the next is

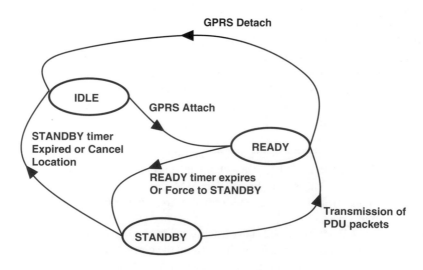

Figure 3.4   State transitions of a mobile station

dependent on the current state, that is, IDLE, STANDBY, or READY. The state transitions are described below:

- *GPRS Attach (IDLE → READY):* In IDLE state, the MS is not reachable and hence no location updating is performed. An MS in READY state informs its SGSN about every movement to a new cell. The MS requests access and a logical link to an SGSN initiated. Performing a GPRS attach, the MS gets into READY state. MM contexts are established at the MS and SGSN.
- *GPRS Detach (READY → IDLE):* With a GPRS detach, it may disconnect from the network and return back to IDLE state. The MS or the network requests that the MM contexts should roll back to IDLE state and that the PDP contexts return to *inactive* state. The SGSN deletes the MM and PDP contexts. The SGSN receives a cancel location message from the HLR, and all PDP contexts will be removed.
- *READY Timer Expiry (READY → STANDBY):* The STANDBY state will be reached when a MS does not send any packets for a long period of time until the READY timer expires. For READY timer expiry, the MS and the SGSN MM contexts return to STANDBY state. For Force to STANDBY, the SGSN indicates an immediate return to STANDBY state before the READY timer expires.
- *Transmission of PDU Packet (STANDBY → READY):* For PDU transmission, the MS sends a logical link control (LLC) PDU to the SGSN, possibly in response to a page. For PDU reception, the SGSN receives an LLC PDU from the MS.
- *Implicit Detach and Cancel Location (STANDBY → IDLE):* The MM and PDP contexts in the SGSN should be returned to IDLE and *inactive* state. The MM and PDP contexts in the SGSN are deleted. The SGSN PDP contexts should be deleted. As for Cancel Location, the SGSN receives a maximum a posteriori (MAP) Cancel Location message from the HLR, and removes the MM and PDP contexts.

3.2.6 Packet Mobility Management (I_u Mode)

Figure 3.5 shows the PMM state transitions from one PMM state to the other as described in the following:

- *PMM-DETACHED State → PMM-CONNECTED State:* In the PMM-detached state, there is no communication between the MS and 3G-SGSN. The MS and SGSN contexts hold no valid location or routing information for the MS. Notice here that 3G-SGSN refers to all functionalities of an SGSN, which serves an MS in I_u mode. In order to establish MM contexts in the MS and the SGSN, the MS should perform the GPRS Attach procedure. The MM context should move to the PMM-connected state when a PS signaling connection is established between the MS and 3G-SGSN for performing a GPRS-PS attach. If the PS attach is accepted, an MM context is created in the MS or the 3G-SGSN.
- *PMM-CONNECTED State → PMM-DETACHED State:* In the PMM-connected state, the MS location is known in 3G-SGSN and is tracked by the serving Radio Network Controller (RNC). The MS will perform the routing area update procedure when the Routing Area Identity (RAI) is changed. When an MS and a 3G-SGSN are in the PMM-connected state, a PS signaling connection is established between the MS and the 3G-SGSN.

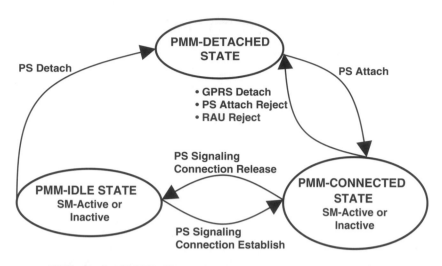

Figure 3.5 PMM State transitions mode

— *GPRS Detach:* The MM context should move to the PMM-detached state when the PS signaling connection is released between the MS and the 3G-SGSN after the MS has performed a GPRS detach.

— *GPRS Attach Reject:* The MM context should move to the PMM-detached state when the PS signaling connection is released between the MS and the 3G-SGSN after a GPRS attach is rejected by the 3G-SGSN.

— *RAU Reject:* The MM context should move to the PMM-detached state when the PS signaling connection is released between the MS and the 3G-SGSN after an RAU is rejected by the 3G-SGSN. The MM context may be deleted.

• *PMM-CONNECTED State → PMM-IDLE State:* The MM context should move to the PMM-idle state when the PS signaling connection is released. The MS should enter the PMM-idle state when its PS signaling connection to the 3G-SGSN has been released or broken. This release or failure is explicitly indicated by the RNC to the MS or detected by the MS.

• *PMM-IDLE State → PMM-CONNECTED State:* For PS signaling connection establishment, the MM context should move to the PMM-connected state when the PS signaling connection is established between the MS and the 3G-SGSN.

• *PMM-IDLE State → PMM-DETACHED State:* For the implicit GPRS detach, the MM context should locally move to the PMM-detached state after expiry of the MS reachable timer, in the case of the 3G-SGSN, but removal of the battery, the User Service Identity Module (USIM), or the GSM Service Identity Module (GSIM) from the TE, in the MS case. Thus GPRS detach changes the state to PMM-DETACHED in both cases.

3.3 Layered Protocol Architecture of Transmission Plane

This section deals with each layered protocol structure providing for both the user and the control plane. The user plane is used to provide user information transfer and its associated signaling, whereas the control plane is used for controlling and supporting the user plane functions. The layered protocol architecture of the user plane for A/G$_b$ mode is illustrated in Figure 3.6 and the control plane for A/G$_b$ mode is shown in Figure 3.7.

3.3.1 User Plane for A/G$_b$ Mode

In this section, the user plane used in A/G$_b$ mode is described with associated layered protocol procedures.

- *Internet Protocol (IP):* IP is the backbone network protocol employed in the network layer to route user data and control signaling. The backbone network was initially based on IPv4, but ultimately IPv6 should be used to route packets through the backbone.
- *Subnetwork Dependent Convergence Protocol (SNDCP):* SNDCP is used to transfer data packets between the SGSN and the MS. Its transmission functionality includes:
  — Multiplexing of several connections of the network layer onto one virtual logical connection of the underlying LLC layer.
  — Compression and decompression of user data and header information.

Next, the data link layer (LLC, RLC, and MAC) and the physical layer (PLL and RFL) at the air interface U$_m$ are considered. The data link layer between the MS and the network is divided into

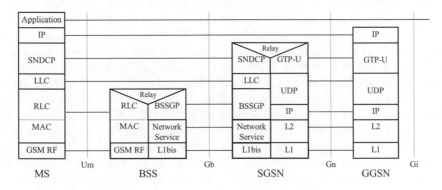

BSS: Base Station System
BSSGP: Base Station System GPRS Protocol
GGSN: Gateway GPRS Support Node
GTP: GPRS Tunneling Protocol
GTP-U: GPRS Tunneling Protocol for the user plane
LLC: Logical Link Control
MAC: Medium Access control

MS: Mobile Station
PDU: Protocol Data Unit
RLC: Radio Link Control
SGSN: Serving GPRS Support Node
SNDCP: Subnetwork Dependent Convergence Protocol
UDP: User Datagram Protocol

Figure 3.6   User plane for A/G$_b$ mode (© 2003. 3GPP™ TSs and TRs are the property of ARIB, ATIS, CCSA, ETSI, TTA and TTC who jointly own the copyright in them. They are subject to further modifications and are therefore provided to you "as is" for information purposes only. Further use is strictly prohibited.)

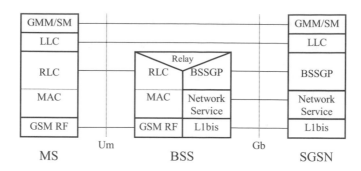

GMM/SM: GPRS Mobility Management and Session Management
BSSGP: Base Station System GPRS Protocol

Figure 3.7   Control plane MS-SGSN in A/G$_b$ mode (© 2003. 3GPP™ TSs and TRs are the property of ARIB, ATIS, CCSA, ETSI, TTA and TTC who jointly own the copyright in them. They are subject to further modifications and are therefore provided to you "as is" for information purposes only. Further use is strictly prohibited.)

two sublayers: the LLC layer between MS-SGSN and the RLC/MAC layer between MS-BSS, as shown in Figure 3.6.

- *Logical Link Control (LLC):* This layer provides a highly ciphered logical link between an MS and its assigned SGSN. Its functionality includes sequence control, in-order delivery, flow control, detection of transmission errors, and retransmission by ARQ. LLC is independent of the underlying radio interface protocols in order to allow introduction of alternative GPRS radio solutions with minimum changes to the Network Subsystem (NSS). The data confidentiality is totally ensured by encryption algorithms. This protocol is mainly an adapted version of the LAPDm protocol used in GSM-Data Link Layer (see Section 1.3.2 of Chapter 1).
- *Radio Link Control/Medium Access Control (RLC/MAC):* The RLC/MAC layer at the air interface contains two functions: The main purpose of the RLC function is to provide for establishing a reliable link between the MS and the BSS. This includes the segmentation and reassembly of LLC frames into RLC data blocks and ARQ of uncorrectable codewords. The MAC function controls the access attempts of an MS on the radio channel shared by several MSs. In general, the MAC function controls the access signaling procedures relating to the request and grant for the radio channel, and mapping of LLC frames onto the GSM physical channel. In the RLC/MAC layer, both the acknowledged and unacknowledged modes of operation are supported.
- *Relay Function:* In the BSS, this function relays LLC PDUs (Protocol Data Units) between the U$_m$ and G$_b$ interfaces. U$_m$ is the interface between the MS and the BSS in the user plane for A/G$_b$ mode. Accordingly, the U$_m$ interface provides GPRS services over the radio to the MS. The A/G$_b$ mode of operation has a functional division that is in accordance with the use of an A or a G$_b$ interface between the radio access network and the core network. Note that an SGSN in A/G$_b$ mode uses only the G$_b$ interface.

  In the SGSN, this function relays PDP PDUs between the G$_b$ and G$_n$ interfaces, where PDP denotes the packet data protocol (i.e. IP). G$_b$ is the interface between a BSS and an SGSN. G$_n$ is the interface between two GSNs (SGSN and GGSN) in the user plane within the same PLMN. Finally, G$_i$ is a reference point between the GPRS user plane and a PDN.

- *Base Station System GPRS Protocol (BSSGP):* This layer conveys routing information as well as QoS-related information between the BSS and the SGSN. BSSGP does not perform error correction.
- *Network Service (NS):* This layer transports BSSGP PDUs. NS is based on the Frame Relay connection between the BSS and the SGSN, and may multi-hop and traverse a network of Frame Relay switching nodes.
- *Physical Layer:* Consider the physical layer at the air interface U_m. The physical layer between the MS and the BSS is divided into the two sublayers: the Physical Link layer (PLL) and the physical RF layer (RFL).

  The PLL provides a physical channel between the MS and the BSS. Its tasks include detection of transmission errors, random or burst error correction, interleaving, and detection of physical link congestion. The physical RF layer operates below the PLL. The function of RFL includes mainly modulation and demodulation as defined in GSM RF (see GSMO5 series).
- *GPRS Tunneling Protocol for the User Plane (GTP-U):* This protocol tunnels user data between GPRS support nodes in the backbone network. The GPRS Tunneling Protocol, which is specified in 3GPP TS 29.060, should encapsulate all PDP PDU. Notice that the user data is transferred transparently between the MS and the PDNs with a method known as encapsulation and tunneling.
- *User Datagram Protocol (UDP):* UDP carries GTP PDUs for protocols that do not need a reliable data link (i.e. IP), and provides protection against corrupted GTP PDUs. This protocol is defined in RFC 768.

3.3.2 Control Plane for A/G$_b$ Mode

The control plane consists of protocols relating to control and support of the user functions.

- *GPRS Mobility Management and Session Management (GMM/SM):* This protocol supports mobility management functionality and PDP context functions:
  - To control the GPRS network access connections such as GPRS attach/detach.
  - To control a PDP context activation and deactivation by controlling a PDP address.
  - To control the routing path by location update in order to support user mobility.
  - To control the assignment of network resources to meet changing user demands.

- *LLC:* This protocol provides a highly reliable ciphered logical link.
- *RLC:* This protocol provides a logical reliable link control over the radio interface.
- *MAC:* The MAC protocol controls the access signaling procedures of request and grant for the radio channel.
- *BSSGP:* This layer conveys routing information and QoS-related information between the BSS and the SGSN.

3.3.3 Control Plane for I$_u$ Mode

Figure 3.8 depicts the control plane MS-SGSN in I_u mode. The layers and protocols in the control plane MS-SGSN in I_u mode are described below.

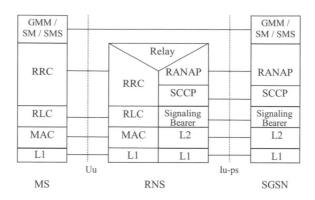

GMM: GPRS Mobility Management   RRC: Radio Resource Control
MAC: Medium Access control   SCCP: Signaling Connection Control Part
MS: Mobile Station   SGSN: Serving GPRS Support Node
RANAP: Radio Access Network Application   SM: Session Management
Protocol   SMS: Short Message Service
RLC: Radio Link Control

Figure 3.8   Control plane MS-SGSN in I_u mode

- *GPRS Mobility Management and Session Management (GMM/SM):* The GMM protocol supports mobility management functionality such as attach, detach, security, and routing area update. The SM protocol supports PDP context deactivation.
- *RRC:* This protocol is concerned with the allocation and maintenance of radio communication paths.
- *Short Message Service (SMS):* This protocol supports the mobile-terminated short message service as described in 3GPP TS 23.040.
- *RLC:* The RLC protocol offers the logical link control over the radio interface for the transmission of higher-layer signaling messages and SMS.
- *MAC:* The MAC protocol controls the access signaling procedures for the radio channel.
- *Radio Access Network Application Protocol (RANAP):* This protocol encapsulates and carries higher-layer signaling, handles signaling between the 3G-SGSN and I_u mode RAN, and manages the GTP connections (i.e. GPRS tunneling connections) on the I_u interface.

3.4 GPRS Ciphering Algorithm

This section covers the specification for the ciphering algorithm that is used to protect the GPRS, introduced by the ETSI Security Algorithm Group of Experts (SAGE). The ETSI SAGE is the design authority for the GPRS ciphering algorithm and is responsible for providing the required information in order to design and deliver a technical specification. The GPRS security algorithm is used to provide confidentiality and integrity protection of user data used for Point-to-Point (P2P) or Point-to-Multipoints (P2M) mobile originated and mobile terminated data transmission. The GPRS ciphering algorithm will be restricted to the MS-SGSN

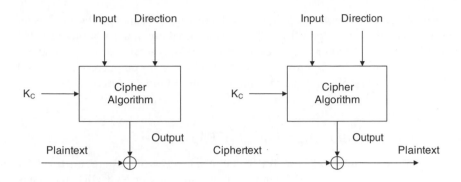

K_C : 64 bits   Direction : 1 bit
Input : 32 bits   Output   : Max 1600 octets, Min 5 octets

Figure 3.9   Basic GRPS stream ciphering scheme

encryption. The encryption algorithm is thus installed in the SGSN and MS. The use of GPRS algorithm should provide GPRS security features that define a service capability to be addressed the security requirements.

3.4.1 Parameters for Algorithm Design

The ETSI SAGE recommended the design of a symmetric stream cipher algorithm that satisfies the functional requirements as illustrated in Figure 3.9. The algorithm specifications relating to the input and output parameters are now described. The inputs consist of the ciphering key (K_c, 64 bits), the frame dependent input (INPUT, 32 bits), and transfer direction (DIRECTION, 1 bit). The output of the algorithm is indicated by the output string (OUTPUT, 1600 octets). The parameters of the algorithm are as follows:

- K_c: The ciphering key K_c is generated in the GPRS authentication and key management procedure. The length of key size is 64 bits. The key is unique to the MS when P2P traffic is used. It may be also common to several MSs when SGSN sends the same data to several MSs in P2M transmission. The K_c is never transmitted over the radio interface and is defined as K_c: K[0], K[1], ..., K[63] where K[i] is the K_c bit with label i.
- INPUT: This is the LLC frame dependent input parameter (32 bits) for the ciphering algorithm. For I-frame carrying user data, the input value is set to a random initial value at LLC connection set-up and incremented by 1 bit for each new frame. For UI-frame carrying user data and signaling messages, the input parameter is a non-repeating 32-bit value derived from the LLC header. The INPUT (32 bits) is denoted as X[i]: X[0], X[1], ..., X[31] where X[i] is the INPUT bit with label i.
- DIRECTION: This defines the direction (1 bit) of the data transmission (uplink/downlink). Z[0] is the DIRECTION bit with label 0.
- OUTPUT: This is the output of the ciphering algorithm (see Figure 3.9). The maximum length of the output string is 1600 octets, which represents the maximum length of the

payload of the LLC frame, including the Frame Check Sequence (FCS), 3 octets. The minimum length of the output string is 5 octets. Normal use of the algorithm is either short packets (25 to 50 octets) or long packets (500 to 1000 octets). However, for an optimum implementation it needs to be possible to generate just as many output octets as needed. W[i]: W[1], W[2], ..., W[1599] for the maximum length of output string.

3.4.2 GPRS Encryption Algorithm 3 (GEA3)

The encryption algorithm GEA3 for GPRS is defined in terms of the core function KGCORE, by mapping the GEA3 inputs onto the inputs of the core function KGCORE and mapping the output of KGCORE onto the outputs of GEA3 algorithm. The GEA3 algorithm is a stream cipher that is used to encrypt/decrypt blocks of data under a ciphering key K_c. The function KGCORE is based on the block cipher KASUMI which is used in a form of output-feedback mode (OFM) and generates the output bitstream in multiples of 64 bits. The GEA3 algorithm architecture is given in Figure 3.10. Referring to Figure 3.11, the keystream generator KGCORE should be initialized with the various inputs before the generation of keystream bits as output. The GEA3 algorithm produces an M-byte keystream where M is number of octets of the output required, but will never exceed $2^{16} = 65536$, that is, $1 \leq M \leq 65536$ inclusive. To generate each keystream block (KSB), we define the inputs to GEA3 as follows:

```
CA = CA[0]...CA[7] = 11111111 = an 8-bit input to the KGCORE func-
tion.
CB = CB[0]...CB[4] = 00000 = an 5-bit input to the KGCORE function.
CC = CC[0]...CC[31] = INPUT[0]...INPUT[31] =32 bit frame dependent
input, CD=CD[0] = DIRECTION[0]=a 1-bit input to the GEA3 algorithm,
indicating the direction of transmission (uplink or downlink)
CE = CE[0] ... CE[15] = 0000000000000000 = a 16-bit input to the
KGCORE function
CK = CK[0] ... CK[KLEN-1] = Kc[KLEN-1], where KLEN is in the range
64...128 bits inclusive.
```

Note that the GEA3 only allows KLEN to be a value of 64 bits, but it must be assumed that K_c is unstructured data. If KLEN < 128, then CK[KLEN] ... CK[KLEN127] = K_c[0]...K_c[127-KLEN].

If KLEN = 64, then CK = K_c[127-KLEN] = K_c[63]. Thus, the ciphering key CK is computed as the binary concatenation of K_c and K_c such that CK = K_c || K_c for producing CK = 128 bits.

Finally, application of these inputs to the core function KGCORE will derive the output CO (CO[0]...C[8M-1]) whose size is 8 M bits. For $0 \leq I \leq M\text{-}1$, we have

OUTPUT {i} = CO[8i]...CO[8i + 7], where CO[8i] is the most significant bit (MSB) of the octet.

No detailed functionalities on KGCORE and KASUMI are presented here, but they are covered in detail in Section 5.7 of Chapter 5.

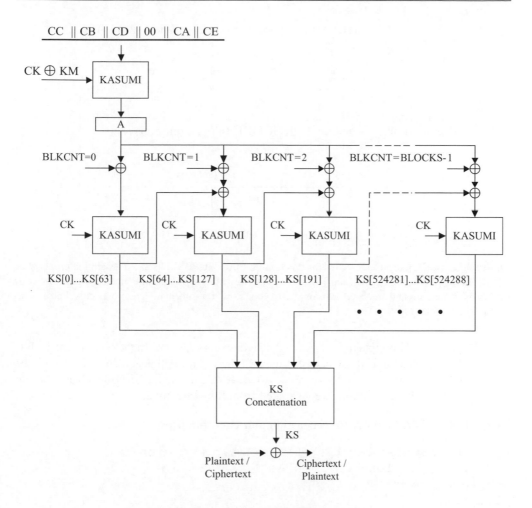

CC: Frame dependent 32-bit input
CB: 5-bit input initialized as 00000
CD: 1-bit input indicating the direction of transmission
CA: 8-bit input initialized as 1111 1111
CE: 16-bit input initialized as 0000 0000 0000 0000

Figure 3.10   GEA3 algorithm architecture

3.4.3 Ciphering and Deciphering

In the sending end, the output string is bit-wise XORed with the plaintext and the resulting ciphertext is sent over the radio interface. In the receiving end, the output string is bit-wise XORed with the ciphertext and the original plaintext is recovered.

- PLAINTEXT: This plaintext consists of the payload of the LLC frame (the information field) and the FCS (or CRC). The maximum length of the payload is 1600 octets. Note that the header part of the LLC frame should not be ciphered.

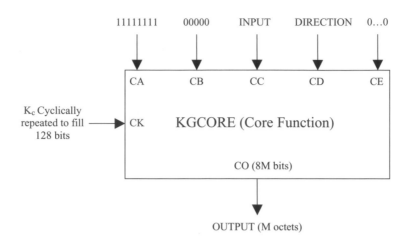

Figure 3.11   GEA3 keystream generator function (© 2003. 3GPP™ TSs and TRs are the property of ARIB, ATIS, CCSA, ETSI, TTA and TTC who jointly own the copyright in them. They are subject to further modifications and are therefore provided to you "as is" for information purposes only. Further use is strictly prohibited.)

- CIPHERTEXT: This ciphered version of the plaintext is generated in the sending end by bit-wise XORing the plaintext and output strings. Ciphering for uplink and downlink should be independent from each other. This contrasts to the A5 algorithm for GSM enciphering where keystreams for both directions are generated from the same input.

Example 3.1. GEA3 keystream generation and encryption

```
Key = 4035c6680af8c6d1d3c5d592327fb11c (128 bit)
Downlink Direction = 0
Length = 51 bits
Plaintext:
59b40a3a
```

In binary format:

```
Key =
    0100000000110101110001100110100000000101011110001100011011010001
    11010011110001011101010110010010001100100111111110110001000
0100011100.
Direction = 0
```

Plaintext:

```
101100110110100000010100011101010
```

Thus the inputs for KGCORE are

```
CA: 11111111.
CB: 00000.
```

```
CC: 10110011011010000001010000111010 (input).
CD: 0 = downlink direction
CE: 0000 0000 0000 0000
CK:
01000000001101011100011001101000000010101111000110001101101
0001
11010011110001011101010110010010001100100111111101100010001
1100
```

First of all, a 64-bit value A should be generated in KGCORE.

To generate A, all the above parameters are concatenated in the following way to form a 64-bit input for a KASUMI block:

```
CC || CB || CD || 00 || CA || CE
= 10110011011010000001010000111010 || 00000 || 0 || 00 || 1111 1111 ||
0000 0000
0000 0000
= 1011001101101000000101000111010000000000111111110000000000000
000
= 59b40a3a00ff0000 (64 bits)
```

XORing the key-modifier with CK computes the key for this KASUMI block. The key modifier (KM) is a constant.

```
Modified CK = CK ⊕ KM
= 0x4035c6680af8c6d1d3c5d592327fb11c ⊕
0x55555555555555555555555555555555
= 0x1560933d5fad9384869080c7672ae449 (128 bits)
```

All the computations in KASUMI block are described below.

First, the 128-bit modified key is subdivided into eight 16-bit values K1...K8 such that

```
K = K1 || K2 || K3||...||K8.
i.e. K = 1560 933d 5fad 9384 8690 80c7 672a e449
```

A second array of subkeys, K'_j is derived from K_j by applying

$$K_j' = K_j \oplus C_j$$

where C is the constant value defined as

```
0x 0123 4567 89ab cdef fedc ba98 7654 3210
And Kⱼ' = 1443 d65a d606 5e6b 784c 3a5f 117e d659 (See key scheduling
example)
```

The round subkeys are then derived from K_j and K'_j in the same manner as defined in Table 5.4 in Chapter 5.

Computed subkeys for KASUMI:

	1	2	3	4	5	6	7	8
$KL_{i,1}$	2ac0	267b	bf5a	2709	0d21	018f	ce54	c893
$KL_{i,2}$	d606	5e6b	784c	3a5f	117e	d659	1443	d65a
$KO_{i,1}$	67b2	f5ab	7092	d210	18f0	e54c	893c	ac02
$KO_{i,2}$	c780	2a67	49e4	6015	3d93	ad5f	8493	9086
$KO_{i,3}$	4ce5	3c89	02ac	b267	abf5	9270	10d2	f018
KI_{i1}	784c	3a5f	117e	d659	1443	d65a	d606	5e6b
KI_{i2}	5e6b	784c	3a5f	117e	d659	1443	d65a	d606
KI_{i3}	d659	1443	d65a	d606	5e6b	784c	3a5f	117e

L_0 and R_0 are initialized as follows:

```
L₀ = 59b4 0a3a
R₀ = 00ff 0000
```

Then an 8-round computation of KASUMI functions is executed. At the first round R_1 and L_1 are computed as follows:

```
R₁ = L₀ = 59b40a3a
L₁ = FO₁(FL₁(L₀, KL₁,₁, KL₁,₂), KO₁,₁, KO₁,₂, KO₁,₃, KI₁,₁, KI₁,₂, KI₁,₃) ⊕
R₀
   = FO₁(FL₁(59b40a3a, 2ac0, d606), KO₁,₁, KO₁,₂, KO₁,₃, KI₁,₁, KI₁,₂,
KI₁,₃) ⊕
   00ff0000
= FO₁(e7c91b3a, 67b2, c780, 4ce5, 784c, 5e6b, d659) ⊕ 00ff0000
= f317d856 ⊕ 00ff0000
= f3e8d856
```

Computation in FL_1 function:

```
FL₁(L₀, KL₁,₁, KL₁,₂) = FL₁(59b40a3a, 2ac0, d606)
The 32-bit input 59b40a3a is divided into two 16-bit values
FLL₀ = 59b4
FLR₀ = 0a3a
```

Then compute

```
FLR₁ = ((FLL₀^ KL₁,₁) <<< 1) ⊕ FLR₀
= ((59b4^ 2ac0) <<< 1) ⊕ 0a3a
    = 1b3a
```

$FLL_1 = FLL_0 \oplus ((FLR_1 \mid KL_{1,2}) <<< 1)$
$= 59b4 \oplus ((1b3a \mid d606) <<< 1)$
$= e7c9$
Output of $FL_1 = FLL_1 \mid\mid FLR_1 = e7c91b3a$

Computation in FO_1 function:

$FO_1(FL_1(L_0, KL_{1,1}, KL_{1,2}), KO_{1,1}, KO_{1,2}, KO_{1,3}, KI_{1,1}, KI_{1,2}, KI_{1,3})$
$= FO_1(e7c91b3a, 67b2, c780, 4ce5, 784c, 5e6b, d659)$

Input e7c91b3a is divided into two 16-bit values:

$FOL_0 = e7c9$
$FOR_0 = 1b3a$

Then compute

$FOL_1 = FOR_0 = 1b3a$
$FOR_1 = FI_{1,1}(FOL_0 \oplus KO_{1,1}, KI_{1,1}) \oplus FOR_0$
$\quad\quad = FI_{1,1}(e7c9 \oplus 67b2, 784c) \oplus 1b3a$
$\quad\quad = d7d2 \oplus 1b3a$ (Calculation of $FI_{1,1}$ function is given later)
$\quad\quad = cce8$

Similarly FOL_2, FOR_2 and FOL_3, FOR_3 are calculated as follows:

$FOL_2 = FOR_1 = cce8$
$FOR_2 = FI_{1,2}(FOL_1 \oplus KO_{1,2}, KI_{1,2}) \oplus FOR_1$
$\quad\quad = f317$
$FOL_3 = FOR_2 = f317$
$FOR_3 = FI_{1,3}(FOL_2 \oplus KO_{1,2}, KI_{1,2}) \oplus FOR_2$
$\quad\quad = d856$

Final output for FO_1 is

$FOL_3 \mid\mid FOR_3 = f317\ d856$

Computation in $FI_{1,1}$ function:

$FI_{1,1}(807b, 784c)$

The input is split into two unequal components: a 9-bit left half FIL_0 and a 7-bit right half FIR_0.

$FIL_0 = 807b >> 7 = 100$
$FIR_0 = 807b \verb|^| 7f = 7b$

The input subkey 784c is split into two unequal components: a 7-bit left half $KI_{1,1,1}$ and a 9-bit right half $KI_{1,1,2}$

$KI_{1,1,1} = 784c >> 9 = 3c$
$KI_{1,1,2} = 784c \wedge 1ff = 4c$

$$
\begin{aligned}
FIL_1 &= S9[FIL_0] \oplus ZE(FIR_0) \\
&= S9[100] \oplus ZE(7b) \\
&= 23 \oplus ZE(7b) \text{ [See Table SBox(9)]} \\
&= 00010 0011_b \oplus 001111011_b \text{ [Zero extended on left side]} \\
&= 58
\end{aligned}
$$

$$
\begin{aligned}
FIR_1 &= S7[FIR_0] \oplus TR(FIL_1) \\
&= S7[7b] \oplus (FIL_1 \wedge 7f) \\
&= 29 \oplus (58 \wedge 7f) \\
&= 29 \oplus 58 \\
&= 71
\end{aligned}
$$

$$
\begin{aligned}
FIL_2 &= FIL_1 \oplus KI_{1,1,2} \\
&= 58 \oplus 4c \\
&= 14
\end{aligned}
$$

$$
\begin{aligned}
FIR_2 &= FIR_0 \oplus KI_{1,1,1} \\
&= 71 \oplus 3c \\
&= 4d
\end{aligned}
$$

$$
\begin{aligned}
FIL_3 &= S9[FIL_2] \oplus ZE(FIR_2) \\
&= S9[14] \oplus ZE(4d) \\
&= 19f \oplus 4d \\
&= 1d2
\end{aligned}
$$

$$
\begin{aligned}
FIR_3 &= S7[FIR_2] \oplus TR(FIL_3) \\
&= S7[4d] \oplus (1d2 \wedge 7f) \\
&= 39 \oplus 52 \\
&= 6b
\end{aligned}
$$

The output of $FI_{1,1}$ function is finally calculated as

$$
\begin{aligned}
FIR_3 \mid\mid FIL_3 &= (FIR_3 << 9) \mid FIL_3 \\
&= (6b << 9) \mid 1d2 \\
&= d7d2
\end{aligned}
$$

Similarly, in the second round of KASUMI, L_2 and R_2 are computed as

$R_2 = L_1 = f3e8d856$
$L_2 = FL_2(F0_2(L_1, KO_{2,1}, KO_{2,2}, KO_{2,3}, KI_{2,1}, KI_{2,2}, KI_{2,3}), KL_{2,1}, KL_{2,2}) \oplus R_1$

```
= 32be701a ⊕ 59b40a3a
= 6b0a7a20
```

Finally, the values of L_8 and R_8 are computed as

```
L₈ = 326bcca0
R₈ = b77f38de
```

Therefore, KASUMI's output b77f 38de 326b cca0 will be considered as the value of A, which will be the input for the subsequent KASUMI blocks of KGCORE.

```
As the length of the output is 51 * 8 = 408, the number of required
KASUMI block
=⌈408/64⌉=7
Thus the values of BLKCNT are 0, 1, ..., 6
For all blocks the confidential key, CK =
4035c6680af8c6d1d3c5d592 327fb11c
```

Now, the input for the first KASUMI block is

```
A ⊕ BLKCNT = b77f 38de 326b cca0 ⊕ 0 = b77f 38de 326b cca0
The output CO[0...63] is calculated in the same manner described
before,
CO[0...63] = e7dabbc73003ff96
```

For the second KASUMI, block input is

```
A ⊕ BLKCNT ⊕ CO[0...63] = b77f38de326bcca0 ⊕ 1 ⊕ e7dabbc73003ff96
                        = 50a5831902683337
The output CO[64...127] = ddae4f7db09aadbd
```

For the third KASUMI, block input is

```
A ⊕ BLKCNT ⊕ CO[64...127] = b77f38de326bcca0 ⊕ 2 ⊕ ddae4f7db09aadbd
                          = 6ad177a382f1611f
The output CO[128...191] = c485f9c36e1a7669
```

Similarly,

```
CO[192...255] = 289a0c0fc93aed4e
CO[256...319] = 78e6464ddae06589
CO[320...383] = 9e20664a437750b7
CO[384...448] = 9b876525ebc40bff
```

Finally, the keystream KS is computed by concatenating the all output bits up to 253. If necessary the least significant bits are discarded.

So the generated keystream is

```
KS = CO[0...63] || CO[64...127] || CO[128...191] || CO[192...255] ||
CO[256...319] || CO[320...383] || CO[384...407]

= e7dabbc73003ff96ddae4f7db09aadbdc485f9c36e1a7669289a0c0fc
93aed4e78e6464ddae065899e20664a437750b79b8765
```

After truncation of the generated keystream if necessary, the ciphertext is computed by XORing the plaintext with the keystream.

☐

4

Third-generation Partnership Projects (3GPP and 3GPP2)

First-generation (1G) technologies were based on analog voice transmissions. Even though digital networks are widely deployed nowadays, analog networks are still in service in more than 40% of cellular phones in the United States. In the early 1990s, second-generation (2G) wireless technologies were introduced. After 1G systems, 2G cellular telephony systems began to incorporate digital technologies in the USA (TDMA and CDMA), Europe (GSM), and Japan (PDC). Most 2G technologies are digital in nature and provide improved system performance and enhanced security. Data rates available for 2G wireless technologies vary from 9.6 to 14.4 Kbps.

Apart from 2G wireless technology, there is a mid-specification group referred to as 2.5G technology. All intermediary systems that provide theoretical data bandwidths on a packet basis of at least 14.4 Kbps are classified as 2.5G. All 2.5G specifications enhance the existing 2G technologies and make a transition to the third-generation (3G) technologies. The technologies classified as 2.5G promise much improved bandwidths and the capability to transmit more complex contents to mobile devices. The 2.5G technologies include High Speed Circuit Switched Data (HSCSD), General Packet Radio Service (GPRS), cdmaOne IS-95B, Enhanced Data rates for Global Evolution (EDGE), and Cellular Digital Packet Data (CDPD). However, NTT DoCoMo's i-mode of Japan is bypassing directly into deployment of 3G FOMA technology without going through the 2.5G technologies.

All European operators that adopted the GSM standard have largely begun to switch their network to 2.5G GPRS. The first GPRS networks were launched at the end of 2000, but the launch of commercial services was delayed until the beginning of 2002 due to several problems encountered in the fine-tuning of terminals. The number of GRPS network service users in Western Europe was estimated at 28 million at the end of 2003. Commercial 2.5G EDGE services are already operational. EDGE technology is going to be used by several European operators including Bouygues Telecom, Orange France, TIM, and TeliaSonera. But these operators are more likely biding their time and waiting on EDGE as a complement of 3G UMTS.

Since the launch of the 2G systems, the following organizations have initiated work to design the 3G communications capabilities: UMTS (Europe); WCDMA (Europe, Japan, Korea, and

Mobile Communication Systems and Security   Man Young Rhee
© 2009 John Wiley & Sons (Asia) Pte Ltd

USA); DoCoMo's FOMA (Japan); CDMA2000 1xEV (USA); and CDMA2000 1xEV-DO and 1xEV-DV (Korea, USA, and Japan):

- *In Western Europe:* Hutchison was a pioneer of the supply of high-performance terminals and announced 20 000 new subscribers per day in Europe in June 2004. Vodafone and T-Mobile launched their first 3G offering for PCMCIA cards.
- *In the United States:* Aggressive 3G deployment plans by Verizon Wireless were followed by those of Sprint and Cingular Wireless. Verizon Wireless, a leader in the USA in terms of number of subscribers, covers 40 of the 50 key markets in the USA. Verizon launched its CDMA2000 1xEV-DO Broadband Access in Washington DC and San Diego in 2003.
- *In Japan:* Since the notably success of DoCoMo's i-mode services at the beginning of 1999, NTT DoCoMo's FOMA services launched its first commercial 3G services in September 2001, followed by equivalent services offered by KDDI and Vodafone KK.
- *In Republic of Korea:* Korea was the world leader in mobile and fixed broadband networks, enabling a theoretical maximum bandwidth of 144 Kbps, with 21.3 million 3G CDMA2000 1xEV-DO subscribers at the end of 2003. Korean operators were in fact the first in the world to offer commercial 3G CDMA2000 1xEV-DO services capable of providing a maximum bandwidth of 2.4 Mbps, from January 2002 for SK Telecom and May 2002 for KTF. In addition, LG Telecom is the holder of a 3G license based on the CDMA2000 standard. SK Telecom and KTF also deployed 3G WCDMA networks and launched their first services at the end of the year 2003.

One of the remarkable features for 3G technologies was to realize the full-scale global roaming services in various countries around the world. 3G throughput data rates range from 144 Kbps to 2 Mbps.

Over the past decade, mobile communications technologies have made tremendous progress, moving rapidly from 1G analog voice-only communications to 2G digital voice and data communications. Now, the world has a revolutionary 3G mobile communications platform.

4.1 3G Partnership Projects

The 3G Partnership Project (3GPP) is a collaboration agreement on crucial technical decisions that was established by five standards organizations in December 1998. These organizations decided to cooperate in their combined efforts to accelerate the joint project aiming at guaranteed global communications interoperability. The 3GPP organizational partners, responsible for determining its general policy and strategy, are Association of Radio Industries and Business (ARIB, Japan), T1 Committee of the Telecommunications of the American National Standards Institute (ANSI T1, USA), European Telecommunications Standard Institute (ETSI, Europe), Telecommunications Technology Association (TTA, Korea), and Telecommunications Technology Committee (TTC, Japan). As a sixth partner, China Wireless Telecommunication Standard Group (CWTS, China) joined the 3GPP project, soon after the five collaborators were organized.

The original scope of 3GPP was to produce globally applicable Technical Specifications (TSs) and Technical Reports (TRs) for a 3G mobile system based on evolved GSM core networks and the radio access technology on Wideband Code Division Multiple Access

(WCDMA). Almost one year later the American National Standards Institute (ANSI) decided to establish 3GPP2, a 3G partnership project for evolved ANSI/TIA/EIA-41 networks.

The International Mobile Telecommunications-2000 (IMT-2000) group within the International Telecommunication Union (ITU) focuses its work on defining interfaces between 3G networks evolved from GSM and ANSI-41 networks in order to enable the realization of seamless harmonization between 3GPP and 3GPP2 networks.

3GPP2 is the international organization in charge of the standardization of CDMA 2000, that is, the 3G evolution of the IS-95 standard. It came from the fruitful results of combined research carried out by the TR45.5 (North America) and TTA (Republic of Korea) standardization groups. Like 3GPP, 3GPP2 has organizational partners and market representation partners. 3GPP2's organizational partners are: ARIB and TTC from Japan, TIA from North America, and CWTS from China. Its market representation partners are: the CDMA Development Group (CDG) and the IPv6 Forum. The task of producing 3GPP2 specifications rests with the project's four Technical Specification Groups (TSGs), comprised of representatives from the project's individual member company. The TSGs are complied with TSG-A (Access Network Interfaces), TSG-C (cdma2000), TSG-S (Services and Systems Aspects), and TSG-X (Intersystem Operations). All TSGs report to the project's steering committee, which is responsible for managing the overall work process and adopting the technical specifications forwarded from each of the TSGs. Thus, the global wireless industry has created two new partnership projects to address the following issues: 3GPP is responsible for developing 3G standards for GSM-based systems; and 3GPP2 is responsible for development of IS-95-based CDMA systems.

There are 3GPP market representation partners with organizations such as the GSM Association, UTMS Forum, Global Mobile Suppliers Association, IPv6 Forum, 3G Americas, and 3G.IP. Their role is to provide markets inputs and requirements for 3GPP standardization groups. 3GPP also includes observers that are standard Development Organizations qualified to become future Organization Partners. 3GPP currently has three observers: the Telecommunications Industries Association (TIA), Telecommunications Standards Advisory Council of Canada (TSACC), and Australian Communications Industry Forum (ACIF).

3GPP officially provides its activities with the preparation of large volumes of Release 99 specifications. The project is run by the Project Coordination Group (PCG). Five TSGs should report to the PCG: they are Radio Access Network, Core Network, Service and Systems Aspects, Terminal, and GSM EDGE Radio Access Network (GERAN). Each of these TSGs has a number of working groups composed of individuals from companies that are members of one or more of the organizational partners, market representation partners, or observers.

The prime purpose of 3GPP partners is to elaborate standards for technologies and applications that are interoperable at both the network and terminal levels. 3GPP objectives, created from an IMT-2000 initiative of ITU and 3GPP2 originated for the CDMA2000 services, aim at the development and implementation of

- realization of the spread of WCDMA-based 3G services;
- high-speed, broadband, and high-capacity communications;
- diversified non-voice visual communications;
- increasing packet transmission volume;
- simultaneous voice and data transmissions;
- multimedia content delivery services;
- global roaming across dissimilar network;

- seamless communications services independent of locality;
- enhanced security against wireless threats for data subscribers' authentication and data privacy for packet data services.

4.2 Evolution of Mobile Radio Technologies

This section presents the evolution and migration from 1G to 3G mobile radio technologies, with an emphasis on wireless security. 1G, or circuit-switched analog systems, consist of voice-only communication; 2G and beyond systems comprise both voice and data communications, largely relying on packet-switched wireless mobile technologies.

This section covers technological development of wireless mobile communications in compliance with each iterative generation over the past decade. Currently, mobile data services have been rapidly transforming to facilitate and ultimately profit from the increased demand for non-voice services. Through aggressive 3G deployment plans, the world's major operators boast attractive and homogeneous portal offerings in all their markets, notably in music and video multimedia services. Despite the improbability of any major changes in the next 4–5 years, rapid technological advances have already bolstered talks for 3.5G and even 4G systems. New All-IP wireless systems will come into a position to compete with current cellular networks; Wi-Fi technology, along with WiMax and FDD/TDD mode, may make it possible for new entrants to compete with incumbent mobile operators.

4.2.1 2G Mobile Radio Technologies

Two major 2G mobile telecommunications standards have been dominating the global wireless market: GSM, developed at the beginning of the 1990s by the ETSI in Western Europe, and TDMA-136/CDMA IS-95, developed by TIA in North America. Since the GSM standard was originally designed for voice, GSM was ill-suited to data transmission. Although GSM is the most widely used circuit-switched cellular system for voice communications, GSM networks were not optimized for high-speed data, image, and other multimedia applications and services. The ETSI hence upgraded the GSM standard, albeit still in circuit-switched mode. The HSCSD technology was first deployed to enable higher data rates. GPRS soon followed, introducing packet-switched mode. Finally, EDGE was introduced to further increase the data speeds provided by GPRS. Currently, most GSM-based networks are expected to evolve 3G UMTS.

Qualcomm developed a mobile communication technology based on the CDMA spectrum-sharing technique. This network technology, which is modulated by codes, is called cdmaOne IS-95A/B and was standardized by TIA. Under the influence of Qualcomm, the IS-95 technology continues to evolve steadily to provide higher data rates, such as 3G CDMA2000 1x networks that were standardized by 3GPP2.

The IS-54, based on the TDMA access mode, was the first North American digital telephony standard. This standard was also adapted for use in wideband PCS networks under the name TDMA-136. The IS-54 was primarily used by the US operators, but limitation in data transfers due to the use of a relatively narrowband forced its demise in December 2001.

In addition, two other proprietary technologies classified as 2G systems include NTT DoCoMo's i-mode in Japan and WAP protocol, the de facto standard created by the WAP Forum, which was founded in 1997 through the initiative of Nokia, Motorola, Ericsson, Phone. com, etc. NTT DoCoMo's i-mode is the mobile Internet access system, which provides service over the packet-switched network. The i-mode gives users a new range of capabilities, offering voice and data cellular service in one convenient package. The WAP Forum's initial aim was to establish a universal and open standard to provide wireless users with access to the Internet. WAP was designed to deliver Internet content by adapting to the features and constrains of mobile phones. WAP technology cannot be defined as a 2G system because it was designed to work with all wireless network technologies, beginning with a majority of 2G systems (GSM, GPRS, PCD, IS-95, TDMA-136) and 3G systems. However, with the commercial failure of the launch of WAP at the beginning of 2000, the European industry missed the mobile data service explosion. Only the SMS service appeared to offer access to the WAP portal via GSM networks. The MMS technology will make it possible to overcome the technological constrains of SMS and to further enhance existing services with SMS. The i-mode service started in February of 1999; WAP 2.0 was released from WAP Forum in August 2001; and the M-service Phase 2 started in GSM-A.

4.2.2 2.5G Mobile Radio Technologies

2G mobile radio technologies enable voice traffic and limited data traffic, such as SMS, to transmit over wireless. Improvements must be made in order to facilitate high data rate services that ultimately allow the transmittal and receiving of high quality data and video to and from the Internet. However, the data handling capabilities of 2G mobile systems are limited.

For 2.5G systems, HSCSD in circuit-switched mode was the first step for GSM's evolution in increased data transmission rates, reaching maximum speeds of about 43 Kbps (3 simultaneous GSM circuits running at 14.4 Kbps). The drawback of HSCSD, when compared to GPRS, is the several time slots used in *circuit mode*, whereas GPRS uses several time slots in *packet mode*. HSCSD is considered an interim technology to GPRS, which offers instant connectivity at higher speeds.

GPRS is the evolution of GSM for higher data rates within the GSM carrier spacing. GPRS introduces packet transmission for data services, replacing GSMs circuit-switched mode. EDGE (an upgraded version of GPRS) was designed for a network to evolve its current 2G GSM system to support faster throughput and to give operators the opportunity to understand the new technology prior to the complete 3G rollout. EDGE is a higher-bandwidth version of GPRS, with transmission rates up to 384 Kbps. Such high speeds can aptly support wireless multimedia applications.

Western Europe has been the leader in terms of GSM mobile penetration since the 1990s. The popularity of these services with subscribers made it possible to create a market representing a volume of almost 170 billion messages in 2003. All European operators that adopted the GSM standard have largely begun to switch their network, first GPRS/EDGE and then to UMTS. The number of GPRS network service users in Western Europe was estimated at 28 million at the end of 2003. EDGE technology is going to be used by several European operators including Bouygues Telecom (France), Orange France, Telecom Italia Mobile (Italy), TeliaSonera (Finland), T-Mobile (Germany), Netcom (Norway), and Orange (UK). Most European operators launched or announced the launch of 3G services in 2004.

ITU defined the IMT 2000 program for the 3GPP as well as 3GPP2 as a main part of the 3G technical framework. The primary objective of the standardization activities for IMT2000 is to develop a globally unified standard for worldwide roaming and mobile multimedia services. In order to achieve these goals, the ITU has placed maximum effort in creating harmonized recommendations backed up by such technical forums as the 3GPP and 3GPP2. The ITU-R and ITU-T are the main bodies that produce recommendations for IMT2000. The 3GPP, created in late 1998, is the group responsible for standardizing UMTS with WCDMA technology.

The 3GPP has so far released: Release 99, Release 4, Release 5, and Release 6. Release 99 includes the basic capabilities and functionalities of UMTS. Release 4 was contributed by CWTS and incorporated as WCDMA/TDD, but it was frozen in 2003. Since Release 5 was successfully completed in March 2002, 3GPP is moving toward the next release, Release 6, to further improve performance and enhance capabilities.

The 3GPP2 specifies an air interface based on cdmaOne technology and the CDMA2000 interface to increase capability and enable faster data communication. The technical area of 3GPP2 is similar to that of the 3GPP.

cdmaOne IS-95B is enhanced through the migration from cdmaOne IS-95A. In the non-GSM regions (notably the USA and South Korea), network operators are preparing next-generation wireless systems based on cdmaOne IS-95A/B. The first phase with IS-95A was fully covered in TIA/EIA/IS-95A + TSB74. In late 1997, the second phase with IS-95B brought about improvements in terms of capacity, allowing data transmission at 64 Kbps. The 2.5G equivalent for CDMA operators is a technology called CDMA2000 1x. The IS-95 technology continues to improve to provide higher data rates towards the natural evolution of CDMA2000 1x networks for 3G, that is, CDMA2000 1xEV-DO and 1xEV-DV, both standardized by 3GPP2, which is covered in the next section.

4.2.3 3G Mobile Radio Technologies (Situation and Status of 3G)

3G mobile technologies are cellular radio systems for mobile technology. ITU defined the 3G technical framework as part of the IMT 2000 program. The 3GPP, created in late 1998, is the group responsible for standardizing UMTS at a global level. It is composed of several international standardization bodies involved with defining 3G technologies. TDMA-136 and GSM/GPRS operators plan to use UMTS (3G), which is an advanced version of EDGE(2.5G). GSM/GPRS operators plan to deploy UMTS with WCDMA technology. Unlike CDMA2000, WCDMA will be deployed in the frequency bandwidths identified for 3G, leading some American operators to adopt EDGE. TDMA networks are steadily being replaced with GSM-evolved technology, that is, from GPRS to EDGE, and finally to the 3G WCDMA (UMTS) standard.

The 3GPP2 is the international organization in charge of the standardization of CDMA2000, which in turn is the 3G evolution of the IS-95A/B standards. CDMA2000 represents a family of ITU-approved IMT-2000 (3G) standards including CDMA2000 1x networks, CDMA2000 1xEV-DO, and 1xEV-DV technologies. The first CDMA2000 1x networks were launched in Korea in October 2000 by SK Telecom and LG Telecom. CDMA2000 1xEV-DO was recognized as an IMT-2000 technology at data rates of 2.4 Mbps on 1.25 MHz CDMA carrier. 1xEV-DO makes use of the existing suite of Internet Protocol (IP) operating systems, and software applications, and therefore it builds on the architecture of the CDMA2000 1x network while preserving seamless backward compatibility with IS-95A/B and CDMA2000 1x. CDMA2000 1xEV-DV provides integrated voice with simultaneous high-speed packet data services at speeds up to 3.09 Mbps. But 1xEV-DV is still in the development stage. The CDMA2000 family of air interfaces operates with an IS-4 network, an IP network, or a GSM-WAP network. This allows operators tremendous flexibility with the network and ensures backward compatibility with deployed terminal base.

In June 2000, the Ministry of Postal and Transportation of Japan awarded 3G licenses to three mobile operators, NTT DoCoMo, KDDI, and Vodafone KK (J-Phone), via a comparative bidding process. NTT DoCoMo's i-mode (2G) is the first mobile Internet service in the world with 42 million subscribers at the end of March 2004. NTT DoCoMo commercially launched a WCDMA network based on the UMTS standard in the Tokyo area under the name of FOMA. Since its first launch of the 3G FOMA service in October 2001, new and exclusive services accessible to FOMA subscribers include: video telephony and i-motion's video clip distribution service, as well as i-motion's mail messaging service.

Unlike its two competitors (NTT DoCoMo and Vodafone KK), KDDI opted for 3G CDMA2000 technology. In April 2002, KDDI opened its CDMA2000 1x network by using 3G bandwidths and began deployment of the 3G version of 1x. KDDI launched its commercial CDMA2000 1xEV-DO services nationwide in the first quarter of 2004 and has offered WIN services based on CDMA 1xEV-DO technology from November 2003. Since March 2004, KDDI has offered a BREW (a new application platform) terminal with Bluetooth technology.

The three major South Korean mobile operators (SK Telecom, KTF, and LG Telecom) provide 2G and 2.5G mobile services, using the Qualcomm developed CDMA IS-95 system and its successor CDMA2000 1x. These three mobile operators have been very quick to set up services based on CDMA2000 1x, thus enabling a maximum bandwidth capacity of 144 Kbps. In fact, SK Telecom was the first operator in the world to launch a CDMA2000 1x service in October 2000. As for 3G technologies, the Ministry of Information and Communication of the Korean government decided to grant licenses according to the type of technology, either WCDMA or CDMA2000. The Korean government decided to grant WCDMA licenses to SK telecom and KTF, and a CDMA2000 license to LG Telecom. Thus, SK Telecom and KTF, holders of 3G WCDMA licenses, are deploying networks based on the CDMA2000 1xEV-DO standard in their existing frequency bandwidths. This resulting system is capable of providing 3G services with a maximum bandwidth of 2 Mbps.

SK Telecom was the first in the world to launch a CDMA2000 1xEV-DO network in January 2002, followed by KTF in May 2002. This was one of the very first operators in the world, including KDDI (2003), to launch a 3G service. SK Telecom launched the WCDMA service at the end of 2003.

KTF is the mobile subsidiary of KT Corporation. KTF launched services based on CDMA2000 1x technology in June 2001 and on CDMA2000 1xEV-DO technology in May 2002. Deployment used Qualcomm's BREW platform, but the Java WIPI platform was used by all operators from the end of 2003. Launch of the WCDMA service started at the end of 2003.

LG Telecom is the subsidiary of the LG Corporation. Services using CDMA2000 1x technology were launched in August 2001. LG Telecom is the holder of a 3G license based on the CDMA2000 standard.

TD-SCDMA operating with the TDD mode is the 3G standard developed in China. This Chinese proposal was submitted by the CWTS to the ITU and is standardized by 3GPP. TD-SCDMA supports asymmetric data traffic with peak data rates of 2 Mbps and can be connected to existing GSM networks. However, the commercial launch of services based on this technology is not expected to take place for several years to come.

Seven operators control the market for mobile telephony in the USA. Instead of facing open competition between three or even four major mobile operators, the North American market is now structured around two main operators, that is, AT&T Wireless + Cingular Wireless and Verizon Wireless. This follows the merger between AT&T Wireless and Cingular Wireless announced in February 2004 and the less important merger between Verizon Wireless and Qwest Wireless. In response to this situation, other operators may have to join forces in terms of operations and capital.

Verizon Wireless is a leader in the USA in terms of number of subscribers; it covers 40 of the 50 key markets in the USA. Launched in January 2002, Verizon Wireless was the first major US operator to commercially provide a CDMA2000 1x network. Verizon Wireless launched its CDMA2000 1xEV-DO broadband access in Washington DC and San Diego in 2003, and plans to continue deploying its market on a national level.

Cingular Wireless is the subsidiary of the regional operators SBC and BellSouth. Originally it was the operator of a TDMA-136 network, but Cingular Wireless decided to migrate to GPRS, first launched in March 2001. Cingular Wireless was also the first US operator to launch EDGE in June 2003.

AT&T Wireless is the North America operator acquired by the Cingular Wireless in February 2004. AT&T Wireless launched a GPRS service in mid-2001 and coverage of all markets was complete by the end of 2002. AT&T Wireless launched an i-mode type services called "mMode" on the GPRS network in April 2002. The EDGE deployment plan was established as of mid-2002 and nationally launched in November 2003. AT&T Wireless announced its UMTS deployment plan at the beginning of 2003, and on 26 December 2003 the company announced the four markets (San Francisco, San Diego, Seattle, and Dallas) in which the first WCDMA networks were to be deployed by the end of 2004. In partnership with NTT DoCoMo the first UMTS call between New York and Tokyo was carried out on 12 November, 2002.

The wireless industry worldwide are putting continuous efforts into deriving the technology evolution to support even greater data throughput and better network capacity than offered by 3G. In a 4G environment, aggressive and iterative generation of all wireless mobile communications (a combination of 2G/2.5G/3G) along with Bluetooth and IEEE 802.11 could all coexist for attaining faster data throughput and greater network capacity.

IMT 2000 has just been commercialized and therefore new standardization work should commence for the system beyond IMT 2000. Those new systems will be expected to provide more sophisticated services to meet the further demands of the wireless community. Overall objectives of the future development of IMT 2000 and of systems beyond IMT 2000 will include new radio access capabilities and a new IP-based core network to result in another phase of harmonization.

Since the 3GPP Release-5 of UMTS was completed in March 2002, 3GPP is moving toward Release-6, which aims to further improve performance and enhance capabilities. The interworking between WLAN and UMTS has been proven to be one of the keys for providing flexibility when accessing multiple radio resources and also providing mobility between WLAN and the 3G system in various mobile environments.

One of the major applications, Multimedia Broadcast Multicast Service (MBMS), may pioneer a new service that allows the broadcasting of multimedia messaging and video/music streaming capabilities. HSDPA (High Speed Downlink Packet Access) will represent a change in WCDMA systems and could be compatible with existing networks. HSDPA may enable packet transmission to make speeds of 8–10 Mbps for the downlink in UMTS channels of 5 MHz. With the MIMO function, speeds of 20 Mbps could even be reached for providing the faster throughput. HSDPA systems will be in a position to compete with Wi-Fi/Bluetooth services at certain mobile markets: NTT DoCoMo in Japan, several Western European operators, and Cingular Wireless and Verizon Wireless in the USA.

Orthogonal Frequency Division Multiple Access (OFDMA) is being studied as a radio access technology that may drastically increase data rates by using a large number of orthogonal frequencies. It is also foreseen that the MIMO antenna technology combined with the OFDM access multiplexing principle could be a promising candidate for what is called the 4G mobile system.

For beyond 3G 2000 systems, CDMA and MIMO-OFDM technologies will be key dynamics in the demand for high-capacity broadband services. With the presence of multiple access technologies, 3G CDMA handsets already exist, but CDMA2000 devices in the coming years should include support for WCDMA/HSPA, WiMAX, UMB, WiBro, and LTE. OFDM-based technologies will offer certain economic benefits for most service operators and will provide them with a complement to their services, features, and coverage. However, 3GCDMA will continue to be the core business for well over a decade and play a key role in the future wireless industry.

The 3GPP2 has been working on an evolution of CDMA technology to enhance new features. CDMA2000, backed by the USA (primarily by Qualcomm), is the direct successor to cdmaOne IS-95 A/B networks. There are two phases to deploy CDMA2000, that is, CDMA2000 1x and CDMA2000 1xEV. CDMA2000 1xEV is the final stage in the evolution from cdmaOne network to 3G. The transition from 1x to 1xEV is taking place in two states: CDMA2000 1xEV-DO and 1xEV-DV.

CDMA2000 1xEV-DO is the first phase, using a separate carrier for traffic and data. The 1xEV-DO may function on a bi-mode operation (1x for voice and EV-DO for data only). At the end of 2000, a specification for High Rate Packet Data (HRPD) was issued to enhance downlink/uplink data transmission. HRPD, sometimes called 1xEV-DO (1x Evolution Data Only), allows mobile terminals to access easily on the IP network through a high-speed data communication link. Since the 1xEV-DO was primarily devised for data communication only, another radio channel was required for speech communication, which led to the development of the 1xEV-DV. CDMA2000 1xEV-DV (1x Evolution Data and Voice) builds on the architecture of CDMA2000 1x while preserving seamless backward compatibility with cdmaOne IS-95 A/B and CDMA2000 1x. CDMA2000 1xEV-DV (approved by the 3GPP2 in June 2002 and submitted to ITU for approval in July 2002) provides integrated voice with simultaneous high-speed packet data services such as video, video-conferencing, and other multimedia services at speeds of up to 3.09 Mbps. In order to support multimedia services, it is necessary to provide simultaneous speech and data communication using the same carrier frequency.

Systems beyond IMT-2000 may include new radio access capabilities and a new IP-based core network to be realized around the year 2010. Due to tireless efforts by ITU and 3GPPs, a global consensus has been recognized to further develop a worldwide harmonized standard that makes it easier to improve mobile services and stimulate the mobile market.

4.3 Cryptographic Protocols Applicable to Wireless Security Technologies

In November 1976, the Data Encryption Standard (DES) was adopted as a federal standard and authorized for use on all unclassified US government communications. The official description of DES was published in FIPS PUB 46 on 15 January 1977. The DES algorithm, developed by IBM, was the best proposed standard, even though there was much criticism of the key size and the design criteria on the internal structure of the S-box. Nevertheless, DES came to fame as a popular security algorithm for organizations worldwide. After much debate, DES was reaffirmed as a US government standard until 1992 because there was still no alternative. In fact, DES survived remarkably well despite intensive cryptanalysis, and it remained the worldwide standard for 20 years. The National Institute of Standards and Technology (NIST) solicited another review to assess the continued adequacy of DES to protect data and in 1993 formally solicited comments on the recertification of DES. After reviewing many comments and technical inputs, NIST recommended that the useful lifetime of DES would end in the late 1990s.

In 2001, the Advanced Encryption Standard (AES), developed by Daemen and Rijmen in 1999 and known as the Rijndael algorithm, became a FIPS-proved advanced block cipher algorithm. It will become a strong advanced algorithm in lieu of DES. AES is expected to provide the data security for the communications network and access control systems along with algorithm specifications such as the key expansion routine, encryption by cipher, and decryption by inverse cipher.

The concept of the Elliptic Curve Cryptosystem (ECC) was first introduced by Victor Miller in 1985 and Neal Koblitz 1987. The elliptic curve discrete algorithm problem appears to be substantially more difficult and somewhat harder than the existing discrete logarithm problem. Implementations can exploit this difference when providing both faster speed and smaller key size for a given level of security. Providing an equivalent level of security, ECC uses smaller parameters than the conventional discrete logarithm systems. Elliptic Curves (ECs) have been well studied by mathematicians for many years, and in the latter half of the twentieth century this research yielded some very significant results.

All practical public-key systems, such as Diffie-Hellman, RSA, ElGamal, Schnorr, DSS, and many other public-key algorithms, exploit the arithmetic properties using large finite groups. For those systems, the security depends directly on the relative difficulty of performing group operations such as exponentiation versus discrete logarithm. Computation of exponentiation is much easier than that of its inverse operation, that is, discrete logarithm. In the commonly used groups, discrete log is hard to compute when the modulus is very large, making large exponentiation expensive.

All commercial public-key cryptosystems rely on the difficulty of a discrete log problem. When discrete log gets easier, long bit-lengths are required to keep the algorithms safe. Discrete logs in ordinary prime number fields Z_p are much easier to solve than in elliptic curve fields. Thus, the discrete log problem for ordinary fields has been getting steadily easier due to successive refinement in the Number Field Sieve (NFS) techniques. In contrast, EC discrete log techniques have not seen significant improvement in the past 20 years. This difference accounts for today's reduced key-size requirement for elliptic curves. Cracking RSA has never been proven to be as hard as prime factoring, while factoring has never been proven to be as hard as discrete log. The only way for future breakthroughs is to resort to the best mathematicians for help.

The study of ECs has yielded some significant results that provide for elliptic curve cryptography. When mathematicians study the points where the EC exactly crosses the integer coordinates (x, y), it was found that the EC could provide a version of public-key cryptosystems. ECs over the finite prime field Z_p or the finite binary field $GF(2^m)$ are particularly interesting because they have the potential to provide faster public-key cryptosystems with smaller key sizes.

The Elliptic Curve Digital Signature Algorithm (ECDSA) was first proposed by Scott Vanstone in 1992 and accepted in 1999 as an ANSI standard and in 2000 as IEEE and NIST standards. ECDSA is the elliptic curve signature protocol analog of DSA specified in DSS. ECCs are viewed as EC analogs to the conventional discrete logarithm cryptosystems in which the subgroup of Z_p^* is replaced by the group of points on an EC over a finite field. The security of ECCs is based on the computational intractability of the elliptic curve discrete logarithm problem.

We now deal with one-way hash function (i.e., MD5 and SHA-1), Hash Message Authentication Code (HMAC), master secret computation, data expansion function, and pseudo-random function. The WTLS Record Protocol requires specification of a suite of algorithms, a master secret, and two peers' random values for secure connection. The encryption and MAC algorithms are determined by the cipher suite selected by the server and revealed in the server hello message. The key exchange and authentication algorithms are determined by the key-cxchangc suitc and arc also revealed in the server hello message. The creation of a shared master by means of the key exchange and the generation of cryptographic parameters from the master secret are of interest. For all key exchange methods, the same algorithm is used to convert the premaster secret into the master secret. In order to create the master secret, a premaster secret is first exchanged between two parties and then the master secret is calculated from it. The master secret is hashed into a sequence of secure bytes, which are assigned to the MAC secret, keys, and non-export IVs required by the CipherSpec. The master secret is always exactly 20 bytes in length. The length of the premaster secret will vary depending on key exchange method. The master secret is used to generate shared keys and secrets for encryption and MAC computations. In 1996, Netscape Communications Corporation introduced its own procedure for generating the master secret and the key block.

When RSA is used for server authentication and key exchange, a 20-byte secret value is generated by the client, encrypted under the server's public key, and sent to the server. The server uses its private key to decrypt the secret value. The `pre_master_secret` is the secret value appended with the server's public key. Both parties then convert the `pre_master_secret` into the `master_secret`. There is RSA key exchange with RSA based certificates. The server sends a certificate that contains its RSA public key. The server certificate is signed with RSA by a third party (i.e. CA) trusted by the client. The client extracts the server's public key from received the certificate, generates a secret value, encrypts it with the server's public key, and sends it to the server. The `pre_master_secret` is the secret value appended with the server's public key. If the client is to be authenticated, it then signs some data with its RSA private key and sends its certificate and the signed data. The key size (bits) is unlimited. Keys smaller than 1024 bits should not be used for RSA and DSA signature operations. The general goal of the key exchange process is to create a `pre_master_secret`, which is known to the communicating parties but not to the attackers. The `pre_master_secret` is used to generate the `master_secret`. The `master_secret` is required to generate the certificate verify and finished messages, encryption keys, and MAC secrets.

Completely anonymous sessions can be established using RSA or Diffie-Hellman, or EC Diffie-Hellman for key exchange. With anonymous RSA, the client encrypts a `pre_master_secret` with the server's uncertified public key extracted from the server key exchange message. The result is sent in a client key exchange message. Since eavesdroppers do not know the server's private key, it is infeasible for them to decode the `pre_master_secret`. For the conventional Diffie-Hellman key exchange, both the client and server generate a Diffie-Hellman common secret key that is used as the `pre_master_secret`, and is then converted into the `master_secret`. For the EC Diffie-Hellman key exchange, the negotiated secret key is used as the `pre_master_secret` and is converted into the `master_secret`. For the case of ECDH-ECDSA key exchange, the server sends a certificate that contains its ECDH public key. The server certificate is signed with ECDSA by a third party trusted by the client. Depending whether the client is to be authenticated or not, it sends its certificate containing its ECDH public key signed with ECDSA by a third party trusted by the server, or just its (temporary) ECDH public key. Each party calculates the `pre_master_secret` based on one's own private key and the counterpart's public key received (contained in a certificate).

A WTLS connection state is the operating environment of the Record protocol. An algorithm is required to generate the connection state (encryption keys, IVs, and MAC secrets) from the secure session parameters provided by the handshake protocol. In WTLS, many connection state parameters can be recalculated during a secure connection. This key refresh is performed in order to minimize the need for new handshakes. In the key refresh, the values of MAC secret, encryption key, and IV will change due to the sequence number. The frequency of these updates depends on the key refresh parameter.

A number of operations in the WTLS record and handshake layer require a keyed-HMAC, which is a secure digest of some data protected by a secret. HMAC can be used with a variety of different hash algorithms, namely MD5 and SHA-1, denoting these as HMAC-MD5 (secret, date) and HMAC-SHA-1 (secret, date). Forgery of the HMAC is infeasible without knowledge of the MAC secret.

An HMAC mechanism can be used with any iterative hash functions where data are hashed by iterating a basic compression function on blocks of data. HMAC uses a secret key for computation and verification of the message authentication values. The HMAC is a cryptographic checksum with the highest degree of security against attacks. HMACs are used to exchange information between parties (where both have knowledge of the secret key K), while a digital signature does not require any secret key to be verified for authentication. The security of the HMAC mechanism depends on cryptographic properties of the hash function, the choice of random keys, a secure key exchange mechanism, periodic key refreshment, and good secrecy protection of keys. Since the HMAC construction and its secure use for message authentication are independent from the particular hash function in use, the hash function can be replaced by any other secure iterative hash function. The strongest attack against HMAC is based on the frequency of collisions for the hash function. As an example, consider a hash function such as MD5 whose hash code length equals 16 bytes (128 bits). The attacker will need to acquire the correct message authentication tags computed on about $2^{64} = 18\,446\,744\,073\,709\,551\,616$ plaintexts. This is an impossible task in any realistic scenario for a block length of 64 bytes, because it will take 250 000 years in a continuous 1 Gbps link and without changing the secret key K during all this time. This attack could become realistic only if serious flaws in the *collision* behavior of the hash function are discovered.

The data expansion function, `P_hash (secret, data)`, uses a single hash function to expand a secret and seed into an arbitrary quantity of output. `P_hash (secret, seed)` is iterated as many times as necessary to produce the required quantity of data. Thus, the data expansion function makes use of the HMAC algorithm with either SHA-1 or MD5 as the underlying hash function.

In the TLS standard, two hash algorithms (for example, MD5 and SHA-1) were used in order to make the Pseudo-Random Function (PRF) as secure as possible. TLS's PRF computation is created by splitting the secret into two halves (S1 and S2) and using one half to generate data with P_MD5 and the other half to generate data with P_SHA-1. These two expansion results created from mixing the two pseudo-random streams are then XORed to produce the PRF output. S1 and S2 are the two halves of the secret and each has the same length. In order to save resources, WTLS can be implemented using only one hash algorithm. That is,

$$\mathrm{PRF}(\mathrm{secret}, \mathrm{label}, \mathrm{seed}) = \mathrm{P\_hash}(\mathrm{secret}, \mathrm{label}||\mathrm{seed})$$

which should be agreed during the handshake as part of the cipher spec.

5

Universal Mobile Telecommunication System (UMTS)

The ETSI initiated the pioneering work for UMTS standardization to the 3G mobile systems. After ETSI's initiatives, the 3GPP international organization took over this pioneering work to provide a joint standard for virtually worldwide coverage. The most important consensus was to aim for joint adoption of the WCDMA/FDD technology by Europe, Japan, Korea, China, and the United States.

5.1 UMTS Standardization

UMTS standardization is based on:

- The standardization of UTRA (UMTS Terrestrial Radio Access) in Frequency Division Duplex (FDD) and Time Division Duplex (TDD) mode. UTRA must fulfill the bit rates of 144 Kbps (in high mobility), 384 Kbps (in limited mobility), and 2 Mbps (indoor micro-cellular environments).
- A UMTS core network based on GSM/GPRS core network, allowing easy migration from 2G to 3G for GSM operators.
- The improvement of radio coverage and new 3G high-speed services.

The 3GPP released several versions of the UMTS standard:

- *Release 99:* This release, also known as Release 3, defines the WCDMA interface and relies on the GSM/GPRS core network. Release 99 was ratified in March 2000.
- *Release 4:* This release brings modifications and new functionalities to Release 99, but does not entail any radical changes. Specifications for Release 4 were ratified in March 2001.

Mobile Communication Systems and Security   Man Young Rhee
© 2009 John Wiley & Sons (Asia) Pte Ltd

- *Release 5:* This release introduced the network's migration to all-IP and adds functions proper to the world of multimedia IP. The switch to IP entails collaboration with IETF (Internet Engineering Task Force, standardization body) and the adoption of its protocols, including Session Initiation Protocol (SIP) and IPv6. The specifications of Release 5 were frozen between March and June 2002. The first Release 5-compatible equipment was expected to become available in mid-2003, but major deployments using Release 5 are not likely to take place for some time, given the size of the investments required and the technological alternations needed to complete the migration. However, in 2007 HSDPA was introduced in Release 5 to realize high-speed data rate on downlink. Some new technologies have been proposed for further enhancement on HSDPA in Releases 6 and 7.
- *Release 6:* Work is currently ongoing for 3GPP Release 6. It is planned that Release 6 will contain Multimedia Broadcast/Multicast service (MBMS), network sharing, priority service, wireless LAN/UMTS internetworking, IMS phase2, push services, and presence. The evolutionary aspects of subsequent 3GPP Releases is expected in the near future.

5.2 FDD/TDD Modes for UTRA Operation

2G mobile technologies are largely classified into TDMA-based GSM (Europe), AMPS (North America) based on IS-136, CDMA-based IS-95A (USA), and PDC (Japan).

The restricted data rates from 96 to 144 Kbps mean that 2G mobile systems are limited to voice and low-rate data services. To cope with these shortcomings, 3G mobile radio technologies have been developed for packet- switched and circuit-switched services with a maximum data rate of 2 Mbps.

ETSI SMG defines two different modes of operation for UTRA:

- *UTRA FDD:* The FDD mode uses different carrier frequencies for uplinks (MS to BS) and downlinks (BS to MS). A maximum of two 5-MHz spectrums are allocated to the FDD mode.
- *UTRA TDD:* The TDD mode uses the same carrier frequency for the uplink and downlink, multiplexed in time.

The SMG agreement assigns WCDMA to the paired bands for UTRA FDD and TD-CDMA to the unpaired bands for UTRA TDD. TD-CDMA is based on a combination of TDMA and CDMA, whereas WCDMA is a pure CDMA-based system. The duplex scheme of UTRA FDD/TDD has been harmonized with respect to the same basic system parameters such as chip rate (3.84 Mcps), bandwidth (5 MHz), modulation (QPSK), frame length (10 ms), and number of time slots per frame (15).

The TDD mode is well suited for environments with high traffic densities, indoor coverage, and asymmetric traffic distributions. It facilitates efficient use of the unpaired spectrum and supports data rates up to 2 Mbps. The TDD mode is beginning to appeal to a certain number of operators in Western Europe as well as in Asia. The FDD mode is well suited for applications in public macro- and microcell environments with typical data rates up to 384 Kbps and high mobility. 3G networks in Europe now use the UTRA FDD mode or WCDMA. The combined development of UTRA FDD/TDD enables efficient use of the whole available spectrum that will be the basis for optimal network design.

5.3 UMTS Architecture

UMTS consists of a number of logical network elements that have their own functionalities. The network elements are grouped into the User Equipment (UE), the UTRAN, and the CN. A simplified UMTS architecture, with the external reference points as well as the UMTS interfaces, is illustrated in Figure 5.1.

UTRAN (Universal Terrestrial Radio Access Network) is a conceptual network, which consists of Radio Network Controllers (RNCs) and node Bs between I_u and U_u. The RNC is a logical node in the Radio Network Subsystem (RNS) in charge of controlling the use and the integrity of the radio resources. An RNS offers the allocation and release of specific radio resources to establish means of connection in between an UE and the UTRAN. An RNS contains one RNC and is responsible for the resources and transmission or reception in a set of cells. The UE is a device allowing a user access to network services via U_u interface. U_u denotes radio interface between the UE and UTRAN. I_u denotes an interface node between RNC and an MSC, SGSN, or CBC, providing an interconnection point between the RNC and the CN. I_{ub} denotes the interface between the RNC and the node B. I_{ur} is a reference point of logical interface between two RNCs. Node B (Base Station) denotes a logical node in the RNS responsible for radio transmission or reception in one or more cells to or from the UE. This logical node terminates the I_{ur} interface towards the RNC.

A Node B can support FDD mode, TDD mode, or dual-mode operation. A Node B, which supports TDD cells, can support one chip-rate option either 3.84 or 1.28 Mcps TDD, or both options. A RNS consists of an RNC, one or more Node Bs, and optionally one Stand-Alone SMLC (SAS). A Node B is connected to the RNC through the I_{ub} interface. An RNC supports one chip-rate option or both options. The RNC is responsible for the handover decisions that require signaling to the UE.

The UMTS system consists of a number of logical network elements that are grouped into the UE, the UTRAN, and the CN. The UTRAN handles all radio related functionality; the CN is

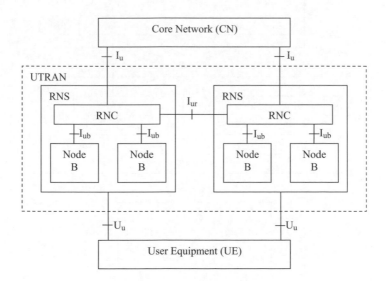

Figure 5.1   Basic UMTS architecture

responsible for switching and routing cells and data connection to the external networks (PSTN, ISDN, Internet, etc.); and the UE is the radio terminal used for radio communication over the U_u interface.

5.4 UTRAN Architecture

The UTRAN consists of a set of RNSs connected to the CN through the I_u interface, as shown in Figure 5.2. An RNS consists of an RNC. The UTRAN architecture is presented here, with a short introduction of all the elements below:

- *UE:* The User Equipment is the radio terminal used for radio communication over the I_u interface. The UMTS Subscriber Identity Module (USIM) is a smartcard, which holds the subscriber identity, performs authentication algorithms, and stores authentication and encryption keys and some subscription information needed at the terminal.
- *UTRAN:* The UTRAN consists of two distinct elements:
  - The Node B converts the data flow between the I_{ub} and U_u interfaces. It also participates in radio resource management. Note that the term "Node B" specified from the 3GPP is used in the more generic term "Base Station" elsewhere.
  - The RNC is the service access point for all services provided for the CN and for management of connections to the UE through the access stratum.
- *CN:* The Core Network can be divided into two groups:
  - CS networks provide circuit-switched connections, such as ISDN and PSTN, which are examples of CN networks.

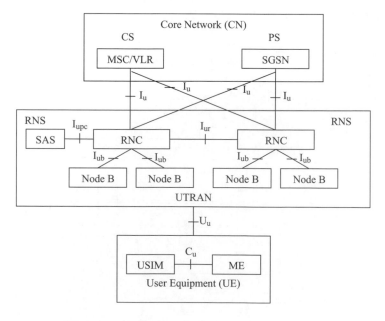

Figure 5.2 UTRAN architecture with the I_{upc} option

— PS networks provide connections for packet-switched data services, such as the Internet, which is one example of a PS network.

The UMTS standards are structured so that the interfaces between the logical network elements have been defined, but internal functionality of the network elements is not defined in detail.

The UTRAN consists of a set of RNSs connected to the CN through the I_u. An RNS is a subnetwork within the UTRAN and consists of one RNC, one (or more) Node B, and optionally one SAS. A Node B is connected to one RNC through the I_{ub} interface. A Node B can support FDD mode, TDD mode and dual-mode operation. There are two chip-rate options in the TDD mode: 3.84 and 1.28 Mcps TDD. Each TDD cell supports either of these options. A RNC, which supports TDD cells, can support one chip-rate option only, or both options.

The RNC is responsible for the handover decisions that require signaling to the UE. A RNC may include a combining/splitting function to support combination/splitting of information streams.

Inside the UTRAN, the RNCs of the RNS can be interconnected together through the I_{ur}. I_u and I_{ur} are logical interfaces. I_{ur} can be conveyed over a direct physical connection between RNCs or virtual networks using any suitable transport network.

The RNC may have full internal support for the EU positioning and/or may be connected to one SAS via the I_{upc} interface. Figure 5.2 illustrates the resulting UTRAN architecture when the I_{upc} interface is adopted.

The main requirements for the design of UTRAN architecture will be: soft handover and the WCDMA-specific radio resource management algorithms as the major impact on the UTRAN design; maximization of the commonalities in the handling of packet-switched and circuit-switched data flow from UTRAN to both the PS and CS domains of the core network; and use of ATM transport mechanism for the UTRAN connection.

5.5 UTRAN Terrestrial Interface

The general protocol mode for UTRAN interface is depicted in Figure 5.3 and will be described in detail in the following subsections. The structure is based on the principle that the layers and planes are logically independent of each other. Therefore, as and when required, the standardization body can easily alter protocol stacks and planes to fit future requirements.

5.5.1 Horizontal Layers

The protocol structure consists of two main layers: Radio Network Layer and Transport Network Layer. All UTRAN related issues are visible only in the Radio Network Layer, and the Transport Network Layer represents standard transport technology that is selected to use for UTRAN, but without any UTRAN specific requirements.

5.5.2 Vertical Planes

The Control Plane is used for all UMTS specific control signaling. The User Plane includes the Data Stream(s) and the Data Bearer(s) for the Data Stream(s).

- *Control Plane:* The Control Plane includes the Application protocol, that is, RANAP (Radio Access Network Application Part) in I_u, RNSAP (Radio Network Subsystems Application

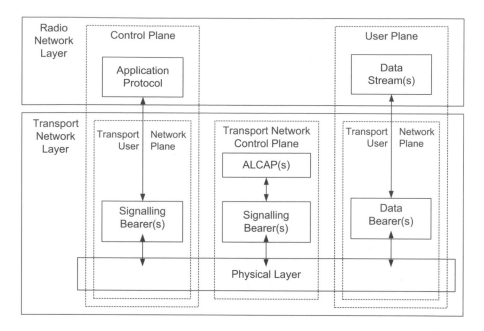

Figure 5.3   General protocol model for UTRAN interfaces

Part) in I_{ur} or NBAP (Node B Application Part) in I_{ub}, and the Signaling Bearer for transporting the Application protocol messages.

Among other things, the Application protocol is used for setting up bearers (i.e. Radio Access Bearer in I_u and subsequently Radio Link in I_{ur} and I_{ub}) in the Radio Network Layer. In the three planes structure, the bearer parameters in the Application protocol are not directly tied to the User Plane technology, but rather are general bearer parameters.

The Signaling Bearer for the Application protocol may or may not be of the same type as the Signaling protocol for the ALCAP (Access Link Control Application Part). The Signaling Bearer is always setup by O&M (Operation and Maintenance) actions.

• *User Plane:* All information sent and received by the user, such as the coded voice in the voice call or the packet in an Internet connection, are transposed via the User Plane. The User Plane includes the Data Stream(s) and the Data Bearer(s) for Data Stream(s). The Data Stream(s) is/ are characterized by one or more frame protocols specified for that interface.

• *Transport Network Control Plane:* The Transport Network Control Plane is a plane that acts between the Control Plane and the User Plane. It does not include any Radio Network Layer information. The Transport Network Control Plane is used for all control signaling within the Transport Network Layer. The Transport Network Control Plane includes the ALCAP protocol that is needed to set up the transport bearers (Data Bearer) for the User Plane. It also includes the appropriate Signaling Bearer(s) needed for the ALCAP protocol(s).

5.6 UTRAN-CN Interface via I_u

The I_u interface, which divides the system into UTRAN and CN, connects UTRAN and the CN. The UTRAN performs radio transmission and reception and the CN handles switching, routing, and service control. Referring to Figure 5.2, the I_u interface has two occasions, that is, one for I_u circuit-switched case connecting UTRAN to CS and the other for I_u packet-switched case connecting UTRAN to PS. Thus, the design standardization should be realized so that the optimized User Plane transport for CS and PS services can be achieved.

5.6.1 I_u CS Protocol Structure

The I_u CS protocol structure is illustrated in Figure 5.4. The three planes in the I_u interface share a common Asynchronous Transfer Mode (ATM) transport, which is used for all planes. The physical layer is the interface to the physical medium: optical fiber, radio link, or copper cable.

- I_u CS Control Plane Protocol Stack: The Control Plane protocol stack consists of RANAP, on top of Broad Band (BB) SS7 (Signaling System #7) protocols. RANAP is the signaling protocol in I_u that contains all the control information specified for the Radio Network Layer.

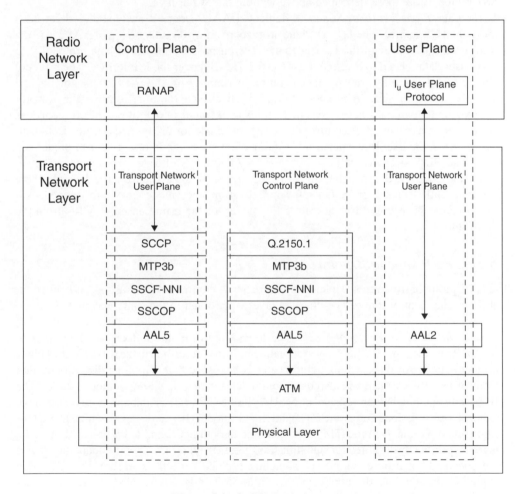

Figure 5.4 I_u CS protocol structure

Various RANAP Elementary Procedures implement the functionality of RANAP. Each RANAP function may require the execution of one or more Elementary Procedures (EPs). Each EP consists of just one request message (class 2 EP), the request and response massage pair (class 1 EP), or one request message and one or more response messages (class 3 EP). The applicable layers are the Signaling Connection Control Part (SCCP), the Message Transfer Part (MTP3b), and Signaling ATM Adaptation Layer for Network-to-Network Interfaces (SAAL-NNI). SAAL-NNI is further divided into Service Specific Coordination Function (SSCF), Service Specific Connection Oriented Protocol (SSCOP), and ATM Adaptation Layer 5 (AAL5) layers. SSCF and SSCOP layers are specifically designed for signaling transport in ATM networks, and take care of such functions as signaling connection management. AAL5 is used for segmenting the data to ATM cells.

- *I_u CS Transport Network Control Plane Protocol Stack:* The Transport Network Control Plane protocol stack consists of the Signaling protocol for setting up AAL5 connections (Q.2630.1 and adaptation layer Q.2150.1), on top of BB SS7 protocols. The applicable BB SS7 protocols are those described above without the SCCP layer.

- *I_u CS User Plane Protocol Stack:* A dedicated AAL2 connection is reserved for each individual CS service. The I_u User Plane protocol residing directly on top of AAL2 is in the Radio Network Layer of the I_u User Plane. The purpose of the User Plane protocol is to carry user data related to Radio Access Bearers (RABs) over the I_u interface. The protocol performs either a fully transparent operation or framing for the user data segments and some basic control signaling to be used for initialization and online control. The protocol has two modes based on these cases: In the transparent mode of operation, the protocol does not perform any framing or control; in the support mode for predefined Service Data Unit (SDU) sizes, the User Plane performs framing of the user data into segments of predefined size. The SDU sizes typically correspond to Adaptive Multirate size and to Adaptive Multirate Code (AMR) speech frames, or to the frame sizes derived from the data rate of the CS data call. Control procedures for initialization and rate control are defined, and functionality is specified for indicating the quality of the frame based on CRC from the radio interface.

5.6.2 I_u PS Protocol Structure

The I_u PS protocol structure is depicted in Figure 5.5. A common ATM transport is applied for both User and Control Plane. The physical layer is specified in the same way as I_u CS.

- *I_u PS Control Plane Protocol Stack:* The I_u PS Control Plane protocol stack consists of RANAP, and the same BB SS7-based signaling bearer as described in the I_u CS Control Plane protocol stack. An IP-based signaling bearer is also specified. The Signaling Connection Control Part (SCCP) layer is also used commonly for both I_u CS/PS protocol stacks. The IP-based signaling bearer consists of M3UA (SS7 MTP3-User Adaptation Layer), Simple Control Transmission Protocol (SCTP), Internet Protocol (IP), and AAL5, which is used for segmentation of the data to ATM cells (common to both I_u CS and I_u PS cases). The SCTP layer is specifically designed for signaling transport in the Internet. Specific adaptation layers are specified for different kinds of signaling protocols, such as M3UA for SS7-based signaling. In addition, the applicable layers are MTP3-B, SAAL-NNI (which is further divided into SSCF and SSCOP), and AAL5 layers.

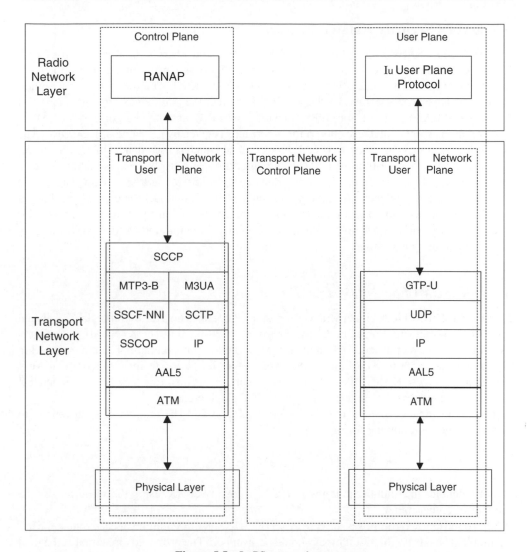

Figure 5.5 Iu PS protocol structure

- I_u *PS Transport Network Control Plane Protocol Stack:* The Transport Network Control Plane is not applied to I_u PS, as shown in Figure 5.5. The setting up of GPRS tunneling protocol requires only an identifier for the tunnel, and the IP addresses for both directions, and they are already included in the RANAP Radio Access Bearer assignment messages. The same information elements that are used in I_u CS for addressing and identifying the AAL2 signaling are used only for the User Plane data in I_u CS.
- I_u *PS User Plane Protocol Stack:* In the I_u PS User Plane, multiple packet data flows are multiplexed on one or several AAL5 Pre-defined Virtual Connections (PVCs). The GTP-U (User Plane part of the GPRS Tunneling protocol) is a multiplexing layer that provides identities for individual packet data flow. Each flow uses UDP connectionless transport and IP addressing.

5.7 UMTS Security Related Features

In Spring 2002, SAGE initiated the task of designing a new encryption algorithm for GSM, ECSD, GPRS, and EGPRS encryptions. Those new algorithms were intended to implement dual-mode handsets for operating with both GSM and UMTS modes. Three encryption algorithms have been specified by the 3GPP Task Force: the A5/3 algorithms for GSM and ECSD; the GEA3 algorithms for GPRS (including EGPRS); and the f8 algorithm for UMTS. The common aspect of all these encryption algorithms is given by the name of KGCORE. The KGCORE function is based on KASUMI block cipher, which is used in a form of OFB mode, specified for the f8 algorithm for UMTS. The algorithms are stream ciphers that are used to encrypt/decrypt blocks of data under a ciphering key K_c. KASUMI, as a keystream generator, is a block cipher that produces a 64-bit output from a 64-bit input under the control of a 128-bit key. The f8 function should be a synchronous stream cipher and must be fully standardized for interoperability within UMTS. This section will deal with requirements for UMTS integrity and confidentiality algorithms.

5.7.1 KASUMI Encryption Function

KASUMI is Feistel block cipher with eight rounds. It operates on a 64-bit input data block and produces a 64-bit output block under the control of a 128-bit key. The KASUMI algorithm is used in a form of output-feedback mode (OFM) as keystream generator. The three ciphering algorithms (i.e. A5/3 for GSM and ECSD, GEA3 for GPRS, and f8 for UMTS) are all very similar and they use KASUMI as a keystream generator. As mentioned before, the KASUMI keystream generator is an enhancement of NISTY block cipher.

The 64-bit input is divided into 32-bit block L_o and R_o. The outputs of each round are generated by the following formulas:

$$R_i = L_{i-1}, L_i = R_{i-1} \oplus f_i(L_{i-1}, RK_i), 1 = i = 8$$

where f_i denotes the round function with L_{i-1} and round key RK_i using the subscript i as the number of cipher round.

The round key RK_i comprises the subkey set (KL_i, KO_i, KI_i). At the end of each round, the left 32-bit block and the right 32-bit block should be swapped. The produced ciphertext at the final eight rounds should be the 64-bit string derived from concatenation of L_8 and R_8. As shown in Figure 5.6, the round function f_i is constructed from two associated subfunctions FL_i and FO_i with their respective subkeys KL_i for FL_i and (KO_i, KI_i) for FO_i, followed by a bitwise XOR operation with the previous branch stream (R_i or L_i). These two subfunctions will have two different orders depending on an even round or an odd round. In the odd round, the FL subfunction is performed first and followed by the FO subfunction. In the even round, the FO subfunction is performed first and followed by the FL subfunction, providing the process of the reverse order. The FO function consists of a 3-round network with a 16-bit nonlinear function FI. This FI function consists of 4-round operations with two S-boxes of S9 and S7. These two S-boxes perform in the binary extension field $GF(2^m)$. Specifically, the S9 performs the x^5 in GF (2^9) operation and the S7 performs the x^{81} in $GF(2^7)$ operation.

The FL function transforms the 32-bit data with two 16-bit subkeys KL_{i1} and KL_{i2} by means of AND, OR, and 1-bit cyclic left sift ($\ll 1$) operations. The FO function divides the 32-bit

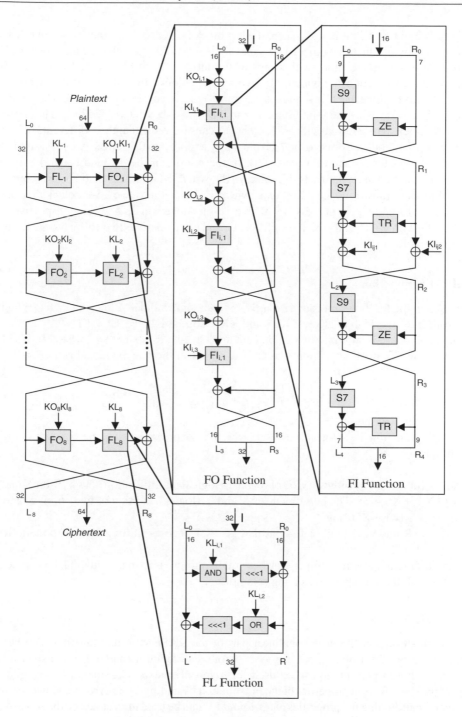

Figure 5.6 The KASUMI encryption data path

input data into two 16-bit blocks. Then, the left block is XORed with the 16-bit subkey KO_{ij}, transformed by FI function with a 16-bit subkey KI_{ij}, and XORed with the right block R_0. This routine is iterated three times with swaps of the left and right blocks. A 16-bit data block entering the FI function is divided into two smaller blocks for S-box transformation. The leftmost 9 bits become one block, and the rightmost 7 bits become another block. Then, they are transformed twice using the 9-bit S-box S9 and the 7-bit S-box S7, respectively. However, the two data blocks are XORed with each other. Because the bit length is different, the Zero-extension (ZE) should be done to the 7-bit blocks by adding two 0s, and the two most significant bits of the 9-bit blocks are truncated (TR). In the middle of the 4-round network, an XOR operation will be done with the 16-bit subkey KI_{ij}, where KI_{ij1} is 7 bits and KI_{ij2} is 9 bits. The 32-bit function transforms the 32-bit data, which is divided into two 16-bit blocks L_0 and R_0 each. These L_0 and R_0 are processed with two 16-bit subkeys KL_{i1} and KL_{i2}, respectively, by using AND, OR, XOR, and 1-bit left shift ($\lll 1$), and finally derives two 16-bit blocks L' and R' from L_0 and R_0.

5.7.1.1 FL Function

The input to the FL_i function comprises a 32-bit data input I and a 32-bit subkey KL_i, which is also split into two 16-bit subkeys, KL_{i1} and KL_{i2}, so that $KL_i = KL_{i1} \parallel KL_{i2}$.

The input data I is split into two 16-bit data units L_0 and R_0 so that $I = L_0 \parallel R_0$. If the 32-bit output of the FL function is defined as L' and R', it can then be expressed as

$$L' = L_0 \oplus ((R' \odot KL_{i2}) \lll 1)$$
$$R' = R_0 \oplus ((L' \boxplus KL_{i1}) \lll 1)$$

The FL function is a linear function, which is expressed using the following logical definitions:

$X \odot Y$ (or X Y): The bitwise AND of X and Y (logical multiplication of two 16-bit X and Y).
$X \boxplus Y$ (or X + Y): The bitwise OR of X and Y (logical addition of two 16-bit X and Y).
$\oplus$: The bitwise XORing.
$X \lll Y$: Rotation X to the left by Y, that is, $\lll 1$ denotes shifting one bit to the left.

The FL function is a linear function whose main purpose is to make individual bits harder to track through the rounds by executing scrambling.

5.7.1.2 FO Function

The FO function constitutes the nonlinear part of the KASUMI round function. The FO is a permutation of 32-bit blocks, but due to its 3-round structure it can be distinguished from a randomly chosen permutation when used for given plaintexts. Consideration was given to improving the diffusion properties of the FO function by adding a fourth round. However, there are no indications that the properties of a 3-round FO can be used in an attack on the full 8-round KASUMI.

The input to the function FO_i comprises a 32-bit data input I and two sets of subkeys, that is, a 48-bit KO_i and 48-bit KI_i. The 32-bit data input is split into two halves, L_0 and R_0, so that $I = L_0 \parallel R_0$, whereas the 48-bit subkeys are subdivided into three 16-bit subkeys, respectively.

$$KO_i = KO_{i1}||KO_{i2}||KO_{i3}$$
$$KI_i = KI_{i1}||KI_{i2}||KI_{i3}$$

For each integer j with $1 \leq j \leq 3$, the operation of the jth round of the function FO_i is defined as:

$$R_1 = FI_{i1}(L_o \oplus KO_{i1}, KI_{i1}) \oplus R_0$$
$$L_1 = R_0$$

Generally,

$$R_j = FI_{ij}(L_{j-1} \oplus KO_{ij}, KI_{ij}) \oplus R_{i-1}$$
$$L_j = R_{j-1}$$

The output from the FO_i function is defined as the 32-bit data block $L_3 \parallel R_3$.

5.7.1.3 FI Function

The FI function is depicted in the rightmost block of Figure 5.6.

The FI-function FI_{ij} takes a 16-bit data input I and a 16-bit subkey KI_{ij}. The input I is split into two unequal components, a 9-bit left half L_0 and a 7-bit right half R_0, so that $I = L_0||R_0 = 16$ bits. Similarly, the subkey KI_{ij} is split into a 7-bit component KI_{ij1} and 9-bit component KI_{ij2}, so that $KI_{ij} = KI_{ij1}||KI_{ij2}$.

Each FI-function FI_{ij} uses two S-boxes, S7 and S9, respectively. S7 maps a 7-bit input R_0 to a 7-bit output, while S9 maps a 9-bit input to a 9-bit output. The FI-function also uses two additional functions, which are designated by ZE and TR. These functions are defined as follows:

- ZE(d) takes a 7-bit data string d and converts it to a 9-bit data string by appending two zero bits to the most significant end of d.
- TR(d) takes a 9-bit data string d and converts it to a 7-bit value by discarding the two most significant bits of d.

The function FI_{ij} is defined by the following series of operations:

$$L_1 = R_0$$
$$R_1 = S9[L_0] \oplus ZE(R_0)$$
$$L_2 = R_1 \oplus KI_{ij2}$$
$$R_2 = S7[L_1] \oplus TR(R_1) \oplus KI_{ij1}$$
$$L_3 = R_2$$
$$R_3 = S9[L_2] \oplus ZE(R_2)$$
$$L_4 = S7[L_3] \oplus TR[R_3]$$
$$R_4 = R_3$$

Finally, the output of the FI_{ij} function is the 16-bit data block $L_4 \parallel R_4$. The FI function is the basic randomizing function of KASUMI with 16-bit input and 16-bit output. It is composed of a four-round structure using two nonlinear substitution boxes S7 and S9. The S-boxes S7 and S9 have been designed to avoid linear structures in the FI functions.

5.7.1.4 S-boxes

The two S-boxes (S7 and S9) have been designed so that they could be easily implemented using combinational logic or by a look-up table. The input x comprises either seven or nine bits with a corresponding number of bits in the output y. Both output formulas are given for each S-box as shown below:

For the S7-box
The input x and output y are expressed by

$$x = x_6||x_5||x_4||x_3||x_2||x_1||x_0$$

and

$$y = y_6||y_5||y_4||y_3||y_2||y_1||y_0,$$

where the x_6 and y_6 bits apply to the S7-box and the x_0 and y_0 bits are the least significant bits.

The output y_i for S-box S7 is computed using the following logical gate operations of x_i, where i stands for $0 \leq i \leq 6$:

$$
\begin{aligned}
y_0 &= x_1x_3 \oplus x_4 \oplus x_0x_1x_4 \oplus x_5 \oplus x_2x_5 \oplus x_3x_4x_5 \oplus x_6 \oplus x_0x_6 \oplus x_1x_6 \oplus x_3x_6 \oplus x_2x_4x_6 \oplus x_1x_5x_6 \\
&\quad \oplus x_4x_5x_6 \\
y_1 &= x_0x_1 \oplus x_0x_4 \oplus x_2x_4 \oplus x_5 \oplus x_1x_2x_5 \oplus x_0x_3x_5 \oplus x_6 \oplus x_0x_2x_6 \oplus x_3x_6 \oplus x_4x_5x_6 \oplus 1 \\
y_2 &= x_0 \oplus x_0x_3 \oplus x_2x_3 \oplus x_1x_2x_4 \oplus x_0x_3x_4 \oplus x_1x_5 \oplus x_0x_2x_5 \oplus x_0x_6 \oplus x_0x_1x_6 \oplus x_2x_6 \oplus x_4x_6 \oplus 1 \\
y_3 &= x_1 \oplus x_0x_1x_2 \oplus x_1x_4 \oplus x_3x_4 \oplus x_0x_5 \oplus x_0x_1x_5 \oplus x_2x_3x_5 \oplus x_1x_4x_5 \oplus x_2x_6 \oplus x_1x_3x_4x_6 \\
y_4 &= x_0x_2 \oplus x_3 \oplus x_1x_3 \oplus x_1x_4 \oplus x_0x_1x_4 \oplus x_2x_3x_4 \oplus x_0x_5 \oplus x_1x_3x_5 \oplus x_0x_4x_5 \oplus x_1x_6 \oplus x_3x_6 \\
&\quad \oplus x_0x_3x_6 \oplus x_5x_6 \oplus 1 \\
y_5 &= x_2 \oplus x_0x_2 \oplus x_0x_3 \oplus x_1x_2x_3 \oplus x_0x_2x_4 \oplus x_0x_5 \oplus x_2x_5 \oplus x_4x_5 \oplus x_1x_6 \oplus x_1x_2x_6 \oplus x_0x_3x_6 \\
&\quad \oplus x_3x_4x_6 \oplus x_2x_5x_6 \oplus 1 \\
y_6 &= x_1x_2 \oplus x_0x_1x_3 \oplus x_0x_4 \oplus x_1x_5 \oplus x_3x_5 \oplus x_6 \oplus x_0x_1x_6 \oplus x_2x_3x_6 \oplus x_1x_4x_6 \oplus x_0x_5x_6
\end{aligned}
$$

The following conventions are used for evaluation of the output y_i, $0 \leq i \leq 6$: The logical AND for any two bits u and v is denoted by uv and the XOR operation is denoted by $u \oplus v$.

The lookup table for S7 is shown in Table 5.1 in which the output decimal numbers are contained from 0 to 127.

Example 5.1. If an input decimal number 45 is given, the value found at the position 45 is S7 [45] = 70. Using the gate logic formulas the output value can be computed as follows:

Table 5.1 The lookup table for S7-box

PoDN	0	1	2	3	4	5	6	7	8	9	10	11	12	13	14	15	PoDN
0	54	50	62	56	22	34	94	96	38	6	63	93	2	18	123	33	**15**
16	55	113	39	114	21	67	65	12	47	73	46	27	25	111	124	81	**31**
32	53	9	121	79	52	60	58	48	101	127	40	120	104	70	71	43	**47**
48	20	122	72	61	23	109	13	100	77	1	16	7	82	10	105	98	**63**
64	117	116	76	11	89	106	0	125	118	99	86	69	30	57	126	87	**79**
80	112	51	17	5	95	14	90	84	91	8	35	103	32	97	28	66	**95**
96	102	31	26	45	75	4	85	92	37	74	80	49	68	29	115	44	**111**
112	64	107	108	24	110	83	36	78	42	19	15	41	88	119	59	3	**127**

PoDN: Position of Decimal Number.

First the input decimal value should be converted to a binary string

$$45_{(10)} = 0101101_{(2)}$$

from which the bit values are derived as

$$x_6 = 0, x_5 = 1, x_4 = 0, x_3 = 1, x_2 = 1, x_1 = 0, \text{ and } x_0 = 1$$

Substituting the input bits to the gate logic equations for S-box S7 yields the following output.

$$y_0 = 0 \oplus 0 \oplus 0 \oplus 1 \oplus 1 \oplus 0 \oplus 0 \oplus 0 \oplus 0 \oplus 0 \oplus 0 \oplus 0 \oplus 0 = 0$$
$$y_1 = 0 \oplus 0 \oplus 0 \oplus 1 \oplus 0 \oplus 1 \oplus 0 \oplus 0 \oplus 0 \oplus 0 \oplus 0 \oplus 1 = 1$$
$$y_2 = 1 \oplus 1 \oplus 1 \oplus 0 \oplus 0 \oplus 0 \oplus 1 \oplus 0 \oplus 0 \oplus 0 \oplus 0 \oplus 1 = 1$$
$$y_3 = 0 \oplus 0 \oplus 0 \oplus 0 \oplus 1 \oplus 0 \oplus 1 \oplus 0 \oplus 0 \oplus 0 = 0$$
$$y_4 = 1 \oplus 1 \oplus 0 \oplus 0 \oplus 0 \oplus 0 \oplus 1 \oplus 0 \oplus 0 \oplus 0 \oplus 0 \oplus 0 \oplus 1 = 0$$
$$y_5 = 1 \oplus 1 \oplus 1 \oplus 0 \oplus 0 \oplus 1 \oplus 1 \oplus 0 \oplus 0 \oplus 0 \oplus 0 \oplus 0 \oplus 1 = 0$$
$$y_6 = 0 \oplus 0 \oplus 0 \oplus 0 \oplus 1 \oplus 0 \oplus 0 \oplus 0 \oplus 0 \oplus 0 = 1$$

Hence $y = 1\,000\,110_{(2)} = 70_{(10)}$. It is thus proved that S7[45] = 70 in Table 5.1 turns out to be the identical value 70 computed by the gate logic equations.

For the S9-box
The input x and output y are expressed by

$$x = x_8||x_7||x_6||x_5||x_4||x_3||x_2||x_1||x_0$$

and

$$y = y_8||y_7||y_6||y_5||y_4||y_3||y_2||y_1||y_0, \text{ respectively.}$$

The gate logic equations for S9-box are expressed as:

$$y_0 = x_0x_2 \oplus x_3 \oplus x_2x_5 \oplus x_5x_6 \oplus x_0x_7 \oplus x_1x_7 \oplus x_2x_7 \oplus x_4x_8 \oplus x_5x_8 \oplus x_7x_8 \oplus 1$$

$$y_1 = x_1 \oplus x_0x_1 \oplus x_2x_3 \oplus x_0x_4 \oplus x_1x_4 \oplus x_0x_5 \oplus x_3x_5 \oplus x_6 \oplus x_1x_7 \oplus x_2x_7 \oplus x_5x_8 \oplus 1$$

$$y_2 = x_1 \oplus x_0x_3 \oplus x_3x_4 \oplus x_0x_5 \oplus x_2x_6 \oplus x_3x_6 \oplus x_5x_6 \oplus x_4x_7 \oplus x_5x_7 \oplus x_6x_7 \oplus x_8 \oplus x_0x_8 \oplus 1$$

$$y_3 = x_0 \oplus x_1x_2 \oplus x_0x_3 \oplus x_2x_4 \oplus x_5 \oplus x_0x_6 \oplus x_1x_6 \oplus x_4x_7 \oplus x_0x_8 \oplus x_1x_8 \oplus x_7x_8$$

$$y_4 = x_0x_1 \oplus x_1x_3 \oplus x_4 \oplus x_0x_5 \oplus x_3x_6 \oplus x_0x_7 \oplus x_6x_7 \oplus x_1x_8 \oplus x_2x_8 \oplus x_3x_8$$

$$y_5 = x_2 \oplus x_1x_4 \oplus x_4x_5 \oplus x_0x_6 \oplus x_1x_6 \oplus x_3x_7 \oplus x_4x_7 \oplus x_6x_7 \oplus x_5x_8 \oplus x_6x_8 \oplus x_7x_8 \mathring{A}_1$$

$$y_6 = x_0 \oplus x_2x_3 \oplus x_1x_5 \oplus x_2x_5 \oplus x_4x_5 \oplus x_3x_6 \oplus x_4x_6 \oplus x_5x_6 \oplus x_7 \oplus x_1x_8 \oplus x_3x_8 \oplus x_5x_8 \oplus x_7x_8$$

$$y_7 = x_0x_1 \oplus x_0x_2 \oplus x_1x_2 \oplus x_3 \oplus x_0x_3 \oplus x_2x_3 \oplus x_4x_5 \oplus x_2x_6 \oplus x_3x_6 \oplus x_2x_7 \oplus x_5x_7 \oplus x_8 \oplus 1$$

$$y_8 = x_0x_1 \oplus x_2 \oplus x_1x_2 \oplus x_3x_4 \oplus x_1x_5 \oplus x_2x_5 \oplus x_1x_6 \oplus x_4x_6 \oplus x_7 \oplus x_2x_8 \oplus x_3x_8$$

The lookup table for S9 is shown in Table 5.2, in which the output decimal numbers from 0 to 155 are contained.   □

Example 5.2. For an input value 304, the value is found as S9[304] = 185 from Table 5.2. Using the gate logic formulas the output value can be computed as follows:

First the input decimal value should be converted to a binary string

$$304_{(10)} = 100110000_{(2)}$$

from which the bit values are derived as

$$x_8 = 1, x_7 = 0, x_6 = 0, x_5 = 1, x_4 = 1, x_3 = 0, x_2 = 0, x_1 = 0, \text{ and } x_0 = 0$$

Substituting the input bits x_i ($0 \le i \le 8$) to the gate logic equations for S-box S9 yields the following output.

$$y_0 = 0 \oplus 0 \oplus 0 \oplus 0 \oplus 0 \oplus 0 \oplus 0 \oplus 1 \oplus 1 \oplus 0 \oplus 1 = 1$$

$$y_1 = 0 \oplus 0 \oplus 0 \oplus 0 \oplus 0 \oplus 0 \oplus 0 \oplus 0 \oplus 0 \oplus 1 \oplus 1 = 0$$

$$y_2 = 0 \oplus 0 \oplus 0 \oplus 0 \oplus 0 \oplus 0 \oplus 0 \oplus 0 \oplus 0 \oplus 1 \oplus 0 \oplus 1 = 0$$

$$y_3 = 0 \oplus 0 \oplus 0 \oplus 0 \oplus 1 \oplus 0 \oplus 0 \oplus 0 \oplus 0 \oplus 0 \oplus 0 = 1$$

$$y_4 = 0 \oplus 0 \oplus 1 \oplus 0 \oplus 0 \oplus 0 \oplus 0 \oplus 0 \oplus 0 \oplus 0 = 1$$

$$y_5 = 0 \oplus 0 \oplus 1 \oplus 0 \oplus 0 \oplus 0 \oplus 0 \oplus 0 \oplus 1 \oplus 0 \oplus 0 \oplus 1 = 1$$

$$y_6 = 0 \oplus 0 \oplus 0 \oplus 0 \oplus 1 \oplus 0 \oplus 0 \oplus 0 \oplus 0 \oplus 0 \oplus 1 \oplus 0 = 0$$

$$y_7 = 0 \oplus 0 \oplus 0 \oplus 0 \oplus 0 \oplus 0 \oplus 1 \oplus 0 \oplus 0 \oplus 0 \oplus 0 \oplus 1 \oplus 1 = 1$$

$$y_8 = 0 \oplus 0 \oplus 0 \oplus 0 \oplus 0 \oplus 0 \oplus 0 \oplus 0 \oplus 0 \oplus 0 \oplus 0 = 0$$

Thus $y = 010111001_{(2)} = 185_{(10)}$. Hence S9[304] = 185 in Table 5.2 is proved to be correct because the value computed using the gate logic equations is identical.   □

Table 5.2 The lookup table for S9-box

PoDN	0	1	2	3	4	5	6	7	8	9	10	11	12	13	14	15	PoDN
0	167	239	161	379	391	334	9	338	38	226	48	358	452	385	90	397	15
16	183	253	147	331	415	340	51	362	306	500	262	82	216	159	356	177	31
32	175	241	489	37	206	17	0	333	44	254	378	58	143	220	81	400	47
48	95	3	315	245	54	235	218	405	472	264	172	494	371	290	399	76	63
64	165	197	395	121	257	480	423	212	240	28	462	176	406	507	288	223	79
80	501	407	249	265	89	186	221	428	164	74	440	196	458	421	350	163	95
96	232	158	134	354	13	250	491	142	191	69	193	425	152	227	366	135	111
112	344	300	276	242	437	320	113	278	11	243	87	317	36	93	496	27	127
128	487	446	482	41	68	156	457	131	326	403	339	20	39	115	442	124	143
144	475	384	508	53	112	170	479	151	126	169	73	268	279	321	168	364	159
160	363	292	46	499	393	327	324	24	456	267	157	460	488	426	309	229	175
176	439	506	208	271	349	401	434	236	16	209	359	52	56	120	199	277	191
192	465	416	252	287	246	6	83	305	420	345	153	502	65	61	244	282	207
208	173	222	418	67	386	368	261	101	476	291	195	430	49	79	166	330	223
224	280	383	373	128	382	408	155	495	367	388	274	107	459	417	62	454	239
240	132	225	203	316	234	14	301	91	503	286	424	211	347	307	140	374	255
256	35	103	125	427	19	214	453	146	498	314	444	230	256	329	198	285	271
272	50	116	78	410	10	205	510	171	231	45	139	467	29	86	505	32	287
288	72	26	342	150	313	490	431	238	411	325	149	473	40	119	174	355	303
304	185	233	389	71	448	273	372	55	110	178	322	12	469	392	369	190	319
320	1	109	375	137	181	88	75	308	260	484	98	272	370	275	412	111	335
336	336	318	4	504	492	259	304	77	337	435	21	357	303	332	483	18	351
352	47	85	25	497	474	289	100	269	296	478	270	106	31	104	433	84	367
368	414	486	394	96	99	154	511	148	413	361	409	255	162	215	302	201	383
384	266	351	343	144	441	365	108	298	251	34	182	509	138	210	335	133	399
400	311	352	328	141	396	346	123	319	450	281	429	228	443	481	92	404	415
416	485	422	248	297	23	213	130	466	22	217	283	70	294	360	419	127	431
432	312	377	7	468	194	2	117	295	463	258	224	447	247	187	80	398	447
448	284	353	105	390	299	471	470	184	57	200	348	63	204	188	33	451	463
464	97	30	310	219	94	160	129	493	64	179	263	102	189	207	114	402	479
480	438	477	387	122	192	42	381	5	145	118	180	449	293	323	136	380	495
496	43	66	60	455	341	445	202	432	8	237	15	376	436	464	59	461	511

PoDN: Position of Decimal Number.

5.7.2 User and Signaling Data Confidentiality

User data and some signaling information should be protected by confidentiality. To ensure user identity confidentiality, which is applied on dedicated channels between UE and the RNC, a protected mode of transmission should be fulfilled by a confidentiality function.

5.7.2.1 Ciphering Method over Radio Link

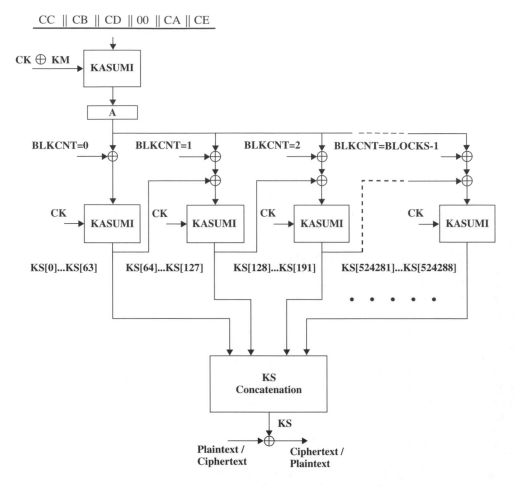

$$CC \parallel CB \parallel CD \parallel 00 \parallel CA \parallel CE$$

CC: Frame dependent 32 bit input
CB: 5 bit bearer input
CD: 1 bit input indicating the direction of transmission
CA: 8 bit input initialized as 0000 0000
CE: 16 bit input initialized as 0000 0000 0000 0000

Figure 5.7   The f8 algorithm architecture

Figure 5.7 illustrates the use of the ciphering algorithm f8 to encrypt the plaintext by applying a keystream using a bit-by-bit binary addition of the plaintext and keystream. The plaintext can be recovered by generating the same keystream using the same input parameters and applying a bit-by-bit binary addition with the ciphertext. The f8 algorithm makes use of the KASUMI key-dependent function, which operates on 64-bit data block and produces 64-bit blocks under control of a 128-bit key CK.

The input parameters to the algorithm f8 are the cipher key (CK), a time dependent input (COUNT-C), the bearer identity (BEARER), the direction of transmission (DIRECTION), and the length of the keystream required (LENGTH).

The input parameters are specified as shown below:

- *CK:* The cipher key is 128 bits long. CK_{cs} denotes one cipher key for circuit switched connections established between the CS service domain and the user. CK_{ps} is denoted one CK for PS connections established between the PS service domain and the user. The CK to be used for a particular radio bearer is described as follows:
  — The radio bearers for CS user data are ciphered with CK_{cs}.
  — The radio bearers for PS user data are ciphered with CK_{ps}.
  — The signaling radio bearers are used for transfer of signaling data for services delivered by both CS and PS service domains. These signaling radio bearers are ciphered by the CK of the service domain for which the most recent security mode negotiation took place.

  For UMTS subscribers, CK is established during UMTS AKA (Authentication and Key Agreement), as the output of the cipher key derivation function f3, available in the USIM and in HLR/AuC. For GSM subscribers that access the UTRAN, CK is established following GSM AKA and is derived from GSM cipher key (K_c). CK is stored in the USIM and a copy is stored in the ME. CK is sent from the USIM to the ME upon request of the ME. The USIM should send CK under the condition that a valid CK is available. CK is sent from HLR/AuC to VLS/SGSN and stored in the VLR/SGSN as part of the quintet. It is sent from the VLR/SGSN to the RNC in the RANAP security mode command. At handover, the CK is transmitted within the network infrastructure from the old RNC to the new RNC, to enable the communication to proceed. The cipher key CK remains unchanged at handover.

- *COUNT-C:* The ciphering sequence number COUNT-C is 32 bits long. There is one COUNT-C value per up-link radio bearer and one COUNT-C value per down-link radio bearer using RLC AM or RLC UM. For all transmission mode RLC radio bearers of the same CN domain, COUNT-C is the same and COUNT-C is also the same for uplink and downlink.

  COUNT-C is composed of two parts: a *short* sequence number that forms the least significant bits of COUNT-C and a *long* sequence number that forms the most significant bits of COUNT-C. The update of COUNT-C depends on the transmission mode as described below:
  — For RLC AM on DCH, the short sequence number is 8-bit CFN (Connection Frame Number) of COUNT-C. It is independently maintained in the ME MAC-d entity and the SRNC MAC-d entity. The long sequence number is the 24-bit MAC-d HFN (Hyper Frame Number), which is incremented at each CFN cycle.
  — For RLC UM mode, the short sequence number is the 7-bit RLC sequence number (RLC SN) and this is part of the RLC UM PDU header. The long sequence number is the 25-bit RLC UM HFN, which is incremented at each RLC SN cycle.

  The hyperframe number (HFN) is initialized by means of the parameter START. The ME and the RNC then initialize the 20 most significant bits of the RLC AM HFN (Radio Link Control Acknowledged Mode Hyper Frame Number), RLC UM HFN (Radio Link Control Unacknowledged Mode Hyper Frame Number), and MAC-d HFN to START. The remaining bits of the RLC AM HFN, RLC UM HFN, and MAC-d HFN are initialized to zero. When a new radio bearer is created during a Radio Resource Control (RRC) connection in ciphered mode, the HFN is initialized by the current START value.

- *BEARER:* The radio bearer identifier (BEARER) is 5 bits long. There is one BEARER parameter per radio bearer associated with the same user and multiplexed on a single 10 ms physical layer frame. The radio bearer identifier is input to avoid different keystreams being used for an identical set of input parameter values. To avoid using the same keystream to encrypt more than one bearer, the algorithm should generate the keystream based on the identity of the radio bearer.

- *DIRECTION:* The direction identifier DIRECTION is 1 bit long. The value of the DIREC-TION is 0 for the uplink (messages from UE to RNC) and 1 for the downlink (messages from RNC to UE). The direction identifier is input to avoid the keystreams for the uplink and downlink using an identical set of input parameter values. The purpose of the DIRECTION bit is to avoid using the same keystream to encrypt both uplink and downlink transmissions.
- *LENGTH:* The length indicator LENGTH is 16 bits long. LENGTH is an integer between 1 and 20 000. LENGTH shall affect only the length of the keystream block, not the actual bits in it.

5.7.3 KGCORE (Core Keystream Generation Function)

This section describes a core keystream generator KGCORE, which is applicable to the 3GPP f8 confidentiality algorithm. The function KGCORE is based on the block cipher KASUMI that is used in a form of OFM that generates the output bitstream in multiples of 64 bits, as illustrated in Figure 5.8.

Following specification of the 2G-security algorithm, an alternative specification of the standard 3GPP f8 algorithm shares the same core function of KGCORE. The original f8 mode has been extended to provide the mechanism for UMTS integrity and the confidentiality algorithm. Thus the mechanism for data confidentiality of user data and signaling data in UMTS requires a cryptographic function called f8. The f8 function should be a synchronous stream cipher algorithm and for interoperability within UMTS it must be fully standardized. The specification of algorithm f8 is given in terms of the core function KGCORE.

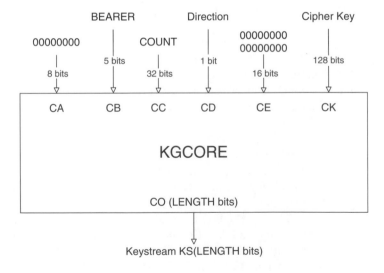

Figure 5.8   3GPP f8 keystream generator function

The definitive specification of f8 specifies the entire encryption procedure, which includes not only the generation of the keystream, but also the encryption procedure. In this section, only the keystream generator part of f8 is described for closer comparison with A5/3 and GEA3.

Mapping the f8 inputs onto the input of KGCORE and mapping the output of KGCORE onto the outputs of f8 define the f8 function. The input parameters to the KGCORE function are defined as

```
CA || BEARER || COUNT || DIRECTION || 0...0
```

obtained as the concatenation of a new 8-bit field CA, the 5 bit BEARER, the 32-bit COUNT, the 1-bit DIRECTION values, and a string of 2-byte field CE. The 8-bit CA is used to specify the mode of encryption and is taken from the 26 bits that were set to all zeros in the f8 specification. In addition, a 2-byte field CE is specified as a variable string of bits reserved for possible future use of KGCORE. The CE field replaces the last two bytes of the all-zero field. After specification of these fields, 2 bits remain unspecified and they are set equal to 0.

The output of KGCORE is the keystream CO of CL bits, denoted by CO[0]...CO[CL-1]. The field CL corresponds to what the f8 specification denoted by LENGTH. By applying KGCORE to the variable inputs described above, it is defined as

```
CA = CA[0]...CA[7] = 00000000 = 8 bits
CB = CB[0]...CB[4] = BEARER[0]...BEARER[4] = 5 bits
CC = CC[0]...CC[31] = COUNT[0]...COUNT[31] = 32 bits
CD = CD[0] = DIRECTION[0] = 1 bit
CE = CE[0]...CE[15] = 0000000000000000 = 16 bits
CK = CK[0]...CK[127] = 128 bits
CO = CO[0]...CO[LENGTH-1] = KS[0]...KS[LENGTH-1]
```

The KGCORE function is applied to the variable inputs to derive the output CO[0]...CO [LENGTH − 1] that is identical to the keystream output of f8 algorithm.

5.7.4 Summary of Four Confidentiality Functions

The four confidentiality algorithms GSM A5/3, ECSD A5/3 (2G), GEA3 (2.5G), and f8 (3G) in terms of KGCORE are summarized in Table 5.3.

Encryption/Decryption functions are a pure computation function based on the usage of a session key whereby the radio-transmitted data can be protected against an unauthorized third party. This function is located in the UE and in the UTRAN.

5.7.5 Key Scheduling

KASUMI operates on a 64-bit data block and uses a 128-bit key K. Before the round keys are calculated, two array of 16-bit values K_j and K'_j ($1 \leq j \leq 8$) are derived as follows:

- First array entry K_j ($j = 1, 2, \ldots, 8$) is derived by subdivision of K into eight 16-bit subblocks:

$$K = K_1 \| K_2 \| \ldots \| K_8$$

Table 5.3 The input/output parameters of KGCORE for four confidentiality functions

Algorithm parameter	GSM A5/3	ECSD A5/3	GEA3	F8
CA	00001111	11110000	11111111	00000000
CB	0000	0000	0000	BEARER
CC	0...0 ‖ COUNT	0...0 ‖ COUNT	INPUT	COUNT
CD	0	0	DIRECTION	DIRECTION
CE	0 0...0	0 0...0	0 0...0	0 0...0
CK	Kc repeated to fill 128 bits	Kc repeated to fill 128 bits	Kc repeated to fill 128 bits	Kc repeated to fill 128 bits
CO	BLOCK 1 ‖ BLOCK 2	BLOCK 1 ‖ BLOCK 2	OUTPUT	KS

ECSD: Enhanced Circuit Switched Data.
CA, CB, CC, CD, CE = KGCORE inputs.
KGCORE = Core Keystream Generator.
CK = Cipher key; CO = KGCORE output.

- Second array entry K'_j ($j = 1, 2,\ldots, 8$) is derived from the first array by adding an array of 16-bit constants C_j:

$$
\begin{aligned}
K &= K'_1 \| K'_2 \| \ldots \| K'_8 \\
K'_j &= K_j \oplus C_j, j = 1, 2, \ldots, 8
\end{aligned}
$$

where the constants C_j are defined as:

$$
\begin{aligned}
C_1 &= 0x\ 0123, C_2 = 0x\ 4567, C_3 = 0x\ 89ab, C_4 = 0x\ cdef \\
C_5 &= 0x\ fedc, C_6 = 0x\ ba98, C_7 = 0x\ 7654, C_8 = 0x\ 3210
\end{aligned}
$$

Next, the subkeys KL, KO, and KI in KASUMI are derived using cyclic left shift as defined by Table 5.4.

Table 5.4 Definition of subkeys in KASUMI

Subkey round	KL_{i1}	KL_{i2}	KO_{i1}	KO_{i2}	KO_{i3}	KI_{i1}	KI_{i2}	KI_{i3}
1	$K_1 \lll 1$	K'_3	$K_2 \lll 5$	$K_6 \lll 8$	$K_7 \lll 13$	K'_5	K'_4	K'_8
2	$K_2 \lll 1$	K'_4	$K_3 \lll 5$	$K_7 \lll 8$	$K_8 \lll 13$	K'_6	K'_5	K'_1
3	$K_3 \lll 1$	K'_5	$K_4 \lll 5$	$K_8 \lll 8$	$K_1 \lll 13$	K'_7	K'_6	K'_2
4	$K_4 \lll 1$	K'_6	$K_5 \lll 5$	$K_1 \lll 8$	$K_2 \lll 13$	K'_8	K'_7	K'_3
5	$K_5 \lll 1$	K'_7	$K_6 \lll 5$	$K_2 \lll 8$	$K_3 \lll 13$	K'_1	K'_8	K'_4
6	$K_6 \lll 1$	K'_8	$K_7 \lll 5$	$K_3 \lll 8$	$K_4 \lll 13$	K'_2	K'_1	K'_5
7	$K_7 \lll 1$	K'_1	$K_8 \lll 5$	$K_4 \lll 8$	$K_5 \lll 13$	K'_3	K'_2	K'_6
8	$K_8 \lll 1$	K'_2	$K_1 \lll 5$	$K_5 \lll 8$	$K_6 \lll 13$	K'_4	K'_3	K'_7

Example 5.3. If the given 128 bit key is 869080c7672ae4491560933d5fad9384, the KL, KO, and KI subkeys of KASUMI are derived as follows:

The 128-bit key is subdivided into eight 16-bit values K_j ($j = 1...8$) where

$$K = K_1||K_2||K_3|| \ldots ||K_8.$$
$$= 8690||80c7||672a||e449||1560||933d||5fad||9384$$

A second array of subkeys, K' is derived from K by applying:

$$K'_j = K_j \oplus C_j (j = 1...8)$$

where C is the constant value defined as

```
0x 0123 4567 89ab cdef fedc ba98 7654 3210
```

So we have, K' = 8690 $\oplus$ 0123 || 80c7 $\oplus$ 4567 || 672a $\oplus$ 89ab || e449 $\oplus$ cdef || 1560 $\oplus$ fedc || 933d $\oplus$ ba98 || 5fad $\oplus$ 7654 || 9384 $\oplus$ 3210
= 87b3 c5a0 ee81 29a6 ebbc 29a5 29f9 a194

The round subkeys are then derived from K_j and K'_j in the manner defined in Table 5.4; the computed values are given in Table 5.5.

```
For instance, KL₁₁ = K₁ <<< 1
                   = 8690 <<< 1
                   = 1000011010010000 <<< 1
                   = 0000110100100001 = 0d21
              KO₁₁ = K₂ <<< 5
                   = 80C7 <<< 5
                   = 1000000011000111 <<< 5
                   = 0001100011110000 = 18f0
              KI₁₁ = K₅' = ebbc
```

Table 5.5 Computed subkeys for KASUMI

	1	2	3	4	5	6	7	8
$KL_{i,1}$	0d21	018f	ce54	c893	2ac0	267b	bf5a	2709
$KL_{i,2}$	ee81	29a6	ebbc	29a5	29f9	a194	87b3	c5a0
$KO_{i,1}$	18f0	e54c	893c	ac02	67b2	f5ab	7092	d210
$KO_{i,2}$	3d93	ad5f	8493	9086	c780	2a67	49e4	6015
$KO_{i,3}$	abf5	9270	10d2	f018	4ce5	3c89	02ac	b267
KI_{i1}	ebbc	29a5	29f9	a194	87b3	c5a0	ee81	29a6
KI_{i2}	29a6	ebbc	29a5	29f9	a194	87b3	c5a0	ee81
KI_{i3}	a194	87b3	c5a0	ee81	29a6	ebbc	29a5	29f9

□

Example 5.4. *UMTS Encryption (f8 function)*

```
Key = d3c5d592327fb11c4035c6680af8c6d1
Frame Count = 398a59b4
Bearer = 15
Uplink Direction = 1
Length = 253 bits
Plaintext:
981ba6824c1bfb1a b485472029b71d80 8ce33e2cc3c0b5fc 1f3de8a6d
c66b1f0
In binary format:
Key = 1101001111000101110101011001001000110010011111111110110001
00011100
01000000001101011100011001101000000010101111100011000110110 1 0001
Count = 00111001100010100101100110110100
Bearer = 10101
Direction = 1

Plaintext:
1001100000011011101001101000001001001100000110111111011000 11010
1011010010000101010001110010000000010100110110111000111011 00 00000
1000110011100011001111100010110011000011110000001011010111 1 11100
0001111100111101111010001010011011011100011001101011000111110

Length = 253 bits

Thus the inputs for KGCORE are:

CA: 0000 0000 (fixed for f8)
CB: 10101 (Bearer)
CC: 00111001100010100101100110110100 (Count)
CD: 1 (Uplink direction)
CE: 0000 0000 0000 0000
CK: 1101001111000101110101011001001000110010011111111011000100011 100
01000000001101011100011001101000000010101111100011000110110100 01
```

First of all, a 64-bit value A should be generated in KGCORE.

To generate A, all the above parameters are concatenated in the following way to form a 64-bit input for a KASUMI block.

```
CC || CB || CD || 00 || CA || CE
= 00111001100010100101100110110100 || 10101 || 1 || 00 || 0000 0000 ||
0000 0000
0000 0000
```

= 0011 1001 1000 1010 0101 1001 1011 0100 1010 1100 0000 0000 0000 0000
0000
0000
= 398a59b4ac000000 (64 bits)

XORing the key-modifier with CK computes the key for this KASUMI block. The key modifier (KM) is a constant.

Modified CK = CK ⊕ KM
= 0xd3c5d592327fb11c4035c6680af8c6d1 ⊕
0x55555555555555555555555555555555
= 0x869080c7672ae4491560933d5fad9384 (128 bits)

All the computations in KASUMI block are described below.
At first the 128-bit modified key is subdivided into eight 16-bit values $K_1 \ldots K_8$ so that

$K = K_1 \parallel K_2 \parallel K_3 \parallel \ldots \parallel K_8$.
i.e. K = 8690 80c7 672a e449 1560 933d 5fad 9384

A second array of subkeys K'_j is derived from K_j by applying:

$$K_j' = K_j \oplus C_j$$

where C is the constant value defined as

0X 0123 4567 89ab cdef fedc ba98 7654 3210

Thus K'_j can be computed as

K_j' = 8690 80c7 672a e449 1560 933d 5fad 9384 ⊕
0123 4567 89ab cdef fedc ba98 7654 3210
= 87b3 c5a0 ee81 29a6 ebbc 29a5 29f9 a194

The round subkeys are then derived from K_j and K'_j in the same manner defined in Table 5.5.
The 64-bit plaintext is then divided into two 32-bit values and L_0 and R_0 are initialized as follows:

L_0 = 398a 59b4
R_0 = ac00 0000

Then 8-round computation of KASUMI functions is executed. At the first round R_1 and L_1 are computed as follows:

$R_1 = L_0$ = 398a59b4
$L_1 = FO_1 (FL_1 (L_0, KL_{1,1}, KL_{1,2}), KO_{1,1}, KO_{1,2}, KO_{1,3}, KI_{1,1}, KI_{1,2}, KI_{1,3}) \oplus R_0$

$= FO_1(FL_1(398a59b4, 0d21, ee81), KO_{1,1}, KO_{1,2}, KO_{1,3}, KI_{1,1}, KI_{1,2},$
$KI_{1,3}) \oplus ac000000$
$= FO_1(e6e14bb4, 18f0, 3d93, abf5, ebbc, 29a6, a194) \oplus ac000000$
$= 13e76fc5 \oplus ac000000$
$= bfe76fc5$

Computation in *FL₁* function

$FL_1(L_0, KL_{1,1}, KL_{1,2}) = FL_1(398a59b4, 0d21, ee81)$

The 32-bit input 398a59b4 is divided into two 16-bit values:

$FLL_0 = 398a$
$FLR_0 = 59b4$

Then compute

$FLR_1 = ((FLL_0 \wedge KL_{1,1}) <<< 1) \oplus FLR_0$
$\quad = ((398a \wedge 0d21) <<< 1) \oplus 59b4$
$\quad = 4bb4$
$FLL_1 = FLL_0 \oplus ((FLR_1 \mid KL_{1,2}) <<< 1)$
$\quad = 398a \oplus ((4bb4 \mid ee81) <<< 1)$
$\quad = e6e1$
Output of $FL_1 = FLL_1 \parallel FLR_1 = e6e14bb4$

Computation in *FO₁* function

$FO_1(FL_1(L_0, KL_{1,1}, KL_{1,2}), KO_{1,1}, KO_{1,2}, KO_{1,3}, KI_{1,1}, KI_{1,2}, KI_{1,3})$
$= FO_1(e6e14bb4, 18f0, 3d93, abf5, ebbc, 29a6, a194)$

Input e6e14bb4 is divided into 16-bit values:

$FOL_0 = e6e1$
$FOR_0 = 4bb4$
Then compute,
$FOL_1 = FOR_0 = 4bb4$
$FOR_1 = FI_{1,1}(FOL_0 \oplus KO_{1,1}, KI_{1,1}) \oplus FOR_0$
$\quad = FI_{1,1}(e6e1 \oplus 18f0, ebbc) \oplus 4bb4$
$\quad = FI_{1,1}(fe11, ebbc) \oplus 4bb4$
$\quad = b9d5 \oplus 4bb4$ (Calculation of $FI_{1,1}$ function is given later)
$\quad = f261$

Similarly, FOL_2, FOR_2 and FOL_3, FOR_3 are calculated as follows:

$FOL_2 = FOR_1 = f261$

$FOR_2 = FI_{1,2}(FOL_1 \oplus KO_{1,2}, KI_{1,2}) \oplus FOR_1$
$\quad = 13e7$
$FOL_3 = FOR_2 = 13e7$
$FOR_3 = FI_{1,3}(FOL_2 \oplus KO_{1,2}, KI_{1,2}) \oplus FOR_2$
$\quad\quad = 6fc5$

The final output for FO_1 is

$FOL_3 \parallel FOR_3 = 13e7\ 6fc5$

Computation in FI$_{1,1}$ function

$FI_{1,1}(fe11, ebbc)$

The input fe11 is split into two unequal components: a 9-bit left half FIL_0 and a 7-bit right half FIR_0.

$FIL_0 = fe11 \gg 7 = 1fc$
$FIR_0 = fe11 \wedge 7f = 11$

The input subkey ebbc is split into two unequal components: a 7-bit left half $KI_{1,1,1}$ and a 9-bit right half $KI_{1,1,2}$.

$KI_{1,1,1} = ebbc \gg 9 = 75$
$KI_{1,1,2} = ebbc \wedge 1ff = 1bc$

$FIL_1 = S9[FIL_0] \oplus ZE(FIR_0)$
$\quad\quad = S9[1fc] \oplus ZE(11)$
$\quad\quad = 1b4 \oplus ZE(11)\ \text{[See Table SBox(9)]}$
$\quad\quad = 110110100_b \oplus 000010001_b\ \text{[Zero extended on left side]}$
$\quad\quad = 110100101$
$\quad\quad = 1a5$

$FIR_1 = S7[FIR_0] \oplus KI_{1,1,1}$
$\quad\quad = S7[11] \oplus 75$
$\quad\quad = 71 \oplus 75$
$\quad\quad = 04$

$FIL_2 = FIL_1 \oplus KI_{1,1,2}$
$\quad\quad = 1a5 \oplus 1bc$
$\quad\quad = 19$

$FIR_2 = FIR_1 \oplus TR(FIL_1)$
$\quad\quad = 04 \oplus (FIL_1 \wedge 7f)$
$\quad\quad = 04 \oplus (1a5 \wedge 7f)$
$\quad\quad = 04 \oplus 25$
$\quad\quad = 21$

```
FIL₃ = S9[FIL₂] ⊕ ZE(FIR₂)
     = S9[19] ⊕ ZE(21)
     = 1f4 ⊕ 21
     = 1d5

FIR₃ = S7[FIR₂] ⊕ TR(FIL₃)
     = S7[21] ⊕ (1d5^ 7f)
     = 9 ⊕ 55
     = 5c
```

The output of $FI_{1,1}$ function is finally calculated as

$$FIR_3 \| FIL_3 = (FIR_3 \ll 9) \| FIL_3$$

```
= (5c << 9) || 1d5
= b9d5
```

Similarly, in the second round of KASUMI, L_2 and R_2 are computed as

```
R₂=L₁ = bfe76fc5
L₂=FL₂(F0₂(L₁, KO₂,₁, KO₂,₂, KO₂,₃, KI₂,₁, KI₂,₂, KI₂,₃), KL₂,₁, KL₂,₂) ⊕R₁
= 23342140 ⊕ 398a59b4
= 1abe78f4
```

Finally, the values of L_8 and R_8 are computed as

$$L_8 = c67b9e63$$
$$R_8 = e78f967c$$

Therefore, KASUMI's output c67b 9e63 e78f 967c is considered as the value of A that will be the input for the subsequent KASUMI blocks of KGCORE.

As the length of the output is 253, the number of the required KASUMI block is

$$= \lceil 253/64 \rceil = 4$$

Thus the values of BLKCNT are 0, 1, 2, and 3.

For all blocks, CK = d3c5d592327fb11c4035c6680af8c6d1.

Now, the input for the first KASUMI block is

```
A ⊕ BLKCNT = c67b 9e63 e78f 967c ⊕ 0 = c67b 9e63 e78f 967c
```

The output CO[0...63] is calculated in the same manner described before:

```
CO[0...63] = 5211c6366585924e
```

For the second KASUMI block, input is

$$A \oplus BLKCNT \oplus CO[0...63] = c67b\ 9e63\ e78f\ 967c \oplus 1 \oplus 5211c6366585924e$$
$$= 946a5855820a0433$$

The output CO[64...127] = 6f722f4e6f435c10.
For the third KASUMI block, input is

$$A \oplus BLKCNT \oplus CO[64...127] = c67b\ 9e63\ e78f\ 967c \oplus 2 \oplus 6f722f4e6f435c10$$
$$= a909b12d88ccca6e$$

The output CO[128...191] = 50628e58c788a649.
For the last KASUMI block, input is

$$A \oplus BLKCNT \oplus CO[128...191] = c67b\ 9e63\ e78f\ 967c \oplus 3 \oplus$$
$$50628e58c788a649$$
$$= 9619103b20073036$$

The output CO[192...255] = 158c16e0851d12c9.
Finally, the keystream is computed by concatenating the all output bits up to 253. If necessary the least significant bits are discarded.
So the generated keystream is

```
KS = CO[0...63] || CO[64...127] || CO[128...191] || CO[192...252]
= 0101001000010001110001100011011001100101100001011001001001001110
0110111101110010001011110100111001101111010000110101110000010000
0101000001100010100011100101100011000111100010001010011001001001
00010101100011000001011011100000100001010001110100010010110012ᵦ
```

```
So the Ciphertext = Plaintext ⊕ KS =
1001100000011011101001101000001001001100000110111111110110001 1010
1011010010000101010001110010000000010100110110111000111011000 0000
1000110011100011001111100001011001100001111000000101101011111100
00011111001111011110100001010011011011100011001101011000111110
    ⊕
0101001000010001110001100011011001100101100001011001001001001110
0110111101110010001011110100111001101111010000110101110000010000
0101000001100010100011100101100011000111100010001010011001001001
00010101100011000001011011100000100001010001110100010010110012ᵦ
    =
1100101000001010011000001011010000010100110011110011010010101010100
11011011111101110110100001101110010001101111010001000001100100000
1101110010000001101100000111010000000010001001000000100111011010101
00001010101011000111111110010001100101100101111011101000110011
    =
ca0a60b4299e6954dbf7686e46f44190dc81b074044813b50ab1fe46597ba339
```

□

5.8 UTRAN Overall Functions

This section summarizes the overall functions of UTRAN:

- The transfer function of user data provides user data transfer capability across the UTRAN between the I_u and U_u reference points.
- System access is the means by which a UMTS user is connected to the UTRAN in order to use UMTS services and facilities. User system access can be initiated from either the mobile side (an originated call) or the network side (a transmitted call).
- The purpose of the admission control is to admit or deny new users, new radio access bearers, or new radio links. The admission control function based on uplink interface and downlink power is located in the controlling RNC. The serving RNC performs admission control towards the I_u interface.
- The task of congestion control is to monitor, detect, and handle situations when the system is reaching an overload or near overload situation with the already connected users. The congestion control should bring the system back to a stable state as seamlessly as possible. Congestion control is performed within UTRAN.
- The information broadcasting function provides the mobile station with the access stratum and non-access stratum information that are needed by the UE for its operation within the network. The basic control and synchronization of this function is located in UTRAN.
- The Enciphering/Deciphering function is a pure computation function whereby the radio-transmitted data can be protected against unauthorized third-parties. This function is located in the UE and in the UTRAN.
- The handover function manages the mobility of radio interface. It is used to maintain the QoS requested by the core network. The handover function is either controlled by the network or independently by the UE. Therefore, this function may be located in the SRNC, the UE, or both. Handover may be directed to/from another system, that is, UMTS to GSM handover.
- The paging function provides the capability to request a UE to contact the UTRAN when the UE is in idle, CELL-PCH, or URA PCH states.
- Radio resource management is concerned with the allocation and maintenance of radio communications resources. UMTS radio resources must be shared between circuit transfer mode services and packet transfer mode services.
- The SRNS relocation function coordinates the activities when the SRNS's role is to be taken over by another RNS. The SRNS relocation function manages the I_u interface connection mobility from one RNS to another. The SRNS relocation is initiated by the SRNC. This function is located in the RNC and the CN.
- The radio resource configuration and operation function configures the radio network resources (cells and common transport channels), and takes the resources into or out of operation.
- The combining/splitting control function controls the combining/splitting of information streams to receive/transmit the same information through multiple physical channels from a single mobile terminal.
- The connection set-up and release function is responsible for the control of connection setup and release of the radio access subnetwork for the processing of the end-to-end connection.

5.9 UTRAN I_{ub} Interface Protocol Structure

The logical interface between a RNC and a Node B is called the I_{ub} interface. I_{ub} is based on a logical mode of Node B, which controls a number of cells. I_{ub} can also be ordered to add/remove radio link in those cells.

The role of an RNS (Serving or Drift) is on a per connection basis between a UE and the UTRAN via the U_u interface. The UTRAN is layered into a Radio Network Layer and a Transport Network Layer. The I_{ub} interface allows the RNC and Node B to negotiate about radio resources for adding and deleting cells controlled by the Node B to support communication of the dedicated connection between UE and SRNC. Figure 5.9 depicts UTRAN radio network layer.

The I_{ub} interface characteristics and capabilities are discussed below:

- *I_{ub} DCH Frame Protocol:* The I_{ub} interface provides the means for transport of uplink and downlink DCH transport frames between RNC and Node B. An I_{ub} DCH data stream corresponds to the data carried on one DCH transport channel. One DCH data stream always corresponds to a bi-directional transport channel in the UTRAN. Therefore, two uni-directional U_u DCH transport channels with opposite directions can be mapped to one or two DCH transport channels in the UTRAN. Thus, one I_{ub} DCH data stream is carried on the one transport bearer. For each DCH data stream a transport bearer must be established over I_{ub}.

- *I_{ub} RACH Frame Protocol:* The I_{ub} interface provides the means for transport of uplink Random Access Channel (RACH) transport frames between Node B and RNC. An I_{ub} RACH data stream corresponds to the data carried on one RACH transport channel. One I_{ub} RACH data stream is carried on one transport bearer. For each RACH in a cell, a transport bearer must be established over the I_{ub} interface.

- *I_{ub} FACH Frame Protocol:* The I_{ub} interface provides the means for transport of downlink Forward Access Channel (FACH) transport frames between RNC and Node B. An I_{ub} FACH data stream corresponds to the data carried on one FACH transport channel. One I_{ub} FACH data stream is carried on one transport bearer. For each FACH in a cell, a transport bearer should be established over the I_{ub} interface.

- *I_{ub} PCH Frame Protocol:* The I_{ub} interface provides the means for transport of Paging Channel (PCH) transport frames between RNC and Node B. An I_{nh} PCH data stream corresponds to the data carried on one PCH transport channel. One I_{ub} PCH data stream is carried on one transport bearer.

- *I_{ub} DSCH Frame Protocol:* The I_{ub} interface provides the means for transport of Downlink Shared Channel (DSCH) data frames between RNC and Node B. An I_{ub} DSCH data stream corresponds to the data carried on one DSCH transport channel for one UE. A UE may have multiple DSCH data streams. One I_{ub} DSCH data stream is carried on one transport bearer. For each DSCH data stream, a transport bearer must be established over the I_{ub} interface.

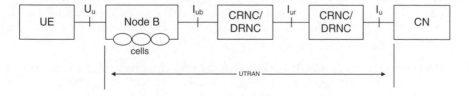

Figure 5.9   UTRAN radio network layer

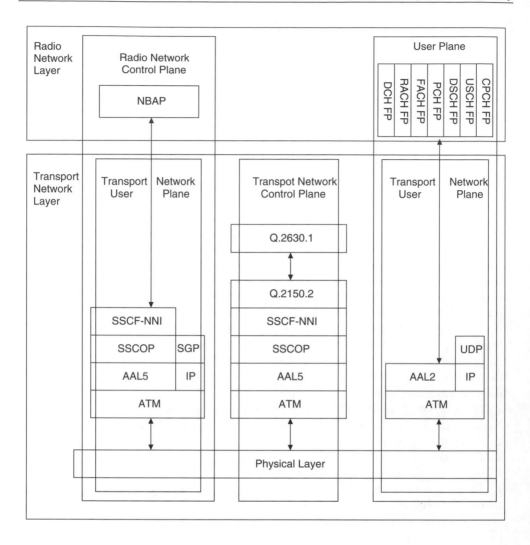

Figure 5.10 Protocol stack of the I_{ub} interface (© 2002. 3GPP™ TSs and TRs are the property of ARIB, ATIS, CCSA, ETSI, TTA and TTC who jointly own the copyright in them. They are subject to further modifications and are therefore provided to you "as is" for information purposes only. Further use is strictly prohibited.)

- *I_{ub} USCH Frame Protocol (TDD):* The I_{ub} interface provides the means for transport of Uplink Shared Channel (USCH) data frames between Node B and RNC. An I_{ub} USCH data stream corresponds to the data carried on one USCH transport channel for one UE. A UE may have multiple USCH data streams. One I_{ub} USCH data stream is carried on one transport bearer. For each USCH data stream, a transport bearer must be established over the I_{ub} interface.

- I_{ub} *CPCH Frame Protocol (FDD):* The I_{ub} interface provides the means for transport of uplink Common Packet Channel (CPCH) transport frames between Node B and RNC. One I_{ub} CPCH data stream is carried on one transport bearer. For each CPCH in a cell, an I_{ub} CPCH data stream must be established over the I_{ub} interface.

The protocol structure for I_{ub} interface is depicted in Figure 5.10.

5.10 UTRAN I_{ur} Interface Protocol Structure

The logical connection that exists between any two RNCs within the UTRAN is referred to as the I_{ur} interface (see Figure 5.11). The I_{ur} interface should support the exchange of signaling

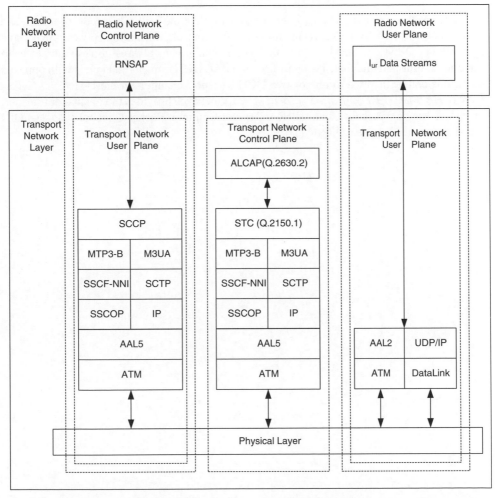

RNSAP: Radio Network Subsystem Application Part
SCCP: Signaling Connection Control Part
ALCAP: Access Link Control Application Part
STC: Signaling Transport Converter
M3UA: SS7 MTP3 User Adaptation Layer

SSCOP: Service Specific Connection Oriented Protocol
SSCF-NNI: Service Specific Co-ordination Function-
Network Node Interface
SCTP: Simple Control Transmission Protocol

Figure 5.11   Protocol stack of the I_{ur} interface

information between two RNCs, and in addition the interface may need to support one or more I_{ur} data streams. From a logical standpoint, the I_{ur} is a point-to-point interface between two RNCs within the UTRAN. The I_{ur} interface facilitates support of continuation between RNCs of the UTRAN services offered via the I_u interface. The I_{ur} interface specification should allow an RNC to address any other RNC within the PLMN for establishing a signaling bearer over I_{ur} and user data bearers for I_{ur} data streams. The I_{ur} interface also provides capability to support radio interface mobility between RNCs when UEs have a connection with UTRAN. This capability includes the support of handover, radio resource handling, and synchronization between RNSs.

The I_{ur} interface characteristics and capabilities are presented below:

- *I_{ur} DCH Frame Protocol:* The I_{ur} interface provides the means for transport of uplink and downlink I_{ub}/I_{ur} DCH frames carrying user data and control information between SRNC and Node B via DRNC. In the UTRAN, one DCH data stream always corresponds to a bi-directional transport channel. Two uni-directional U_u DCH transport channels with opposite directions can be mapped to one or two DCH transport channels in the UTRAN.
- *I_{ur} RACH/CPCH (FDD) Frame Protocol:* The I_{ur} interface provides the means for transport of uplink Random Access Channel (RACH) and Common Packet Channel (CPCH-FDD) transport frames between DRNC and SRNC.

6

High Speed Downlink Packet Access (HSDPA)

This chapter describes HSDPA operational features relating to Adaptive Modulation and Coding, Hybrid ARQ, Fast Cell Selection, Scheduling at the Node B, and MIMO Antenna techniques. These operational techniques are primarily aimed at increasing throughput, reducing delay, and achieving high peak rates. These functionalities rely on a new type of HS-DSCH (High Speed Downlink Shared Channel) transport channel to which the Node B is terminated. HS-DSCH is applicable only to packet switched domain Radio Access Bearers (RABs).

6.1 Basic Structure of HS-DSCH

The HS-DSCH functionality should be able to operate in an environment where certain cells are not updated with HS-DSCH functionality. Two architectures have been considered as part of the study items: an RNC-based architecture consistent with Release 99 and a Node B-based architecture for scheduling. The moving schedule to the Node B provides a more efficient implementation enabling the scheduler to work with the most recent channel information. The scheduler can adapt the modulation to better match the current channel conditions and fading environment. Moreover, the scheduler can exploit the multi-user diversity by scheduling only those users in constructive fades. Furthermore, the HSDPA proposal has the additional potentiality to improve the RNC-based HARQ architecture for both UE memory requirements and transmission delay. A protocol architecture for the HS-DSCH has been proposed, in order specifically to facilitate the introduction of MAC and HARQ, with the aim of maximizing throughput and peak rates.

6.1.1 Protocol Structure

Since the HSDPA functionality should be able to operate in an environment where certain cells are not updated with HSDPA functionality, there is a need to keep, if possible, the Release 99 functional split between layers. The PDCP (Packet Data Coverage Protocol) and MAC-d layers are unchanged from the Release 99 and Release 4 architecture. In addition to the RLC layer from Release 99, it is possible to use Release 7 RLC layer, which is modified to support flexible

RLC PDU sizes for RLC AM, when MAC-ehs is configured. A MAC-d flow is a flow of MAC-d PDUs, which belong to logical channels, which are MAC-d multiplexed. MAC-d is retained in the SRNC. Transport channel type switching is therefore feasible.

The RLC layer can operate in AM or UM mode (but not TM mode due to ciphering). The PDCP can be configured either to perform or not to perform header compression.

The new functionalities of hybrid ARQ, segmentation (MAC-ehs only), and HS-DSCH scheduling are included in the MAC layer. In the UTRAN these functions are included in new entities called MAC-hs and MAC-ehs located in Node B. The MAC entity controlling the transport channel is called MAC-hs, which is located in the Node B. Upper layers configure which of the two entities, MAC-hs or MAC-ehs, is to be applied to handle HS-DSCH functionality. HS-DSCH is the transport channel that exhibits its own functionality.

Two MAC protocol configurations on the UTRAN side can be classified into configuration with MAC-c/sh and configuration without MAC c/sh. Both configurations are transparent to both the UE and Node B.

Figure 6.1 shows the radio interface protocol architecture with termination points of the configuration with MAC-c/sh. In this case, the MAC-hs or MAC-ehs in the Node B is located below MAC-c/sh in CRNC. MAC-c/sh should provide functions to HS-DSCH identical to those provided for the DSCH in the Release 99. The HS-DSCH FP (Frame Protocol) will handle the data transport from SRNC to CRNC (if the I_{ur} interface is involved) and between CRNC and the Node B.

In the case of configuration without MAC-c/sh (as shown in Figure 6.2), the CRNC does not have any user plane function for the HS-DSCH. MAC-d in SRNC is located directly above MAC-hs or MAC-ehs in the Node B; that is, in the HS-DSCH user plane by which the SRNC is directly connected to the Node B, thus bypassing the CRNC. Both configurations are transparent to both the UE and Node B. The same architecture supports both FDD and TDD modes of operation even though some details of the associated signaling for HS-DSCH are different. The MAC-hs (UTRAN side) is responsible for handling the data transmitted on the HS-DSCH. It has also responsibility for managing the physical resources allocated to HSDPA. The MAC-hs receives configuration parameters from the RRC (Radio Resource Control) layer

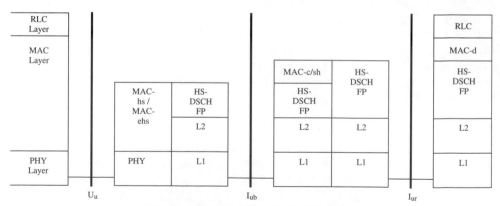

L1: Physical Layer, L2: Data Link Layer, FP: Frame Protocol

Figure 6.1 Protocol architecture of HS-DSCH, configuration with MAC-c/sh (© 2007. 3GPP™ TSs and TRs are the property of ARIB, ATIS, CCSA, ETSI, TTA and TTC who jointly own the copyright in them. They are subject to further modifications and are therefore provided to you "as is" for information purposes only. Further use is strictly prohibited.)

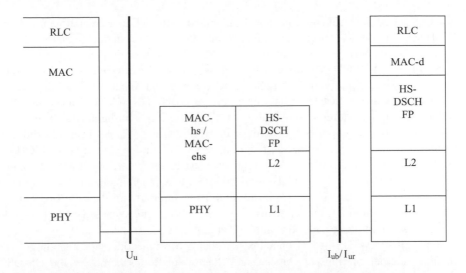

Figure 6.2   Protocol architecture of HS-DSCH, configuration without MAC-c/sh

via the MAC-control SAP (Service Access Point). There should be priority handling per MAC-d PDU in the MAC-hs. The MAC-hs, which is added to the MAC architecture of Release 99, is located in the Node B. If an HS-DSCH is assigned to the UE, then the MAC-hs SDUs (that is, the MAC-d PDUs to be transmitted) are transferred from the MAC-c/sh to MAC-hs via the I_{ub} interface in the case of configuration with MAC-c/sh, or from the MAC-d via I_{ur}/I_{ub} in the case of configuration without MAC-c/sh. Figure 6.3 depicts a simplified schematic illustration for MAC multiplexing in UTRAN side.

6.1.2 HS-DSCH Physical Layer Model

When operating in CELL-DCH state, the basic downlink channel configuration consists of one or several HS-PDSCHs along with an associated DPCH (Dedicated Physical Channel) combined with a number of separate shared physical control channels, High Speed Shared Control Channels (HS-SCCHs).

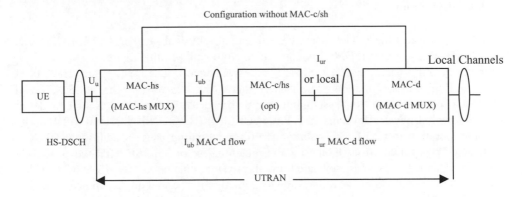

Figure 6.3   UTRAN side of MAC multiplexing

The set of shared physical control channels allocated to the UE at a given time is called an HS-SCCH set. The UTRAN may use more than one HS-SCCH set in one given cell. There is a fixed time offset between the start of the HS-SCCH information and the start of the corresponding HS-PDSCH subframe. The UE is provided with one HS-SCCH set of HS-PDSCH configuration/re-configuration via RPC signaling. What a UE is allocated at a given time is called an HS-SCCH set. The maximum number of HS-SCCHs in an HS-SCCH set is four. The UE should monitor continuously all the HS-SCCHs in the allocated set.

The two-step signaling approach is used for indicating which UE has been scheduled and for signaling the necessary information required for the UE to decode the HS-PDSCHs.

The HS-PDSCH channelization codes that are used in a given cell are not sent to the UE using RRC signaling. The HS-SCCH signals the set of HS-PDSCH channelization codes that are allocated to a UE for a given TTI (Transmission Time Interval). In the case of HS-DSCH transmission to the same UE in consecutive HS-DSCH TTIs, the same HS-SCCH should be used for corresponding associated downlink signaling.

The first part of the HS-SCCH contains the channelization code set, precoding weight information, number of transport blocks, and the modulation scheme for the HS-DSCH allocation with the second part containing the transport block size and Hybrid-ARQ related information. One CRC is calculated over both parts and the UE identity, and attached to the HS-SCCH information.

When operating in CELL-DCH state, the upper layer signaling on the DCCH can be mapped to the DCH mapped to the associated DPCH or the HS-DSCH. For each HS-DSCH TTI, each HS-SCCH carries downlink signaling for one UE. The following information is carried on the HS-SCCH:

- *Transport Format and Resource Indicator (TFRI):* The TFRI includes information about the dynamic part of the HS-DSCH transport format, including transport block size. The HS-SCCH also includes information about the modulation scheme and the set of physical channels (channelization codes) onto which HS-DSCH is mapped in the corresponding HS-DSCH TTI. If MIMO mode is configured, it also contains the number of transport blocks and the precoding weight information, which informs the UE of which precoding weight is applied to the primary transport block.
- *Hybrid-ARQ-related Information (HARQ information):* This includes the HARQ protocol related information for the corresponding HS-DSCH TTI and information about the redundancy version.

Figure 6.4 illustrates the model of the UE's downlink physical layer in CELL-DCH state. HS-PDSCH with associated DPCH is transmitted from cell 1 in this figure.

When operating CELL-FACH (Forward Access Channel), CELL-PCH, and URA-PCH (UTRAN Registration Area-Paging Channel) states, the UE obtains the HS-SCCH and HS-PDSCH configuration from system information broadcast. Figure 6.5 shows the model of the UE's downlink physical layer in CELL-FACH, CELL-PCH, and URA-PCH state (FDD only).

When operating in CELL-DCH state, the uplink signaling uses an additional DPCCH (Dedicated Physical Control Channel) with Spreading Factor 256 (SF = 256), that is, code multiplexed with the existing dedicated uplink physical channels. The HS-DSCH related uplink signaling consists of HARQ acknowledgement and channel quality indicator (CQI). The DCH model of the UE's uplink physical layer in CELL-DCH state is illustrated in

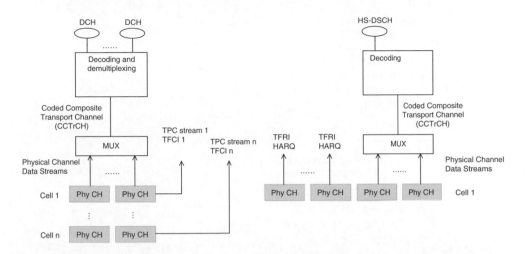

Figure 6.4   Model of the UE's downlink physical layer in CELL-DCH state. HS-PDSCH with associated DPCH is transmitted from cell 1 in this figure (© 2007. 3GPP™ TSs and TRs are the property of ARIB, ATIS, CCSA, ETSI, TTA and TTC who jointly own the copyright in them. They are subject to further modifications and are therefore provided to you "as is" for information purposes only. Further use is strictly prohibited.)

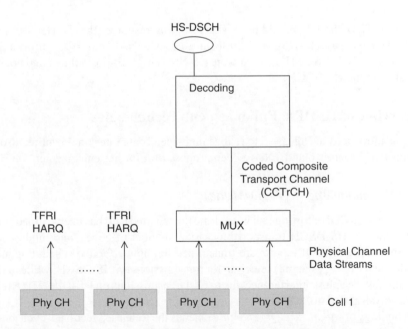

Figure 6.5   Model of the UE's downlink physical layer in CELL-FACH, CELL-PCH, and URA-PCH state (FDD only) (© 2007. 3GPP™ TSs and TRs are the property of ARIB, ATIS, CCSA, ETSI, TTA and TTC who jointly own the copyright in them. They are subject to further modifications and are therefore provided to you "as is" for information purposes only. Further use is strictly prohibited.)

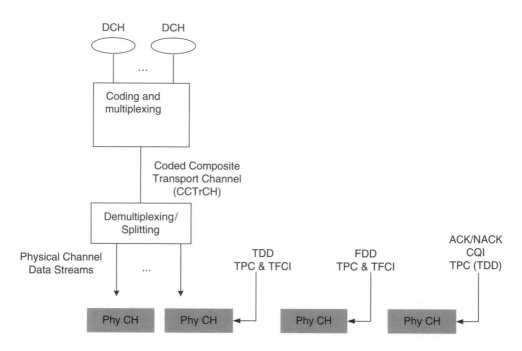

Figure 6.6   Model of the UE's uplink physical layer in CELL-DCH state

Figure 6.6. In TDD, the UE should use a shared uplink resource, that is, HS-SICH (Shared Information Channel for HS-DSCH) for transmitting ACK/NACK and CQI information. The relation between the HS-SCCH in DL and the HS-SICH in UL is pre-defined and not signaled dynamically on the HS-SCCH.

6.2 Overview of HSDPA Enhancement Technologies

HSDPA was introduced in Release 5 to realize high-speed data rates on downlink. (Some new technologies, as Releases 6 and 7, have been proposed for further enhancement on HSDPA.)

6.2.1 CQI Enhancement (FDD Mode)

In Release 5, a fixed CQI reporting rate is used and the reported CQI has been defined to relate to the combination of HS-DSCH transport block size, number of codes, and modulation. It is desirable that sufficient CQI reports are transmitted to allow effective scheduling and MCS (Modulation and Coding Scheme) selection for transmissions on HS-DSCH, while at the same time minimizing the uplink interference due to CQI transmissions on HS-DPCCH (High Speed Dedicated Physical Control Channel).

The technology of *enhanced CQI reporting* extends the Release 5 feedback cycle based CQI scheme by introducing tunable reporting rates through additional CQI reports during periods of downlink activity and fewer reports at other times. The additional CQI reports are initiated implicitly with every ACK and/or NACK and through the possibility of requesting them on demand using fast Layer 1 (Physical Layer) signaling. Requesting CQI reports by means of fast

Layer 1 signaling is especially advantageous prior to the first packets of a packet cell, whereas implicit ACK/NACK based CQI reports are efficient after the transmission of packets has already started. By fast Layer 1 messages, it may be possible to improve the performance of the first packet of a packet cell and to reduce the number of retransmissions. Fast Layer 1 signaling is realized by using the redundant area of channelization code set mapping of HS-SCCH.

The motivation for this technology is to improve the performance of HSDPA. The uplink-signaling overhead may be able to be reduced, while ensuring that up-to-date CQI information is available at the time of downlink activity for efficient scheduling and rate selection.

6.2.2 Multiple Simultaneous Transmission to a UE Within Sub-Frame

Release 5 of the HSDPA specifications allows only one transmission to a UE within an HSDPA subframe. The restriction of a single transmission to a UE within a subframe may lead to scheduling and transmission inefficiencies when one or more retransmissions for a given UE are pending on one or more HARQ processes and/or when a new transmission needs to be code-multiplied with one or more retransmissions within a subframe. Providing the flexibility to support multiple simultaneous transmissions to a UE within a subframe has the potential to enhance HSDPA system performance through better exploitation of multi-user diversity. Thus, the enhancement that allows multiple simultaneous HARQ transmissions to a UE can improve scheduling flexibility and system performance.

6.2.3 Code Reuse for Downlink HS-DSCH

Downlink OVSF (Orthogonal Variable Spreading Factor) codes are a critical resource factor in defining the capacity of the network. The capacity of HS-DSCH can be limited due to a shortage of available orthogonal codes. The shortage of codes can result in inefficient code space usage by the associated DPCH (Dedicated Physical Channel) and other dedicated channels. For Release 99 data services, an inactivity timer is typically employed to ensure that code resources are released for other users. However, there are certain data applications for which long inactivity timers may be needed in order to ensure low delay. As a result, these applications tend to use up significant fractions of the code space but have very low power requirements. Voice users can also make inefficient use of code space because codes remain assigned during periods of inactivity. This leads to power code imbalance, an effect further compounded by soft handoff on the downlink. Enhancements that provide power benefit, such as beamforming, also need corresponding improvements in the code dimension so that the system capacity benefits can be realized. These effects can result in a large amount of power available for HS-DSCH as compared to codes. An enhancement that allows OVSF code reuse without requiring multiple receive antennas can improve downlink capacity in code limited situations.

6.2.4 Fast Signaling Between Node B and UE

In the Release 5 specifications, the UE can be signaled about MAC-hs and control channel reconfiguration through RRC signaling. However, the RRC signaling is slow due to delays in the radio access network and use of longer TTI for transmission. Moreover, when a control message related to MAC-hs (scheduler, AMC or HARQ, etc.) needs to be carried to the UEs, the information is first sent from MAC-hs in the Node B to the RRC in the RNC and only then can RRC forward the signaling message to UE.

This enhancement will help to reduce the delays involved due to RRC signaling between UE and RNC. The objective is to devise first signaling schemes that allow Node B to signal various information such as control channel reconfiguration and power offsets over the air interface to the UE, without having to use RRC signaling.

6.2.5 Fast Adaptive Emphasis

The aim of Fast Adaptive Emphasis is to offer the UE a seamless gain of closed loop transmission diversity when it enters in a soft handover region. In fact, in such a region the UE loses transmission diversity gain because the weights are not optimized anymore for the radio link from the HSDPA serving cell.

According to the Release 5 specifications, during operation of HSDPA in closed loop transmission diversity with an associated DPCH in soft handover, it is possible to emphasize the radio link from the serving cell that carries HS-PDSCH and HS-SCCH. With Fast Adaptive Emphasis, the UE puts emphasis on the serving cell only when it is scheduled for transmission on the HS-PDSCH. This enhancement is expected to improve performance of HSDPA and reduce downlink transmission power for DPCH when closed loop transmission diversity is used in the soft handover region.

6.2.6 ACK/NACK Transmit Power Reduction for HS-DPCCH

In Release 5, the UE always uses Discontinuous Transmission (DTX) in the ACK/NACK field of the HS-DPCCH except when an ACK or NACK is being transmitted in response to an HS-DSCH transmission. This means that if the UE fails to detect typical probability 0.01 of the HS-SCCH, the UE will use DTX in the corresponding ACK/NACK field. The Node B must avoid decoding this DTX as ACK if it is to avoid loss of the HS-DSCH TTI at the physical layer. If an HS-DSCH TTI is lost at the physical layer, the only means of recovering the TTI is by higher-layer transmission, which may be too slow for the delay requirements and buffering capabilities.

The transmission power for ACK messages must therefore be set high enough to reduce the probability of such misinterpretations to a sufficiently low level. It is then desirable to find means by which the Node B can set its ACK detection threshold closer to TDX without resulting in misinterpretations.

6.2.7 Fractional Dedicated Physical Channel (F-DPCH)

The current specifications mandate the set up of Dedicated Physical Channels (DPCHs) both in the uplink and in the downlink for any user operating in HSDPA. However, with the development of data-only applications for low to medium bit rates, the cost of the dedicated channel may restrict a wider use of HSDPA.

When a user only wants to have streaming, interactive, or background service, that is a data-only service, there is still a need from the system perspective to set up a DPCH in the downlink. It is generally foreseen that this downlink dedicated channel will be mainly used to carry RRC signaling and that all the traffic will go through the HSDPA channel. RRC signaling has a minimum data rate however because transmission of RRC signaling is rather infrequent; the physical channel carrying this signaling will be DTXed most of time except for TPC (Transmit Power Control) and pilot bits transmission.

As the current standard allows the signaling to be carried on the HS-DSCH transport channel, the dedicated physical channel may be set up in the downlink to carry only Layer 1 (physical layer) signaling.

Providing the flexibility to share the dedicated code channels associated to data-only users has the potential to allow for a wider use of the DSDPA system by reducing the code limitation problem. The fractional dedicated physical channel implements the concept of code sharing between data-only HSDPA users to carry Layer 1 control information as described earlier.

6.3 HS-DSCH MAC Architecture—UE Side

This section describes the MAC architecture and functional split required to support HSDPA (i.e. HS-DSCH) on the UE side. Both MAC-hs and MAC-ehs handle the HS-DSCH specific functions. Upper layers configure which of the two entities, MAC-hs or MAC-ehs, is to be applied to handle HS-DSCH functionality.

6.3.1 Overall Architecture

Figure 6.7 shows the overall MAC architecture (UE side). In the case of HSDPA, the data received on HS-DSCH is mapped to MAC-hs or MAC-ehs. The MAC-hs or MAC-ehs is configured via the MAC Control SAP by RRC similar to the MAC-c/sh and MAC-d, to set the parameters in the MAC-hs or MAC-ehs, such as allowed transport format combinations for HS-DSCH.

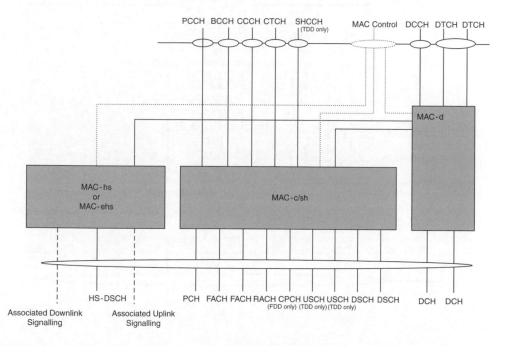

Figure 6.7 UE side MAC architecture with HS-DSCH (MAC-hs or MAC-chs) (© 2007. 3GPP™ TSs and TRs are the property of ARIB, ATIS, CCSA, ETSI, TTA and TTC who jointly own the copyright in them. They are subject to further modifications and are therefore provided to you "as is" for information purposes only. Further use is strictly prohibited.)

The associated downlink signaling carries information for support of HS-DSCH, while the uplink signaling carries feedback information.

In FDD, when operating in CELL-FACH, CELL-PCH, and URA-PCH defined in TS 25.308 v7.3.0 and the transmission of HS-DPCCH for HS-DSCH, related ACK/NACK and CQI signaling is not supported. The direct UE to Node B uplink CQI signaling on RACH is for further study (FFS).

6.3.2 MAC-d Entity

The MAC-d entity is modified with the addition of a link to the MAC-hs or MAC-ehs entity. The links to MAC-hs, MAC-ehs and MAC-c/sh cannot be configured in one UE simultaneously.

The mapping between C/T MUX entity in MAC-d and the reordering buffer in MAC-hs is configured by higher layers. One recording buffer maps to one C/T MUX entity and many recording buffers can map to the same C/T MUX entity. If MAC-ehs is configured, C/T MUX toward MAC-ehs is not used. Figure 6.8 depicts the MAC-d architecture for MAC-hs/MAC-ehs. It is noted that for DCH, DSCH, and HS-DSCH, different scheduling mechanisms apply; and ciphering is performed in MAC-d only for transparent RLC mode.

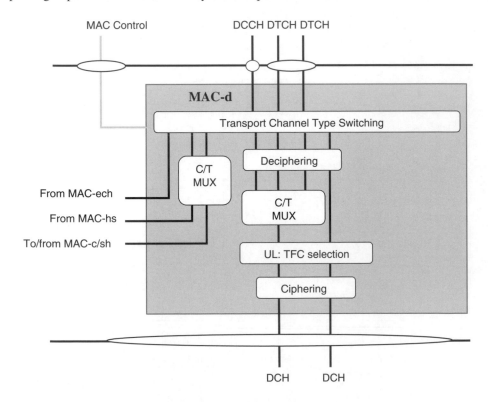

Note 1: For DCH, DSCH and HS-DSCH, different scheduling mechanisms apply
Note 2: Ciphering is performed in MAC-d only for transparent RLC mode

Figure 6.8   MAC-d architecture for MAC-hs/MAC-ehs

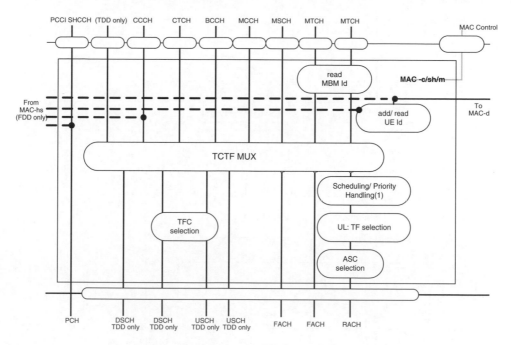

Note: Dashed lines are FDD only

Figure 6.9   UE side MAC architecture/MAC-c/sh/m detail (© 2007. 3GPP™ TSs and TRs are the property of ARIB, ATIS, CCSA, ETSI, TTA and TTC who jointly own the copyright in them. They are subject to further modifications and are therefore provided to you "as is" for information purposes only. Further use is strictly prohibited.)

6.3.3 MAC-c/sh Entity

The MAC-c/sh/m on the UE side is not modified for HS-DSCH operation in CELL-DCH state and the MAC-c/sh details are depicted in Figure 6.9.

6.3.4 MAC-hs Entity

The MAC-hs entity handles the HSDPA specific functions. In the model below the MAC-hs comprises the following entities:

* *HARQ:* The HARQ entity is responsible for handling the HARQ protocol. There will be one HARQ process per HS-DSCH per TTI. The HARQ functional entity handles all the tasks that are required for hybrid ARQ. It is responsible for generating ACKs or NACKs. The detailed configuration of the hybrid ARQ protocol is provided by RRC over the MAC-Control SAP (Service Access Point).
* *Reordering:* The reordering entity organizes received data blocks according to the received TSN (Transmission Sequence Number). Data blocks with consecutive TSNs are delivered to higher layers upon reception. A timer mechanism determines delivery of non-consecutive data blocks to higher layers. There is one reordering entity for each priority class.

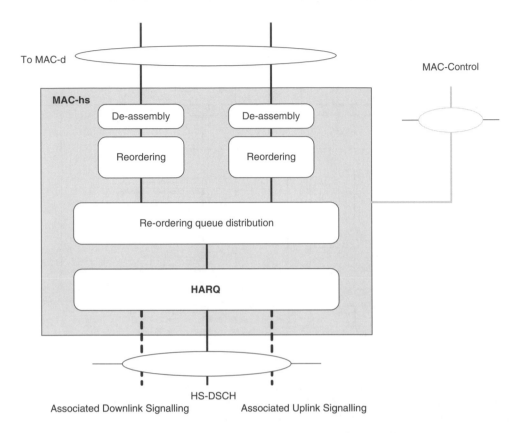

Figure 6.10   UE side MAC architecture/MAC-hs details

- The following is allowed:
  — One MAC-hs PDU contains only MAC-d PDUs with the same priority, and from the
    same MAC-d flow.
  — Different MAC-d PDU sizes can be supported in a given MAC-hs PDU (Protocol Data
    Unit).

The UE side MAC architecture/MAC-hs details are illustrated in Figure 6.10.

6.3.5 MAC-ehs Entity

The model for MAC-ehs entity comprises the following entities:

- *HARQ:* The HARQ entity is responsible for handling the HARQ protocol. There should be
  one HARQ process per HS-DSCH per TTI for single stream transmission and two HARQ
  processes per HS-DSCH per TTI for dual stream transmission. The HARQ functional entity
  handles all the tasks that are required for hybrid ARQ. It is responsible for generating ACKs
  or NACKs. The detailed configuration of the hybrid ARQ protocol is provided by RRC over
  the MAC-Control SAP.

- *Disassembly:* The disassembly entity disassembles the MAC-ehs PDUs.
- *Reordering queue distribution:* The reordering queue distribution function routes the received MAC-hs SDUs or segments of MAC-ehs SDUs to correct reordering queues based on the received logical channel identifier.
- *Reordering:* The reordering entity organizes received MAC-ehs SDUs or segments of MAC-ehs SDUs according to the received TSN. Data blocks with consecutive TSNs are delivered to higher layers upon reception. A timer mechanism determines delivery of non-consecutive data blocks to higher layers. There is one reordering entity for each priority class.
- *Logical Channel Identifier (LCH-ID) demultiplexing:* The demultiplexing entity routes the MAC-ehs SDUs or segments of MAC-ehs SDUs to the correct reassembly entity based on the received logical channel identifier.
- *Reassembly:* The reassembly entity reassembles segmented MAC-ehs SDUs to MAC PDUs and forwards the MAC PDUs to upper layers.
- The following is allowed:
  - The MAC-ehs SDUs included in a MAC-ehs PDU can have a different size and a different priority and can be mapped to different MAC-d flows.

Figure 6.11 illustrates the UE side MAC architecture/MAC-ehs details.

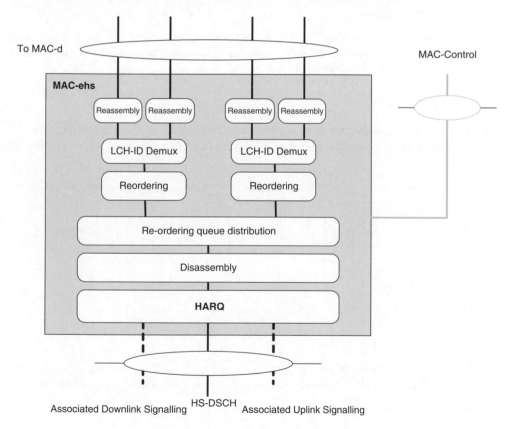

Figure 6.11 UE side MAC architecture/MAC-ehs details

6.4 HS-DSCH MAC Architecture—UTRAN Side

This section describes the modifications to the MAC model with respect to the Release 99 model for supporting the features for HS-DSCH on the UTRAN side. Both MAC-hs and MAC-ehs are responsible for handling the data transmitted to the HS-DSCH. Furthermore, they are responsible for the management of the physical resources allocated to HS-DSCH. Upper layers configure which of the two entries, MAC-hs or MAC-ehs, is to be applied to handle HS-DSCH functionality.

6.4.1 Overall MAC Architecture

New MAC functional entities, the MAC-hs and MAC-ehs that are located in the Node B, were added to the MAC architecture of Release 99. An HS-DSCH is assigned to the MAC-hs and MAC-ehs SDUs, that is MAC-d PDUs to be transmitted from MAC-c/sh to MAC-hs or MAC-ehs via the I_{ub} interface in the case of configuration with MAC-c/sh, or from MAC-d via I_{ur}/I_{ub} in the case of configuration without MAC-c/sh. Figure 6.12 illustrates the overall MAC architecture on the UTRAN side, MAC-hs or MAC-ehs. The MAC multiplexing chain for MAC-hs on the UTRAN side is illustrated in Figure 6.13.

6.4.2 MAC-c/sh Entity

The data for HS-DSCH is subject to flow control between the serving and the drift RNC. As shown in Figure 6.14, a new flow control function is included to support the data transfer between MAC-d and MAC-hs/MAC-ehs.

6.4.3 MAC-hs Entity

The MAC-hs is responsible for handling the data transmitted on the HS-DSCH. MAC-hs receives configuration parameters from the RRC layer via the MAC-Control SAP. There should

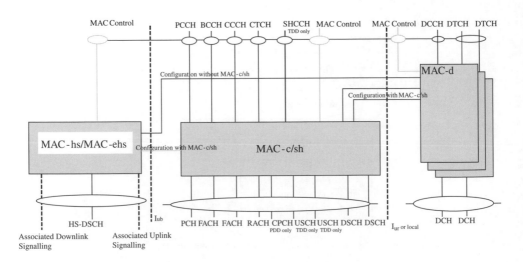

Figure 6.12   UTRAN side overall MAC architecture MAC-hs/MAC-ehs

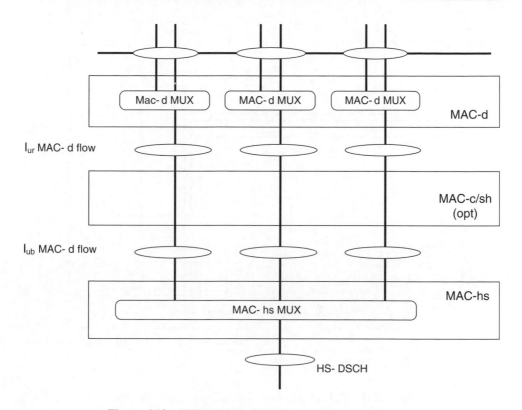

Figure 6.13 UTRAN side of MAC multiplexing for MAC-hs

be priority handling per MAC-d PDU in the MAC-hs. The MAC-hs is comprised of four different functional entities as described below:

- *Flow Control:* This is the companion flow control function to the flow control function in the MAC c/sh in the case of configuration with MAC-c/sh and MAC-d in the case of configuration without MAC-c/sh. Both entities together provide a controlled data flow between the MAC-c/sh and the MAC-hs or the MAC-d and the MAC-hs taking the transmission capabilities of the air interface into account in a dynamic manner. This function is intended to limit Layer 2 signaling latency and reduce discarded and retransmitted data as a result of HS-DSCH congestion. Flow control is provided independently per priority class for each MAC-d flow.
- *Scheduling/Priority Handling:* This function manages HS-DSCH resources between HARQ entities and data flows according to their priority class. Based on status reports from associated uplink signaling either new transmission or retransmission is determined when operating in CELL-DCH state. In FDD, when operating in CELL-FACH, CELL-PCH, and URA-PCH state, the MAC-hs can perform retransmission without uplink signaling. Further, it sets the priority class identifier and TSN for each new data block being serviced. To maintain proper transmission priority a new transmission can be initiated on a HARQ process at any time. The TSN (Transmission Sequence Number) is unique to each priority class within a HS-DSCH, and is incremented for each new data block. It is not permitted to

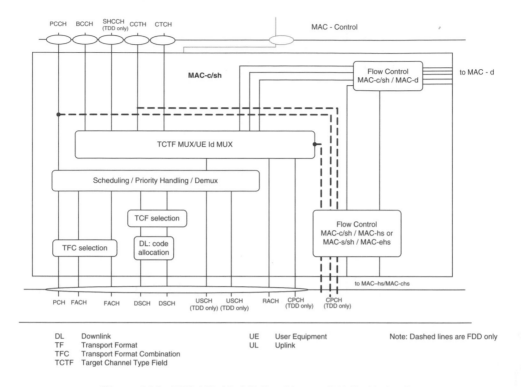

Figure 6.14   UTRAN side MAC architecture/MAC-c/sh details

schedule new transmissions, including retransmissions originating in the RLC layer, within the same TTI, along with retransmission originating from the HARQ layer.

• *HARQ:* One HARQ entity handles the hybrid ARQ functionality for one user. One HARQ entity is capable of supporting multiple instances (HARQ process) of stop-and-wait HARQ protocols. There should be one HARQ process per TTI.

• *Transport Format and Resource Combination (TFRC) Selection:* This entity performs the selection of an appropriate TFRC for the data to be transmitted on HS-DSCH.

Figure 6.15 depicts the MAC architecture/MAC-hs details on the UTRAN side.

6.4.4 MAC-ehs Entity

MAC-ehs on the UTRAN side receives configuration parameters from the RRC layer via the MAC-control SAP. There should be priority handling per MAC-ehs SDU in the MAC-ehs. The MAC-ehs comprises of the following six different functional entities:

• *Flow Control:* The flow control for MAC-ehs is identical to the flow control for MAC-hs described above.

• *Scheduling/Priority Handling:* This function manages HS-DSCH resources between HARQ entities and data flows according to their priority class. In FDD, the scheduler determines for each TTI whether single or dual stream transmission should be used. Based on status reports

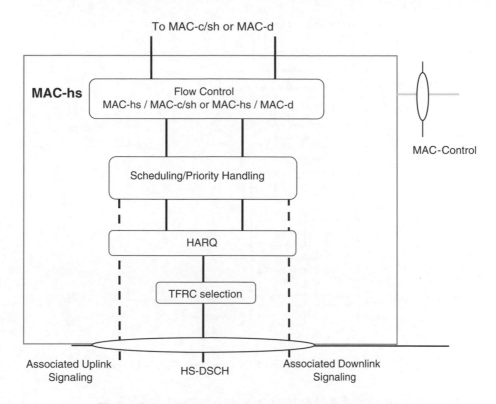

Figure 6.15   UTRAN side MAC architecture/MAC-hs details

from associated uplink signaling, either new transmission or retransmission is determined when operating in CELL-DCH state. In FDD, when operating in CELL-FACH, CELL-PCH, and URA-PCH state HS-DSCH reception, the MAC-ehs can perform retransmission without uplink signaling. Furthermore, it sets the logical channel identifiers and TSNs for each new data block being serviced. To maintain proper transmission priority, a new transmission can be initiated on a HARQ process at any time. The TSN is unique to each priority class within a HS-DSCH. It is not permitted to schedule new transmissions, including retransmissions originating in the RLC layer, within the same TTI, along with retransmissions originating from HARQ layer.

- *HARQ:* One HARQ entity handles the hybrid ARQ functionality for one user. One HARQ entity is capable of supporting multiple instances (HARQ process) of stop-and-wait HARQ protocols. There should be one HARQ process per TTI for single stream transmission and two HARQ processes per TTI for dual stream transmission.
- *TFRC Selection:* The TFRC selection for MAC-ehs is identical to the TFRC selection of the MAC-hs described above.
- *LCH-ID MUX:* This function determinates the number of octets to be included to MAC-ehs PDU from each logical channel based on the scheduling decision and available TFRC.
- *Segmentation:* This function performs necessary segmentation of MAC-ehs SDUs.

Figure 6.16 illustrates the MAC architecture/MAC-ehs details on the UTRAN side.

to MAC-c/sh or MAC-d

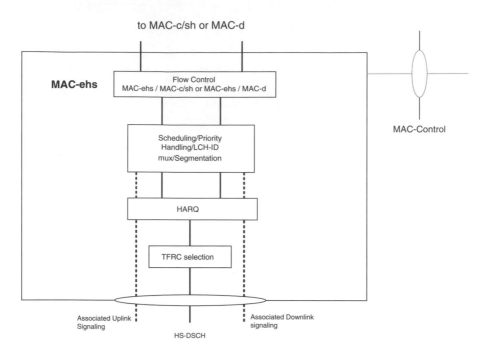

Figure 6.16   UTRAN side MAC architecture/MAC-ehs details

6.5 Overview of HSDPA Techniques to Support UTRA

The HSDPA enhancement technology applied to UTRA is based on techniques such as AMC, HARQ, FCS, MIMO antenna processing, and other features to increase throughput, reduce delay, and achieve high pack rates.

6.5.1 Adaptive Modulation and Coding (AMC)

Adaptation of the transmission parameters against the changing channel conditions is benefi-cial to the implementation of reliable communications. Link adaptation is generally known as the process of modifying the transmission parameters to compensate for variable channel conditions. Another method that falls under this category of link adaptation is the Adaptive Modulation and Coding (AMC) technique. In other words, the AMC principle is to change the modulation and coding format in accordance with the changing channel conditions.

In a system with AMC, users close to the cell site in favorable positions are typically assigned higher order modulation and code rates (i.e. 64 QAM and $R = 3/4$ turbo codes), whereas users close to the cell boundary in unfavorable positions are assigned lower order modulation code rates (i.e. QPSK with $R = 1/2$ turbo codes). Thus, the main benefits of AMC are that it enables users in favorable positions to increase the average throughput of the cell with reduced interference due to the variations in the modulation and coding scheme. AMC is also effective when combined with fat-pipe scheduling techniques such as those enabled by DSCH.

HS-DSCH with AMC and HARQ can provide substantially higher peak rates and average throughput than Release 99 DSCH. Although the use of the AMC scheme is feasible, the sophisticated complex demodulation techniques are still required that lead to higher receiver complexity compared to a Release 99 UE. It may also need a more refined synchronizing tracking mechanism and more sophisticated channel estimation means, required by utilization of 64 QAM.

The implementation of AMC offers several challenges. First, AMC is sensitive to measurement error and delay. In order to select the appropriate modulation, the scheduler must be aware of the channel quality. Errors in the channel estimate will cause the scheduler to select the wrong data rate and either transmit it at too high a power, wasting system capacity, or too low a power, raising the block error rate. Delay in reporting channel measurements also reduces the reliability of the channel quality estimate due to the constantly varying mobile channel. Furthermore, changes in the interference add to the measurement errors. Hybrid ARQ enables the implementation of AMC by reducing the number of required MCS (Modulation and Coding Scheme) levels and the sensitivity to measurement error and traffic fluctuations.

6.5.2 Hybrid ARQ (HARQ)

HARQ is an implicit link adaptation technique. The link adaptation in AMC is used to set the modulation and coding format for improving system capacity. In HARQ, link layer acknowledgements are used for retransmission decisions.

There are many schemes for implementing HARQ: Chase combining (also called HARQ-type-III with one redundancy version), rate compatible punctured turbo codes, and incremented redundancy. HARQ-type-III belongs to the class of incremental redundancy ARQ schemes. However, with HARQ-type-III, each transmission is self-decodable, which is not the case with HARQ-type-II. HARQ-type-II (also incremental redundancy) is another implementation of the HARQ technique wherein instead of sending simple repeats of the entire coded packet, additional redundant information is incrementally transmitted if the decoding fails on the first attempt. Chase combining involves the retransmission by the transmitter of the same coded data packet. The decoder at the receiver combines multiple copies of the transmitted packet weighted by the received SNR. Thus, diversity gain is obtained. In the HARQ-type-III with multiple redundancy version, different puncture bits are used in each retransmission.

AMC by itself does provide some flexibility to choose an appropriate MCS for the channel conditions, either based on measurement of UE measurement reports or network determination. However, an accurate measurement is required and there is an effect of delay. However, an ARQ mechanism is still required. HARQ autonomously adapts to the instantaneous channel conditions and is insensitive to error and delay. Combining AMC with HARQ leads to the best results: AMC provides the coarse data rate selection, while HARQ provides for fine data rate adjustment based on channel conditions.

The choice of HARQ mechanism is also important. Window based Selective Repeat (SR) is a common type of ARQ protocol employed by many systems including RLC R99. SR is generally insensitive to delay and has the favorable property of repeating only those blocks that have been received in error. To accomplish this feat, the SR ARQ transmitter must employ a sequence number to identify each block it sends. SR may fully use the available channel capacity by ensuring that the maximum block sequence number (MBSN) exceeds the number of blocks transmitted in one round trip feedback delay. The greater the feedback delay the longer the maximum sequence number must be. However, when HARQ is partnered with SR, two difficulties in particular are observed:

- UE memory requirements are high. A large MBSN requires significant storage in the UE adding to the unit's cost.
- Hybrid ARQ requires the receiver to determine reliably the sequence number of each transmission.

Stop-and-wait is one of the simplest forms of ARQ requiring very little overhead. In stop-and-wait, the transmitter operates on the current block until the block has been received successfully. Protocol correctness is ensured with a simple one-bit sequence number that identifies the current or the next block. As a result, the control overhead is minimal. Acknowledgement overhead is also minimal, because the indication of a successful/unsuccessful decoding using ACK/NACK may be signaled concisely with a single bit. Furthermore, because only a single block is in transit at a time, memory requirements at the US are also minimized. Therefore, HARQ using a stop-and-wait mechanism offers significant improvements by reducing the overall bandwidth required for signaling and the UE memory. However, one major drawback exists: acknowledgements are not instantaneous and therefore after every transmission, the transmitter must wait to receive the acknowledgement prior to transmitting the next block. This is a well-known problem with stop-and-wait ARQ. In the interim, the channel remains idle and system capacity is wasted. In a slotted system, the feedback delay will waste at least half the system capacity while the transmitter is waiting for acknowledgements. As a result, at least every other timeslot must go idle even on an error free channel.

N channel stop-and-wait HARQ offers a solution by parallelizing the stop-and-wait protocol and in effect running a separate instantiation of the HARQ protocol when the channel is idle. As a result, no system capacity goes wasted because one instance of the algorithm communicates a data block on the forward link at the same time that another communicates an acknowledgement on the reverse link. However, the receiver has to store N blocks for this scheme.

As the HSDPA study item, an AMC scheme using 7MCS levels were simulated using a symbol level link simulator. The AMC scheme uses QPSK, 8-PSK, and 16/64 QAM modulation using $R = 1/2$ and $R = 3/4$ turbo code and can support a maximum peak data rate of 10.8 Mbps.

The simplest form of HARQ scheme was proposed by Chase. The basic idea in Chase's combining scheme (also called HARQ-type-III) is to send a number of repeats of each coded data packet and allow the decoder to combine multiple received copies of the coded packet weighted by the SNR prior to decoding. This method provides diversity gain and is very simple to implement.

Incremental redundancy is another HARQ technique wherein instead of sending simple repeats of the entire coded packet, additional redundant information is incrementally transmitted if the decoding fails on the first attempt. Incremental redundancy is called HARQ-type-II, or HARQ-type-III if each retransmission is restricted to be self-decodable.

6.5.3 Fast Cell Selection

Fast Cell Selection (FCS) has been proposed for HSDPA. Using FCS, the UE indicates the best cell that should serve it on the downlink, through uplink signaling. Thus, although multiple cells may be members of the active set, only one of them transmits at any time, potentially decreasing interference and increasing system capacity.

The effect of FCS on throughput was studied using a dynamic system simulation tool for a data only system. The system simulator tool models Rayleigh and Rician channel fading, time evolution with discrete steps (0.66 ms for example), AMC, fast HARQ, FCS, and open loop transmits diversity (STTD).

FCS improves throughput and residual FER (Frame Error Rate) for UE in multi-coverage regions. This is because a UE in a multi-coverage region typically has a weaker channel to any single serving cell compared to UEs closer to their serving cell. With FCS, the multi-coverage UE has more opportunities to select a better link to one of the serving cells and be scheduled. The overall system benefit due to FCS is more significant with fair schedulers compared to maximum C/I schedulers, because the users with a weak link are scheduled more often. With a maximum C/I scheduler the larger the load the less impact FCS has on performance. Without FCS, it takes longer for UEs with weak links to finish a packet call and hence longer to release the dedicated control channel, which results in further overhead and reduced system capacity. Open issues include how much larger the FCS benefit is with motion and allowing the MCS changes between retransmissions.

FCS allows a UE to choose rapidly in its active set for downlink transmission. The potential benefit is that for each frame interval the active set cell with the best faded link can be chosen for frame transmission to the UE. The UE chooses the best cell by comparing each active set cell's estimated Common Pilot Channel (CPICH) E_c/I_o and transmits a cell indicator to be detected by the desired cell on an uplink dedicated control channel. Frame retransmissions can therefore take place by any active set cell if chosen and the resulting received signal energy from each frame is accumulated to model a Chase combining process. Note that the active set evolves with time as a UE's position changes.

As part of the AMC schemes proposed for HSDPA, the UE also estimates CPICH E_c/N_t (C/I) for each cell in its pilot active set for the current slot. The pilot E_c/N_t information for the selected cell is then also fed back on an uplink dedicated control channel. The Node B then uses the C/I estimate to determine the modulation and coding level for that user's subsequent frame and possibly also for setting scheduling priorities. The cell selection update is once very 3.33 ms. The FCS and CPICH measurement delay is about 10 ms, as shown in Figure 6.17.

6.5.4 Multiple Input Multiple Output Antenna Processing

Diversity techniques based on the use of multiple downlink transmit antennas are well known and these diversity techniques have been applied in the UTRA Release 99 specification. Such techniques exploit spatial and/or polarization decorrelations over multiple channels to achieve fading diversity gains.

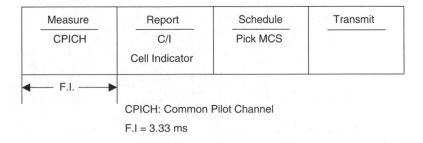

Figure 6.17   Time diagram of FCS and CPICH measurement delay

Multiple Input Multiple Output (MIMO) processing employs multiple antennas at both the base station transmitter and terminal receiver, providing several advantages over transmit diversity techniques with multiple antenna only at the transmitter and over conventional single antenna systems. If multiple antennas are available both at the transmitter and receiver, the peak throughput can be increased using a technique known as code reuse. With code reuse, each channelization/scrambling code pair allocated for HS-DSCH transmission can modulate up to M distinct data streams, where M is the number of transmit antennas. Data streams that share the same channelization/scrambling code must be distinguished based on their spatial characteristics, requiring a receiver with at least M antennas. In principle, the peak throughput with code reuse is M times the rate achievable with a single transmit antenna. Some intermediate data rates with code reuse can be achieved with a combination of code reuse and smaller modulation constellations of 16 QAM instead of 64 QAM. Compared with a single antenna transmission scheme with a larger modulation constellation to achieve the same rate, the code reuse technique may have a smaller required E_b/N_o, resulting in overall improved system performance.

The technique discussed so far is an *open-loop* MIMO technique, because the Node B transmitter does not require feedback from the UE other than the conventional HSDPA information required for rate determination. Further performance gains can be achieved using *closed-loop* MIMO techniques whereby the Node B transmitter employs feedback information from the UE. For example, with knowledge of channel realizations, the Node B can transmit on orthogonal eigen modes, eliminating the spatial multi-access interference.

In a conventional single antenna HSDPA transmission, a set of N downlink physical channels (codes) is shared among many users. Using an open-loop MIMO architecture with M transmit antennas, the same set of codes is used, but each code is reused M times and each modulates distinct data substreams. More specifically, a high-rate data source is coded, rate-matched, and interleaved. As shown by the block diagram of a MIMO transmitter in Figure 6.18, the coded data stream is then demultiplexed into MN substreams, and the nth group ($n = 1, 2, \ldots, N$) of M substreams is spread by the spreading code. The mth substream ($m = 1, 2, \ldots, M$) of each group is summed and transmitted over the mth antenna so that the substreams sharing the same code

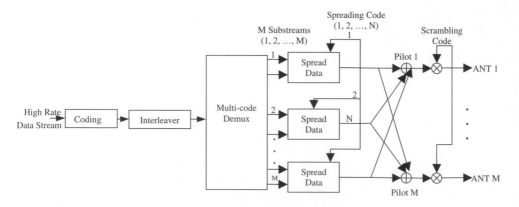

Figure 6.18   Block diagram of a MIMO transmitter (© 2007. 3GPP™ TSs and TRs are the property of ARIB, ATIS, CCSA, ETSI, TTA and TTC who jointly own the copyright in them. They are subject to further modifications and are therefore provided to you "as is" for information purposes only. Further use is strictly prohibited.)

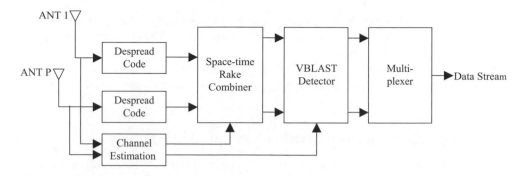

Figure 6.19   Block diagram of a MIMO receiver (© 2007. 3GPP™ TSs and TRs are the property of ARIB, ATIS, CCSA, ETSI, TTA and TTC who jointly own the copyright in them. They are subject to further modifications and are therefore provided to you "as is" for information purposes only. Further use is strictly prohibited.)

are transmitted over different antennas. Mutually orthogonal dedicated pilot symbols are also added to each antenna's CPICH to allow for coherent detection.

To distinguish the M substreams sharing the same code, the UE uses multiple antennas and spatial signal processing. A representative MIMO receiver with P antennas is shown in Figure 6.19. For coherent detection at the UE, complex amplitude channel estimates are required for each Tx/Rx antenna pair.

The purpose of MIMO techniques in UTRA is to improve system capacity and spectral efficiency by increasing the data throughput in the downlink within the existing 5 MHz carrier. This is achieved by deploying multiple antennas at both the UE and the Node-B side. The antenna configurations at the Node-B and UE, with regard to the number of antennas and antenna spacing/polarization, are important items to be considered for MIMO effects in UTRA.

If the notation (x, y) denotes a system with x Node B antennas and y UE antennas, the following antenna configurations covering $(1, 1), (1, 2), (2, 1), (2, 2), (1, 4), (4, 1), (2, 4), (4, 2)$, or $(4, 4)$ should be supported by the proposed MIMO technique.

The operation of a MIMO technique should be described in sufficient detail to determine straightforwardly the changes necessary to UTRA to include the technique. Detailed descriptions of aspects that are specific to the technique should be provided, including transmit and receive algorithms, physical layer signaling, and control. Higher-level signaling on both uplink and downlink should be described. An analysis of its complexity should be provided compared to existing solutions in terms of RF complexity, memory requirements, UE size, computational complexity, hardware reusability, and signaling requirements. MIMO techniques should demonstrate significant incremental gain over the best performing systems supported in the current release with reasonable complexity. The operation of MIMO techniques should be described in sufficient detail to enable realistic calibration and system level performance studies such as delay, channel estimation error, signaling error, and pilots.

6.5.5 Handling for Error Cases

The following list describes the most frequent error handling cases:

- If NACK is detected as an ACK, the network starts afresh with new data in the HARQ process. The data block is discarded in the network and lost. Retransmission is left up to higher layers.

- If ACK is detected as NACK, the transmitter at the network will send an abort indicator by incrementing the New Packet Indicator, and the receiver at the UE will continue to process the data block as in the normal case.
- If a CRC error on the HS-SCCH is detected, UE receives no data and sends out no status report. When the absence of status report is detected, the network can retransmit the data block.

6.6 Orthogonal Frequency Division Multiplexing (OFDM)

Frequency Division Multiplex (FDM) transmits multiple signals simultaneously over a single transmission channel and each signal travels over its own 5 MHz carrier, which is individually modulated by the data stream. FDM thus divides the bandwidth into many channels so that each channel is allocated to a single user in a multi-user environment.

Orthogonal Frequency Division Multiplexing (OFDM) is a multicarrier transmission technique. OFDM transmits the data stream over a large number of subcarriers that are closely spaced at their precise frequencies. Thus, OFDM divides the frequency available into many closely-spaced carriers, modulated individually by low-rate data streams. This orthogonal spacing prevents the demodulators from recognizing frequencies other than their own. In this sense, OFDM is similar to FDM, but the difference lies in the fact that the carriers chosen in OFDM are much more closely spaced than in FDM. Therefore, the benefits of OFDM are high spectral efficiency and lower multipath distortion. The OFDM spread spectrum technique is useful in a typical terrestrial broadcasting scenario because the transmitted signal arrives at the receiver via various paths of different length through multipath-channels.

Multiple Input Multiple Output-Orthogonal Frequency Division Multiplexing (MIMO-OFDM) is a technology that uses multiple antennas to transmit and receive radio signals. MIMO-OFDM allows service providers to deploy a Broadband Wireless Access (BWA) system that has Non-Line-of-Sight (NLOS) functionality. Particularly, MIMO-OFDM takes advantage of the multipath properties of environments using base station antennas that do not have Line-of-Sight (LOS).

A number of parameters have to be considered in OFDM system design: the number of subcarriers, guard time, symbol duration, subcarrier spacing, modulation type per subcarrier, and the encoder type for error correction. These design issues will be influenced by the system requirements such as available bandwidth, required bit rate, tolerable delay spread, and Doppler values.

6.6.1 OFDM Modulation Scheme

An OFDM signal consists of a sum of subcarriers that are modulated by using Phase Shift Keying (PSK) or Quadrature Amplitude Modulation (QAM).

Figure 6.20 shows the schematic diagram of an OFDM modulator. One OFDM data symbol starting at $t = t_s$ can be written as

$$s(t) = \sum_{k=-N_{s/2}}^{N_{s/2}-1} d_{k+N_{s/2}} \exp\left(j2\pi \frac{k}{T}(t-t_s)\right) \quad \text{for } t_s \leq t < t_s + T$$

$$s(t) = 0 \quad \text{for } t < t_s \text{ and } t > t_s + T$$

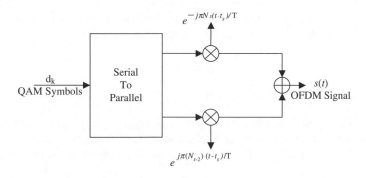

Figure 6.20   Schematic diagram of OFDM modulator

where d_k = QAM symbols, f_i = carrier frequency, N_s = number of subcarriers, t_s = symbol starting time, T = symbol duration, and $s(t)$ = OFDM signal.

In this equation, the real and imaginary parts correspond to the in-phase and quadrature parts of the OFDM signal, which have to be multiplied by a cosine and sine of desired carrier frequency to produce the final OFDM signal.

The spectrum required should be chosen based on the input data and modulation scheme (BPSK, QPSK, or QAM). Amplitudes and phases of the carriers are evaluated based on the chosen modulation scheme. The required spectrum is then converted back to its time-domain signal by employing Inverse Fourier Transform (IFT) algorithms such as IFFT (Inverse Fast Fourier Transform). In order to ensure robustness against multipath delay spread, a guard interval between successive symbols should be added to the transmitted symbol for minimizing intersymbol interference. The most effective method against multipath intersymbol interference is the addition of a cyclic prefix, which is a copy of the last part (cyclic postfix) of the OFDM symbol. A cyclic prefix should be precoded to the transmitted symbol.

A mathematical model of a baseband OFDM system depending on the continuous-time OFDM model or the discrete-time OFDM model is now considered.

6.6.1.1 Continuous Time OFDM Model

Consider an OFDM system with N carriers, a bandwidth of W Hz and a symbol length of T seconds, of which T_{cp} seconds is the length of the cyclic prefix. Starting with the waveforms $\varphi_k(t)$ at the transmitter, the mathematical waveforms are modeled as

$$\varphi_k(t) \quad = \frac{1}{\sqrt{T-T_{cp}}} e^{j2\pi \frac{W}{T} k(t - T_{cp})} \quad \text{for all } t \in [0, T]$$

$$= 0 \text{ otherwise}$$

$$\text{where } T = \frac{N}{W} + T_{cp}$$

In fact, $\varphi_k(t) = \varphi_k\left(t + \frac{N}{W}\right)$ is true when t is within the cyclic prefix. The waveforms $\varphi_k(t)$ are used in the modulation and transmitted baseband signal such that

$$s_\lambda(t) = \sum_{k=0}^{N-1} x_{k,\lambda}\, \varphi_k(t - \lambda T)$$

where λ is OFDM symbol number and $x_{k,\lambda} = (x_{0,\lambda}, x_{1,\lambda}, \ldots, x_{n-1,\lambda})$ represents input signal symbols obtained from a set of signal constellation points. Considering an infinite sequence of OFDM symbols, the output signal, $s(t)$, from the transmitter will be expressed as

$$s(t) = \sum_{t=-\infty}^{\infty} s_\lambda(t) = \sum_{t=-\infty}^{\infty} \sum_{k=0}^{N-1} x_{k,\lambda}\, \varphi_k(t - \lambda T)$$

If the effect of the impulse response of physical channel, $h(\tau)$, is restricted to the time period $\tau \in [0, T_{cp}]$, then the received signal becomes

$$r(t) = h(\tau) * s(t) = \int_{0}^{T_{cp}} h(\tau, t) s(t - \tau) d\tau + n(t)$$

where $*$ denotes the cyclic convolution and $n(t)$ is additive white Gaussian noise.

At the receiver, a filter bank to be matched to the last part $[T_{cp}, T]$ of the transmitter waveforms $\varphi_k(t)$ will remove the cyclic prefix effectively:

$$\Phi_k(t) = \varphi_k(T - t) \quad \text{for all } t \in [0, T - T_{cp}]$$
$$= 0 \text{ otherwise}$$

The sampled output obtained at the kth matched filter is calculated as

$$y_k = r(t) * \Phi_k(t)|_{t=T} = \int_{-\infty}^{\infty} r(t)\Phi_k(T - t) dt$$

The channel, denoted by $h(\tau)$, fixed over the OFDM symbol interval will be expressed after simplification:

$$y_k = H\left(\frac{kW}{N}\right) x_k + n_k$$

where $H(f)$ designates the Fourier transform of $h(\tau)$ and n_k is additive white Gaussian noise.

6.6.1.2 Discrete Time OFDM Model

$\varphi_k(t)$ is a rectangular pulse modulated on the carrier frequency $k\left(\frac{W}{N}\right)$. According to the Fourier Transform theorems, the rectangular pulse shape will lead to a $\sin(x)/x$ sinc type of spectrum of the subcarriers. $\Phi_k(t)$ is the sampled output obtained at the receiver filter bank. The integrals in the continuous-time domain should be changed to summations when in the discrete-time domain. Also, the modulation and demodulation in the continuous-time OFDM model are replaced by the Inverse Discrete Fourier Transform and Discrete Fourier Transform respectively, and the channel is a Discrete-Time convolution. The whole Discrete-Time OFDM system can be expressed by the following equation:

$$\begin{aligned} y_\lambda &= \mathrm{DFT}(\mathrm{IDFT}(x_\lambda) * (h_\lambda + x_\lambda)) \\ &= \mathrm{DFT}(\mathrm{IDFT}(x_\lambda) * h_\lambda) + \mathrm{DFT}(x_\lambda) \end{aligned}$$

where $y_\lambda = N$ received data points, $x_\lambda = N$ transmitted constellation points, $h_\lambda =$ channel impulse response, and $n_\lambda =$ channel noise.

Since the channel noise n_λ is assumed to be white Gaussian, $n'_\lambda = \mathrm{DFT}(n_\lambda)$ also represents uncorrelated Gaussian noise; and because the DFT of two cyclically convolved signals is generally equivalent to the product of their individual DFTs, the equation expressed above can be written as

$$y_\lambda = x_\lambda \cdot \mathrm{DFT}(h_\lambda) + n'_\lambda$$

where the symbol "$\cdot$" denotes element-by-element multiplication and $n'_\lambda = \mathrm{DFT}(n_\lambda)$ is uncorrelated Gaussian noise.

In order for the orthogonality to be preserved, the receiver and transmitter should be perfectly synchronized. Synchronization is a key issue in the design of a robust OFDM receiver. Time and frequency synchronization are paramount for identifying the start of the OFDM symbol and aligning the modulators' and demodulators' local oscillator frequencies. If any of these synchronization tasks is not performed with sufficient accuracy, the orthogonality of the subcarriers is lost. To eliminate intersymbol interference (ISI) almost completely, a guard time is introduced for each OFDM symbol. The guard time should be chosen larger than the expected delay spread, so that multipath components from one symbol would not interfere with the next symbol. The guard time may consist of no signal at all. In that case, the problem of intercarrier interference (ICI) arises. ICI is defined as crosstalk between different subcarriers, which means they are no longer orthogonal. To eliminate ICI, the OFDM symbol should be cyclically extended in the guard time. This ensures that delayed replicas of the OFDM symbols always have an integer number of cycles within the FFT interval, as long as the delay is smaller than the guard time. As a result, multipath signals with delays smaller than the guard time cannot cause ICI.

The equalization for symbol demapping required for detecting the data constellations is an element-wise multiplication of the DFT output by the inverse of the estimated channel estimation.

The frequency-selective radio channel may severely attenuate the data symbols transmitted to one or several subchannels, leading to bit errors. Spreading the coded bit over the bandwidth of the transmitted system, and an efficient coding scheme, can correct for the erroneous bits and thereby exploit the wideband channel's frequency diversity. OFDM systems using error-correction coding are often referred to as coded OFDM systems.

A rectangular pulse has in general a very large bandwidth due to the sidelobs of its Fourier Transform being a sinc function. Windowing is a well-known technique to reduce the level of these sidelobes and consequently reduce the signal power transmitted out of band. In an OFDM system, the applied window should not influence the signal during its effective period. In fact, cyclically extended parts of the symbol are pulse shaped. This additional cyclic prefix extends the guard interval (GI) to some extent so that the delay-spread robustness is slightly enhanced. On the other hand, the efficiency will be further reduced because the window part is discarded by the receiver. The orthogonality of the subcarriers of the OFDM is restored by the rectangular receiver filter implemented by the DFT, requiring the correct estimation of the DFT start time kT, where T denotes the OFDM symbol period.

6.6.2 Signal Processing Over OFDM Transceiver

Figure 6.21 illustrates an OFDM modem system that consists of the upper path, the center block, and the lower path. The upper path is the transmitter chain and the lower path corresponds to the receiver chain. The center block modulates a block of QAM values onto a number of subcarrier.

The previous section describes how the basic OFDM signal is formed using the IFFT, adding a cyclic extension and performing windowing to get a steeper spectral rolloff. In order to build a

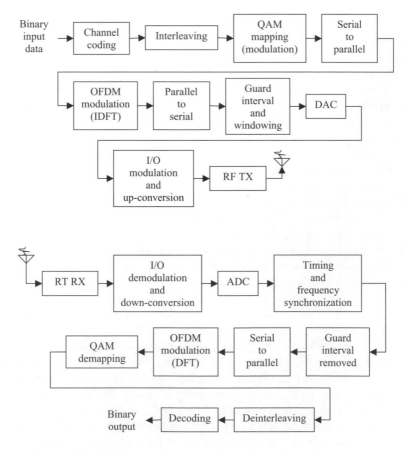

Figure 6.21   OFDM transceiver for simplex point-to-point transmission

complete OFDM modem, the subcarrier in the receiver should be demodulated by a Fast Fourier Transform (FFT), which performs the reverse operation of an IFFT in the transmitter. Referring to Figure 6.21, binary input data is first encoded by a forward error correction code. The encoded data is then interleaved and mapped onto QAM values. In the receiver path, after passing the RF (Rx) part and the analog-to-digital conversion (ADC), the digital signal processing starts with a training phase to determine symbol timing and frequency offset. An FFT is used to demodulate all subcarriers. The output of the FFT contains N_s QAM values, which are mapped onto binary values and decoded to produce binary output data. To map the QAM values onto binary values successfully, first the reference phases and amplitudes of all subcarriers have to be acquired. Note also that the IDFT and DFT are used in an OFDM transceiver for modulating and demodulating the data constellations on the orthogonal subcarriers. The signal processing algorithms replace the bank of I/Q-modulators and demodulators. At the input of the IDFT, N data constellation points $\{x_{k,\lambda}\}$ are present, where N is the number of DFT points (k is an index on the SC and λ is an index on the OFDM symbol).

6.7 Prospect of OFDM-based Applications

Existing 3G mobile broadband technologies, including WCDMA/HSPA and CDMA2000 1x EV-DO, use CDMA technology as the core radio access. Since their first launch in 2000, 3G cellular technologies have evolved to deliver high capabilities, throughputs, and efficiencies to support growing usage of broadband-intensive data services. The next-generation of IMT 2000 systems based on CDMA and OFDM technologies, along with OFDM-based broadcast technologies, will be key enablers of this transition. With significant momentum of markets and large scale of economics, 3G CDMA technologies will continue to be the leading platform for mobile communications and broadband services for the next generation.

A single operator's ability to deliver voice, video, broadband Internet access, position location by GPS, and other mobile data services will maximize its revenue opportunity across a diverse subscriber base. However, it is impossible for these services to fulfill all these multi-varied requirements with a single telecommunications delivery mechanism. In order for service providers to deliver large revenue-generating services across different types of networks, including concurrent voice and data services, seamlessly and with greater flexibility, they need the combination of network and device convergence.

In narrow bandwidth allocations up to 5 MHz, CDMA2000 1x EV-DO Revision A in 2006 and multi-carrier EV-DO Revision B in 2008, HSPA and HSPA+ can achieve some of the highest data throughputs possible in a given amount of spectrum. With wider radio channels more than 10 MHz, OFDM-based technologies such as UMB, LTE, WiBro, and Mobile WiMax have emerged as viable options for delivering wider-bandwidth mobile broadband services.

Since both 3G CDMA and OFDM-based technologies use different air interfaces and antenna techniques to achieve their respective performance characteristics, their combined technologies will provide the spectral efficiencies, network capabilities, and latencies necessary to support the mobile television and rich broadband services of the future.

For service providers who want to offer mobile broadcast or multicasting services, OFDM-based technologies such as DMB, DVB-H, or MediaFlo are better suited to simultaneously delivering rich multimedia content from one base station transceiver to many people in a single sector. In this case, the supplementary OFDM-base broadcast network will overlay the 2G/3G mobile network to deliver premium video content and television-like services. For service providers with existing or a limited amount of spectrum, CDMA-based technologies are the best option because they are more spectrally efficient in bandwidths up to 5 MHz. For a service provider that has access to a large amount of bandwidth more than 10 MHz, OFDM-based technologies may be a suitable option for introducing new bandwidth-intensive broadband services or complementing existing 2G or 3G solutions with additional broadband capacity in densely-populated metro-zones. At bandwidth scales beyond 2×5 MHz FDD or 10 MHz TDD, OFDM-based technologies offer a simpler implementation than CDMA technologies. Outside of high traffic metro areas, OFDM-based systems are not economical because the spectrum and network will most likely remain under utilized. It is expected that the coexistence of CDMA and OFDM-based solutions will persist well beyond 2020. Until then, 3G CDMA-based solutions will remain the core business for hundreds of operators. Once the adoption criteria are met, the 3–4 billion wireless subscribers in the world will begin migrating to the new generation of wireless technologies.

Present wireless technologies include 1G Analog, cdmaOne IS-95, GSM, GPRS, EDGE, WCDMA, HSDPA/HSPA, CDMA2000 1x EV-DO and 1x EV-DV for wide area network access, Wi-Fi for the local area network access, GPS for location-based services, NFC (Near Field Communication) for mobile commerce, and Bluetooth for personal area network connectivity. By the end of 2009, it is hoped that some multimode worldwide devices will include WCDMA/HSPA, UMB, LTE, WiBro, and Mobile WiMAX.

OFDM-based solutions will be built out over time as the demand for broadband services grows and spectrum becomes available. Mass adoption of these wide-bandwidth OFDM-based solutions will take years, as coverage is expanded and economics are built in larger scale. Meanwhile, CDMA2000 systems will continue to be the core business for hundreds of operators for well over a decade and play a key role in the future of the wireless industry.

7

CDMA2000 1x High Rate Packet Data System (1xEV-DO)

CDMA2000 is the registered trademark for the technical nomenclature of certain specifications and standards of 3GPP2. This air interface standard provides high rate packet data services. The technical requirements for CDMA2000 1x series ensure that a compliant access terminal can obtain service through any access network conforming to this standard.

7.1 Architectural Reference Protocol Model

The architectural reference model, as shown in Figure 7.1, includes the air interface between the access terminal (AT) and the access network (AN). Any AT can obtain service through any AN (typically the Internet). The protocols used over the air interface have been layered in Figure 7.2 as specified in this chapter. The layered architecture for the air interface consists of one or more protocols that provide the layer's functionality as shown below:

- *Application Layer:* This layer provides multiple applications that provide both the default signaling application and default packet application for transporting air interface protocol messages or user data.
  - *Default Signaling Application:* This application encompasses the Signaling Network Protocol (SNP) and the Signaling Link Protocol (SLP). The SNP provides message transmission services for signaling messages. The SNP uses the header to route the message to the appropriate protocol instance. The SLP provides message fragmentation mechanisms, along with reliable and best-effort delivery mechanisms for signaling messages. When used in the context of the default signaling application, SLP carries SNP packets.
  - *Default Packet Application:* The Radio Link Protocol (RLP) provides retransmission and duplicate detection for an octet data stream. The Location Update Protocol (LUP) defines location update procedures and messages in support of mobility management for the Default Packet Application. The Flow Control Protocol (FCP) defines flow control procedures to enable and disable the Default Packet Application data flow.

Mobile Communication Systems and Security   Man Young Rhee
© 2009 John Wiley & Sons (Asia) Pte Ltd

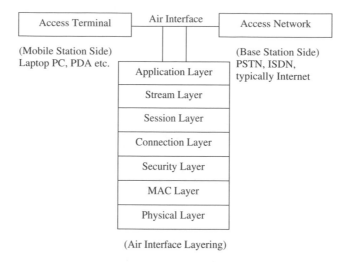

(Air Interface Layering)

Figure 7.1   Reference model architecture

- *Stream Layer:* This layer provides multiplexing of distinct application streams. The stream protocol adds the stream header to application packets prior to transmission; and after reception, removes the stream header and forwards application packets to the correct application.

  The default stream protocol provides four streams:
  — Stream 0 is dedicated to signaling and defaults to the Default Signaling Application.
  — Stream 1 defaults to the default packet service (RLP).
  — Streams 2 and 3 are not used by default.

  The generic virtual stream protocol provides 127 virtual streams to which applications can be bound. Applications bound to a stream send messages using the SNP. Applications bound to a virtual stream send messages using the stream layer.
- *Session Layer:* This layer provides address management, protocol negotiation, protocol configuration, and state maintenance services.
  — *Address Management Protocol:* This protocol provides access terminal identifier (ATI) management.
  — *Session Configuration Protocol:* This protocol provides negotiation and configuration of the protocols used in the session.
  — *Session Management Protocol:* This protocol provides the means to control the activation and deactivation of the address management protocol and the session configuration protocol. It also provides a session keep-alive mechanism.
- *Connection Layer:* This layer provides air link connection establishment and maintenance services.
  — *Air Link Management Protocol:* This protocol provides the overall state machine management that an AT and AN follow during a connection.
  — *Initialization State Protocol:* This protocol provides the procedures that an AT follows to acquire a network and that an AN follows to support network acquisition.

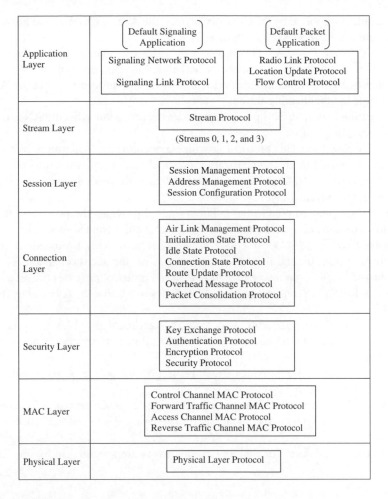

Figure 7.2   Default protocols defined for each one of the reference layers

— *Idle State Protocol:* This protocol provides the procedures that an AT and AN follow when a connection is not open.
— *Connected State Protocol:* This protocol provides the procedures that an AT and AN follow when a connection is open.
— *Route Update Protocol:* This protocol provides a means to maintain the route between the AT and AN.
— *Overhead Messages Protocol:* This protocol provides broadcast messages containing information that is mostly used by connection layer protocols.
— *Packet Consolidation Protocol:* This protocol provides transmit prioritization and packet encapsulation for the connection layer.

• *Security Layer:* This layer provides authentication and encryption services. The air interface supports a security layer, which can be used for authentication and encryption of access terminal traffic transported by the Control Channel, the Access Channel, the Forward Traffic Channel, and the Reverse Traffic Channel.

— *Key Exchange Protocol:* This protocol provides the procedures that the AN and AT follow to exchange security keys for authentication and encryption.

— *Authentication Protocol:* This protocol provides the procedures that the AN and AT follow for authenticating traffic.

— *Encryption Protocol:* This protocol provides the procedures that the AN and AT follow for encrypting traffic.

— *Security Protocol:* This protocol provides procedures for generation of a cryptosync that can be used by the authentication protocol and encryption protocol.

• *Medium Access Control (MAC) Layer:* This layer defines the procedure used to receive and to transmit over the physical layer.

— *Control Channel MAC Protocol:* This protocol provides the procedure that the AN follows to transmit and the AT follows to receive the control channel.

— *Access Channel MAC Protocol:* This protocol provides the procedure that the AT follows to transmit and the AN follows to receive the access channel.

— *Forward Traffic Channel MAC Protocol:* This protocol provides the procedures that the AN follows to transmit and the AT follows to receive the forward traffic channel (FTC).

— *Reverse Traffic Channel MAC Protocol:* This protocol provides the procedure that the AT follows to transmit and the AN follows to receive the reverse traffic channel (RTC).

• *Physical Layer:* This layer provides structure, frequency, power output, modulation, and encoding specifications for FTC and RTC.

— *Physical Layer Protocol:* This protocol provides channel structure, frequency, power output, and modulation specifications for the forward and reverse links.

The default protocols defined for each one of the layers are specified in Figure 7.2.

7.2 Air Interface Layering Protocol

The protocols and layers over the air interface are specifically defined in this chapter. Each layer in the reference model consists of one or more protocols that perform the layer's functionality. Default protocols defined for each one of the layers provide the means to carry messages between a protocol in one entry and the same protocol in other entry.

7.2.1 Application Layer Protocols

The application layer protocols consist of the default signaling application and the default packet application.

• *Default Signaling Application:* The default signaling application consists of the SNP and the SLP. The relationship between SNP and SLP is illustrated in Figure 7.3.

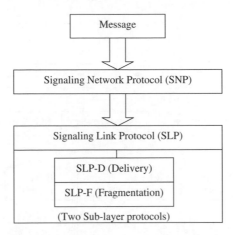

Figure 7.3   Default signaling layer protocols

— *Signaling Network Protocol (SNP):* The SNP provides message transmission services for signaling messages. Protocols that control each layer use SNP to exchange their messages between the peer-to-peer. SNP is also used by application specific control messages. SNP provides a single octet header that defines the Type of protocol and the protocol instance, such as InConfiguration or InUse, with which the message is associated. SNP is a message-routing protocol and uses the header to route the messages to the appropriate protocol instance.

The InConfiguration protocol field in the SNP header determines whether the encapsulated message corresponds to the InUse protocol instance or the InConfiguration protocol instance.

The actual protocol indicated by the Type is negotiated during session set-up. As an example, Type 0×01 is associated with the Control Channel MAC. The specific Control Channel MAC protocol used must be negotiated when the session is setup.

The protocol data unit for this protocol is an SNP packet. Each SNP packet consists of a SNP header and one message sent by a protocol using SNP. The protocol constructs an SNP packet by adding the SNP header in front of the payload, as shown in Figure 7.4, and forwards it for transmission to SLP.

SNP routes messages to their associated protocols according to the value of the InConfiguration protocol and Type field in the SNP header. If the InConfiguration protocol field in the SNP header is set to "1," the SNP should route the message to the InConfiguration instance of the protocol that is identified by the Type field, otherwise

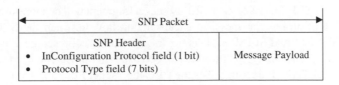

Figure 7.4   SNP packet structure

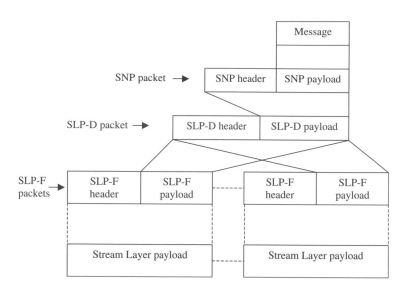

Figure 7.5   Message fragmented encapsulation

the SNP should route the message to the InUse instance of the protocol identified by the Type field. The protocol Type field should be set the Type value for the protocol associated with the encapsulated message.

Figure 7.5 illustrates the relationship between a message, SNP packets, SLP packets, and stream Layer payloads. Figure 7.5 also shows a case in which the SLP fragments the SNP packet into more than one SLP-F payload.

— *Signaling Link Protocol (SLP):* The SLP is a protocol associated with the default signaling application, as shown in Figure 7.3. SLP has two sub-layers: the SLP delivery layer (SLP-D) and the SLP fragmentation layer (SLP-F).

The purpose of SLP-D is to provide the best effort and reliable delivery for SNP packets. SLP-D also provides duplicate detection and retransmission for messages using the reliable delivery. On the other hand, the purpose of SLP-F is to provide fragmentation for SLD-D packets. The protocol data units of SLP are an SLP-D packet and an SLP-F packet. The packet taken from the upper layer should add the SLP-D header. The resulting SLP-D packet should forward to the SLP-F sublayer. The SLP-F packet(s) should be constructed by adding the SLP-F header in front of each SLP-F payload. The SLP-F payload(s) should be constructed from an SLP-D packet. If the SLP-D packet exceeds the current maximum SLP-F payload size, the sender should fragment the SLP-D packet. If the sender does not fragment the SLP-D packet, the SLP-D will be the SLP-F payload. If the sender does fragment the SLP-D packet, each SLP-D packet fragment will become an SLP-F payload. Figure 7.5 illustrates the message encapsulation flow between the SLP-D packet for reliable delivery and SLP-F packets for fragmentation as presented above.

• *Default Packet Application:* This provides an octet stream that can be used to carry packets between the AT and AN. It encompasses the RLP, the Packet Location Update protocol, and the FCP. It also provides functionality for those three protocols. The relationship among the protocols associated with the Default Packet Application is illustrated in Figure 7.6.

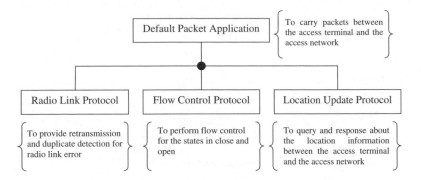

Figure 7.6 Default packet application protocols

— *Radio Link Protocol (RLP):* RLP is a protocol associated with the default packet application. This protocol provides Nak-based retransmission, and duplicate detection for an octet data stream such that the radio link error can be reduced with an acceptably low erasure rate.

Transmission unit of this protocol is an RLP packet. RLP receives octets for transmission from the higher layer and forms an RLP packet by concatenating the RLP packet header with a number of received contiguous octets. However, an RLP packet should not exceed the maximum payload length that can be carried by a Stream Layer packet.

— *Location Update Protocol (LUP):* This protocol defines location update procedures and messages for mobility management for the default packet application. The transmission unit of this protocol is a message. The AN uses this message to query the AT of its location information. The AN should set the MessageID field (8 bits) to 0x 03. The AT sends the LocationNotification message in response to the Location-Request message.

If the AT has a stored value for the LocationValue parameter, it should set the LocationType (8 bits), LocationLength (0 or 8 bits), and LocationValue (0 or 8 × LocationLength bits) fields in this message to its stored value of these fields.

If the AT receives a LocationAssignment message, then a LocationComplete message should be sent to the AN and the AT also stores the value of the LocationType (8 bits), LocationLength (8 bits), and LocationValue (8 × LocationLength bits) variables, respectively.

The AN uses the LocationRequest message to query the AT of its location information. The AT sends the LocationNotification message in response to the LocationRequest message. The AN uses the LocationAssignment message to update the location information of the AT. The AT sends the LocationComplete message in response to the LocationAssignment message.

— *Flow Control Protocol (FCP):* This protocol provides procedures and messages used by the AT and AN to perform flow control for the default packet application. In the close state, the default packet application does not send or receive any RLP packets; while in the open state, the default packet application can send or receive any RLP packets. Figure 7.7 illustrates the state transition diagram, (a) at the AT and (b) at the AN.

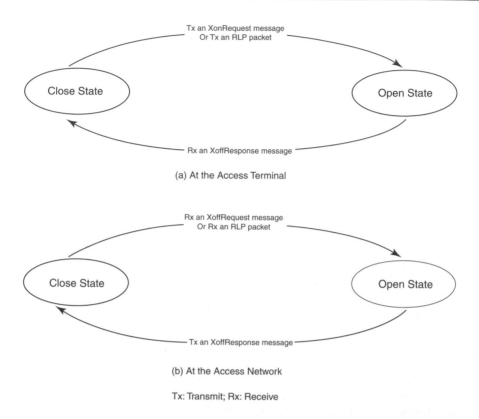

(a) At the Access Terminal

(b) At the Access Network

Tx: Transmit; Rx: Receive

Figure 7.7   Flow control state transition diagram

The protocol data unit of this protocol is a message. Since this is a control protocol, it does not carry payload on behalf of other layers or protocols. The AN may send a DataReady message to indicate that there is data awaiting to be transmitted. The AT should send a DataReadyAck message within the specific time period after reception of the DataReady message to acknowledge reception of the message. Figure 7.8 shows the diagram for control message flow.

The flow control message's functionality is summarized as shown in Table 7.1.

7.2.2 *Multi-Flow Packet Application*

The multi-flow packet application provided for multiple octet streams is used to carry octets between the AT and AN. The functionality of the multi-flow application consists of the following:

- *Default Packet Application:* This protocol provides one or more octet streams from the higher layer and uses Nak-based retransmissions. The transmission unit of this protocol is an RLP packet, as shown in Figure 7.9.
- *Data Over Signaling Protocol:* This protocol provides transmission and duplicate detection of higher layer packets using signaling messages. Message sequence numbers are provided for duplicate detection. The AT and AN send the DataOnlySignaling message to transmit a

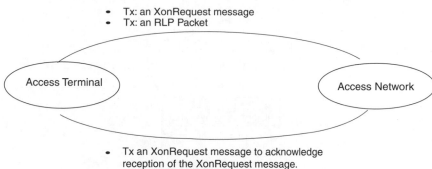

- Tx: an XonRequest message
- Tx: an RLP Packet

- Tx an XonRequest message to acknowledge
  reception of the XonRequest message.
- Tx an XonRequest message to request to stop
  sending RLP packets.
- Tx the DataReadyAck message to acknowledge
  reception of a DataReady message.

Tx: Transmit

Figure 7.8 Sending or receiving any RLP packets between the access terminal and access network

higher layer packet. The transmission unit of the protocol is a DataOverSignaling message that carries payload on behalf of the higher layer.

- *Location Update Protocol:* This protocol defines location update procedures and messages for mobility management for the Multi-Flow Packet Application. This protocol is a control protocol and therefore does not carry payload on behalf of other layers or protocol.

Table 7.1   Functionality of flow control messages

Close state	Open state
Neither sending nor receiving any RLP packets	Sending or receiving any RLP packets
Access Terminal Requirements	
Send an XonRequest message or an RLP packet for transition to the open state.	Send an XonResponse message to acknowledge reception of the XonRequest message to the close state.
Transmit an XonRequest message or RLP packet when it is ready to exchange RLP packets with access network.	Send an XoffRequest message to request the access network to stop sending RLP letters.
Send an XonRequest message or an RLP packet when it receives a DataReady message from the access network.	Re-send an XoffRequest message if it does not receive an XoffResponse message within the specified time period.
	Send the DataReadyAck message to acknowledge reception of a DataReady message.
Access Network Requirement	
Send an XonResponse message after reception of the XonRequest message to acknowledge reception of the message.	Send an XoffResponse message after reception of an XoffRequest message to acknowledge reception of the message.
Transmit to the open state if it receives an RLP packet.	Transmit acknowledgement of message reception to the close state.

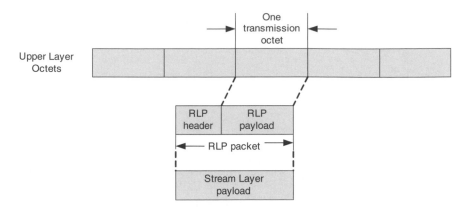

Figure 7.9   Multi-flow packet application encapsulation for one transmission unit

If the AT receives a LocationRequest message, it should send a LocationNotification message. If the AT receives a LocationAssignment message, it should send a LocationComplete message. If the AT receives a ServiceNetworkIDRequest message, it should send a ServiceNetworkIDNotification message. If the AT receives a ServiceNetworkAssignment message, it should send a ServiceNetworkIDComplete message.

- *Flow Control Protocol:* This protocol provides procedures and messages used by the AT and AN to perform flow control for the Multi-Flow Packet Application. In the close state, the AT and AN should not send or receive any RLP packet or DataOverSignaling message. In the open state, the Multi-Flow Packet Application can send or receive RLP packets or DataOverSignaling messages.

7.3 Stream Layer Protocol

The Stream Layer protocol provides the stream layer functionality for multiplexing of application streams for one access terminal. The generic virtual stream protocol provides the ability to multiplex up to 127 application streams. The air interface can support up to four parallel application streams. Stream 0 is always assigned to the signaling application. The other three streams are assigned to different QoS (Quality of Service) requirements. The stream layer protocol is responsible for defining the format and processing of the configuration messages that map applications to streams.

The relationship among an application layer packet, a stream layer packet, and a session layer payload is illustrated as shown in Figure 7.10. The protocol data unit for this protocol is a stream layer packet. The header added by 2 bits and the 6-bit payload will become the session layer payload.

The Stream Layer protocol carries data from the higher layer in an ApplicationData message. This protocol also carries signaling messages from the higher layer in an ApplicationSignaling message. This protocol uses the Signaling Application to transmit and receive messages.

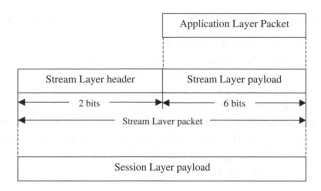

Figure 7.10   Stream layer encapsulation

7.3.1 Protocol Initialization

Upon protocol initialization, the InConfiguration instance or the InUse instance of this protocol in the AT and AN should perform the following specifications:

- The fall-back values of the attributes should be set to the default values specified for each attribute.
- If the InUse protocol has the same protocol subtype as the InConfiguration protocol instance, the fall-back values of the attributes defined by the InConfiguration protocol instance should be set to the values of the corresponding attributes associated with the InUse protocol instance.
- The value of the attributes for the InUse instance should be set to the default values specified for each attribute.

The format for the ConfigurationRequest and ConfigurationResponse messages consists of the following fields:

- MessageIndicator Field (8 bits): Set this field to 0x 00 for indicating a message.
- MessageID field (8 bits): Set this field to 0x 50 for the ConfigurationRequest and to 0x 51 for the ConfigurationResponse.
- TransmissionID field (8 bits): The sender increments this value for each new ConfigurationRequest message sent for the ConfigurationRequest, and the sender sets this value to the TransmissionID field of the corresponding ConfigurationRequest message.
- AttributeRecord field (Attribute dependent): This field defines a set of suggested values for a given attribute. The attribute record format is defined such that if the recipient does not recognize the attribute, it can discard it and parse attribute records that follow this record. For the ConfigurationResponse message, an attribute record field contains a single attribute value. If this message selects a complex attribute, only the ValueID field of the complex attribute should be included in the message. The sender should not include more than one attribute record with the same attribute identifier.

7.3.2 Procedures and Messages for the InConfiguration and InUse Instances

The AT and AN may use the ConfigurationRequest and ConfigurationResponse messages to select the application carried by each virtual stream. Once the AT and AN agree upon the mapping of a new application layer protocol to a virtual stream, they should create an InConfiguration instance of the agreed upon application and replace the InConfiguration instance with the agreed upon application instance.

If the InUse protocol instance has the same subtype as the InConfiguration protocol instance, the AT and AN should set the attribute values associated with the InUse instance to the attribute values associated with the InConfiguration instance. If the InUse protocol instance doesn't have the same subtype as the InConfiguration protocol instance, the AT and AN should perform as follows:

- The public data values of the InConfiguration protocol instance should be copied to the corresponding public data of the InUse protocol.
- The InConfiguration protocol instance becomes the InUse protocol instance for the session configuration protocol at the AT and AN.

The virtual stream protocol receives an application packet for transmission from up to 127 different applications. This protocol carries payload for the higher layer using virtual streams. A virtual stream is identified by its virtual stream number in the range "0000001"–"1111111" inclusive. If x bits is the length of data payload presented to the Stream Layer by an application, x should satisfy x modulo $8 = 6$. If S bits is the length of the signaling message payload presented to the stream layer by an application, S should satisfy S modulo $8 = 0$.

7.4 Session Layer Protocol

The Session Layer protocol is used to negotiate a session between the AT and AN. The AT is a device providing connectivity to a user, which may be a computing device such as a laptop, personal computer, or a personal digital assistant, and the AN is the network equipment providing data connectivity between a packet switched data network (PSDN such as the Internet) and the ATs.

A session contains information such as UATI (Unicast Address Terminal Identifier) assigned to the AT and configuration settings for authentication keys and parameters for connection layer and MAC layer protocols.

The session layer contains the following protocols:

- *Session Management Protocol:* This protocol provides the management means to control the activation of the other session layer protocol.
- *Address Management Protocol:* This protocol specifies the procedure for the initial UATI assignment and maintains the AT addresses as the AT moves between subnets.
- *Session Configuration Protocol:* This protocol provides the means to negotiate and provision the protocols used during the session, and negotiate the configuration parameters for these protocols.

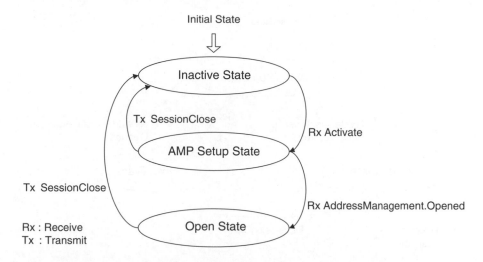

Figure 7.11   Session management protocol state diagram based on the AT

7.4.1 Default Session Management Protocol (SMP)

The default session management protocol provides for the negotiation and configuration of the set of protocols used during a session.

The actual behavior and message exchange in each state of the default session management protocol are mainly governed by protocols that are activated by the default session management protocol. Figure 7.11 illustrates an overview of the AT states and state transitions. Each state shown in Figure 7.11 is explained by the following.

- *Inactive state:* This state applies only to the AT. In this state, there are no communications between the AT and AN.
- *AMP Setup State:* In this state, the AT and AN perform message exchange governed by the Address Management protocol and the AN assigns a UATI to the AT. In the AMP setup state, the Session Management protocol in the AT sends an AddressManagement.Activate command to the Address Management protocol and waits for the Address Management protocol to respond.
- *Open State:* A session is open in this state. In the open state, the AT has an assigned UATI and then AT and AN have a session.

Figure 7.12 provides an overview of the AN states and the state transitions. The inactive, AMP setup, and open states are similar to the protocol state diagram for the AT. However, the close state applies only to the AN. In this case, the AN waits for the close procedure to complete.

- *Close State:* The close state is only associated with the protocol in the AN. In this state, the protocol in the AN waits for a SessionClose message from the AT or an expiration of a timer.

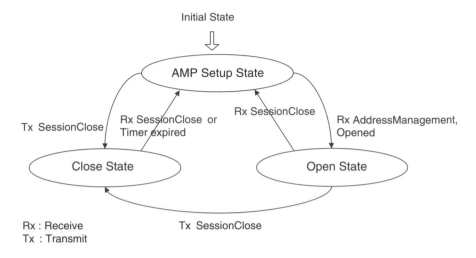

Figure 7.12   Session management protocol state diagram based on the AN

The session management protocol should periodically ensure that the session is still valid and manages the closing of the session. In addition, it is recommended to learn that the AT and AN should perform the procedure in the order specified in the protocol initialization for InConfiguration Protocol Instance and InUse Protocol Instance of this protocol.

7.4.2 Default Address Management Protocol (AMP)

This protocol provides the means to control the activation of the Address Management Protocol (AMP) and then the Session Configuration Protocol (SCP) before a session is established. This protocol operates in the following three states:

- *Inactive State:* The inactive state applies only to the AT. In this state there are no communications between the AT and AN.
- *Setup State:* In this state, the AT and AN perform a UATIRequest/UATIAssignment/ UATIComplete exchange to assign the AT with a UATI.
- *Open State:* In this state, the AT has been assigned a UATI. The AT and AN may also perform a UATIAssignment/UATIComplete exchange so that the AT obtains a new UATI.

The protocol states and messages and events causing the transition between the states are shown in Figures 7.13 and 7.14.

If the AT receives an Activate command in the Inactive State, the AT will transition the command to the Setup State. If the AT receives the Deactivate command in the Inactive State, the command should be ignored.

In the Setup State, the AT sends a request to the AN asking for a UATI and waits for the AN's response. Accordingly, upon entering the Setup State, the AT should send a UATIRequest message to the AN.

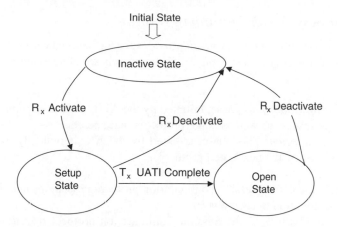

Figure 7.13   Address management protocol state diagram (AT)

The AN may send a UATIAssignment message at any time in the open state. When the AN sends a UATIAssignment message, the AN should assign a UATI to the AT for the session and include it in a UATIAssignment message.

In the Open State, the AT has been assigned a UATI. The AT sends a UATIComplete message to notify the AN that it has received the UATIAssignment message.

When the AN receives a UATIComplete message, it should return a UATIAssigned indication. If the AN does not receive the UATIComplete message in response to the corresponding UATIAssignment message within a certain time interval that is specified by the AN, it should retransmit the UATIAssignment message. If the AN does not receive the UATIComplete message after an implementation specific number of re-transmissions of the UATIAssignment message, it should return a Failed indication and transition to the Inactive State.

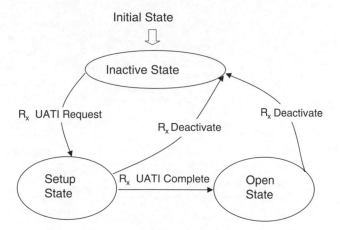

Figure 7.14   Address management protocol state diagram (AN)

7.4.3 Default Session Configuration Protocol

The Default Session Configuration protocol provides for the negotiation and configuration of the set of protocols used during a session. There are two phases of negotiation supporting this protocol:

- *AT Initiated Negotiation:* This phase, initiated by the AT, is used to negotiate the protocols employed in the session as well as some protocols' parameters.
- *AN Initiated Negotiation:* This phase, initiated by the AN, is typically used to override default values used by the negotiated protocols.

This protocol uses the default session configuration protocol procedures and messages when performing the negotiation in each phase.

As illustrated in Figure 7.15 the Session Configuration protocol operates in one of the following four states:

- *Inactive State:* In this state, the protocol waits for an activate command. Upon entering this state, the protocol should set the SessionConfigurationToken to 0x 0000.
- *AT Initiated State:* In this state, negotiation is performed at the initiative of the AT. During the AT Initiated State of the Default Session Configuration protocol, the AT and AN use the Generic Configuration protocol with the AT being the initiator of each exchange. The AT and AN use the ConfigurationRequest/ConfigurationResponse exchange to select the protocols and configure their associated parameters that will be used for the session.
- *AN Initiated State:* In this state, negotiation is performed at the initiative of the AN. During the AN Initiated State, the AN and AT execute the AN-initiated configuration procedures specified by each negotiated protocol.

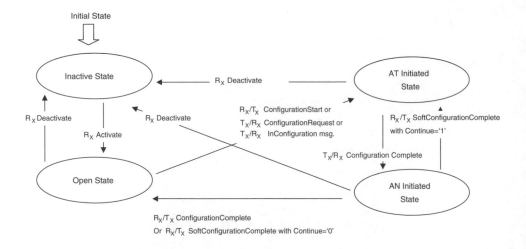

Figure 7.15   Session configuration protocol state diagram based on AT/AN

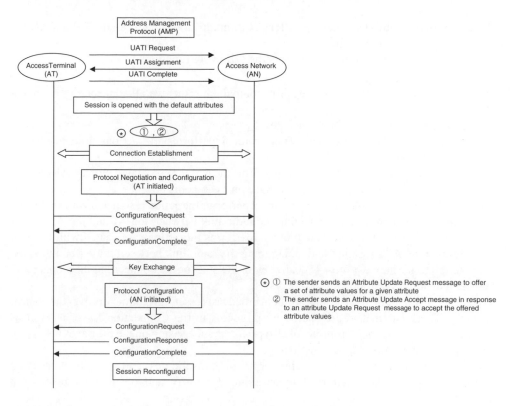

Figure 7.16   Negotiation procedure with key exchange for default session configuration protocol

- *Open State:* In this state, the AT may initiate the session configuration at any time during the Open State to start the negotiation process and the AN may request the AT to initiate the session configuration at any time.

Thus the AT and AN can use the negotiated protocols to exchange data and signaling in accordance with the requirements of each protocol. Figure 7.16 shows the default session configuration protocol, illustrating the negotiation procedure with key exchange.

7.5 Connection Layer Protocol

The Connection Layer provides air link connection establishment and maintenance services. This section presents default protocols for the connection layer. The connection layer not only controls the state of the air-link, but also prioritizes the traffic that is sent over it. The connection layer encapsulates transmitted data received from the session layer and forwards it to the security layer. This layer also de-capsulates data received from the security layer and forwards it to the session layer.

The AT and AN maintain a connection whose state dictates the form in which communications between them can take place. When a connection is closed, the AT is not assigned any dedicated air link resources. When a connection is open, the AT can be assigned the FTC, a Reverse Power Control Channel, and a RTC. Communications between the AT and AN are connected over the assigned channels.

The Connection Layer protocol performs the following functions:

- *Air Link Management Protocol:* This protocol maintains the overall connection state in the AT and AN.

  The protocol can be in one of these states:
  - — *Initialization State Protocol:* This protocol performs the actions associated with acquiring an AN.
  - — *Idle State Protocol:* This protocol performs the actions associated with an AT that has acquired the network, but does not have an open connection.
  - — *Connected State Protocol:* This protocol performs the action associated with an AT that has an open connection, managing the radio link between the AT and the AN.
  - — *Route Update Protocol:* This protocol performs the actions associated with keeping track of an AT's location and maintaining the radio link between the AT and the AN.

  In addition to the above protocols dealing with the state of the connection, the Connection Layer also contains the following protocols:
- *Overhead Messages Protocol:* This protocol broadcasts essential parameters over the control channel. These parameters are shared by protocols in the Connection Layer as well as protocols in other layers. This protocol also performs supervision on the messages necessary to keep the Connection Layer functioning.
- *Packet Consolidation Protocol:* This protocol consolidates and prioritizes packets for transmission as a function of their assigned priority and the target transmission channel.

7.5.1 Data Encapsulation for InUse Protocol Instance

In the transmit direction, the Connection Layer receives Session Layer packets, adds Connection Layer header(s), concatenates them in the order to be processed on the receive side, adds padding (where applicable), and forwards the resulting packet for transmission to the Security Layer. In the receive direction, the Connection Layer receives Security Layer packets from the Security Layer, and forwards the Session Layer packets to the Session Layer in the order received after removing the Connection layer headers and padding. Figure 7.17 illustrates the relationship between Session Layer packets, Connection Layer packets, and Security Layer payloads.

7.5.2 Air-Link Management Protocol

The Air-Link Management protocol provides the status and state-transition rules to be followed by an AT and an AN for the connection layer. All transitions are caused by indications returned from protocols activated by the default air-link management protocol. The message exchange in each state is mainly governed by the following three protocols:

- *Initialization State:* In this state, the AT acquires an AN. The protocol activates the initialization state protocol to execute the procedures relevant to this state.

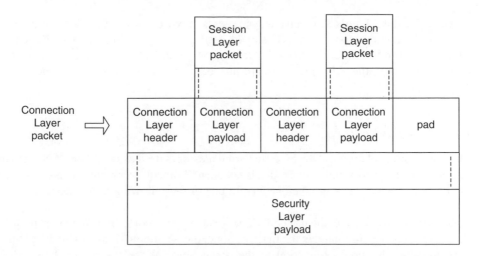

Figure 7.17   Connection layer encapsulation

- *Idle State:* In this state, the connection is closed. The protocol activates the idle state protocol to execute the procedures relevant to this state.
- *Connection State:* In this state, the connection is open. The protocol activates the connection state protocol to execute the procedures relevant to this state.

7.5.3 Initialization State Protocol

The Initialization State protocol provides the procedures and messages required for an AT to acquire a serving network. At the AT, protocol states and events causing transition between states will operate in one of the following four states:

- *Inactive State:* In this state the protocol waits for an activate command. If the protocol receives a activate command in the Inactive State, the AT should transition to the Network Determination State. If the protocol receives a deactivate command in the Inactive State, the AT should ignore it.
- *Network Determination State:* In this state, the AT chooses an AN on which to operate. Specifically, the AT selects a CDMA channel on which to try and acquire the AN. Upon selecting a CDMA channel, the AT should enter the Pilot Acquisition State.
- *Pilot Acquisition State:* In this state, the AT acquires a Forward Pilot Channel of the selected CDMA channel. Upon entering the Pilot Acquisition State, the AT should tune to the selected CDMA channel and should search for the pilot. If the AT acquires the pilot, it should enter the Synchronization State. If the AT fails to acquire the pilot within $T_{PA} = 60$ seconds, it should enter the Network Determination State. T_{PA} denotes the time to acquire the pilot in an AT.
- *Synchronization State:* In this state, the AT synchronizes to the control channel cycle, receives the Sync message, and synchronizes to system time. Upon entering this state, the AT completes timing synchronization and should issue the ControlChannelMAC. Activate

command. If the AT fails to receive a sync message within $T_{PA} = 5$ seconds of entering the synchronization state, the AT should issue a ControlChannelMAC.Deactivate command and should enter the Network Determination State. While attempting to receive the sync message, the AT should discard any other messages received on the control channel.

7.5.4 Idle State Protocol

The Idle State protocol provides the procedures and messages used by the AT and AN when the AT has acquired a network and a connection is not open. Protocol States and events causing the transition between the states will be operated in one of the following four states:

- *Inactive State:* In this state, the protocol waits for an activate command. When the protocol receives an activate command in the Inactive State, the AT should transition to the Monitor State and the AN should transition to the Sleep State. When the protocol receives a Deactivate command in the Inactive State, it should be ignored. When the protocol received a Deactivate command in any other state, both the AT and AN should transition to the Inactive State.
- *Sleep State:* In this state, the AT may shut down part of its subsystems to conserve power. The AT does not monitor the Forward Channel, and the AN is not allowed to transmit unicast packets to it. When the AT is in the Sleep State it may stop monitoring the control channel by issuing the OverheadMessages.Deactivate command and ControlChannelMAC.Deactivate command.

  When the AN is in the Sleep State, it is prohibited from sending unicast packets to the AT. If the AN receives a ConnectionRequest message, it will transition to the Connection Setup State.

  The AN and AT should transition from the Sleep State to the Monitor State in time to send and receive, respectively, the sub-synchronous capsule or the synchronous capsule sent at a time satisfying the following condition:

$$[T + 256 \times R] \bmod \text{Period} = \text{Offset}$$

  where T is the current CDMA system Time in slots and R is the result of applying the hash function to compute an index and period depending on PreferredControlChannelEnabled being equal to "0" or "01," respectively.
- *Monitor State:* In this state, the AT monitors the control channel, listens for a Page message and, if necessary, updates the parameters received from the Overhead Message protocol. The AN may transmit unicast packets to the AT in this state.
- *Connection Setup State:* In this state, the AT and AN setup a connection.

7.5.5 Connected State Protocol

The AT and AN use the Connected State protocol while a connection is open. This protocol defines the activate command, the deactivate command, and the close connection command. The Connected State protocol can be in one of the following three states:

- *Inactive State:* In this state, the protocol waits for an activate command. When the protocol receives an activate command in the Inactive State, the AT and AN should transition to the Open State. When the protocol receives a deactivate command in the Inactive State, it should return a ConnectionClosed indication. When the protocol receives this command in the Open State, the AT sends a ConnectionClose message to the AN and performs the close-up procedures as below:
  — Issue RouteUpdate.Close command, return a ConnectionClosed indication, and transition to the Inactive State.
  — The AN should send a ConnectionClose message to the AT, perform the cleanup procedures, and transition to the Closed State.
- *Open State:* In this state, the AT can use the RTC and the AN can use the FTC and the control channel to send application traffic to each other. Upon entering the Open State, the AT should issue the following commands: OverheadMessage.Active and ControlChannelMAC. Activate.

  The AT should comply with the following requirements when in the Open State:
  — The AT receives the Control Channel and Forward Traffic Channel.
  — The AT should not transmit on the Access Channel.
  — The AT should monitor the overhead messages.
  — If the AT receives a ConnectionClose message, it will send a ConnectionClose message with CloseReason set to "CloseReply" and execute the cleanup procedures. If the AT sends a ConnectionClose message, it may advertise, as part of the ConnectionClose message, that it should be monitoring the ControlChannel continuously, until a certain time following the closure of the connection. This period is called a suspend period, and can be used by the AN to accelerate the process of sending a unicast packet to the AT.
  — If the AT sends a ConnectionClose message in response to a ConnectionClose message with CloseReason field "100," then the AT should advertise a non-zero suspend period in the ConnectionClose message.
- *Close State:* This state is associated only with the AN, which will wait for connection resources to be safely released. Upon entering this state, the AN waits for a replying ConnectionClose message from the AT. If the AN receives a ConnectionClose message or if the time expires, it should execute the cleanup procedures, may close all connection-related resources assigned to the AT, and should transition to the Inactive State.

7.5.6 *Route Update Protocol*

The Route Update protocol provides the procedures and messages used by the AT and AN to keep track of the AT's approximate location and to maintain the radio link as the AT moves between the coverage areas of different sectors.

Transitions between states are driven by commands received from Connection Layer protocols and the transmission and reception of the TrafficChannelAssignment message. The protocol states, messages, and commands causing the transition between states are shown in Figure 7.18.

This protocol can be in one of the following three states.

- *Inactive State:* In this state, the protocol waits for an activate command. If the protocol receives an activate command in the Inactive State, the AT and AN will transition to

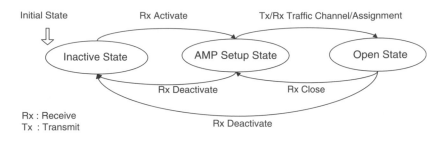

Figure 7.18   Route Update protocol state diagram

the Idle States. If the protocol receives a deactivate command in the Inactive State, it should be ignored. If the protocol receives this command in any other state, the AT and AN should:

— Issue a ReverseTrafficChannelMAC.Deactivate command.
— Issue a ForwardTrafficChannelMAC.Deactivate command.
— Transition to the Inactive State.

• *Idle State:* This state corresponds to the Air-Link Management Protocol Idle State. In this state, the AT autonomously maintains the Active Set. Route update messages from the AT to the AN are based on the distance between the AT's current serving sector and the serving sector at the time the AT last sent an update. The AT sends the RouteUpdate message to notify the AN of its current location and provides it with an estimate of its surrounding radio link conditions.

• *Connected State:* In this state, the AN dictates the AT's Active Set. The active set is the set of pilots associated with the sectors currently serving the AT. The AN sends the TrafficChannelAssignment message to manage the AT's Active Set. Route Update messages from the AT to the AN are based on changing radio link conditions, which can be obtained through pilot measurements at the AT. The AT estimates the strength of the Forward Channel transmitted by each sector in its neighborhood. This estimate is based on measuring the strength of the Forward Pilot Channel. When this protocol is in the Connected State, the AT uses pilot strengths to decide when to generate RouteUpdate messages. The AT sends the RouteUpdate message to notify the AN of its current location and provide it with an estimate of its surrounding radio link conditions.

7.5.7 Packet Consolidation Protocol

This protocol provides packet consolidation on the transmit side and packet demultiplexing on the receive side. The protocol data unit for this protocol is a Connection Layer packet. Connection Layer packets contain Session Layer packets to or from the same AT address. All transmitted packets are forwarded to the Security Layer. All transmitted packets are forwarded to the Session Layer after the Connection Layer headers are removed. There are two types of Connection Layer packets:

• *Format A:* Format A packets contain one Session Layer packet and do not have Connection Layer headers or padding, as shown in Figure 7.19(a). Format A provides an extra octet of payload per packet. The AN should create a Format A Connection Layer packet only if the length of the highest priority pending Session Layer packet will fill the security layer payload.

- *Format B:* Format B packets contain one or more Session Layer packets and have a Connection Layer header(s). The protocol places the Connection Layer header in front of each Session Layer packet and concatenates enough padding to create a maximum length packet, as illustrated in Figure 7.19(b). All concatenated Connection Layer packets should be transmitted on the same Physical Layer Channel.

The protocol uses the priority order to determine which Session Layer packets should be included in the Connection Layer packet. The protocol should concatenate and encapsulate Session Layer packets into a Connection Layer packet in priority order. The AN should forward the Connection Layer packet for transmission to the Security Layer.

If a FTC session layer packet is marked as forced single encapsulation, the AN should encapsulate it without any other Session Layer packets in a Connection Layer packet. Forced single encapsulation applies only to the FTC MAC Layer packets.

The control channel carries broadcast transmissions as well as directed transmissions to multiple ATs. If the AN transmits a unicast packet to an AT over the control channel, it should transmit this packet at least from all the sectors in the AT's active set.

If the data is carried in a synchronous capsule, the transmission should occur simultaneously at least once. Transmitting asynchronous capsules on the control channel is optional, because all data marked for transmission in these capsules can also be transmitted in a synchronous capsule. If the AN chooses to transmit Connection Layer packets in an asynchronous capsule of the control channel, the AN should transmit, in priority order, packets marked for transmission on the FTC or the control channel.

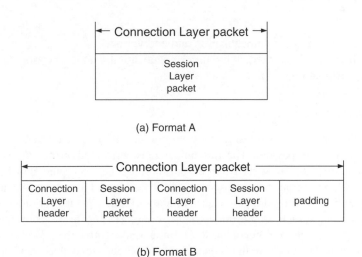

Figure 7.19   Connection layer packet structures of (a) Format A and (b) Format B

If two packets have different priority numbers, the packet with the lower priority number has priority. If two packets have the same priority number, the packet received first by the protocol has priority.

7.5.8 Overhead Messages Protocol

The overhead messages are broadcast by the AN over the control channel. The Overhead Messages protocol provides procedures related to transmission, reception, and supervision of these messages.

- *Inactive State:* This state corresponds only to the AT and occurs when the AT has not acquired an AN or is not required to receive overhead messages. In this state, the protocol waits for an activate command. If this protocol receives an activate command in the Inactive State, the AT should transit to the Active State and the AN should ignore it. If this protocol receives a deactivate command in the Inactive State, it should be ignored. If this protocol receives a deactivate command in the Active State, the AT should transit to the Inactive State and the AN should ignore it.
- *Active State:* In this state, the AN transmits and the AT receives overhead messages.

7.6 Security Layer Protocols

The air interface supports a security layer. This layer can be used for encryption and authentication of AT traffic transported by the control channel, the access channel (AC), the FTC, and the RTC.

The Security Layer provides the key exchange protocol, authentication protocol, and encryption protocol. The security protocol provides the following functions between the AN and AT:

- exchange security keys for authentication and encryption;
- provide public variables needed by the authentication and encryption protocols (i.e. cryptosync, time-stamp, etc.).

7.6.1 Security Layer Encapsulation

Figure 7.20 illustrates the relationship between a connection layer packet, a security layer packet, and a MAC layer payload. The authentication is performed on the encryption protocol packet and the portions of security layer packet may be encrypted and authenticated. Specifically, the encryption protocol should add a trailer to hide the actual length of the plaintext or padding to be used by the encryption algorithm. The encryption protocol header may contain variables such as the initialization vector (IV) to be used by the encryption protocol. The authentication protocol header or trailer may contain the digital signature that is used to authenticate the portion of the authentication protocol packet that is authenticated. The security protocol header or trailer may contain variables needed by the authentication and encryption protocols. However, the security layer headers or trailers may not present if the session configuration establishes the default security layer or if the configured security protocol

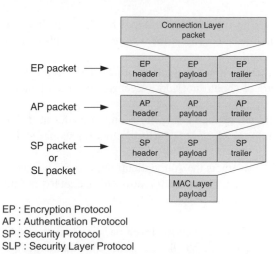

EP : Encryption Protocol
AP : Authentication Protocol
SP : Security Protocol
SLP : Security Layer Protocol

Figure 7.20   Security layer encapsulation

does not require a header or trailer. The fields added by the MAC layer indicate presence or absence of the security layer headers and trailer.

7.6.2 Default Security Protocol

The Default Security protocol provides a mechanism for transferring packets between the authentication protocol and the MAC layer. As observed from Figure 7.20, the protocol data unit for this protocol is a Security Layer packet. Each security layer packet consists of an authentication protocol packet. The default security protocol will set the security layer packet (i.e. Security protocol packet) to the authentication protocol packet and forward it to transmit to the MAC Layer. The default security protocol does not add a security protocol header and trailer. The protocol for the InUse instance should set the authentication protocol packet to the Security Layer packet received for the MAC Layer and should forward the packet to the authentication protocol. The default security protocol adds neither a header nor a trailer.

The Generic Security protocol on the transmission side provides a cryptosync that may be used by the negotiated authentication protocol and encryption protocol, while on the receiving side, this protocol computes the cryptosync using the information provided in the security protocol header and makes the cryptosync publicly available. The protocol data unit for this protocol is a security layer packet. Each security layer packet consists of an authentication protocol packet and a security packet header. The protocol constructs a security layer packet out of the authentication protocol packet and should pass the packets for transmission to the MAC Layer as follows:

- When the protocol receives an authentication protocol packet that is either authenticated or encrypted, it should set CryptosyncShort in the security protocol header to the least significant 16-bits of the Cryptosync that is used by the Authentication protocol or the Encryption protocol to authenticate or encrypt this packet. The sender should include the

CryptosyncShort field in the security protocol header only if the authentication protocol packet is either authenticated or encrypted. The sender sets this field to the 16 least significant bits of the Cryptosync.

- When the protocol receives an authentication protocol packet that is neither authenticated nor encrypted, the protocol should not add a Security protocol header to the Authentication protocol packet.
- This protocol should not append a security protocol trailer to the Authentication protocol packet. Note thus that the Generic Security protocol does not add a trailer.

The Security protocol constructs the Authentication protocol packet using the security layer packet received from the MAC Layer and forwards the packet to the Authentication protocol as follows:

- When the protocol receives a security layer packet from the MAC Layer that is either authenticated or encrypted, it should construct the authentication protocol packet by removing the security layer header.
- When the protocol receives a security layer packet from the MAC Layer that is neither authenticated nor encrypted, it should set the authentication protocol packet to the security layer packet.

When the Security Layer receives a Connection Layer packet that is to be either authenticated or encrypted, the Security protocol should choose a value for the Cryptosync based on the current 64-bit representation of the CDMA System Time in units of 80 ms. Cryptosync does not specify a time later than the time that the security layer packet will be transmitted by the Physical Layer or earlier than the current CDMA System Time. The protocol should then set CryptosyncShort in the security protocol header to Cryptosync [15 : 0].

When the Security protocol receives a security layer packet from the MAC Layer that is either authenticated or encrypted, it should compute the 64-bit Cryptosync using Cryptosync-Short given in the security protocol header as follows:

$$Cryptosync = (SystemTime - (SystemTime[15 : 0] - CryptosyncShort) \bmod 2^{16}) \bmod 2^{64},$$

where SystemTime is the current CDMA System Time in units of 80 ms, System Time[15 : 0] is the 16 least significant bits of the SystemTime, and CryptosyncShort is the 16-bit security protocol header.

7.6.3 Diffie-Hellman Key Exchange Protocol

The Diffie-Hellman (DH) key exchange protocol provides a method for session key exchanges based on the DH key exchange algorithm.

The session key being created from the DH key exchange protocol can be applied for use on channels as shown below:

- *Authentication Key:* This key is used on the FTC, RTC, Forward Control Channel (FCC), and Reverse Access Channel (RAC), respectively.

- *Encryption Key:* This key is also used for FTC, RTC, FCC, and RAC, respectively. Upon the protocol initialization, the InConfiguration or InUse instances of this protocol in the AT and AN should perform the following:
  — set SKey (Session Key) to zero and its length to the value of the KeyLength attribute;
  — set FTCAuthKey or RTCAuthKey to zero and its length to 160 bits;
  — set FTCEncKey or RACEncKey to zero and its length to 160 bits;
  — set FCCAuthKey or RACAuthKey to zero and its length to 160 bits;
  — set FCCEncKey or RACEncKey to zero and its length to 160 bits.

The Key Exchange protocol uses the KeyRequest and KeyResponse messages for exchanging public session Keys, and the ANKeyComplete and ATKeyComplete messages for indicating that the secret session keys have been calculated.

The key exchange procedures performed in AT and AN during session configuration are described in the following sections.

7.6.4 Access Terminal (AT) Requirements

Upon receiving the KeyRequest message, the AT should perform as follows:

- Choose a random number ATRand between $1 \leq ATRand \leq p - 2$ and set the ATPubKey field of the KeyResponse message as follows:

$$ATPubKey = g^{ATRand}(\bmod\, p)$$

  where g and p are KeyLength dependent protocol constants for the DH Key Exchange protocol.
- The AT sends a KeyResponse message with the ATPubKey filed.
- The AT computes the SKey as follows:

$$SKey = ANPubKey^{ATRand}(\bmod\, p)$$

  After receiving an ANKeyComplete message with a TransactionID field (8 bits) that matches the TransactionID field (8 bits) of the associated KeyRequest message, the AT should perform the following:

- AT computes the 64-bit variable TimeStampLong as follows:

$$TimeStampLong = (SystemTime - (SystemTime[15:0] - TimeStampShort)\bmod\, 2^{16})$$

  where SystemTime is the current CDMA System Time (80 ms), SystemTime[15:0] is the least significant bits of the SystemTime, and TimeStampShort is the 16-bit field received in the ANKeyComplete message.
- AT pads the message bits and computes the 160-bit message digest.
- If the message digest matches the KeySignature field of ANKeyComplete message, the AT sends an ATKeyComplete message with the Result field set to "1."
- Otherwise, the AT declares failure and sends an ATKeyComplete message with the Result field set to "0."

7.6.5 Access Network (AN) Requirements

The AN sends the KeyRequest message to initiate the Session key exchange. The AT sends the KeyResponse message in response to the KeyRequest message. The AN chooses a random number ANRand between $1 < \text{ANRand} \leq p - 2$ and set the ANPubKey field of the KeyRequest message as follows:

$$\text{ANPubKey} \equiv g^{\text{ANRand}}(\text{mod}\, p)$$

where, g, p and, KeyLength are specified during session configuration of the DH Key Exchange protocol.

The random number ANRand has the following properties:

- The generated ANRand should have a uniform distribution over its range and should be statistically uncorrelated in formulating different KeyRequest messages.
- Formulation of each KeyRequest message should not be derivable from the random number used previously.
- The numbers used in formulating KeyRequest messages sent by different ANs should be statistically uncorrelated.

After receiving a KeyResponse message with a TransactionID field (8 bits) that matches the TransactionID field (8 bits) of the associated KeyRequest message, the AN shall perform the following.

- The AN computes the SKey such that

$$\text{SKey} = \text{ATPubKey}^{\text{ANRand}}(\text{mod}\, p)$$

- The AN constructs the message bits that consist of the computed SKey, TransactionID (8 bits), a 16-bit pseudorandom value (Nonce), and TimeStampLong (64 bits). The 64-bit TimeStampLong is set based on the current 64-bit representation of the CDMA System Time in units of 80 ms.
- The AN pads the message bits and computes the 160-bit message digest.
- The AN sends an ANKeyComplete message with the KeySignature field of the message set to the message digest computed in the previous step and the TimeStampShort field of the message set to the 16 least significant bits of the CDMA System Time. The AN then starts the AT Signature Computation Timer with a timeout value of the AN Key Signature Computation Timer.

The AN should declare failure and stop performing the rest of the key exchange procedure if both AT Key Computation and the AT Key Signature Computation Timers are expired, or the AN receives an ATKeyComplete message with Request field set to "0."

The session key exchange procedures during session configuration between the AT and AN are summarized as shown in Table 7.2.

Table 7.2   DH key exchange algorithm for SKey computation

AN	AT
Choose ANRand, $1 \leq$ ANRand $\leq p-2$	Choose ATRand, $1 \leq$ ATRand $\leq p-2$
Compute the AN public key:	Compute the AT public key:
$ANPubKey \equiv g^{ANRand} \pmod p$	$ATPubKey \equiv g^{ATRand} \pmod p$
Send ANPubKey to the AT $\rightleftharpoons$	Send ATPubKey to the AN
Finally compute the session key:	Compute the session key:
$SKey \equiv ATPubKey^{ANRand} \pmod p$	$SKey \equiv ANPubKey^{ATRand} \pmod p$
$\quad\quad \equiv g^{(ATRand)(ANRand)} \pmod p$	$\quad\quad \equiv g^{(ANRand)(ATRand)} \pmod p$

If the SKey computed at AN and AT is equal, the session key is established.

ANRand: Access network random number
ATRand: Access terminal random number
ANPubKey: Access network public key
ATPubKey: Access terminal public key
SKey: Session key
g: a primitive element of a prime module p
p: a prime number
g and p are specified during session configurations of the DH Key Exchange Protocol.
The parameters g and p are called cryptosync.

Example 7.1. A simple example will be presented for the reader to help understand how to compute the SKey.

Choose protocol constants $g=6$ and $p=11$ for the DH key exchange protocol. Pick $ATRand=3$, $1 \leq ATRand \leq 9$. Pick $ANRand=7$, $1 \leq ANRand \leq 9$.

Compute first the public keys of the AT and AN, respectively:

$$ATPubKey \equiv g^{ATRand} \pmod p, \quad ANPubKey \equiv g^{ANRand} \pmod p,$$
$$\equiv 63 \pmod{11} = 7 \quad\quad\quad\quad \equiv 67 \pmod{11} = 8$$

AT sends $ATPubKey$ to AN $\rightleftharpoons$ AN sends $ANPubKey$ to AT

$$\text{Finally, AT computes the SKey}: \quad \text{AN computes the SKey}:$$
$$SKey \equiv ANPubKey^{ATRand} \pmod p, \quad SKey \equiv ATPubKey^{ANRand} \pmod p,$$
$$\equiv 83 \pmod{11} = 6 \quad\quad\quad\quad \equiv 77 \pmod{11} = 6$$

Thus, the SKey between AT and AN is computed.

$\square$

The SKey computation by means of Elliptic Curve (EC) cryptography over either Z_p or $GF(2^m)$ can be done using the EC DH key exchange algorithm as shown in Tables 7.3 and 7.4.

Example 7.2. Suppose G(5,2) is a generator point on the EC curve $y^2 + xy = x^3 + x + 6$ for $a=1$ and $b=6$. Choose a prime module $p=11$.

Table 7.3 SKey computation using EC DH Key Exchange protocol over the prime field Z_p

Choose a generator $G = (x, y)$ on the EC curve and two random numbers of $ATRand$ and $ANRand$.

AT	AN
Compute $ATPubKey \equiv (ATRand)G$ AT sends its publicKey to AN	Compute $ANPubKey \equiv (ANRand)G$ AN also sends its publicKey to AT
Then, compute the SKey as follows: $SKey = (ATRand)(ANPubKey) \equiv (ATRand)$ $(ANPand)G$	$SKey = (ANRand)(ATPubKey) \equiv (ANRand)$ $(ATPand)G$
Thus, the session key is computed.	

When choosing $ATRand = 2$ and $ANRand = 3$, $ATPubKey = (ATRand)G$ and $ANPubKey = (ANRand)G$ can be computed as shown below.

Computation of the public key at AT:

$$ATPubKey = 2G = 2(5,2) = (5,2) + (5,2) = (x_3, y_3)$$

This is the case in which two points are doubling.

$$\beta = \frac{3x_1^2 + 1}{2y_1}(\bmod\, p) = \frac{3 \times 25 + 1}{2 \times 2}(\bmod\, 11) = \frac{76}{4} = 19(\bmod\, 11) = 8$$

$$x_3 = \beta^2 - 2x_1 = 64 - 2 \times 5 = 54(\bmod\, 11) = 10$$

$$y_3 = \beta(x_1 - x_3) - y_1 = 8(5 - 10) - 2 = -42(\bmod\, 11) = 2$$

Thus, $ATPubKey = (10,2)$.

Computation of the public key at AN:

$$ANPubKey = 3G = (5,2) + (5,2) + (5,2) = (x_3, y_3)$$

Table 7.4 The SKey computation using the EC DH key exchange algorithm over the binary extension field GF(2^m)

Select the base point (a generator) G on the EC curve $y^2 + xy = x^3 + ax^2 + b$. Let n denote the order of GF(2^m).

AT	AN
Choose ATRand, $1 \le ATRand \le n$ Compute ATPubKey $= (ATRand)G$	Choose ANRand, $1 \le ANRand \le n$ Compute ANPubKey $= (ANRand)G$
Computation of the SKey (session key) ATSKey $= (ATRand)(ANPubKey) = (ATRand)$ $(ANRand)G$	ANSKey $= (ANRand)(ATPubKey) = (ANRand)$ $(ATRand)G$
Thus the session key between AT and AN can be computed.	

This is the case where scalar multiplication of a point on EC is simply a repeated addition of that point.

Since $2G = (10,2)$, $ANPubKey = (10,2) + (5,2) = (x_3, y_3)$

$$\alpha = \frac{y_2 - y_1}{x_2 - x_1} = \frac{2-2}{5-10} = 0$$

$$x_3 = \alpha^3 - x_1 - x_2 = -10-5 = -15(\text{mod } 11) = 7$$

$$y_3 = \alpha(x_1 - x_3) - y_1 = -2(\text{mod } 11) = 9$$

Hence, $ANPubKey = (7, 9)$.

Next, consider the computation of SKeys at AT and AN, respectively.

$ATSKey = (ATRand)(ANPubKey) = 2(7,9) = (7,9) + (7,9)$

$$\beta = \frac{3 \times 49 + 1}{2 \times 9} = \frac{74}{9} = \frac{8}{9}(\text{mod } 11) = 8 \times 9^{-1}(\text{mod } 11) = 8 \times 5(\text{mod } 11) = 7$$

$$x_3 = \beta^2 - 2x_1 = 49 - 2 \times 7 = 35(\text{mod } 11) = 2$$

$$y_3 = \beta(x_1 - x_3) - y_1 = 7(7 - 2) - 9 = 26(\text{mod } 11) = 4$$

Thus, $ATSkey = (2,4)$

$$\begin{aligned} ANSKey &= (ANRand)(ATPubKey) = 3(10, 2) \\ &= (10, 2) + (10, 2) + (10, 2) \end{aligned}$$

The addition of the first two terms becomes: Since $\beta = 1$, it results in $x_3 = 3$ and $y_3 = 5$. Hence, $ANSKey = (3,5) + (10,2)$

$$\alpha = \frac{2 - 5}{10 - 3} = -3 \times 7^{-1} = -3 \times 8 = -24(\text{mod } 11) = 9$$

$$x_3 = \alpha^2 - x_1 - x_2 = 81 - 3 - 10 = 68(\text{mod } 11) = 2$$

$$y_3 = \alpha(x_1 - x_3) - y_1 = 9(3 - 2) - 5 = 4$$

Thus, $ANSKey = (2,4)$. Since computed results at AT and AN are identical, the session key is established.   □

Example 7.3. Consider $GF(2^4)$ whose primitive polynomial is $p(x) = x^4 + x + 1$ of degree 4. If α denotes a root of $p(x)$, then the field elements α^i, $1 \leq i \leq 15$, of $GF(2^4)$ can be generated by using $\alpha^4 = \alpha + 1$ (see Table 9.7 of Chapter 9).

Suppose the EC equation over $GF(2^4)$ is $y^2 + xy = x^3 + \alpha^4 x^2 + 1$ by $a = \alpha^4$ and $b = 1$. Let the base point (a generator) be $G = (\alpha^5, \alpha^3)$.

Computation at AT (Access Terminal)

Pick ATRand $= 3$ from the range of $1 \leq$ ATRand ≤ 15.

Compute ATPubKey $= ($ATRand$)G = 3(\alpha^5, \alpha^3) = (\alpha^5, \alpha^3) + (\alpha^5, \alpha^3) + (\alpha^5, \alpha^3)$.

Find the solution of the first two doubling points as follows:

$$(\alpha^5, \alpha^3) + (\alpha^5, \alpha^3) = (x^3, y^3) = (1, \alpha^{13})$$

where

$$x_3 = x_1^2 + \frac{b}{x_1^2} = (\alpha^5)^2 + \frac{1}{(\alpha^5)^2} = \alpha^{10} + \alpha^5 = 1$$

$$y_3 = x_1^2 + \left(x_1 + \frac{y_1}{x_1} \right) x_3 + x_3 = (\alpha^5)^2 + \left(\alpha^5 + \frac{\alpha^3}{\alpha^5} \right) + 1$$

$$= \alpha_{10} + \alpha_5 + \alpha^{13} + 1 = a^{13}$$

Thus, ATPubKey $= (1, \alpha^{13}) + (\alpha^5, \alpha^3) = (x_3, y_3)$.

This is the case for adding two distinct points on the elliptic curve.

$$\lambda = \frac{y_1 + y_2}{x_1 + x_2} = \frac{\alpha^{13} + \alpha^3}{1 + \alpha^5} = \frac{1 + \alpha^2}{\alpha^{10}} = \alpha^{13}$$

$$x_3 = \lambda^2 + \lambda + x_1 + x_2 + a = \alpha^{26} + \alpha^{13} + 1 + \alpha^5 + \alpha^4 = \alpha^2 + 1 + \alpha = \alpha^{10}$$

$$y_3 = \lambda(x_1 + x_3) + x_3 + y_1 = \alpha^{13}(1 + \alpha^{10}) + \alpha^{10} + \alpha^{13} = \alpha^{10} + \alpha^8 = \alpha$$

Finally, ATPubKey $= (\alpha^{10}, \alpha)$.

AT sends ATPubKey $= (\alpha^{10}, \alpha)$ to AN.

Computation at AN (Access Network)

Let us pick ANRand $= 2$ from the range of $1 \leq$ ANRand ≤ 15.

Compute ANPubKey $= ($ANRand$)G = 2(\alpha^5, \alpha^3) = (\alpha^5, \alpha^3) + (\alpha^5, \alpha^3) = (1, \alpha^{13})$.

AN sends ANPubKey $= (1, \alpha^{13})$ to AT.

Lastly, the SKey is computed as follows:

$$\text{AT computes the ATSKey} = ($ATRand$)($ANPubKey$)$$
$$= 3(1, \alpha^{13}) = (1, \alpha^{13}) + (1, \alpha^{13}) + (1, \alpha^{13})$$

For the first two doubling points, we have

$$(1, \alpha^{13}) + (1, \alpha^{13}) = (x_3, y_3) = (0, 1)$$

where

$$x_3 = x_1^2 + \frac{b}{x_1^2} = 1 + 1 = 0$$

$$y_3 = x_1^2 + \left(x_1 + \frac{y_1}{x_1}\right)x_3 + x_3 = 1 + 0 + 0 = 1$$

ATSKey $= (0, 1) + (1, \alpha^{13}) = (x_3, y_3)$.
Applying the formula for the addition of two distinct points on EC, we have

$$\lambda = \frac{y_1 + y_2}{x_1 + x_2} = \frac{1 + \alpha^{13}}{0 + 1} = \alpha^2 + \alpha^3 = \alpha^6$$

$$x_3 = \lambda^2 + \lambda + x_1 + x_2 + a = (\alpha^6)^2 + \alpha^6 + 0 + 1 + \alpha^4 = \alpha^{12} + \alpha^6 + 1 + \alpha^4 = 1$$

$$y_3 = \lambda(x_1 + x_3) + x_3 + y_1 = \alpha^6(0 + 1) + 1 + 1 = \alpha^6$$

The SKey computed at AT is

$$\text{ATSKey} = (1, \alpha^6)$$

AN computes the ANSKey $= (\text{ANRand})(\text{ATPubKey})$

$$= 2(\alpha^{10}, \alpha) = (\alpha^{10}, \alpha) + (\alpha^{10}, \alpha)$$

$$x_3 = x_1^2 + \frac{b}{x_1^2} = (\alpha^{10})^2 + \frac{1}{(\alpha^{10})^2} = \alpha^{20} + \alpha^{-20} = \alpha^5 + \alpha^{10} = 1$$

$$y_3 = x_1^2 + \left(x_1 + \frac{y_1}{x_1}\right)x_3 + x_3 = (\alpha^{10})^2 + \left(\alpha^{10} + \frac{\alpha}{\alpha^{10}}\right) + 1 = \alpha^5 + \alpha^{10} + \alpha^6 + 1 = \alpha^6$$

The SKey computed at AN is

$$\text{ANSKey} = (1, \alpha^6)$$

Since ATSKey $=$ ANSKey $= (1, \alpha^6)$, it is proved that the SKey was established between AT and AN. $\square$

7.6.6 Authentication Key and Encryption Key Generation

The keys used for authentication and encryption are generated from the SKey, using the procedures specified as follows. SKey contains eight sub-fields (K0, K1, ..., K7) that are of equal length (Keylength/8). The AT and AN construct their message bits using the computed SKey, the 8-bit TransactionID, and a 64-bit TimeStampLong (based on the current 64-bit

Table 7.5 Message bits for generation of authentication and encryption keys

	MSB		LSB
Message bits for generation of four AuthKeys and four EncKeys	$K_i\ (0 \le i \le 7)$	Nonce $(0 \le i \le 7)$	TimeStampLong $(0 \le i \le 7)$
	Keylength/8	16-bits	64-bits

representation of CDMA System Time in units of 80 ms). The AN and AT should pad the message bits and compute the 160-bit message digests for each of the following 8 subKeys of SKey. The AN and AT then set the FACAuthKey, RACAuthKey, FACEncKey, RACEncKey, FPCAuthKey, RPCAuthKey, FPCEncKey, and RPCEncKey to the message digests for the corresponding key as shown in Table 7.5.

AuthKeys and EncKeys corresponding to message digests computed from eight sub-field within the SKey are shown in the following examples.

Example 7.4. Consider the 128-bit SKey that contains eight sub-fields of equal length, that is, K0 = K1 = K2 = K3 = K4 = K5 = K6 = K7 = 16 bits.

The AT and AN pad the message bits to produce a 512-bit padded message and compute the 160-bit message digests by means of the SHA-1 hash function that can be used for computation of authentication and encryption keys.

Suppose the SKey is given by:

SKey =	20a7	4f12	30ae	c5ff	52ad	621b	8601	a901
	K0	K1	K2	K3	K4	K5	K6	K7

First, the FTCAuthKey enables us to compute as shown below:

$$FTCAuthKey = K0||TransactionID||Nonce||TimeStampLong$$
$$= 16bits||8bits||16bits||64bits$$

where
K0 = 0x20a7 (16 bits).
TransactionID = 0x 10 (8 bits).
Nonce = 0xf263 (16 bits).
TimeStampLong = 0xa801 4cff 5106 de22 (64 bits).
The padded message bits, as shown below, should be used for computation of 160-bit message digest as follows:

Append a "1" followed by padding of 312 "0"s.

20a710f2	63a8014c	ff5106de	22800000
00000000	00000000	00000000	00000000
00000000	00000000	00000000	00000000
00000000	00000000	00000000	00000068
		length	field

(bigendian form)

A, B, C, D, and E are initialized by the following initial constants:

A = H0 = 67452301

B = H1 = efcdab89

C = H2 = 98badcfe

D = H3 = 10325476

E = H4 = c3d2e1f0

$Temp = S^5(A) + F_t(B,C,D) + E + W_t + K_t$

$E = D; D = C; C = S^{30}(B); B = A; A = Temp$

t: round number, $0 \leq t \leq 79$; S^i: circular left shift by i bits; K_t: four distinct additive constants over $0 \leq t \leq 79$; W_t: a 32-bit word derived from the current 512-bit input block; $\boxplus$: addition module 2^{32}

Through 80 continuous operations, according to SHA-1 algorithm (see Figure 10.2), the hex values of A, B, C, D, and E are computed as shown below:

Steps	A	B	C	D	E
0	c05ba9a5	67452301	7bf36ae2	98badcfe	10325476
1	d5ce0311	c05ba9a5	59d148c0	7bf36ae2	98badcfe
2	28400a71	d5ce0311	7016ea69	59d148c0	7bf36ae2
3	590e7d61	28400a71	757380c4	7016ea69	59d148c0
4	467a4ecc	590e7d61	4a10029c	757380c4	7016ea69
5	0654be0e	467a4ecc	56439f58	4a10029c	757380c4
6	e8cfca75	654be0e	119e93b3	56439f58	4a10029c
7	ea35e44	e8cfca75	81952f83	119e93b3	56439f58
8	16c6fcf5	ea35e44	7a33f29d	81952f83	119e93b3
9	d0381f75	16c6fcf5	03a8d791	7a33f29d	81952f83
10	4dcd6e6f	d0381f75	45b1bf3d	3a8d791	7a33f29d
11	d21519d4	4dcd6e6f	740e07dd	45b1bf3d	03a8d791
12	e50b2321	d21519d4	d3735b9b	740e07dd	45b1bf3d
13	37b3bcab	e50b2321	34854675	D3735b9b	740e07dd
14	fb797197	37b3bcab	7942c8c8	34854675	d3735b9b
15	ce2ad377	fb797197	cdecef2a	7942c8c8	34854675
16	ddb944aa	ce2ad377	fede5c65	cdecef2a	7942c8c8
17	dd0c56c1	ddb944aa	f38ab4dd	fede5c65	cdecef2a
18	bc6a6b38	dd0c56c1	b76e512a	F38ab4dd	fede5c65
19	d91585e4	bc6a6b38	774315b0	B76e512a	f38ab4dd
20	05fc91ec	d91585e4	2f1a9ace	774315b0	b76e512a
21	646a9fc0	5fc91ec	36456179	2f1a9ace	774315b0
22	05c5151e	646a9fc0	017f247b	36456179	2f1a9ace
23	b3280fa3	05c5151e	191aa7f0	017f247b	36456179
24	ad124529	b3280fa3	81714547	191aa7f0	017f247b
25	2da70a62	ad124529	ecca03e8	81714547	191aa7f0
26	ed42f31b	2da70a62	6b44914a	ecca03e8	81714547
27	c1e39894	ed42f31b	8b69c298	6b44914a	ecca03e8
28	7c496883	c1e39894	fb50bcc6	8b69c298	6b44914a
29	3a269e3c	7c496883	3078e625	fb50bcc6	8b69c298
30	611294c1	3a269e3c	df125a20	3078e625	fb50bcc6

Steps	A	B	C	D	E
31	1d6efe82	611294c1	0e89a78f	df125a20	3078e625
32	bff8d7c8	1d6efe82	5844a530	0e89a78f	df125a20
33	144cde03	bff8d7c8	875bbfa0	5844a530	0e89a78f
34	c4ae5995	144cde03	2ffe35f2	875bbfa0	5844a530
35	f34e286	c4ae5995	c5133780	2ffe35f2	875bbfa0
36	c3fcfe2a	f34e286	712b9665	C5133780	2ffe35f2
37	f50a3003	c3fcfe2a	83cd38a1	712b9665	c5133780
38	e47f696b	f50a3003	b0ff3f8a	83cd38a1	712b9665
39	1b1881b	e47f696b	fd428c00	b0ff3f8a	83cd38a1
40	a3f67ca6	1b1881b	f91fda5a	fd428c00	b0ff3f8a
41	2280ce98	a3f67ca6	c06c6206	F91fda5a	fd428c00
42	1a853ddf	2280ce98	a8fd9f29	C06c6206	f91fda5a
43	c29dbea	1a853ddf	8a033a6	A8fd9f29	c06c6206
44	b9e5504c	c29dbea	c6a14f77	8a033a6	a8fd9f29
45	34bee947	b9e5504c	830a76fa	C6a14f77	8a033a6
46	7cd54db5	34bee947	2e795413	830a76fa	c6a14f77
47	c69f0ffc	7cd54db5	cd2fba51	2e795413	830a76fa
48	ea7e9a7f	c69f0ffc	5f35536d	cd2fba51	2e795413
49	38577dfb	ea7e9a7f	31a7c3ff	5f35536d	cd2fba51
50	111acd7b	38577dfb	fa9fa69f	31a7c3ff	5f35536d
51	20d1a54b	111acd7b	ce15df7e	fa9fa69f	31a7c3ff
52	774c48ad	20d1a54b	c446b35e	ce15df7e	fa9fa69f
53	d394b159	774c48ad	c8346952	C446b35e	ce15df7e
54	e2fd3511	d394b159	5dd3122b	C8346952	c446b35e
55	263848cc	e2fd3511	74e52c56	5dd3122b	c8346952
56	0ef68ed	263848cc	78bf4d44	74e52c56	5dd3122b
57	5acd4cec	0ef68ed	98e1233	78bf4d44	74e52c56
58	d5f5f742	5acd4cec	403bda3b	98e1233	78bf4d44
59	438477c0	d5f5f742	16b3533b	403bda3b	98e1233
60	e1cb3109	438477c0	b57d7dd0	16b3533b	403bda3b
61	f5d1c599	e1cb3109	10e11df0	B57d7dd0	16b3533b
62	293a0d2e	f5d1c599	7872cc42	10e11df0	b57d7dd0
63	2c9a7086	293a0d2e	7d747166	7872cc42	10e11df0
64	a5c7c557	2c9a7086	8a4e834b	7d747166	7872cc42
65	16a0f2bf	a5c7c557	8b269c21	8a4e834b	7d747166
66	97190c09	16a0f2bf	e971f155	8b269c21	8a4e834b
67	94774b92	97190c09	c5a83caf	E971f155	8b269c21
68	f4a8c8d9	94774b92	65c64302	C5a83caf	e971f155
69	88281df4	f4a8c8d9	a51dd2e4	65c64302	c5a83caf
70	e45fe640	88281df4	7d2a3236	A51dd2e4	65c64302
71	a1e6b49	e45fe640	220a077d	7d2a3236	a51dd2e4
72	769795f2	a1e6b49	3917f990	220a077d	7d2a3236
73	c084d7d1	769795f2	42879ad2	3917f990	220a077d
74	c6054c69	c084d7d1	9da5e57c	42879ad2	3917f990
75	ef4439c1	c6054c69	702135f4	9da5e57c	42879ad2

76	43488a63	ef4439c1	7181531a	702135f4	9da5e57c
77	d2bef450	43488a63	7bd10e70	7181531a	702135f4
78	8117f963	d2bef450	d0d22298	7bd10e70	7181531a
79	e9b48aac	8117f963	34afbd14	D0d22298	7bd10e70

Finally,

$H_0 = H_0 + A = 50f9adad$

$H_1 = H_1 + B = 70e5a4ec$

$H_2 = H_2 + C = cd6a9a12$

$H_3 = H_3 + D = e104770e$

$H_4 = H_4 + E = 3fa3f060$

Now the 160-bit message digest is calculated by concatenating the H_0, H_1, H_2, H_3, and H_4:

50f9adad 70e5a4ec cd6a9a12 e104770e 3fa3f060

Thus, the computed FTCAuthkey by using $K_0 = 20a7$ is

50f9adad 70e5a4ec cd6a9a12 e104770e 3fa3f060.

Similarly, the following keys are computed by using K_i for $i = 1, 2, \ldots, 7$, respectively:

```
RTCAuthKey = 8691443d86a13cda16bc75c1e8332b077f287262
FTCEncKey = 2de7b52c5f95473a94a12ad5f44a467c6d65153c
RTCEncKey = ba15cb193662d15c8d7bf18ef71291fe5c783968
FCCAuthKey = bf08eb2aeee452349d418e9e6f7a3ce51cdbe02b
RACAuthKey = 344cb44161dfeccdf8be13b1f1f0955544f1cadd
FCCEncKey = 8e019fab7e5a85668eefa3adbd9cfc64bcee2798
RACEncKey = abb0380ab055c0c47761d14126fad8dbcc5bb01f.
```

□

- *KeyRequest Message Format:* The AN sends the KeyRequest message to initiate the SKey exchange. The AN should perform the following:
  — The AN sets the MessageID field (8 bits) to 0x 00.
  — The AN increments the TransmissionID field (8 bits) for each new KeyRequest message sent.
  — The AN sets the Timeout field (8 bits) to the maximum time required for calculation of the SKey.
  — The AN sets the ANPubKey field to the AN ephemeral public DH key.
- *KeyResponse Message Format:* The AT sends the KeyResponse message in response to the KeyRequest message. The AT should perform the following:
  — The AT sets the MessageID field (8 bits) to 0x 01.
  — The AT sets the TransactionID field (8 bits) to the value of the corresponding TransactionID field (8 bits) of the KeyRequest message.
  — The AT sets the Timeout field (8 bits) to the maximum time (in seconds) required for computation of the SKey.
  — The AT sets the ATPubKey field to the AT ephemeral to the public DH key.

- *ANKeyComplete Message:* The AN sends the ANKeyComplete message in response to the KeyResponse message. The AN should perform as follows:
  — The AN sets the 8-bit MessageID field to 0x 02.
  — The AN sets the 8-bit TransactionID field to the value of the 8-bit TransactionID field of the corresponding KeyRequest message.
  — The AN sets the Nonce field (16 bits) to an arbitrary chosen 16-bit value Nonce that is used to compute the KeySignature.
  — The AN sets the TimeStampShort field (16 bits) to the 16 least significant bits of the SystemTimeLong used in computing the KeySignature.
  — The AN sets the 160-bit KeySignature field to the 20-octet signature of the SKey.
- *ATKeyComplete Message:* The AT sends the ATKeyComplete message in response to the ANKeyComplete message. The AT should perform as follows:
  — The AT sets the 8-bit MessageID field to 0x 03.
  — The AT sets the 8-bit TransactionID field to the value of the 8-bit TransactionID field of the corresponding KeyRequest message.
  — The AT sets the Result field (1 bit) to "1" if the KeySignature field of ANKeyComplete message matches the message digest computed for the KeySignature; otherwise the AT sets this field to "0."
  — The AT sets the Reserved field (7 bits) to zero. The AT should ignore this field.

7.7 MAC Layer Protocols

This section presents the protocols for the MAC layer. The MAC layer contains the rules governing operation of the Control Channel, the Access Channel, the FTC Channel, and the RTC channel.

7.7.1 Data Encapsulation for the MAC Protocols

This section presents the data encapsulation for the In-Use instances of the MAC protocols. In the transmit direction, the MAC layer receives the security header packet, adds layer-related headers and trailers, concatenates them in the order to be processed on the receive side, adds padding where applicable, and forwards the resulting packet for transmission to the Physical Layer. In the receive direction, the MAC layer receives MAC packets from the Physical Layer and forwards the contained Security Layer packets to the Security Layer in the order received after removing the layer-related headers, trailers, and padding.

Figures 7.21–7.24 illustrate the relationship between Security Layer packets, MAC packets, and Physical Layer packets, for the Control Channel, the Access Channel, and the FTC and RTC channels.

7.7.2 Control Channel MAC Protocol

This protocol builds control channel MAC layer packets out of one or more security layer packets, and contains the rules concerning AN transmission and packet scheduling on the

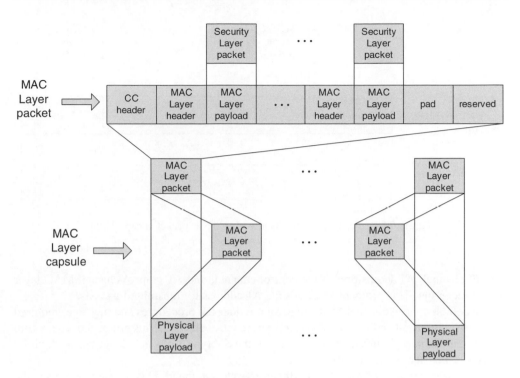

Figure 7.21 Control channel MAC layer packet encapsulation

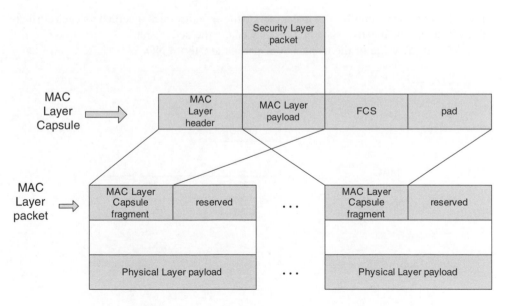

Figure 7.22 Access channel MAC layer packet encapsulation

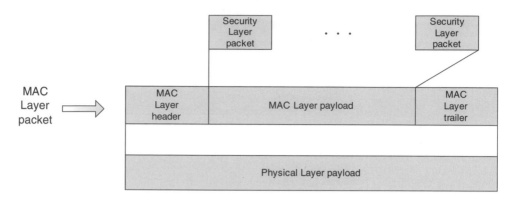

Figure 7.23   Forward traffic channel MAC layer packet encapsulation

control channel, AT acquisition of the control channel, and AT control channel MAC layer packet reception. This protocol also adds the AT address to transmitted packets.

The default control channel MAC protocol provides the procedures and messages required for an AN to transmit and for an AT to receive the control channel. This protocol operates with the default (Subtype 0) physical layer protocol, the Subtype 1 physical layer protocol, or the Subtype 2 physical layer protocol.

In the inactive state, the protocol waits for an activate command. This state applies only to the AT and occurs when the AT has not acquired an AN or is not monitoring the control channel. In the active state, AN transmits and the AT receives the control channel.

Up protocol initialization, the In-Use instant in the AT and AN, should perform as follows:

• The value of the attributes for this instant sets to the default values specified for each attribute.
• The AT enters the inactive state and the AN enters the active state.
• The AT sets the value of the ConnectionOpen parameter to NO.

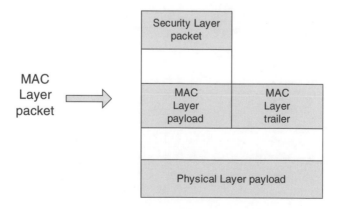

Figure 7.24   Reverse traffic channel MAC layer packet encapsulation

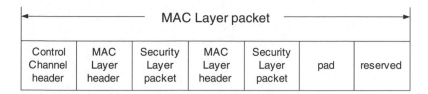

Figure 7.25 Control channel MAC packet structure

7.7.3 Procedures and Messages for the InUse Instance

Referring to Figure 7.25, each control channel MAC layer packet consists of zero or more security layer packets for zero or more ATs.

The protocol constructs a MAC layer packet out of the security layer packets as shown below:

- It adds the MAC layer header in front of every security layer packet (i.e. MAC layer payload).
- It concatenates the control channel header followed by the above formed packets.
- It pads the resulting packet.
- It adds the reserved bits.

The protocol then sends the packet for transmission to the physical layer.

7.7.4 Control Channel Capsules

Control channel MAC layer packets can be transmitted in a synchronous capsule or in a sub-synchronous capsule that is transmitted at a particular time. Control channel MAC layer packets can also be transmitted in an asynchronous capsule that can be transmitted at any time except when a synchronous or sub-synchronous capsule is transmitted. A synchronous capsule consists of one or more control channel MAC layer packets. A sub-synchronous capsule consists of one control channel MAC layer packet. An asynchronous capsule consists of one control channel MAC layer packet.

A control channel synchronous sleep state capsule constitutes the control channel MAC layer packets of a control channel synchronous capsule starting from the beginning of a synchronous capsule up to and including the first MAC layer packet for which the SleepStateCapsuleDone bit in the control channel header is set to "1."

This protocol expects an address and a parameter indicating transmission in a synchronous capsule, synchronous sleep state capsule, sub-synchronous capsule, or an asynchronous capsule with each transmitted security layer packet. For security layer packets that are carried by an asynchronous capsule or by a sub-synchronous capsule this protocol can also receive an optional parameter indicating a transmission deadline. Received packets are parsed into their constituent security layer packets. The packets that are addressed to the AT are then forwarded for further processing to the security layer.

The control channel cycle designates a 256 slot period, synchronous with CDMA system time; that is, an integer multiple of 256 slots between the beginning of a cycle and the beginning of CDMA system time.

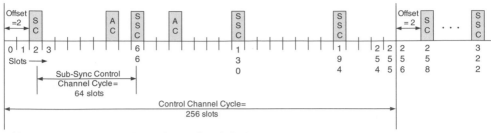

Figure 7.26 Location of control channel capsules

When the AN transmits the control channel, it should do so using the transmission formats defined by payload size (bits), normal transmission duration (slots), and preamble length (chips) with normal data rates of 19.2, 38.4, or 76.8 Kbps.

The timing of synchronous, sub-synchronous, and asynchronous capsules is shown in Figure 7.26.

- *Transmission of Synchronous Capsules:* The AN constructs a synchronous capsule out of all the pending security layer packets that are destined for transmission in a synchronous capsule. The synchronous capsule may contain more than one control channel MAC layer packet. The AN transmits exactly one synchronous control channel capsule per sector during each control channel cycle. The AN should transmit the control channel MAC layer packet of a synchronous capsule as follows:
    - The first MAC layer packet starts transmission at times T where T is CDMA system time in slots, satisfying as such

$$T \bmod 256 = \text{Offset}$$

    - All other MAC layer packets of the capsule start at the earliest time T following the end of transmission of the previous capsule that satisfies

$$T \bmod 4 = \text{Offset}$$

      where Offset is the value specified in the control channel header of the first control channel MAC layer packet of the synchronous capsule.
- *Transmission of Asynchronous Capsules:* The AN transmits asynchronous capsules at any time during the control channel cycle in which it does not transmit a synchronous capsule or a sub-synchronous capsule. The AN sets the FirstPacket and LastPacket bits of the control channel header to "1" for the control channel MAC layer packet of an asynchronous capsule.
- *Transmission of Sub-Synchronous Capsules:* The AN constructs a sub-synchronous capsule out of pending security layer packets that are destined for transmission in a sub-synchronous capsule. The sub-synchronous capsule contains one control channel MAC layer packet. The

Table 7.6 Control channel packets with transmission formats

MAC index	Normal data rates (Kbps)	Payload size (bits)	Normal transmit duration (slots)	Preamble length (bits)	Transmission format
71	19.2	128	4	1024	(128, 4, 1024)
71	38.4	256	4	1024	(256, 4, 1024)
71	76.8	512	4	1024	(512, 4, 1024)
3	38.4	1024	16	1024	(1024, 16, 1024)
2	76.8	1024	8	512	(1024, 8, 512)

AN may transmit zero or more sub-synchronous control channel capsules per sector during each control channel cycle. The AN should set the FirstPacket and LastPacket of the control channel header to "1" for the control channel MAC layer packet of a sub-synchronous capsule.

The AN should not transmit the control channel MAC layer packet of a sub-synchronous capsule except when the MAC layer packet starts transmission at times T where T satisfies

$$T \bmod 4 = \text{Offset}$$
$$T \bmod 256 \neq \text{Offset}$$

- *Access Network Requirements:* When the AN transmits the control channel, it uses the Transmission Formats as shown in Table 7.6.

7.7.5 Access Channel MAC Protocol

The Access Channel MAC protocol provides the procedures and messages required for an AT to transmit and for an AN to receive the access channel. The default access channel protocol operates with the Subtype 0 (default) physical layer protocol, the Subtype 1 physical layer protocol, and the Subtype 2 physical layer protocol, while the enhanced access channel MAC protocol operates with the Subtype 1 physical layer protocol or the Subtype 2 physical layer protocol.

7.7.5.1 Access Channel MAC Packet Structure

The transmission unit of this protocol is the access channel MAC layer packet. Each MAC layer packet contains part or all of a security layer packet. The protocol constructs one or more packets out of the security layer packet as follows:

- The protocol adds the MAC layer header in front of the security layer packet and adds the Frame Check Sequence (FCS) to the security layer packet.
- The protocol pads the security layer packet.
- The protocol splits the result into one or more access control MAC layer capsule fragments.
- The protocol adds the reserved bits to the capsule fragments to construct the access channel MAC layer packets.

This protocol passes the packets for transmission to the physical layer. Figure 7.22 shows an example of the MAC layer packet structure.

7.7.5.2 Access Probe Structure and Sequences

The access channel cycle specifies the time instants at which the AT starts an access probe. An access channel probe only begins at times T such that

$$T \bmod ACD = 0 \text{ for the default case, and}$$
$$(T - Access\ Offset) \bmod ACD = 0 \text{ for the enhanced case,}$$

where T is CDMA system time in slots, and ACD denotes Access Cycle Duration.

The structure of an individual access probe is illustrated in Figure 7.27. In each access probe, the pilot (I-channel) is first enabled and functions as a preamble. For the default case, after PreambleLength frames (i.e. PreambleLength $\times$ 16 slots), the probe data (Q-channel) is enabled for up to CapsuleLengthMax $\times$ 16 slots. For the enhanced case, after Preamble-LengthSlots slots, the probe data (Q-channel) is enabled for up to (CapsuleLengthMax $\times$ 16 slots). If the PreambleLengthSlots field is not included in the AccessParameter message, then the AT sets PreambleLengthSlots to (PreambleLength $\times$ 16 slots).

Figure 7.28 illustrates the access probe sequences. Each probe in a sequence is transmitted at increased power until any of the following conditions are met:

- the AT receives an ACAck message;
- transmission is aborted because the protocol received a deactivate command;
- the maximum number of probes per sequence (ProbNumStep) has been transmitted.

Prior to the transmission of the first probe, the AT performs a persistence test, which is used to control congestion in the access channel. In addition, the AT also performs a persistence test in between probe sequences, as shown in Figure 7.28.

7.7.5.3 Access Channel Long Code Mask

The AT sets two 42-bit AC long code masks. For transmission on the AC, the in-phase and quadrature-phase long codes, MI_{ACMAC} and MQ_{ACMAC}, are given as public data of the access channel MAC protocol, respectively. The 42-bit mask MI_{ACMAC} is specified in Table 7.7.

The 42-bit Mask MQ_{ACMAC} is derived from the mask MI_{ACMAC} as follows:

$$\text{MQ}_{\text{ACMAC}}[k] = \text{MI}_{\text{ACMAC}}[k-1] \text{ for } k = 1, 2, \ldots, 41$$
$$\begin{aligned}
\text{MQ}_{\text{ACMAC}}[0] = {} & \text{MI}_{\text{ACMAC}}[0] \oplus \text{MI}_{\text{ACMAC}}[1] \oplus \text{MI}_{\text{ACMAC}}[2] \oplus \text{MI}_{\text{ACMAC}}[4] \\
& \oplus \text{MI}_{\text{ACMAC}}[5] \oplus \text{MI}_{\text{ACMAC}}[6] \oplus \text{MI}_{\text{ACMAC}}[9] \oplus \text{MI}_{\text{ACMAC}}[15] \oplus \text{MI}_{\text{ACMAC}}[16] \\
& \oplus \text{MI}_{\text{ACMAC}}[17] \oplus \text{MI}_{\text{ACMAC}}[18] \oplus \text{MI}_{\text{ACMAC}}[20] \oplus \text{MI}_{\text{ACMAC}}[21] \\
& \oplus \text{MI}_{\text{ACMAC}}[24] \oplus \text{MI}_{\text{ACMAC}}[25] \oplus \text{MI}_{\text{ACMAC}}[26] \oplus \text{MI}_{\text{ACMAC}}[30] \oplus \text{MI}_{\text{ACMAC}}[32] \\
& \oplus \text{MI}_{\text{ACMAC}}[34] \oplus \text{MI}_{\text{ACMAC}}[41]
\end{aligned}$$

where $\oplus$ denotes the XORing operation, and $\text{MQ}_{\text{ACMAC}}[i]$ and $\text{MI}_{\text{ACMAC}}[i]$ denote the ith least significant bit of MQ_{ACMAC} and MI_{ACMAC}, respectively.

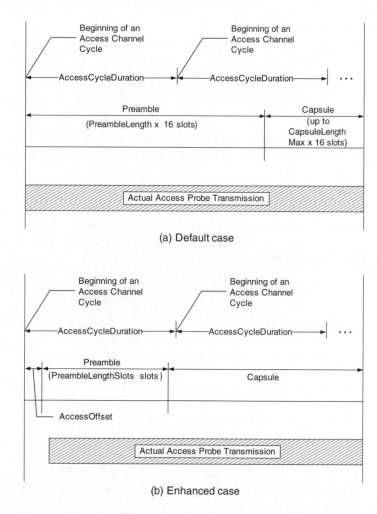

Figure 7.27 Access probe structure (Reproduced under written permission from Telecommunications Industry Association)

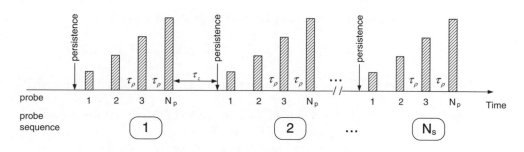

Figure 7.28 Access probe sequences

Table 7.7 Access channel long code masks

Bit	4 1	4 0	3 9	3 8	3 7	3 6	3 5	3 4	3 3	3 2	3 1	3 0	2 9	2 8	2 7	...	0 4	0 3	0 2	0 1	0 0
MI$_{ACMAC}$	1	1	AccessCycleNumber								Permuted(Color Code|SectorID[23:0])										

Referring to Table 7.7, ColorCode and SectorID designate public data of the overhead message protocol and correspond to the sector to which the AT is sending the access probe. AccessCycleNumber is defined as:

$$AccessCycleNumber = (SystemTime - AccessOffset) \bmod 256,$$

where SystemTime is the CDMA system time in slots corresponding to the first access probe preamble sent. Permuted (ColorCode\|SectorID[23 : 0]) is a permutation of the bits in Color-Code\|SectorID [23 : 0] and is defined as:

$$ColorCode|SectorID[23:0]) = (S_{31}, S_{30}, \ldots, S_0).$$

Permuted $(ColorCode|SectorID[23:0]) = (S_0, S_{31}, S_{22}, S_{13}, S_4, S_{26}, S_{17}, S_8, S_{30}, S_{21}, S_{12}, S_3,$ $S_{25}, S_{16}, S_7, S_{29}, S_{20}, S_{11}, S_2, S_{24}, S_{15}, S_6, S_{28}, S_{19}, S_{10}, S_1, S_{23}, S_{14}, S_5, S_{27}, S_{18}, S_9).$
When sending a probe sequence, the AT should perform the following rules:

- Prior to sending the first probe of the sequence, the AT waits until the beginning of the first AC cycle so as not to overlap with the reverse link silence interval. The reverse link silence interval is defined as the time interval of ReverseLinkSilenceDuration frames that starts at time T:

$$T \bmod (2048 \times 2^{ReverseLinkSilencePeriod} - 1) = 0$$

- If the AT receives an ACAck message, it stops the probe sequence and returns a TransmissionSuccessful indication.
- The AN should send an ACAck message to acknowledge receipt of an Access Channel MAC layer capsule it receives.

7.7.5.4 Frame Check Sequence (FCS)

The FCS is calculated using the following CRC-CCITT generator polynomial:

$$g(x) = x^{32} + x^{26} + x^{23} + x^{22} + x^{16} + x^{12} + x^{11} + x^{10} + x^8 + x^7 + x^5 + x^4 + x^2 + x + 1$$

The FCS computation is equal to the value computed by the logic depicted in Figure 7.29.

- The initial contents of the Linear Feedback Shift Register (LFSR) are set to binary "1" in order to make the FCS field to nonzero values even for all-zero data.
- Set the switches S1 and S2 in the up position and the switch S3 to the down position.

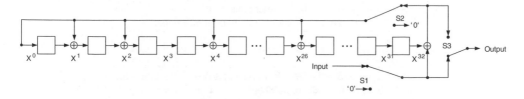

Figure 7.29   Access channel MAC layer capsule FCS

- The input bits of the Access Channel MAC capsule start to transmit to the encoder output as well as the feedback path of the shift register.
- The register should be clocked once for each bit of Access Channel MAC layer capsule, excluding the FCS and padding bits.
- The Access Channel MAC layer capsule is read in order from MSB to LSB, starting with the MSB of the MAC layer header.
- The switches S1 and S2 should be set in the down positions and the switch S3 in the up position so that the output is a module-2 addition with a "0" and the successive shift register inputs are "0."
- Finally, the register should be clocked an additional 32 times for the 32 FCS bits.

7.7.6 Forward Traffic Channel MAC Protocol

The FTC MAC protocol provides the techniques required for an AN to transmit and an AT to receive the FTC. Specifically, this protocol addresses FTC addressing and FTC rate control. The default FTC MAC protocol operates with the Subtype 0 or the Subtype 1 physical layer protocols. The enhanced FTC MAC protocol operates with the Subtype 2 physical layer protocol.

This protocol operates in one of three states:

- *Inactive State:* Since the AT is not assigned an FTC in this state, it waits for an activate command.
- *Variable Rate State:* In this state, the AN transmits the FTC to the AT, in accordance with the Data Rate Control (DRC) channel received from the AT.
- *Fixed DRC State:* In this state, the AN transmits the FTC to the AT, in accordance with the FixedModeEnable message received from the AT.

7.7.6.1 MAC Layer Packet

The MAC layer packet is the basic unit of data provided by the enhanced FTC MAC protocol to the physical layer protocol.

The MAC layer packets are of the following three types:

- single user simplex;
- single user multiplex;
- multi-user.

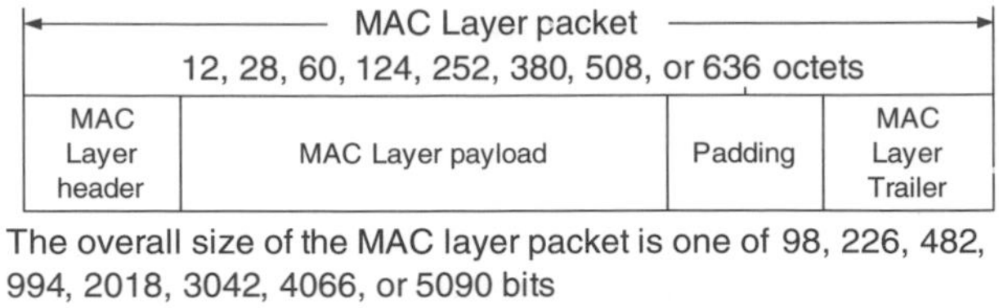

The overall size of the MAC layer packet is one of 98, 226, 482, 994, 2018, 3042, 4066, or 5090 bits

Figure 7.30   FTC MAC layer packet structure

The structure of a MAC layer packet is shown in Figure 7.30. The MAC layer packet consists of the following:

- MAC layer header (which may be empty);
- MAC layer payload;
- padding (if required);
- MAC layer trailer.

The MAC layer payload consists of one or more security layer packets addressed to one or more access terminals. The overall size of the MAC layer packet is one of a discrete set of the following values: 98, 226, 482, 994, 2018, 3042, 4066, or 5090 bits. The MAC layer header and MAC layer trailer are used to specify the type of the MAC layer packet (i.e. single user simplex, single user multiplex, or multi-user) and to provide the information needed to parse the contents of the MAC layer payload. The MAC layer payload is followed by padding, which consists of all "0"s. The size of the padding sequence will depend on the overall size of the MAC layer packet.

A single user simplex MAC layer packet is used to carry one security layer packet in its payload and is addressed to one AT. The MAC layer payload size equals the size of the MAC layer packet minus the size of the MAC layer trailer. A single user simplex packet consists of an empty MAC layer header, a MAC layer payload, and no padding. The security layer packet in a single user simplex packet contains a Format A or Format B connection layer packet.

A single user multiplex MAC layer packet is used to carry in its payload one or more security layer packets addressed to one AT. The combined size of the MAC layer payload is the size of the MAC layer packet minus the size of the MAC layer trailer. A single user multiplex packet consists of a non-empty MAC layer header, MAC layer payload, which consists of one or more security layer packets addressed to one AT, and padding (if required). Each security layer packet in a single user multiplex packet contains a Format A connection layer packet.

A multi-user MAC layer packet is used to carry in its payload one or more security layer packets addressed to one or more ATs. A multi-user packet consists of a non-empty MAC layer header, MAC layer payload consisting of one or more (maximum of eight) security layer packets, and padding (if required). Each security layer packet in a multi-user packet contains a Format A or Format B connection layer packet.

7.7.6.2 Construction of MAC Layer Packets

The structure of a single user simplex MAC packet is shown in Figure 7.31. The MAC layer trailer field (2 bits) indicates a single user simplex packet if it is equal to 01 or 11. A single user

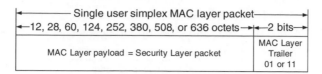

Figure 7.31 FTC single user simplex MAC layer packet

simplex MAC layer packet is used to carry one security layer packet in its payload if the security layer packet contains a Format A or a Format B connection layer packet. The AN sets the MAC trailer field (2 bits) to "11" if the security layer packet contains a Format B connection layer packet. Otherwise, the AN sets the trailer field to "01."

The structure of a single user multiplex packet is shown in Figure 7.32. The AN constructs a single user multiplex MAC packet as follows: The MAC layer trailer field (2 bits) indicates a single user multiplex packet if the AN sets the MAC layer trailer field to "10." The MAC layer payload consists of n security layer packets, $n = 1, 2, 3, \ldots$, where n is an integer greater than or equal to one. Note that a security layer packet is from the Security protocol containing a Format A connection layer packet. The MAC layer header consists of m length fields, in octets, where m is equal to n or $n + 1$. If $m = n + 1$, the AN sets the mth occurrence of this field to "00000000." The AN sets the size of the padding field to the size of the MAC layer packet minus the size of the MAC layer header (m octets), payload, and trailer (2 bits). The AN should set the value of this field to all "0"s.

The AN constructs a multi-user layer packet as shown in Figure 7.33. The MAC layer payload consists of n security layer packet where, $n = 1, 2, \ldots, 8$. The MAC layer header consists of m PacketInfo fields, and n length field, where m is equal to n or $n + 1$. If the MAC layer payload contains eight security layer packets, the AN sets m to n. Otherwise, the AN should set m to $n + 1$.

Specifically, if an ith security layer packet for $1 \le i \le n$ contains a Format B connection layer packet, the AN sets the ith occurrence of the Format field to "1." Otherwise, the AN should set the ith occurrence of this field to "0." The ith occurrence of the MACIndex field should be set to the MACIndex of the AT to which the ith security layer packet is addressed. The AN uses the MACIndex assigned to the AT by the sector transmitting this MAC layer packet. The ith occurrence of the length field should be set to the length, in octets, of the ith security layer packet in this MAC layer payload. The AN sets the size of the padding field to the size of the MAC layer packet minus the size of MAC layer header ($m + n$ octets), payload, and trailer (2 bits). The AN sets the value of padding to all "0." The AN sets the FTC MAC trailer (2 bits) to "00."

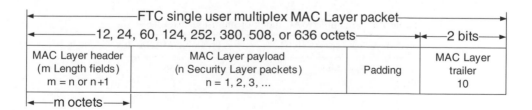

Figure 7.32 FTC single user multiplex MAC Layer packet

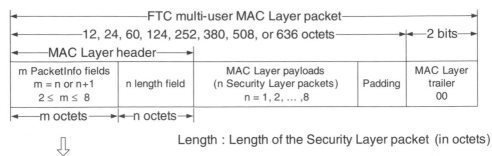

Length : Length of the Security Layer packet (in octets)

Figure 7.33   FTC multi-user MAC Layer packet

7.7.6.3 Preamble MAC Index

The transmission of a MAC layer packet on the FTC is preceded by the transmission of a preamble MAC index. When transmitting a single user MAC layer packet to an AT on the FTC, the AN should set the preamble MAC index to the MACIndex assigned to the AT by transmitting sector. When transmitting a multi-user MAC layer packet on the FTC, the AN sets the preamble MAC index based on the MAC layer packet size in accordance with Table 7.8.

The preamble MAC index preceding a single user (simplex or multiplex) MAC layer packet specifies the address of the single user MAC layer packet. The preamble MAC index preceding a multi-user MAC layer packet specifies the MAC layer packet size of the multi-user MAC layer packet.

A single user MAC layer packet is addressed to one AT, while different security layer packets embedded in a multi-user MAC layer packet are addressed to different ATs. When transmitting a single user MAC layer packet, the MACIndex of the AT is indicated by the preamble MAC index. When transmitting a multi-user packet, the MACIndex of the AT is specified in the PktInfo field of the MAC layer header, corresponding to the security layer packet being addressed to the AT.

Table 7.8   Preamble MAC index of multi-user packets

MAC layer packet size (bits)	Preamble MAC index
98, 226, 482, or 994	66
2018	67
3042	68
4066	68
5090	69

7.7.6.4 DRC and DSC Channels

The AT transmits the Data Rate Control (DRC) channel and the Data Source Control (DSC) channel. The DRC channel transmission consists of a DRC cover and a DRC value. The AT uses the DRC cover to specify the requested sector for packet transmission. The AT transmitting a cell cover on the DSC channel should set the cell cover to the value indicated by one of the DSC fields in the last traffic channel assignment message. If the AN transmits a packet to the AT starting in slot T, it should do so from the sector associated with DRC cover. The AT uses the DRC value to specify the set of requested transmission formats.

The DSC channel transmission is either a null-cover or a cell-cover. If the DSC is a cell-cover, it specifies the Forward Link Data Source (or equivalently, a Data Source) represented by the cell cover. A Data Source of the AT is a group of sectors that maintain a command forward link data queue for the AT. The AT sets the DRC value to a 4-bit DRC index, ranging from 0x0 to 0xe. The DRC index 0 is known as the null-rate DRC. Each DRC index is associated with a Rate Matrix, a span, a list of single user transmission formats, and a list of multi-user transmission formats.

7.7.7 Reverse Traffic Channel MAC Protocol

The Reverse Traffic Channel (RTC) MAC protocol addresses RTC transmission rules and rate control. This protocol operates with the Subtype 0 (default), Subtype 1, Subtype 2, or Subtype 3 Physical layer protocols.

7.7.7.1 Default Subtype 0 RTC MAC Protocol

The transmission unit of this protocol is a RTC MAC layer packet. Each packet contains one security layer packet adding the MAC layer trailer:

RTC MAC layer packet = Security layer packet + MAC layer trailer

The AT sets the ConnectionLayerFormat field to "1" if the MAC layer packet is in Format B and contains valid payload. The protocol then sends the packet for transmission to the physical layer.

If the MACLayerFormat field of the MAC layer trailer is equal to "1," received packets are passed for further processing to the Security layer after removing the layer-related trailer. The AN discards the MAC packet if the MACLayerFormat field of the MACLayer trailer is equal to "0." The ConnectionLayerFormat field in the MAC layer trailer should be passed to the Security layer with the Security layer packet.

Payload sizes used on the RTC are a function of the transmission rate as shown in Table 7.9. The AT and AN should perform the following requirements:

- The AT should set a timer when it enters the Setup State.
- If the protocol is still in the Setup State when the timer expires, the AT should return to a SupervisionFailed indication.
- The AT starts transmission on the RTC upon entering the Setup State, and should obey the Reverse Power Control (RPC) channel. The AT sets the DRC value and cover as specified in the FTC MAC protocol.

Table 7.9   RTC transmission rates and payload sizes

Transmission rate (Kbps)	Minimum payload (bits)	Maximum payload (bits)
9.6	1	232
19.2	233	488
38.4	489	1000
76.8	1001	2024
153.6	2025	4072

- The AN using the current AT Identifier sends the RTCAck message to notify the AT that it has acquired the RTC channel.
- If the AT receives an RTCAck message, it should return a LinkAcquired indication and transmission to the Open State. If the AN acquires the RTC channel, it should send an RTCAck message to the AT, return a LinkAcquired indication, and transition on the Open State.

The AN uses the UnicastReverseRateLimit message to control the transmission rate on the reverse link for a particular AT. The AN uses the BroadReverseRateLimit message to control the transmission rate on the reverse link.

The rate control uses the following variables: MaxRate, CombineBusyBit, and CurrentRateLimit. CurrentRateLimit should be set initially to 9.6 Kbps. After the AT receives a BroadcastReserveRateLimit or the UnicastReserveRateLimit message, the AT should update the CurrentRateLimit value as follows:

- If the RateLimit value in the message is less than or equal to the CurrentRateLimit value, the AT should set CurrentRateLimit to the RateLimit value in the message immediately after the receipt of the message
- If the RateLimit value in the message is greater than the CurrentRateLimit value, the AT should set CurrentRateLimit to the RateLimit value in the message, one frame (16 slots) after the message is received.
- CurrentRate is set to the rate at which the AT was transmitting data immediately before the new transmission time.

The AT should sets the variable MaxRate based on its current transmission rate, the value of the CombinedBusyBit, and a random number. The AT should generate a uniformly distributed random number x, $0 < x < 1$.

7.7.7.2 Subtype 1 RTC MAC Protocol

Subtype 1 RTC MAC protocol provides the procedures and messages required for an AT to transmit, and for an AN to receive, the RTC, in an almost identical manner to the Subtype 0 RTC MAC protocol.

The only difference regarding the Subtype 1 from the Default (Subtype 0) RTC MAC protocol is to offer a set of attribute-values for a given attribute. The AT (sender) sends an

AttributedUpdateAccept message in response to an AttributeUpdateRequest message to accept the offered attribute values. The AN sends an AttributeUpdateReject message in response to an AttributeUpdateRequest message to reject the offered attribute values. This protocol operates with the Subtype (Default) and Subtype 1 physical layer.

7.7.7.3 Subtype 2 RTC MAC Protocol

The Subtype 2 RTC MAC protocol supports intra-access terminal QoS for multiple concurrent active MAC flows at the AT. Rate Control is accomplished via per active MAC flow Traffic-to-Pilot (T2P) power ratio control. The Subtype 2 RTC MAC protocol provides per MAC flow QoS control. This is achieved by distributed rate selection at the AT with centralized resource allocation by the AN. This protocol operates with the Default (Subtype 0) or the Subtype 1 physical layer protocol. It is assumed that the AT has one instant of this protocol for every AT.

This protocol operates in one of three states:

- *Inactive State:* The AT in this state is not assigned a RTC and waits for an activate command.
- *Setup State:* The AT in this state obeys the power control command that it receives from the AN. Data transmission on the RTC is not allowed.
- *Open State:* The AT in this state obeys the power control command from the AN. In this state, the AT may negotiate different Subtype 2 RTC MAC protocol parameters and attributes per MAC flow and transmit data on the RTC.

Each active MAC flow contributes to sector loading and its contribution is strongly correlated with its average transmit T2P. The Subtype 2 RTC MAC protocol supports multiple MAC flows. An AT may transmit multiple active MAC flows simultaneously. This protocol controls the average T2P of an active MAC flow based on the requirements of that MAC flow, requirements of other concurrent active MAC flow, transmit power constraints, and sector loading. The Subtype 2 RTC MAC protocol allows updating RTC MAC attributes using the generic attribute update protocol.

Transmission unit of this protocol is a RTC MAC layer packet that contains one security layer packet. The MAC layer packet is the basic unit of data provided by the Subtype 2 RTC MAC protocol to the physical layer protocol. The structure of a MAC layer packet consists of a security layer packet followed by the MAC layer trailer.

7.7.7.4 Subtype 3 RTC MAC Protocol

The Subtype 3 RTC MAC protocol addresses RTC transmission rules and rate control. This protocol supports intra-access terminal QoS for multiple concurrent active MAC flows at the AT. Rate control is accomplished via per active MAC flow T2P control. This protocol operates with the Subtype 2 physical layer protocol.

- An AT may transmit multiple active MAC flows simultaneously. The Subtype 3 RTC MAC protocol uses the average transmit T2P per active MAC flow as a measure of the air link resource used by that MAC flow. The Subtype 3 RTC MAC protocol treats multiple active MAC flows associated with a single AT in a manner consistent with multiple active MAC flows associated with multiple access MAC access terminals, subject to AT transmit power constraints.

- The MAC layer packet is the basic unit of data provided by the Subtype 3 RTC MAC protocol to the physical layer protocol. The structure of a MAC layer packet consists of a security layer packet followed by the MAC layer trailer. The MAC layer packet is transmitted in one of two modes: low latency or high capacity mode.

7.7.7.5 RTC Long Code Mask

The AT should set the long code masks for the RTChannel (MI_{RTCMAC} and MQ_{RTCMAC}) as shown below. The 42-bit mask MI_{RTCMAC} is specified in Table 7.10.

The MI_{RTCMAC} field (42 bits) should be set to the value of RTC in-phase long code mask associated with the AT's session. The MQ_{RTCMAC} field (42 bits) is set to the value of the RTC quadrature-phase long code mask associated with the AT's session. These two parameters are included in the session state information record only if the session state information is being transferred while the connection is open. The session state information is to be used for transferring the session parameters corresponding to the InUse protocol instances from a source AN to a target AN. Session parameters are the attribute and the internal parameters that define the state of each protocol. The sender should include all the parameter records associated with the ProtocolType and ProtocolSubtype in the same session state information record.

Permuted (ATI_{LCM}) in Table 7.10 is defined as follows:

$$ATI_{LCM} = (A_{31}, A_{30}, A_{29}, \ldots, A_1, A_0)$$

$$Permuted(ATI_{LCM}) = (A_0, A_{31}, A_{22}, A_{13}, A_4, A_{26}, A_{17}, A_8, A_{30}, A_{21}, A_{12}, A_3, A_{25}, A_{16}, A_7,$$
$$A_{29}, A_{20}, A_{11}, A_2, A_{24}, A_{15}, A_6, A_{28}, A_{19}, A_{10}, A_1, A_{23}, A_{14}, A_5, A_{27}, A_{18}, A_9)$$

The 42-bit mask MQ_{RTCMAC} should be derived from the mask MI_{RTCMAC} as follows:

$$MQ_{RTCMAC}[k] = MI_{RTCMAC}[k-1] \, for \, k = 1, 2, \ldots, 41$$
$$MQ_{RTCMAC}[0] = MI_{RTCMAC}[0] \oplus MI_{RTCMAC}[1] \oplus MI_{RTCMAC}[2] \oplus MI_{RTCMAC}[4]$$
$$\oplus MI_{RTCMAC}[5] \oplus MI_{RTCMAC}[6] \oplus MI_{RTCMAC}[9] \oplus MI_{RTCMAC}[15]$$
$$\oplus MI_{RTCMAC}[16] \oplus MI_{RTCMAC}[17] \oplus MI_{RTCMAC}[18]$$
$$\oplus MI_{RTCMAC}[20] \oplus MI_{RTCMAC}[21] \oplus MI_{RTCMAC}[24]$$
$$\oplus MI_{RTCMAC}[25] \oplus MI_{RTCMAC}[26] \oplus MI_{RTCMAC}[30]$$
$$\oplus MI_{RTCMAC}[32] \oplus MI_{RTCMAC}[34] \oplus MI_{RTCMAC}[41]$$

Table 7.10 RTC long code mask

BIT	41	40	39	37	36	35	34	33	32	31	...	22	21	20	...	14	13	...	02	01	00
MI_{RTCMAC}	1	1	1	1	1	1	1	1	1	Permuted (ATI_{LCM})										

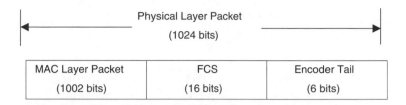

Figure 7.34 Physical layer packet format for the control channel

where the ⊕ denotes the Exclusive OR operation, and $MQ_{RTCMAC}[i]$ and $MI_{RTCMAC}[i]$ denote the ith least significant bit of MQ_{RTCMAC} and MI_{RTCMAC}, respectively.

7.8 Physical Layer Protocol

This section contains the specification for the Default (Subtype 0), the Subtype 1, and the Subtype 2 Physical Layer protocol. The transmission unit of this protocol is the physical layer packet that contains one or more MAC layer packets.

7.8.1 Subtype 0 (Default) and Subtype 1 Physical Layer Protocol

A physical layer packet has its own length of 256, 512, 1024, 2048, 3072, or 4096 bits.

The length of a control channel physical layer packet is 1024 bits. Each control channel packet carries one control channel MAC layer packet. The format of the control channel physical layer packet contains the three fields composed of one control channel MAC layer packet (1002 bits), the FCS (16 bits), and Encoder Tail (6 bits) as shown in Figure 7.34.

The length of an AC physical layer packet of the Default (Subtype 0) physical layer protocol is 256 bits. The length of an AC physical layer packet of the Subtype 1 physical layer protocol will be 256, 512, or 1024 bits. Each AC physical layer packet carries one AC MAC layer packet. AC physical layer packets of Subtype 0 physical layer protocol have the field format consisting of MAC layer packet (234 bits), FCS (16 bits), and Encoder Tail (6 bits) as shown in Figure 7.35. AC physical layer packets of Subtype 1 physical layer protocol use the following fields as illustrated in Figure 7.36: MAC layer packet (234, 490, or 1002 bits), FCS (16 bits), and Encoder Tail (6 bits).

The length of a FTC physical layer packet will be 1024, 2048, 3072, or 4096 bits. A FTC physical layer packet will carry 1, 2, 3, or 4 FTC MAC layer packets depending on the transmission rate. The physical layer packet format for the FTC will use the following fields:

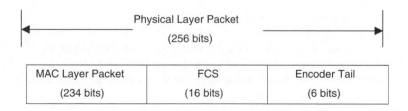

Figure 7.35 Subtype 0 physical layer protocol packet format for the access channel

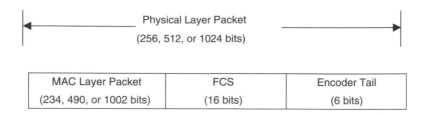

Figure 7.36   Subtype 1 physical layer protocol packet format for the access channel

- *MAC layer packet (1002 bits):* This field is the MAC layer from the FTC MAC protocol.
- *PAD (22 bits):* This padding field should be set to all "0"s. The receiver should ignore this field.
- *FCS (16 bits):* This field is the frame check sequence.
- *TAIL (6 bits):* This field represents Encoder Tail bits and should be set to all "0"s.

Figure 7.37 illustrates the format of FTC physical layer packets.

The length of a RTC physical layer packet will be 256, 512, 1024, 2048, or 4096 bits. Each RTC physical layer packet carries one RTC MAC layer packet. The physical layer packet format for the RTC will use the following fields: MAC layer packet (234, 490, 1002, 20026, or 4074 bits), FCS (16 bits), and Encoder Tail (6 bits).

Figure 7.38 illustrates the format of RTC physical layer packets.

Each field of the physical layer packets is transmitted in sequence such that the most significant bit (MSB) is transmitted first and the least significant bit (LSB) is transmitted last. The MSB is the left-most bit in the figures of this section.

7.8.2 Frame Check Sequence (FCS) Computation

The FCS computation should be used for computing the FCS field in the control channel physical layer packets, the FTC physical layer packets, the AC physical layer packets, and the RTC physical layer packets.

The FCS is a CRC calculated using the standard CRC-CCITT generator polynomial:

$$g(x) = x^{16} + x^{12} + x^5 + 1$$

The FCS bits are equal to the values computed according to the following procedure as shown in Figure 7.39:

- All shift-register elements should be initialized to "0"s.
- Set the switches in the up position.
- Clock the register once for each bit of the physical layer packet except for the FCS and TAIL fields. The physical layer packet should be read from MSB to LSB.
- Set the switches in the down position so that the output is a modulo-2 addition with a "0" and the successive shift-register inputs are "0"s.
- The register should be clocked an additional 16 times for the 16 FCS bits.
- The output bits constitute all fields of the physical layer packets except the TAIL field.

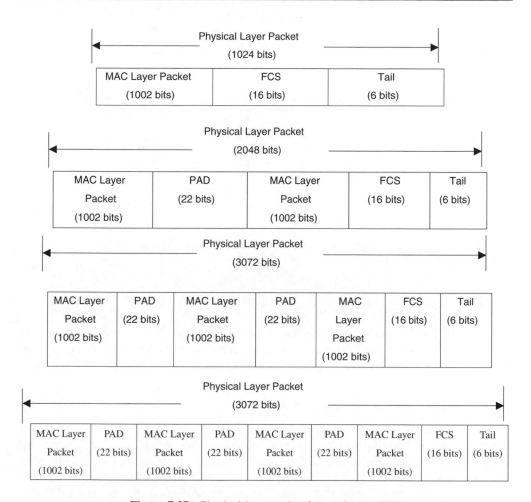

Figure 7.37 Physical layer packet format for the FTC

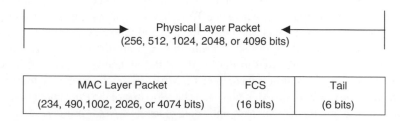

Figure 7.38 Physical layer packet for the RTC

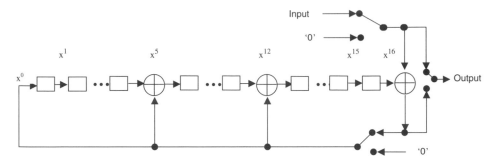

Up position for all the bits of the physical layer packet except for the
FCS and TAIL fields; then down position for the 16-bit FCS field.

Figure 7.39   FCS computation for the physical layer packet

7.8.3 Role of Access Terminal

This section defines requirements specific to the AT operation. When transmitting, the AT transmits access probes until the access attempt succeeds or ends. The normal access probe structure and its transmit power requirements are defined as part of the AC MAC protocol.

7.8.3.1 Output Power of Reverse Channels

The power of the AC relative to that of the pilot channel is defined in 3GPP2 C.S0024-A v1.0. When the AT is transmitting the RTC, the AT transmits the Pilot channel, the DRC channel, the ACK channel when acknowledging received physical layer packets, and the Data channel when transmitting physical layer packets. The transmitted power level of the Data channel should be adjusted depending on the selected data rate and reverse link power control. The traffic data should be transmitted in the form of physical layer packets. When the data rate is changed, the AT output power should be within ±0.5 dB or 20% of the change in dB, whichever is greater.

The AT should disable its transmitter except when it is instructed by a MAC protocol to transmit. The normal access probe structure and its power requirements are defined as part of the AC MAC protocol. Once instructed by RTC MAC protocol, the AT initiates RTC transmission. During the transmission of the RTC, the determination of the output power needed to support the Data channel, the DRC channel, and the ACK channel is an additional open-loop process performed by the AT. During the transmission of the DRC channel, the power of the DRC channel relative to that of the Pilot channel should be as specified by DRCChannelGain, which is public data of the FTC MAC protocol. During the transmission of the ACK channel, the power of the ACK channel relative to that of the Pilot channel should be as specified by ACKChannelGain, which is public data of the FTC MAC protocol. The AC should maintain the power of the Data channel, DRC channel, and ACK channel, relative to that of the Pilot channel, to within ±0.25 dB of the specified values. If the AT is unable to transmit at the requested output power level when the maximum RTC data rate is 9600 bps, the AT should reduce the power of the DRC channel and the ACK channel accordingly.

7.8.3.2 Reverse Channel Structure

The reverse channel consists of the AC and the RTC. The AC consists of a Pilot channel and a Data channel. The RTC channel is composed of a Pilot channel, a reverse rate indicator (RRI) channel, a Dtaa Rate control (DRC) channel, an Aknowledgement (ACK) channel, and a Data channel.

- *Pilot Channel:* The AT transmits un-modulated symbols with a binary value of "0" on the pilot channel.
- *RRI Channel:* This channel is used to indicate the data rate of the data channel being transmitted on the RTC.
- *DRC Channel:* This channel is used by the AT to indicate to the AN the requested FTC channel data rate and the selected serving sector on the forward channel.
- *ACK Channel:* This channel is used by the AT to inform the AN whether or not the physical layer packet transmitted on the FTC channel has been received successfully.

When the AT is transmitting a RTC, it should continuously transmit the pilot channel and the RRI channel. These channels are time-division multiplexed and should be transmitted on Walsh channel W_0^{16}. When the DRC channel is active, it should be transmitted for full slot durations on a Walsh channel W_8^{16}. The AT will transmit an ACK channel bit in response to every FTC slot that is associated with a detected preamble directed to the AT. Otherwise, the ACK channel should be gated off. When the ACK channel bit is transmitted, it should be transmitted on the first half slot on Walsh channel W_4^8.

For the RTC, the encoded RRI symbols will be time-division multiplexed (TDM) with the Pilot channel, and the encoded RRI symbols should be allocated the first 256 chips of every slot as shown in Figure 7.40.

Figure 7.41 shows an example of ACK channel operation for a 153.6 Kbps FTC. The 153.6 Kbps FTC physical layer packets use four slots, and these slots are transmitted with a three-slot interval between them.

Figure 7.41 shows the case of a normal physical layer packet termination. Suppose the AT transmits NAK responses on the ACK channel after the first three slots of the physical layer packet are received. In this case, it will indicate that it was unable to receive the FTC physical layer packet correctly after only one, two, or three of the normal four slots.

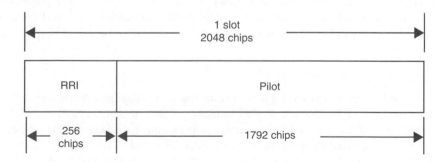

Figure 7.40 Pilot channel and RRI channel TDM allocations for the RTC

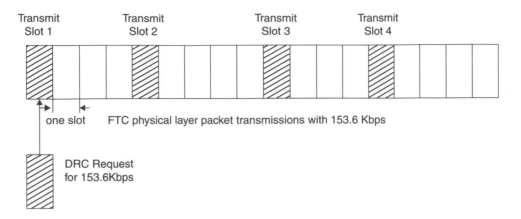

Figure 7.41   A normal physical layer packet transmission

7.8.3.3 Access Channel (AC)

The AT uses the AC to initiate communication with the AN. The AT is also used to respond to an AT directed message. The AC consists of a Pilot channel and a Data channel as shown in Figure 7.42.

The AT transmits information on the access channel at a fixed data rate of 9.6 Kbps if using the Subtype 0 (Default) physical layer protocol. When using the Subtype 1 physical layer protocol, the AT transmits information on the AC at a variable data rate of 9.6, 19.2, or 38.4 Kbps.

Pilot Channel
The AT transmits unmodulated symbols with a binary value of "0" on the Pilot channel. This channel should be transmitted continuously during AC transmission in such a way that it should be transmitted on the I channel using the 16-chip Walsh function number 0 cover (i. e. $W_0^{16} = + + \cdots + + = 16$ chips). Notice here that the signal point mapping is $0 \rightarrow +$ and $1 \rightarrow -$.

Data Channel
One or more AC physical layer packets are transmitted on the Data channel during every access probe. The AC physical layer packets should be transmitted at a fixed data rate of 9.6 Kbps on the Q channel using the 4-chip Walsh function number 2 (i.e. $W_2^4 = + + --$). There, physical layer packets will be preceded by a preamble of PreambleLength frames where only the Pilot channel is transmitted. The PreambleLength parameter is public data from the AC MAC protocol.

7.8.3.4 Reverse Traffic Channel (RTC)

The RTC is used by the AT to transmit user-specific traffic or signaling information to the AN. The RTC consists of a Pilot channel, an RRI channel, a DRC channel, an ACK channel, and a Data channel. The AT will support transmission of information on the Data channel at the rate of 9.6 Kbps for Subtype 0 and at the rate of 9.6, 19.2, 38.4, 76.8, and 153.6 Kbps for Subtype 1, as shown in Figure 7.43. The data rate used on the Data channel is specified by the RTC MAC protocol.

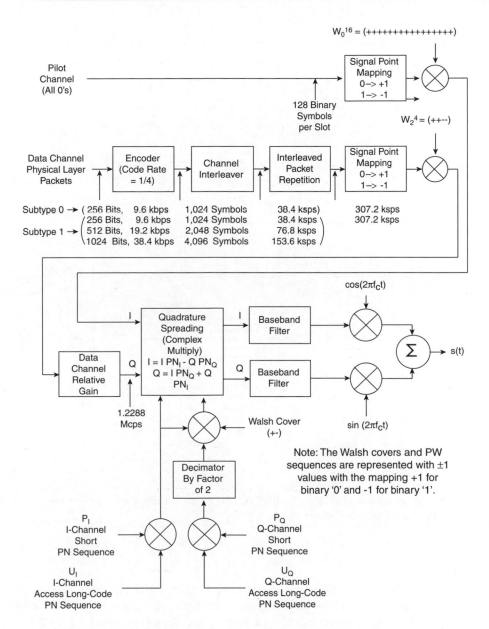

Figure 7.42 Subtype 0 and subtype 1 physical layer protocol reverse channel structure for the access channel (Reproduced under written permission from Telecommunications Industry Association)

Pilot Channel

The AT transmits unmodulated symbols with a binary value of "0" for mapping +1 on the Pilot channel. The transmission of the Pilot channel and the RRI channel should be time-multiplexed on the same Walsh channel as illustrated in Figure 7.43. The Pilot channel and the RRI channel are transmitted at the same power.

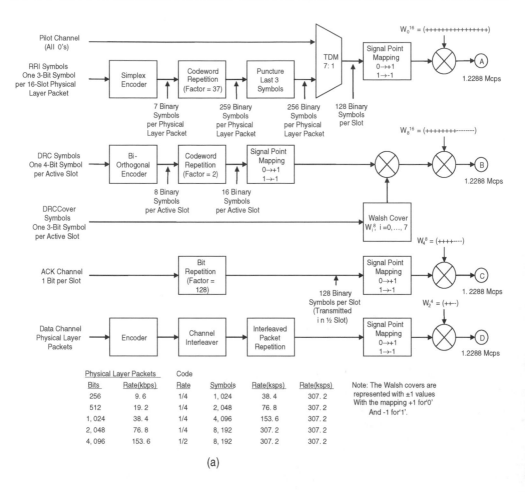

Figure 7.43 Reverse channel structure for the RTC (Reproduced under written permission from Telecommunications Industry Association)

RRI Channel

The AT uses the RRI channel to indicate the data rate at which the Data channel is transmitted. The data rate is represented by a 3-bit RRI symbol per 16-slot physical layer packet. Each RRI symbol is encoded in a 7-bit codeword by a simplex encoder as specified in Table 7.11.

Each codeword is repeated 37 times and the last three symbols should be disregarded, as shown in Figure 7.43. The resulting 256 binary symbols per physical layer packet should be time-division multiplexed with the Pilot channel symbols and span the same time interval as the corresponding physical layer packet. The time-division-multiplexed pilot and RRI channel sequence will be spread with the 16-chip Walsh function W_0^{16} producing 256 RRI chips per slot.

When no physical layer packet is transmitted on the RTC, the AT will transmit the zero data rate RRI codeword on the RRI channel, which should be transmitted on the I channel.

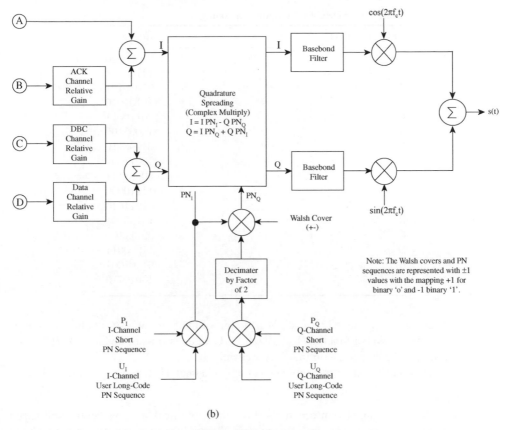

(b)

Figure 7.43 (*continued*)

DRC Channel

The DRC channel is used by the AT to indicate to the AN the selected serving sector and the requested data rate on the FTC. The requested FTC data rate is mapped into a 4-bit DRC value as specified by the FTC MAC protocol. The DRC value is block encoded to yield 8-bit bi-orthogonal codewords, as specified in Table 7.12. Each DRC codeword is transmitted twice per slot. Each bit of a repeated codeword is spread by an 8-ary Walsh function W_i^8 where $i, 0 \le i \le 7,$

Table 7.11   RRI symbol and simplex encoder assignment

Data rate (Kbps)	RRI symbol	RRI codeword
0	000	0000000
9.6	001	1010101
19.2	010	0110011
38.4	011	1100110
76.8	100	0001111
153.6	101	1011010

Table 7.12   DRC bi-orthogonal encoding

DRC value (0x)	Codeword
0	00000000
1	11111111
2	01010101
3	10101010
4	00110011
5	11001100
6	01100110
7	10011001
8	00001111
9	11110000
A	01011010
B	10100101
C	00111100
D	11000011
E	01101001
F	10010110

equals DRCCover (see Table 7.13). Each Walsh chip of the 8-ary Walsh function should be further spread by the Walsh function W_8^{16}. Each DRC value is transmitted over DRCLength slots when the DRC channel is continuously transmitted.

The DRC channel should be transmitted on the Q channel as shown in Figure 7.43.

ACK Channel

The AT uses the ACK channel to inform the AN whether a physical layer packet transmitted on the FTC has been received successfully or not. The AT should transmit an ACK channel bit in response to every FTC slot that is associated with a detected preamble directed to the AT. The ACK channel is BPSK modulated. A "0" bit (ACK) will be transmitted on the ACK channel if a FTC physical layer packet has been successfully received; otherwise, a "1" bit (NAK) should be transmitted. A FTC physical layer packet is considered successfully received if the FCS checks. For a FTC physical layer packet transmitted in slot n on the Forward Channel, the corresponding ACK channel bit should be transmitted in slot $n + 3$ on the reverse channel. The ACK channel uses the Walsh channel identified by Walsh function W_4^8 and should be transmitted on the I channel.

Table 7.13   8-ary Walsh functions

W_0^8	0000 0000
W_1^8	0101 0101
W_2^8	0011 0011
W_3^8	0110 0110
W_4^8	0000 1111
W_5^8	0101 1010
W_6^8	0011 1100
W_7^8	0110 1001

Data Channel

The Data channel will be transmitted at the data rates as indicated in Figure 7.43. Data transmission begins only at slot Frame Offset within a frame. The Frame Offset parameter is public data of the RTC MAC protocol. All data transmitted on the RTC are encoded, block interleaved, sequence repeated, and orthogonally spread by the Walsh function W_2^4.

7.8.3.5 Reverse Link Encoding

The RTC and AC physical layer packets are encoded with code rates of 1/2 and 1/4, depending on the data rate. The encoder discards the 6 bits of the Tail field in the physical layer packet inputs, and then encodes the remaining bits with a turbo encoder.

Turbo Encoder

The turbo encoder encodes the input data and adds an output tail sequence. If the total number of input bits is designated by N_{TB}, the turbo encoder generates N_{TB}/R encoded data output symbols followed by $6/R$ tail output symbols, where R is the code rate of 1/2 or 1/4. The turbo encoder employs two systematic, recursive, convolutional encoders connected in parallel, as shown in Figure 7.44. The two recursive convolutional codes are called the constituent codes of the turbo code. The turbo interleaver should precede the constituent encoder 2. The common constituent code is used for the turbo codes of rate1/2 or 1/4.

The transfer function for the turbo code is

$$G(D) = \left[1 \quad \frac{n_0(D)}{d(D)} \quad \frac{n_1(D)}{d(D)}\right]$$

where $n_0(D) = 1 + D + D^3$, $n_1(D) = 1 + D + D^2 + D^3$, and $d(D) = 1 + D^2 + D^3$.

The turbo encoder will generate an output symbol sequence that is identical to the one generated by the encoder illustrated in Figure 7.44. Initially, the constituent encoder registers are set to zero, and then the constituent encoders are clocked with the switches in positions noted. The encoded data output symbols are generated by clocking the constituent encoders N_{TB} times with the switches in the up positions and puncturing the outputs as specified in Table 7.14. The outputs of the constituent encoders are punctured and repeated to achieve the $(N_{TB} + 6)/R$ output symbols.

Within a puncturing pattern, a "0" means that the symbol should be deleted and a "1" means that the symbol should be passed onwards. The constituent encoder outputs for each bit period will be output in the sequence $X, Y_0, Y_1, X', Y_0', Y_1'$ with the X output first.

The turbo encoder generates $6/R$ tail output symbols following the encoded data output symbols. The tail output symbols are generated after the constituent encoders have been clocked N_{TB} with the switches in the up position. The first $3/R$ tail output symbols are generated by clocking the constituent encoder 1 three times with its switch in the down position, while the constituent encoder 2 is not clocked, and puncturing and repeating the resulting constituent encoder output symbols. The constituent encoder output symbol puncturing and symbol repetition should be as specified in Table 7.15. For rate-1/2 turbo codes, the tail output symbols

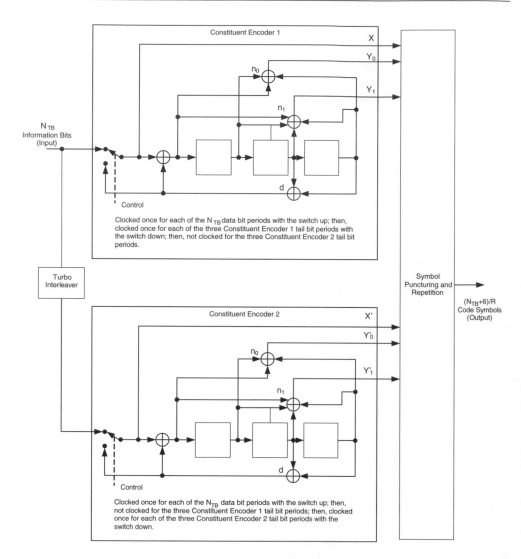

Figure 7.44 Turbo encoder (Reproduced under written permission from Telecommunications Industry Association)

for each of the first three tail bit periods should be XY_0, and the tail output symbols for each of the last three tail bit periods should be $X'Y_0'$. For rate-1/4 turbo codes, the tail output symbols for each of the first three tail bit periods should be XXY_0Y_1, and the tail output symbols for each of the last three tail bit periods should be $X'X'Y_0'Y_1'$.

Turbo Interleavers
The turbo interleaver is part of the turbo encoder and is block interleaved with the turbo encoder input data that is fed to the constituent encoder 2. The entire sequence of turbo encoder input bits are written sequentially into an array at a sequence of input addresses from 0 to $N_{TB} - 1$;

Table 7.14 Puncturing patterns for data bit periods

Output	Code rate	
	1/2	1/4
X	11	11
Y_0	10	11
Y_1	00	10
X'	00	00
Y_0'	01	01
Y_1'	00	11

Note: For each code rate, the puncturing table should be read first from top to bottom and then from left to right.

and then the entire sequence is read out from a sequence of output addresses by the procedure illustrated in Figure 7.45 and described below:

1. Determine the turbo interleaver parameter n that is the smallest integer such that $N_{TB} \leq 2^{n+5}$. Table 7.16 gives the turbo interleaver parameter for the different physical layer packet sizes. For example, for $n = 5$, the physical layer packet size is $2^{n+5} = 2^{10} = 1024$ and the turbo interleaver block size is $N_{TB} = 1024 - 6 = 1018$.
2. Initially set an $(n + 5)$-bit counter to zero "0."
3. Extract the n most significant bits (MSBs) from the counter and add one to form a new value. Then, discard all except the n least significant bits (LSBs) of this new value.
4. Obtain the n-bit output of the lookup table (depending on the value of n) defined in Table 7.17 with a read address equal to the five LSBs of the counter.
5. Multiply the values obtained in Steps 3 and 4, and discard all except the n LSBs (t_{n-1}, $t_{n-2}, \ldots, t_0$).
6. Bit-reverse the five LSBs (i_4, $i_3, \ldots, i_0$) of the counter.
7. Form a tentative output address that has its five MSBs (i_0, $i_1, \ldots, i_4$) obtained in Step 6 and its n LSBs (t_{n-1}, $t_{n-2}, \ldots, t_0$) obtained in Step 5.

Table 7.15 Puncturing patterns for the tail bit periods

Output	Code rate	
	1/2	1/4
X	111 000	111 000
Y_0	111 000	111 000
Y_1	000 000	111 000
X'	000 111	000 111
Y_0'	000 111	000 111
Y_1'	000 000	000 111

Note: For rate-1/2 turbo codes, the puncturing table should be read first from top to bottom and then from left to right. For rate-1/4 turbo codes, the puncturing table should be read first from top to bottom repeating X and X', and then from left to right.

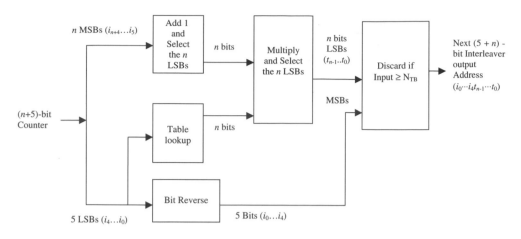

Figure 7.45 Turbo interleaver output address calculation procedure (Reproduced under written permission from Telecommunications Industry Association)

8. Accept the tentative output address as an output address if it is less than N_{TB}, $(i_0, \ldots, i_4, t_{n-1}, \ldots, t_0)$; otherwise, discard it.
9. Increment the counter and repeat Steps 3 through 8 until all N_{TB} interleaver output addresses are obtained.

7.8.3.6 Channel Interleaving

The encoder output of binary symbols is interleaved with a bit-reversal channel interleaver as shown in Figure 7.46. The bit-reversal interleaver is functionally equivalent to an approach so that the entire symbol sequence to be interleaved is written into a linear sequential array with addresses from 0 to $2^L - 1$. The sequence of array addressed from which the interleaved symbols are read out is generated by a bit-reversal address generator. Thus, the binary form of $i = b_{L-1} b_{L-2} \cdots b_1 b_0$ is written into the bit-reversal interleaver and read out the sequence of array addresses $A_i = b_0 b_1 \cdots b_{L-2} b_{L-1}$, where $b_k = 0$ or 1 and b_0 is the LSB and b_{L-1} is the MSB.

7.8.3.7 Interleaved Sequence Repetition

If the data rate is lower than 76.8 Kbps, the sequence of interleaved code symbols should be repeated before being modulated. The number of repeats varies depending on each data rate as specified in Table 7.18.

Table 7.16 Turbo interleaver parameter

Turbo interleaver parameter n	Physical layer packet size $= 2^{n+5}$ (bits)	Turbo interleaver block size $N_{TB} = 2^{n+5} - 6$ (symbols)
3	256	250
4	512	506
5	1024	1018
6	2048	2042
7	4096	4090

Table 7.17   Turbo interleaver lookup table

Table index	Entries				
	$n=3$	$n=4$	$n=5$	$n=6$	$n=7$
0	1	5	27	3	15
1	1	15	3	27	127
2	3	5	1	15	89
3	5	15	15	13	1
4	1	1	13	29	31
5	5	9	17	5	15
6	1	9	23	1	61
7	5	15	13	31	47
8	3	13	9	3	127
9	5	15	3	9	17
10	3	7	15	15	119
11	5	11	3	31	15
12	3	15	13	17	57
13	5	3	1	5	123
14	5	15	13	39	95
15	1	5	29	1	5
16	3	13	21	19	85
17	5	15	19	27	17
18	3	9	1	15	55
19	5	3	3	13	57
20	3	1	29	45	15
21	5	3	17	5	41
22	5	15	25	33	93
23	5	1	29	15	87
24	1	13	9	13	63
25	5	1	13	9	15
26	1	9	23	15	13
27	5	15	13	31	15
28	3	11	13	17	81
29	5	3	1	5	57
30	5	15	13	15	31
31	3	5	13	33	69

7.8.3.8 Orthogonal Walsh Covers

Figure 7.43 illustrates the reverse channel structure relating to the Pilot channel, the RRT channel, the DRC channel, the ACK channel, and the Data channel. The Pilot channel is time-division multiplexed with the RRI channel. These reverse channels are spread with Walsh functions at a fixed chip rate of 1.2288 Mcps. The Walsh function assignments are also show in Figure 7.43. The Pilot channel is covered by the 16-chip Walsh function number 0, that is, $W_0^{16} = (+ + + + + + + + + + + + + + + +)$. The DRC channel should be covered by the 16-chip Walsh function number 8, that is, $W_8^{16} = (+ + + + + + + + - - - - - - - -)$. The ACK channel is covered by the 8-chip Walsh function number

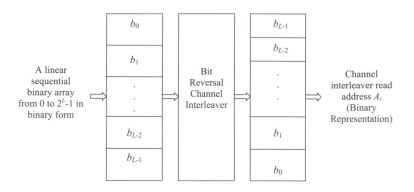

Figure 7.46   Channel interleaver output address generation

4 $W_4^8 = (+ + + + - - - -)$. The data channel is covered by the 4-chip Walsh function number 2 $W_2^4 = (+ + - -)$.

Walsh function time alignment should be such that the first Walsh chip begins at a slot boundary referenced to the AT transmission time.

7.8.3.9 Quadrature Spreading

After the orthogonal spreading with Walsh functions, the ACK, DRC, and Data channel chip sequences will be scaled by a factor that gives the gain of each of these channels relative to that of the Pilot channel.

During the transmission of the RTC, the determination of the output power needed to support the Data channel, the DRC channel, and the ACK channel is an additional open-loop process performed by the AT. During the transmission of the DRC channel or the ACK channel, the power of the DRC or ACK channel relative to that of the Pilot channel is specified as DRCChannelGain or ACKChannelGain, which is public data of the FTC MAC protocol. The AT should maintain the power of the Data channel, DRC channel, and ACK channel, relative to that of the Pilot channel, to within ±0.25 dB of the specified values. If the AT is unable to transmit at the requested output power level when the maximum RTC data rate is 9600 bpc, the AT should reduce the power of the DRC and ACK channels, accordingly. The maximum power reduction for the DRC channel or the

Table 7.18   Interleaved packet repeats

Parameter	Data rate (Kbps)				
	9.6	19.2	38.4	76.8	153.6
Interleaved packet repeats	8	4	2	1	1
Bits per physical layer packet	256	512	1,024	2,048	4,096
Code rate	1/4	1/4	1/4	1/4	1/2
Code symbol per physical layer packet	1,024	2,048	4,096	8,192	8,192
Code symbol rate (Kbps)	38.4	76.8	153.6	307.2	307.2
Physical layer packet duration (ms)	26.66	26.66	26.66	26.66	26.66

ACK channel corresponds to gating off the DRC or ACK channel, respectively. If the ACK channel is active, the ACK channel power reduction occurs only after the DRC channel has been gated off.

After the scaling, the pilot and scaled ACK, DRC, and Data channel sequences are combined to form resultant I-channel and Q-channel sequences, and these sequences are quadrature spread as shown in Figures 7.42 and 7.43(b). The quadrature spreading occurs at the chip rate of 1.2288 Mcps, and it is used for the RTC and ACs. The pilot and scaled ACK channel sequences are added to form the resultant I-channel sequences, and the scaled DRC and Data channel sequences are added to form the resultant Q-channel sequence. The quadrature spreading operation is equivalent to a complex multiply operation of the resultant I-channel and Q-channel sequences by the PN_I and PN_Q PN sequences, as shown in Figures 7.42 and 7.43.

The I and Q PN sequences (PN_I and PN_Q) are obtained from the long code PN sequences (U_I and U_Q), and the AT common short PN sequences (P_I and P_Q). The binary long code PN sequence and short PN sequence values of "O" and "I" should be mapped into values of $+1$ and -1, respectively. The bipolar PN_I sequence values are equivalent to those obtained by multiplying the P_I short code sequence values by the U_I long code sequence values.

The bipolar PN_Q sequence values are equivalent to those obtained with the following procedure (see Figure 7.43(b)):

- Multiply the P_Q values by the U_Q values.
- Decimate the sequence values of $P_Q \times U_Q$ by a factor of two.
- Multiply pairs of decimator output symbols by the Walsh cover sequences $(+ \ -)$.
- Finally, multiply the sequence obtained above by the bipolar PN_I sequence.

Short Code PN Sequences
The AT common short code PN sequences are the zero-offset I and Q PN sequences with a period of 2^{15} chips, based on the following characteristic polynomials $P_I(x)$ and $P_Q(x)$, respectively:

$$P_I(x) = x^{15} + x^{13} + x^9 + x^8 + x^7 + x^5 + 1 \text{ for the in-phase sequence}$$

$$P_Q(x) = x^{15} + x^{12} + x^{11} + x^{10} + x^6 + x^5 + x^4 + x^3 + 1 \quad \text{for the quadrature-phase}$$
sequence. The maximum length linear feedback shift-register (LFSR) sequences $\{I(n)\}$ and $\{Q(n)\}$ based on the above polynomials is $2^{15} - 1$ and can be generated by the following linear recursions:

$$I(n) = I(n-15) \oplus I(n-10) \oplus I(n-8) \oplus I(n-7) \oplus I(n-6) \oplus I(n-2) (based \ on \ P_I(x))$$
$$Q(n) = Q(n-15) \oplus Q(n-12) \oplus Q(n-11) \oplus Q(n-10) \oplus Q(n-9) \oplus Q(n-5)$$
$$\oplus Q(n-4) \oplus Q(n-3) (based \ on \ P_Q(x)),$$

where I(n) and Q(n) are binary valued ("0" and "1") and the additions are module-2.

In order to obtain the I and Q short code PN sequences, a "O" is inserted in the $\{I(n)\}$ and $\{Q(n)\}$ sequences after 14 consecutive "O" outputs. Therefore, the short code PN sequences have one run of 15 consecutive "O" outputs instead 14. The chip rate for the AT common short code PN sequence is 1.2288 Mcps. The short code PN sequence period is 26.666...(32768/1228800) ms and exact 75 PN sequence repetitions occur every 2 seconds.

Long Code Sequences

The in-phase and quadrature-phase long codes designated by U_I and U_Q are generated from a long code generating sequence by using two different masks. The long code generating sequence is specified by the following characteristic polynomial:

$$p(x) = x^{42} + x^{35} + x^{33} + x^{31} + x^{27} + x^{26} + x^{25} + x^{22} + x^{21} + x^{19} + x^{18} + x^{17} + x^{16}$$
$$+ x^{10} + x^7 + x^6 + x^5 + x^3 + x^2 + x + 1$$

The long codes, U_I and U_Q are generated by a module-2 addition of the 42-bit state vector of the LFSR sequence generator and two 42-bit masks (MI and MQ) as shown in Figure 7.47. The masks MI and MQ vary depending on the channel on that the AT is transmitting.

For transmission on the RTC, MI and MQ will be set to MI_{RTCMAC} and MQ_{RTCMAC}, respectively, and the long code sequences are referred to as the user long codes. The long code generator should be reloaded with the hexadecimal value 0x24b91bfd3a8 at the beginning of every period of the short codes. Thus, the long codes are periodic with a period of 2^{15} PN chips.

Baseband Filtering

Following the quadrature spreading, the I and Q impulses are applied to the inputs of the I and Q baseband filters as shown in Figures 7.42 and 7.43. The baseband filters have a frequency response S(f) that satisfies the limits given in Figure 7.48. Specifically, the normalized frequency response of the filter should be contained within $\pm\delta_1$, in the passband $0 \leq f \leq f_p$ and should be less than or equal to $-\delta_2$ in the stopband $f \geq f_s$. The numerical values for parameters are $\delta_1 = 1.5$ dB, $\delta_2 = 40$ dB, $f_p = 590$ KHz, and $f_s = 740$ KHz.

The impulse response of the baseband filter, S(t), should satisfy the following equation:

$$\text{Mean Squared Error} = \sum_{k=0}^{\infty} [\alpha S(KT_s - \tau) - h(k)]^2 \leq 0.03,$$

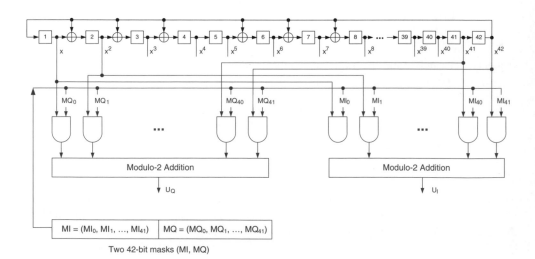

Figure 7.47   Long code generators

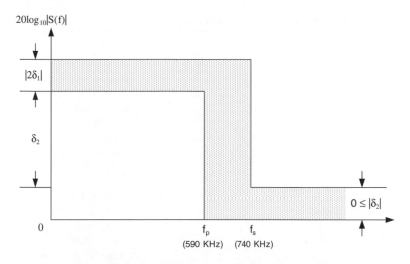

Figure 7.48   Baseband filter frequency response limits

where the constants α and τ are used to minimize the mean square error. The constant T_s is equal to 203.451 ns, which equals one quarter of a PN chip. The values of the coefficients $h(k)$ for $k \leq 48$ are given in Table 7.19. Note that $h(k) = 0$ for $k \geq 48$ and $h(k)$ equals $h(47 - k)$.

7.8.4 Access Network Requirements

This section defines the specific requirements to the AN operation.

7.8.4.1 Forward Channel Structure

The forward channel consists of the following time-multiplexed channels: the Pilot channel, the Forward MAC channel, and FTC channel or control channel. The traffic channel carries user physical layer packets. The control channel carries control messages and it may also carry user traffic. Each channel is further decomposed into code-division-multiplexed quadrature Walsh channels.

The forward link slot structure consists of slots of length 2048 chips (1.666... ms), or 1024 chips in the 1/2 slot. Within each slot, the active 1/2 slot consists of the pilot (96 chips), MAC ($64 \times 2 = 128$ chips), and data ($400 \times 2 = 800$ chips) symbols. The traffic or control channels are time-division multiplexed and should be transmitted at the same power level.

Pilot Channel

The pilot channel is an unmodulated signal that is used for synchronization and other functions by the AT operating within the coverage area of the sector. An additional function of the Pilot channels are used for initial acquisition, phase recovery, timing recovery, and maximal-ratio combining.

The Pilot channel consists of all-"0" symbols transmitted on the I channel with Walsh cover 0. Each slot is divided into two half slots, each of which contains a pilot burst having a duration of 96 chips and is centered at the midpoint of the half slot. The pilot channel within the forward channel structure is shown in Figure 7.49.

Table 7.19   Baseband filter coefficients

k	$h(k)$
0, 47	−0.025288315
1, 46	−0.034167931
2, 45	−0.035752323
3, 44	−0.016733702
4, 43	0.021602514
5, 42	0.064938487
6, 41	0.091002137
7, 40	0.081894974
8, 39	0.037071157
9, 38	−0.021998074
10, 37	−0.060716277
11, 36	−0.051178658
12, 35	0.007874526
13, 34	0.084368728
14, 33	0.126869306
15, 32	0.094528345
16, 31	−0.012839661
17, 30	−0.143477028
18, 29	−0.211829088
19, 28	−0.140513128
20, 27	0.094601918
21, 26	0.441387140
25, 22	0.785875640
23, 24	1.0

Forward MAC Channel

The MAC channel consists of three subchannels: the RPC channel, the DRCLock channel, and the reverse activity (RA) channel. The RA channel transmits a reverse link activity bit (RAB) stream. Each MAC channel symbol is BPSK modulated on one of 64-ary Walsh covers. The MAC symbol Walsh covers should be transmitted immediately preceding each of the pilot bursts in a slot. A burst should also be transmitted immediately following each of the pilot bursts in a slot.

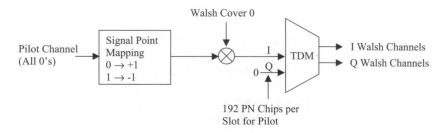

Figure 7.49   Pilot channel

The forward MAC channel is composed of Walsh channels that are orthogonally covered and BPSK modulated on either the in-phase (I) modulation phase or the quadrature (Q) modulation phase. Each Walsh channel is identified by MAC index value i that is between 0 and 63 and defines a unique 64-ary Walsh cover and a unique modulation phase as shown below:

$$W_{i/2}^{64} \text{ for } i = 0, 2, \ldots, 62 (\text{even-numbered MACIndex value})$$

$$W_{(i-1)/2+32}^{64} \text{ for } i = 1, 3, \ldots, 63 (\text{odd-numbered MACIndex value})$$

where i denotes the MAC index value. Even-numbered MAC channels $W_{i/2}^{64}$ are assigned to the I-modulation phase, whereas odd-numbered ones $W_{(i-1)/2+32}^{64}$ are assigned to the Q-modulation phase. The MAC symbol Walsh covers are transmitted four times per slot in bursts of length 64 chips each. These bursts should be transmitted immediately preceding and following the pilot bursts (96 chips each) of each slot.

Reverse Power Control Channel

The RPC channel for each AT with an open connection is assigned to one of the available MAC channels. It is used for the transmission of the RPC bit stream destined to that access terminal. The RPC channel and DRCLock channel are time-division multiplexed and transmitted on the same MAC channel. The RPC channel should be transmitted in DRCLockPeriod-1 slots out of every DRCLockPeriod slots. Each RPC bit should be transmitted four times in a slot in bursts of 64 chips each.

Data Rate Control Lock Channel

The DRC channel is used by the AT to indicate to the AN the selected serving sector and the requested data rate on the FTC. The DRCLock channel for each AT with an open connection is assigned to one of the available MAC channels. It is used for the transmission of the DRCLock bit stream destined to that AT. The RPC channel and the DRCLock channel are time-division multiplexed and transmitted on the same MAC channel. The DRCLock channel should be transmitted in one out of every DRCLockPeriod slots. Each DRCLock bit should be transmitted four times in a slot in bursts of 64 chips each. One burst should be transmitted immediately preceding and following each pilot burst in a slot as shown in Figure 7.50.

Reverse Activity Channel

The RA channel transmits the Reverse Activity Bit (RAB) stream over the MAC channel with MACIndex 4. The RA bit is transmitted over RABLength successive slots. The RA channel data rate is 600/RABLength bps. Each RA bit should be repeated and transmitted over RABLength consecutive slots. The RA bit in each slot is further repeated to form four symbols per slot for transmission.

Forward Traffic Channel

The FTC is a packet-based channel. Since the FTC is also called a variable-rate channel, the user physical layer packets for the AT are transmitted at a data rate that varies from 38.4 Kbps to 2.4576 Mbps.

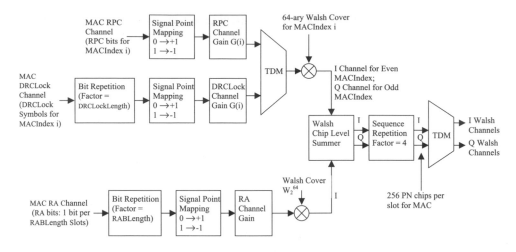

Figure 7.50   Forward MAC channel with RPC, DRCLock, and RA channels

The functionalities of the FTC and control channel are described by the procedure illustrated in Figure 7.51 and described in the following.

1. Channel data is encoded in the physical layer packets on a block-by-block basis.
2. Encoded packets are scrambled and then fed into a channel interleaver.

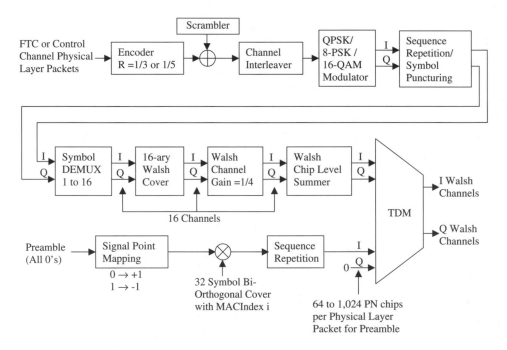

Figure 7.51   Forward Channel Structure for the FTC or control channel and preamble (Reproduced under written permission from Telecommunications Industry Association)

3. The output of the channel interleaver is fed into a QPSK/8-PSK/16-QAM modulator.
4. The modulated symbol sequences should be repeated and punctured.
5. The resulting sequence of modulation symbols should be demultiplexed to form 16 pairs (in-phase and quadrature) of parallel streams.
6. Each of the parallel streams will be covered with a distinct 16-ary Walsh function at a chip rate to yield Walsh symbols at 76.8 Ksps.
7. The Walsh-coded symbols of all the streams are summed together to form a single in-phase stream and a single quadrature stream at a chip rate of 1.2288 Mcps.
8. The resulting chips are time-division multiplexed with the preamble, Pilot channel, and MAC channel chips to form the resultant sequence of chips for the quadrature spreading operation.

FTC and control channel physical layer packets can be transmitted in 1 to 16 slots. When more than one slot is allocated, the transmit slots should use 4-slot interlacing, that is, separated by three interleaving slots, and slots of other physical layer packets should be transmitted in the slots between those transmit slots. If a positive acknowledgment is received on the reverse link ACK channel before all the allocated slots have been transmitted, the remaining untransmitted slots should not be transmitted and the next allocated slot may be used for the first slot of the next physical layer packet transmission.

The multislot interlacing approach for a 153.6 Kbps FTC channel with DRCLength of one slot is illustrated in Figure 7.52. The 153.6 Kbps FTC physical layer packets use four slots, and those slots are transmitted with a three-slot interval between them. The slots from other physical layer packets are interlaced in the three intervening slots. In the case of a normal physical layer packet transmission, the AT transmits NAK responses on the ACK channel after the first three slots of the physical layer packet are received indicating that it was unable to receive the FTC physical layer packet correctly after only one, two, or three of the nominal four slots. When the AN has transmitted all the slots of a physical layer packet or has received a positive ACK response, the physical layer will return a ForwardTrafficCompleted indication.

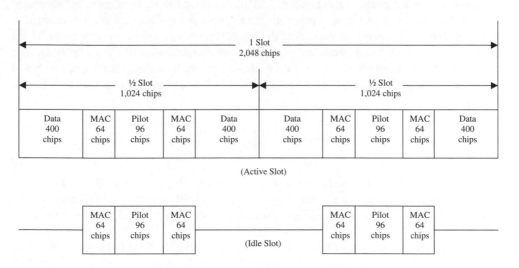

Figure 7.52 Forward link slot structure

The control channel should be transmitted at a data rate of 76.8 Kbps or 38.4 Kbps. The modulation characteristics for the control channel are the same as those of the FTC transmitted at the corresponding rate. The FTC and control channel data symbol should fill the slot as described earlier in Figure 7.52. A slot during which no traffic or control data are transmitted is referred to as an idle slot. During an idle state, the sector should transmit the pilot channel and MAC channel.

The FTC preamble sequence should be transmitted with each FTC and control channel physical layer packet in order to assist the AT with synchronization of each variable-rate transmission. The preamble consists of all-"0" symbols transmitted on the in-phase component only. The preamble sequence is covered by a 32-chip bi-orthogonal sequence and the sequence is repeated several times depending on the transmit mode. The preamble is then time multiplexed into the FTC stream as shown in Figure 7.51.

The bi-orthogonal sequence is specified in terms of the 32-ary Walsh functions and their bit-by-bit complements by expressing

$$W_{i/2}^{32} \quad \text{for} \quad i = 0, 2, \ldots, 62$$

$$\overline{W_{(i-1)/2}^{32}} \quad \text{for} \quad i = 1, 3, \ldots, 63$$

where $i = 0, 1, 2, \ldots, 63$ is the MACIndex value and $\overline{W_i^{32}}$ is the bit-by-bit complement of the 32-chip Walsh function of order i.

Turbo Encoding

The traffic channel physical layer packets are encoded with code nodes of $R = 1/3$ or $1/5$. The encoder discards the 6-bit tail field of the packet inputs and encodes the remaining bits with a parallel turbo encoder, as shown in Figure 7.53. The turbo encoder will add an internally generated tail of 6/R output code symbols, so that the total number of output symbols is 1/R times the number of bits in the input packet. Figure 7.54 illustrates the forward link encoding approach.

The turbo encoder employs two systematic, recursive, convolutional encoders connected in parallel with a turbo interleaver, preceding the second recursive convolutional encoder. These two recursive convolutional codes are called the *constituent* codes of the turbo code. The outputs of the constituent encoders are punctured and repeated to achieve the desired number of turbo encoder output symbols. The transfer function for the constituent code is identical to the RTC case:

$$G(D) = \left[1 \, \frac{n_0(D)}{d(D)} \, \frac{n_1(D)}{d(D)} \right]$$

where $d(D) = 1 + D^2 + D^3$, $n_0(D) = 1 + D + D^3$, and $n_1(D) = 1 + D + D^2 + D^3$.

The turbo encoder generates an output symbol that is identical to the one generated by the turbo encoder shown in Figure 7.53. The constituent encoders are clocked with the switches in the positions noted.

Let N_{TB} denote the number of bits in the turbo encoder after the 6-bit tail field is discarded. Then the encoded data output symbols are generated by clocking the constituent encoders N_{TB} times with the switches in the up positions and puncturing the outputs as specified in Table 7.20.

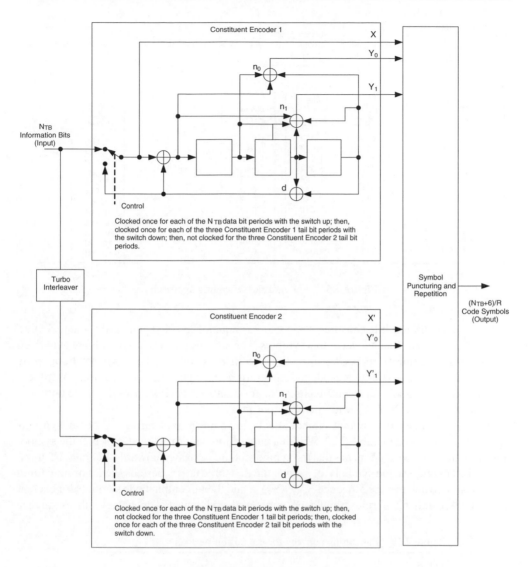

Figure 7.53 Turbo encoder

Within a puncturing pattern, a "0" means that the symbol should be deleted and a "1" means that the symbol should be passed onwards. The constituent encoder output for each bit period should be output in the sequence $XY_0Y_1X'Y'_0Y'_1$ with the X output first. Symbol repetition is not used in generating the encoded date output symbols.

The turbo encoder generates 6/R tail output symbols following the encoded data output symbols. This tail output symbol sequence should be identical to the one generated by the turbo encoder illustrated in Figure 7.53. The tail output symbols are generated after the constituent encoders have been clocked N_{TB} times with the switches in the up position. The first 3/R tail output symbols are generated by clocking the constituent encoder 1 three times with the switch

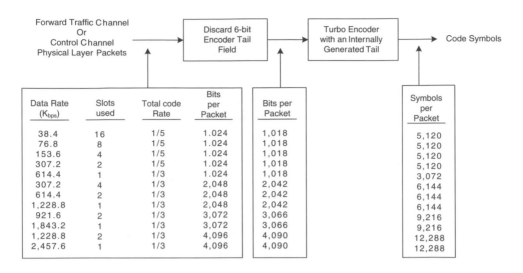

Figure 7.54 Forward link encoding approach

in the down position, while constituent encoder 2 is not clocked, and puncturing and repeating the resulting constituent encoder output symbols. The last 3/R tail output symbols are generated by clocking constituent encoder 2 three times with its switch in the down position, while constituent encoder 1 is not clocked, and puncturing and repeating the resulting constituent encoder output symbols. The constituent encoder outputs for each bit period will be output in the sequence $XY_0Y_1X'Y_0'Y_1'$ with the X output first.

The constituent encoder output symbol puncturing patterns puncturing for the tail symbols should be as specified in Table 7.20. Within a puncturing pattern, a "0" means that the symbol should be deleted and a "1" means that the symbol should be passed onwards. For rate-1/5 turbo codes, the tail output code symbols for each of the first three tail bit periods should be punctured and repeated to achieve the sequence $XXY_0Y_1Y_1$, and the tail output code symbols for each of the last three tail bit periods should be punctured and repeated to achieve the sequence

Table 7.20 Puncturing patterns for the tail bit periods

Output	Code Rate	
	1/3	1/5
X	111000	111000
Y_0	111000	111000
Y_1	000000	111000
X'	000111	000111
Y_0'	000111	000111
Y_1'	000000	000111

Note: For rate-1/3 turbo codes, the puncturing table should be read first from top to bottom repeating X and X', and then from left to right. For rate-1/5 turbo codes, the puncturing table should be read first from top to bottom repeating X, X', Y_1 and Y_1' and then from left to right.

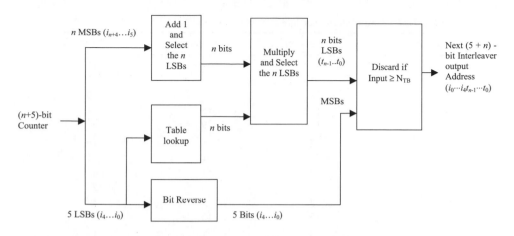

Figure 7.55   Turbo interleaver output address calculation procedure

$X'X'Y'_0Y'_1Y'_1$. For rate-1/3 turbo codes, The tail output symbols for each of the first three tail bit periods should be XXY_0, and the tail output symbols for each of the last three tail bit periods should be $X'X'Y'_0$.

Turbo Interleaver

The turbo interleaver shown in Figure 7.55 is block interleaved from the turbo encoder input data fed to constituent encoder 2. The entire sequence of turbo interleaver input bits are written sequentially into an array at a sequence of addresses from 0 to N_{TB-1}, and then the entire sequence is read out from the sequence of addresses that are defined by the following procedure:

1. Determine the turbo interleaver parameter n such that $N_{TB} \leq 2^{n+5}$. Table 7.21 gives this parameter for the different physical layer packet sizes.
2. Initialize an $(n+5)$-bit counter to 0.
3. Extract n MSBs from the counter and add one to form a new value. Then discard all except the n LSBs of this new value.
4. Obtain the n-bit output of the lookup table defined in Table 7.22 with a real address equal the five LSBs of the counter.
5. Multiply the values obtained in Steps 3 and 4 and discard all except the n LSBs.
6. Bit-reverse the LSBs $(i_4, i_3, \ldots, i_0)$ of the counter.

Table 7.21   Turbo interleaver parameter

Physical layer packet size N_{PS}	Turbo interleaver block size $N_{TB} = N_{PS} - 6$	Turbo interleaver parameter n
1024	1018	5
2048	2042	6
3072	3066	7
4096	4090	7

Choose n such that $N_{TB} \leq 2^{n+5}$ should be met.

Table 7.22   Turbo interleaver lookup table

Table: index	$n = 5$ entries	$n = 6$ entries	$n = 7$ entries
0	27	3	15
1	3	27	127
2	1	15	89
3	15	13	1
4	13	29	31
5	17	5	15
6	23	1	61
7	13	31	47
8	9	3	127
9	3	9	17
10	15	15	119
11	3	31	15
12	13	17	57
13	1	5	123
14	13	39	95
15	29	1	5
16	21	19	85
17	19	27	17
18	1	15	55
19	3	13	57
20	29	45	15
21	17	5	41
22	25	33	93
23	29	15	87
24	9	13	63
25	13	9	15
26	23	15	13
27	13	31	15
28	13	17	81
29	1	5	57
30	13	15	31
31	13	33	69

7. Form a tentative output address that has its MSBs equal to the value obtained in Step 6 and its LSBs equal to the value obtained in Step 5.
8. Accept the tentative output address as an output address if it is less than N_{TB}; otherwise, discard it.
9. Increment the counter and repeat Steps 3−8 until all N_{TB} interleaver output addresses are obtained.

The turbo interleaver output address calculation procedure is shown in Figure 7.55.

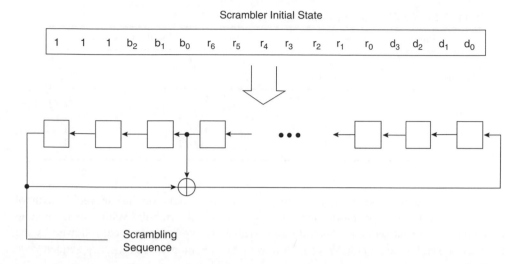

Figure 7.56 Symbol scrambler

Scrambling

The output of the turbo encoder is scrambled to randomize the data prior to modulation. The scrambling sequence is equivalent to one generated with a 17-stage Linear Feedback Shift Register (LFSR) with a characteristic polynomial $h(D) = D^{17} + D^{14} + 1$, as shown in Figure 7.56. At the start of the physical layer packet, the shift register should be initialized to the state $[1\ 1\ 1\ 1\ 1\ 1\ 1\ r_5\ r_4\ r_3\ r_2\ r_1\ r_0\ d_3\ d_2\ d_1\ d_0]$. The $r_5 r_4 r_3 r_2 r_1 r_0$ bits are equal to the 6-bit preamble MACIndex value. The $d_3 d_2 d_1 d_0$ bits are determined by the data rate. The initial state will generate every encoder output code symbol to generate a bit of the scrambling sequence. Every encoder output code symbol should be XORed with the corresponding bit of the scrambling sequence to yield a scrambled encoded bit.

Channel Interleaving

The channel interleaving consists of a symbol reordering followed by symbol permutation:

- *Symbol Reordering:* The scrambled turbo encoder data and tail output symbols generated with the rate-1/3 encoder or the rate-1/5 encoder will be recorded into the following sequences according to the demultiplexed procedure:
  - U, V_0, and V_0' sequences for the rate-1/3 encoder;
  - U, V_0, V_1, V_0', and V_1' sequences for the rate-1/5 encoder.
- *Symbol Permuting:* The reordered symbols should be permuted in three separate bit-reversal interlever blocks with rate-1/5 coding and in two separate blocks with rate-1/3 coding. The permuter input blocks with rate-1/3 coding consist of the U symbol sequences, the V_0 symbol sequence, followed by the V_0' symbol sequence. With rate-1/5 coding, the V_1 symbol sequence is followed by the V_1' sequence of symbols.

Table 7.23 QPSK modulation table

Interleaved symbols		Modulation symbols	
s_1	s_0	$m_I(k)$	$m_Q(k)$
$x(2k + 1)$	$x(2k)$		
0	0	$1/\sqrt{2}$	$1/\sqrt{2}$
0	1	$-1/\sqrt{2}$	$1/\sqrt{2}$
1	0	$1/\sqrt{2}$	$-1/\sqrt{2}$
1	1	$-1/\sqrt{2}$	$-1/\sqrt{2}$

With rate-1/3 coding, the interleaver output sequence should be interleaved U symbol sequence, followed by the interleaved V_0-V_0' sequence of symbols. With rate-1/5 coding the interleaver output sequence should be interleaved U symbol sequence, followed by the interleaved symbol sequence V_0-V_0', followed by the interleaved V_1-V_1' sequence of symbols.

Signal Constellation for Modulation Symbols
The output of the channel interleaver is applied to a modulator that outputs the in-phase and quadrature-phase sequence of modulated symbols. The modulator generates QPSK, 8-PSK, or 16-QAM modulation symbols, depending on the data rate.

- *QPSK Modulation:* For packet size of 1024 or 2048 bits, two successive output symbols of the channel interleaver should be grouped to form QPSK modulation symbols. Each group of two adjacent interleaved symbols, $x(2k)$ and $x(2k + 1)$, $k = 0, 1, 2, \ldots, M - 1$, should be mapped into a complex modulation symbol $(m_I(k), m_Q(k))$ as specified in Table 7.23. Figure 7.57 shows the signal constellation of the QPSK modulation, where $s_0 = x(2k)$ and $s_1 = x(2k + 1)$.

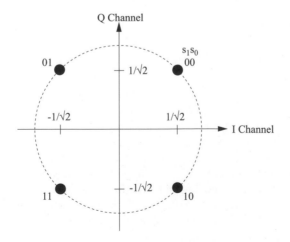

Figure 7.57 Signal constellation for QPSK modulation

Table 7.24 8-PSK modulation table

Interleaved symbols			Modulation symbols	
s_2	s_1	s_0	$m_I(k)$	$m_Q(k)$
$x(3k + 2)$	$x(3k + 1)$	$x(3k)$		
0	0	0	C	S
0	0	1	S	C
0	1	0	−S	C
0	1	1	−C	S
1	0	0	−C	−S
1	0	1	−S	−C
1	1	0	S	−C
1	1	1	C	−S

Note: $C = \cos(\pi/8) = 0.9239$ and $S = \sin(\pi/8) = 0.3827$.

- *8-PSK Modulation:* For the packet size of 3072 bits, groups of three successive channel interleaver output symbols should be grouped to form 8-PSK modulation symbols. Each group of three adjacent output symbols from the block interleaver, $x(3k)$, $x(3k + 1)$, and $x(3k + 2)$, where $k = 0, 1, \ldots, M - 1$, should be mapped into a complex modulation symbol $(m_I(k), m_Q(k))$ as specified in Table 7.24. Figure 7.58 shows the signal constellation of the 8-PSK modulation, where $s_0 = x(3k)$, $s_1 = x(3k + 1)$, and $s_2 = x(3k + 2)$.

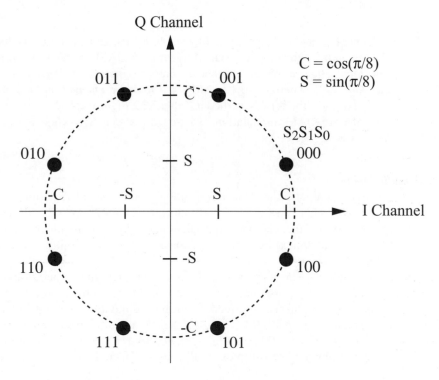

Figure 7.58 Signal constellation for 8-PSK modulation

Table 7.25   16-QAM modulation table

Interleaved symbols				Modulation symbols	
s_3	s_2	s_1	s_0	$m_I(k)$	$m_Q(k)$
$x(4k + 3)$	$x(4k + 2)$	$x(4k + 1)$	$x(4k)$		
0	0	0	0	3A	3A
0	0	0	1	3A	A
0	0	1	1	3A	−A
0	0	1	0	3A	−3A
0	1	0	0	A	3A
0	1	0	1	A	A
0	1	1	1	A	−A
0	1	1	0	A	−3A
1	1	0	0	−A	3A
1	1	0	1	−A	A
1	1	1	1	−A	−A
1	1	1	0	−A	−3A
1	0	0	0	−3A	3A
1	0	0	1	−3A	A
1	0	1	1	−3A	−A
1	0	1	0	−3A	−3A

Note: A $= 1/\sqrt{10} = 0.3162$.

- *16-QAM Modulation:* For physical layer packet size of 4096 bits, groups of four successive channel interleaver output symbols are grouped to form 16-QAM modulation symbols. Each group of four adjacent block interleaver output symbols, $x(4k)$, $x(4k + 1)$, $x(4k + 2)$, and $x(4k + 3)$, where $k = 0,1,\ldots,M - 1$, should be mapped into a complex modulation symbol $(m_I(k), m_Q(k))$ as specified in Table 7.25. Figure 7.59 shows the signal constellation of the 16-QAM modulation, where $s_0 = x(4k)$, $s_1 = x(4k + 1)$, $s_2 = x(4k + 2)$, and $s_3 = x(4k + 3)$.

Symbol Demultiplexing

The in-phase stream at the output of the sequence repetition operation should be demultiplexed into 16 parallel streams labeled I_0, I_1, I_2, $\ldots$, I_{15}. If $m_I(0)$, $m_I(1)$, $m_I(2)$, $m_I(3)$,$\ldots$ denote the sequence of sequence-repeated modulation output values in the in-phase stream, then for each $k = 0,1,\ldots$, 15 the kth demultiplexed stream I_k will consist of the values $m_I(k)$, $m_I(16 + k)$, $m_I(32 + k)$, $m_I(48 + k)$,$\ldots$.

Similarly, the quadrature stream at the output of the sequence repetition operation should be demultiplexed into 16 parallel streams labeled Q_0, Q_1, Q_2,$\ldots$, Q_{15}. If $m_Q(0)$, $m_Q(1)$, $m_Q(2)$, $m_Q(3)$,$\ldots$ denote the sequence of sequence-repeated modulation output values in the quadrature streams, then for each $k = 0,1,2,\ldots$, 15, the kth demultiplexed stream Q_k will consist of the values $m_Q(k)$, $m_Q(16 + k)$, $m_Q(32 + k)$, $m_Q(48 + k)$,$\ldots$. Each demultiplexed stream at the output of the symbol demultiplexer will consist of modulation values at the rate of 76.8 Kbps.

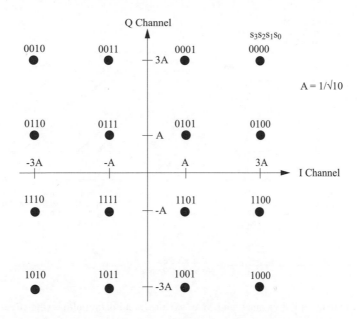

Figure 7.59   Signal constellation for 16-QAM modulation

Walsh Channel Assignment

The individual streams generated by the symbol demultiplexer should be assigned to one of the 16 distinct channels. For each $k = 0, 1, 2, \ldots, 15$, the demultiplexed streams with labels I_k and Q_k should be assigned to the in-phase and quadrature phases, respectively, of the kth Walsh channel W_k^{16}. The modulation values associated with the in-phase and quadrature-phase components of the same Walsh channel are referred to as Walsh symbols.

Walsh Channel Gain

The modulated symbols on each branch of each Walsh channel should be scaled to maintain a constant total transit power independent of the data rate. For this purpose, each orthogonal channel should be scaled by a gain of $1/\sqrt{16} = 1/4$.

Walsh Chip Level

The scaled Walsh chips associated with the 16 Walsh channels should be summed on a chip-by-chip basis.

Control Channel

The control channel transmits broadcast messages and access terminal directed messages. The control channel messages are transmitted at a data rate of 76.8 Kbps or 38.4 Kbps. The modulation characteristics are the same as those of the forward traffic channel at the corresponding data rate. The control channel transmissions should be distinguished from FTC transmissions by having a preamble that is covered by a bi-orthogonal cover sequence with MACIndex 2 or 3. A MACIndex value of 2 is used for the 76.8 Kbps data rate, and a MACIndex of 3 is used for the 38.4 Kbps data rate.

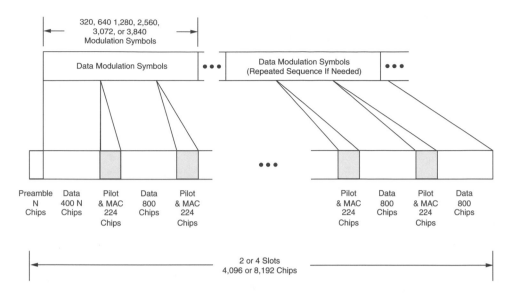

Figure 7.60 Preamble, pilot channel, and MAC channel, and data multiplexing (Reproduced under written permission from Telecommunications Industry Association)

Time-Division Multiplexing

The forward traffic channel or control channel data modulation chips should be time-division multiplexed with preamble, Pilot channel, and MAC channel, and data multiplexing with data rates of 38.4 and 76.8 Kbps, which is shown in Figure 7.60.

Quadrature Spreading

Following orthogonal spreading (*I*-Walsh and *Q*-Walsh covers), the combined modulation sequence should be quadrature spread as shown in Figure 7.61. The spreading sequence will be a quadrature sequence of length $2^{15} = 32768$ PN chips. This sequence is called the Pilot PN

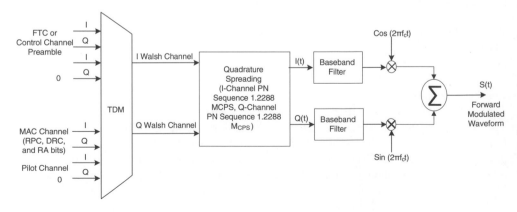

Figure 7.61 Forward channel quadrature spreading, baseband filtering, and QPSK

sequence, which is based on the following characteristic polynomials:

$$P_I(x) = x^{15} + x^{10} + x^8 + x^7 + x^6 + x^2 + 1 \text{ for the in-phase sequence(I)}$$

and

$$P_Q(x) = x^{15} + x^{12} + x^{11} + x^{10} + x^9 + x^5 + x^4 + x^3 + 1 \text{ for the quadrature-phase sequence(Q)}$$

The I-phase and Q-phase Pilot PN sequences are also induced by means of the following reciprocal polynomials:

$$I(x) = x^{15} P_I(x^{-1}) = x^{15} + x^{13} + x^9 + x^8 + x^7 + x^5 + 1$$

and

$$Q(x) = x^{15} P_Q(x^{-1}) = x^{15} + x^{12} + x^{11} + x^{10} + x^6 + x^5 + x^4 + x^3 + 1$$

The maximum length LFSR sequence $\{I(n)\}$ and $\{Q(n)\}$ based on the above reciprocal polynomials is $2^{15} - 1$ and can be generated by the following linear recursions:

$$I(n) = I(n-15) \oplus I(n-13) \oplus I(n-9) \oplus I(n-8) \oplus I(n-7) \oplus I(n-5)$$

$$(\text{based on } P_I(x))$$

and

$$Q(n) = Q(n-15) \oplus Q(n-12) \oplus Q(n-11) \oplus Q(n-10) \oplus Q(n-6) \oplus Q(n-5) \oplus Q(n-4) \oplus Q(n-3)(\text{based on } P_Q(x))$$

where $I(n)$ and $Q(n)$ are binary valued ("0" and "1") and the additions are modulo-2. In order to obtain the I and Q Pilot PN sequences of period $2^{15} = 32768$ PN chips, a "0" is inserted in the $\{I(n)\}$ and $\{Q(n)\}$ sequences after 14 consecutive "0"outputs. Therefore, the Pilot PN sequences have one run of 15 consecutive "0"ouptputs instead of 14. Since the chip rate for the pilot PN sequence is 1.2288 Mcps, the Pilot PN sequence period is $32768/1228800 = 26.666...$ms, and exactly 75 Pilot PN sequence repetitions occur every 2 seconds.

The zero-offset pilot PN sequence is such that the start of the sequence should be output at the beginning of every even second in time. The start of the zero-offset pilot PN sequence for either the I or Q sequences must be defined as the state of the sequence for which the next 15 outputs inclusive are "0." Equivalently, the zero-offset sequence is defined such that the last chip prior to the even-second mark is a "1" prior to the 15 consecutive "0"s. Pilot channels are identified by an offset index in the range from 0 through 511 inclusive. This offset index should specify the offset value (in units of 64 chips) by which the pilot PN sequence lags the zero-offset pilot PN sequence.

8

CDMA2000 1x Evolution-Data and Voice (1xEV-DV)

CDMA2000, originated by the US manufacturer Qualcomm, is a family of technologies allowing seamless evolution from CDMA2000 1x to CDMA2000 1x EV-DO and 1x EV-DV. CDMA2000 1x evolved from cdmaOne IS-95A/B networks. The 1x standard was finalized in late 1999 and its commercial system was launched by SK Telecom (Korea) in October 2000. Compared to IS-95 networks, it provides double voice capacity and offers packet data speeds of 153 Kbps (Release 0) and 307 Kbps (Release 1) in a single CDMA channel with a bandwidth of 1.25 MHz. The ITU (International Telecommunication Union) has designated the 1x standard as the first 3G technology to be commercially deployed. CDMA2000 represents a family of ITU-approved 3G standards and includes CDMA2000 1x and CDMA2000 1x EV technologies. CDMA2000 1x and 1xEV support advanced applications such as e-mail messages, games, video and still images with sound, picture and music downloading, and GPS-based location services.

CDMA2000 1xEV is the final stage in the evolution from CDMA 2G network to 3G. Its standardization within 3GPP2 is mostly driven by Qualcomm, as opposed to 3GPP. CDMA2000 1x EV is backward compatible with CDMA 1x when migrating from 1x to 1x EV. Like 1x technology, 1x EV can be deployed both in the frequency bands used by 2G systems and in IMT-2000 frequency bands. The transition from 1x to 1x EV will take place in two stages: 1x EV-DO and 1x EV-DV.

1x EV-DO is bi-mode (1x for voice and EV-DO for data). 1x EV-DO provides a peak data rate of 2.4 Mbps within a 1.25 MHz CDMA carrier and brings additional capacity to 1x networks. Since the technology functions on a permanent (always on) basis, users are free to send and receive information from the Internet and their corporate Intranets. 1x EV-DO supports applications such as MP3 transfers and video conferencing. The first CDMA2000 1x EV-DO networks were launched by SK Telecom in January 2002 and by KTF in May 2002 in South Korea. The 1x EV-DO system in the 2 GHz band was in operation by KDDI with a license approved in October 2004 in Japan. CDMA2000 1x EV-DO technology was fully covered in Chapter 7.

CDMA2000 1x EV-DV (also called CDMA2000 Releases C/D) builds on the architecture of CDMA2000 1x while preserving seamless backward compatibilities to IS-95 A/B and CDMA 1x networks. The migration required only simple upgrades to the BTS, BSC, PDSN, and the AAA. CDMA2000 1x EV-DV provides integrated voice with simultaneous high-speed packet data services such as video-conferencing and other multimedia services at speeds of up to 3.09 Mbps. The 1x EV-DV standard was approved by the 3GPP2 in June 2002 and was submitted to ITU for approval in July 2002. The 1x EV-DV standard was completed in early 2004, but at the time of this writing it is not deployed because no wireless operators in the world are willingly to launch for commercial reasons.

8.1 UMTS (WCDMA) Versus CDMA2000—Physical Layer Harmonization

It seems that UMTS (WCDMA) and CDMA2000 are continuing to diverge both in IP core network and radio access. 3GPP has been developing HSDPA, while 3GPP2 has developed 1x EV-DV (Revision D), an evolved version of 1x RTT for voice and data. Many similarities exist between HSDPA and 1x EV-DV. There will be an opportunity for harmonization after they were completely standardized. The ultimate goal should be to achieve minimal difference and a single access in the near future. Harmonization should allow mobile operators to choose WCDMA and 1x EV-DV depending on the amount of radio spectrum available to them in point of spectrum commonality in the physical layer. Harmonization for high-speed data services should also analyze possible common technical paths for high-speed data solutions in 3GPP/ 3GPP2. Thus, harmonization must solve some of the issues that were not successful in creating a single common technology.

Both HSDPA and 1x EV-DV should produce a single technology to enhance downlink packet data performances and to improve spectral efficiency for data services by using items such as SDPDCH, high-order modulation (16-QAM, 64-QAM), AMC, HARQ retransmission schemes, fast scheduling for packet data, FCS (Fast Call Setup), and shorter frame size.

8.2 Reverse CDMA Channel

The Reverse CDMA Channel consists of the number of code channels transmitted by a mobile station and its structure as shown in Figure 8.1. All the channels contained in the Reverse CDMA Channel will now be described.

8.2.1 Reverse Pilot Channel (R-PICH)

The Reverse Pilot Channel (R-PICH) is an unmodulated spread spectrum signal used to assist the base station (BS) in detecting a mobile station (MS) transmission. When the Reverse Power Control Subchannel is not inserted, the R-PICH is continuous. When the Reverse Power Control Subchannel is inserted, the R-PICH becomes the Reverse Secondary Pilot Channel (all "0"s) without the power control bit, as shown in Figure 8.2.

The R-PICH should be transmitted when the Reverse Enhanced Access Channel (R-EACH), Reverse Common Control Channel (R-CCCH), or Reverse Traffic Channel (RTC) with Radio Configurations (RCs) 3 through 7 is enabled. The R-PICH should also be transmitted during the EACH preamble, the R-CCCH preamble, and the RTC preamble.

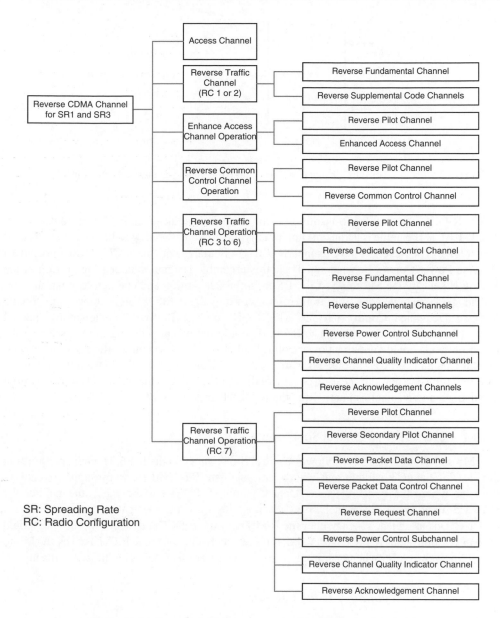

Figure 8.1 Reverse CDMA channels received at the base station

The R-EACH preamble is transmitted to aid the BS in acquiring an R-EACH transmission. The R-EACH preamble is a transmission of only the non-data-bearing R-PICH at an increased power level. The R-PICH associated with R-EACH does not have a Reverse Power Control Subchannel. Thus, R-EACH preamble is a non-data bearing portion of the Enhanced Access probe sent by the MS to assist the BS in acquisition and channel estimation.

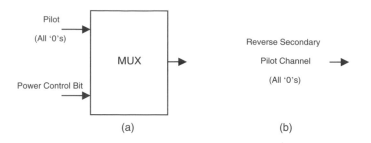

Figure 8.2   (a) Reverse Pilot Channel structure and (b) reverse Secondary Pilot Channel structure

The R-CCCH preamble is transmitted to aid the BS in acquiring an R-CCCH transmission. The R-CCCH preamble is a transmission of only the non-data-bearing R-PICH at an increased power level. The Reverse Pilot Channel (R-PICH) associated with the R-CCCH does not have a Reverse Power Control Subchannel. Thus, this preamble is a non-data bearing portion of the R-CCCH sent by the MS to assist the BS in initial acquisition and channel estimation.

The RTC preamble consists of transmissions of only the R-PICH before transmitting on the Reverse Dedicated Control Channel (R-DCCH) on the Reverse Fundamental Channel (R-FCH) with RCs 3 through 6. The R-PICH should not be gated during transmission of the preamble and should contain a power control subchannel unless neither the Forward Fundamental Channel (F-FCH) nor the Forward Dedicated Control Channel (F-DCCH) is required. This preamble is a non-data bearing portion of the R-PICH sent by the MS to aid the BS in initial acquisition and channel estimation for the R-DCCH and the R-FCH.

8.2.1.1 Reverse Power Control Subchannel

The MS should insert a Reverse Power Control Subchannel on the R-PICH when transmitting the Traffic Channel Preamble, and when operating on the RTC with RCs 3 through 7, except for the case where the Forward Packet Data Channel (F-PDCH) is assigned, and neither the Forward Fundamental Channel (F-FCH) nor the Forward Dedicated Control Channel (F-DCCH) is assigned. When neither the F-FCH nor the F-DCCH is assigned, the MS should transmit a continuous R-PICH. In a word, this is a subchannel on the R-PICH used by the MS to control power of a BS when operating on the Forward Traffic Channel with RCs 3 through 7.

8.2.2 Reverse Secondary Pilot Channel (R-SPICH)

The R-SPICH is an unmodulated, direct-sequence spread spectrum signal transmitted by a CDMA mobile station in conjunction with certain transmissions on the Reverse Packet Data Channel (R-PDCH). The secondary pilot channel provides additional phase reference for the R-PDCH for coherent demodulation and may provide a means for signal strength measurement.

The MAC layer instructs the Physical Layer when to transmit on the R-SPICH. When the Physical Layer receives a PHY-RSPICH.Request from the MAC layer, the MS transmits the R-SPICH for 10 ms starting at System Time equal to SYS-TIME.

The R-SPICH data should be orthogonally spread, quadrature spread, and filtered. It should be gated off unless instructed by the MAC Layer.

8.2.3 Access Channel

Access Channel (AC) is used by the MS to initiate communication with the BS and to response to Paging Channel messages. An R-ACH (Reverse Access Channel) transmission is a coded, interleaved, and modulated spread-spectrum signal. The AC uses a random-access protocol. ACs are uniquely identified by their long codes.

An access probe consists of an access preamble, followed by a series of AC frames, each carrying a Service Data Unit (SDU).

The MS transmits information on the R-ACH at a fixed data rate of 4800 bps. An R-ACH frame should be 20 ms in duration. An AC frame begins only when System Time is an integral multiple of 20 ms.

The MS should delay the transmit timing of the probe by RN PN chips, where the value of RN is supplied by the Common Channel multiplex sublayer. This transmit timing adjustment includes delay of the direct sequence spreading long code and of quadrature spreading I and Q pilot PN sequences, and so it effectively increases the apparent range from the MS to the BS. Note that RN denotes the pseudo-random offset of the access probe from a zero-offset AC frame, while PN designates the Pilot PN sequence offset index for the Forward CDMA Channel.

The Reverse CDMA Channel may contain up to 32 ACs numbered 0 through 31 per supported Paging Channel. At least one AC exists on the Reverse CDMA Channel for each Paging Channel on the corresponding Forward CDMA Channel. Each AC is associated with a signal Paging Channel. The AC preamble consists of 96 zeros that are transmitted at 4800 bps. The AC preamble is transmitted to aid the BS in acquiring an AC transmission. Each AC frame contains 96 bits (20 ms frame at 4800 bps). Each AC frame consists of 88 information bits and 8 Encoder Tail Bits.

The AC data should be convolutionally encoded, repeated once (each code symbol occurs two consecutive times), interleaved, modulated, transmitting all power control groups, spread by the long code, spread by the pilot PN sequences, and filtered.

The AC is used for short signaling message exchanges such as cell originations, response to pages, and registrations.

The overall structure of AC for spreading rate 1 is shown in Figure 8.3.

8.2.4 Enhanced Access Channel (R-EACH)

The EACH is used by the MS to initiate communication with the BS or to respond to a MS directed message. The R-EACH can be used in two possible modes: Basic Access mode and Reservation Access mode. When operating in the Basic Access mode, the MS should not transmit the Enhanced Access header on the R-EACH. In Basic Access mode, the Enhanced Access probe consists of an EACH preamble followed by Enhanced Access data. When operating in the Reservation Access mode, the Enhanced Access probe should consist of an EACH preamble followed by an Enhanced Access header. Enhanced Access data is sent on the R-CCCH upon receiving permission from the BS.

The R-EACH uses a random-access protocol. The long code masks will uniquely identify each of the R-EACHs.

The MS should transmit the Enhanced Access header on the R-EACH at a fixed data rate of 9600 bps. The MS should transmit the Enhanced Access data on the R-EACH channel at a fixed data rate of 9600, 19 200, or 38 400 bps. The frame duration for the Enhanced Access header on the R-EACH channel should be 5 ms. The frame duration for the Enhanced Access data on the

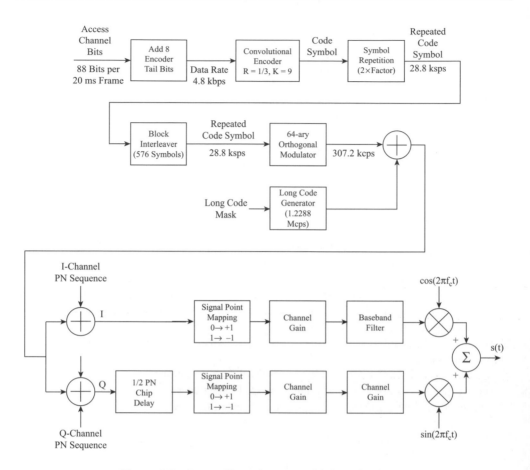

Figure 8.3   Access Channel structure for spreading rate 1

R-EACH should be 20, 10, or 5 ms. The timing of EACH transmissions should start at 1.25 ms increments of System Time.

The Reverse CDMA Channel may contain up to 32 Enhanced Access Channels per supported Forward Common Control Channel (F-CCCH), numbered 0 through 31. There is a Forward Common Assignment Channel (F-CACH) associated with every R-EACH operating in the Reservation Access mode.

The Frame Quality Indicator (also called Cyclic Redundancy Code (CRC)) should be calculated on all bits within the frame, except the CRC itself and the Encoder Tail Bits. The R-EACH should use the following Frame Quality Indicator with the generator polynomial as shown below:

- When transmitting the Enhanced Access header, the R-EACH should use an 8-bit Frame Quality Indicator (CRC) with the generator polynomial:

$$g(x) = x^8 + x^7 + x^4 + x^3 + x + 1$$

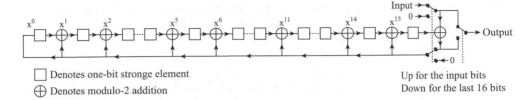

Denotes one-bit stronge element
⊕ Denotes modulo-2 addition

Up for the input bits
Down for the last 16 bits

Figure 8.4 Frame Quality Indicator calculation for the 16-bit Frame Quality Indicator

- When transmitting the Enhanced Access data, the 20 ms R-EACH should use a 12-bit CRC with the generator polynomial for the 9600 bps frame:

$$g(x) = x^{12} + x^{11} + x^{10} + x^9 + x^8 + x^4 + x + 1$$

- When transmitting the Enhanced Access data, a 16-bit CRC with the generator polynomial for the 38 400 and 19 200 bps frames can be computed by

$$g(x) = x^{16} + x^{15} + x^{14} + x^{11} + x^6 + x^5 + x^2 + x + 1$$

The CRC of length 16 can be generated by the shift register structure shown in Figure 8.4.

8.2.5 Reverse Common Control Channel (R-CCCH)

The R-CCCH is used for the transmission of user and signaling information to the BS when RTCs are not in use. The MS transmits information on the R-CCCH at variable data rates of 9.6, 19.2, and 38.4 Kbps, and its frame will be 20, 10, or 5 ms in duration. The timing of R-CCCH transmissions should start on 1.25 ms increments of System Time. The Reverse CDMA Channel may contain up to 32 R-CCCHs numbered 0 through 31 per supported Forward Common Control Channel (F-CCCH) and up to 32 R-CCCH numbered 0 through 31 per supported F-CACH. At least one R-CCCH exists on the Reverse CDMA Channel for each F-CCCH on the corresponding Forward CDMA Channel, which means that each R-CCCH is associated with a single F-CCCH.

The R-CCCH bit allocations depend on the transmission rates (9.6, 19.2, or 38.4 Kbps) of the information bits (17.2, 360, or 744 bit), followed by a Frame Quality Indicator (12, 16, or 16 bit) and 8-bit Encoder Tails. The Frame Quality Indicator (CRC) should be calculated on all bits within the frame, except the CRC itself and the Encoder Tail Bits. The 20 ms R-CCCH should use a 12-bit CRC with the generator polynomial $g(x) = x^{12} + x^{11} + x^{10} + x^9 + x^8 + x^4 + x + 1$ for the 9600 bps frame and a 16-bit Frame Quality Indicator (CRC) with the generator polynomial $g(x) = x^{16} + x^{15} + x^{14} + x^{11} + x^6 + x^5 + x^2 + x + 1$ for the 38.4 and 19.2 Kbps frames. The 10 ms R-CCCH should use a 12-bit CRC with the generator polynomial $g(x) = x^{12} + x^{11} + x^{10} + x^9 + x^8 + x^4 + x + 1$ for the 14.2 Kbps frame and a 16-bit CRC with the generator polynomial $g(x) = x^{16} + x^{15} + x^{14} + x^{11} + x^6 + x^5 + x^2 + x + 1$ for the 38.4 Kbps frame. The 5 ms R-CCCH shall use a 12-bit CRC with the generator polynomial $g(x) = x^{12} + x^{11} + x^{10} + x^9 + x^8 + x^4 + x + 1$ for the 38.4 Kbps frame.

The R-CCCH preamble is transmitted to aid the BS in acquiring a R-CCCH transmission. This preamble is a transmission of only the non-data-bearing R-PICH at an increased power level. The R-PICH associated with the R-CCCH does not have a Reverse Power Control Subchannel. The total preamble duration should be an integer multiple of 1.25 ms. No preamble should be transmitted when operating in the Reservation Access mode.

The R-CCCH data must be convolutionally encoded, repeated (each code symbol output from the encoder), interleaved, modulated and orthogonally spread, quadrature spread, and baseband filtered.

When the physical layer receives a PHY-RCCCH Primitive.Request (FCCCH-ID, RCCCH-ID, BASE-ID) from the MAC Layer, the mobile station should store the arguments FCCCH-ID, RCCCH-ID, and BASE-ID, set the R-CCCH Long Code Mask using FCCCH-ID, RCCCH-ID, and BASE-ID, and transmit a R-CCCH preamble. When the Physical Layer receives a PHY-RCCCH.Request (FCCCH-ID, RCCCH-ID, BASE-ID, SDU, FRAME-DURATON, NUM-BITS) from the MAC Layer, the mobile station should store the arguments FCCCH-ID, RCCCH-ID, BASE-ID, SDU, FRAME-DURATION, and NUM-BITS; set the R-CCCH Long Code Mask using FCCCH-ID, RCCCH-ID and BASE-ID; set the information bits to SDU; and transmit a R-CCCH frame of duration FRAME-DURATION (5 ms, 10 ms, or 20 ms) at a data rate that corresponds to NUM-BITS and FRAME-Duration. Figure 8.5 illustrates the R-EACH data and R-CCCH structure for SR1.

8.2.6 Reserve Packet Data Control Channel (R-PDCCH)

The mobile station uses the R-PDCCH for transmitting control information for the associated Reverse Packet Data Channel (R-PDCH) and the Mobile Status Indicator Bit (MSIB). The MS should transmit information on the R-PDCCH at a fixed data rate of 700 bps. For each frame, the MAC Layer instructs the Physical Layer whether to transmit on the R-PDCCH. When the MS receives PHY-RPDCCH.Request primitives from the MAC Layer, the MS should transmit one R-PDCCH.

A R-PDCCH frame is 10 ms in duration, and begins only when System Time is an integral multiple of 10 ms. Transmission on the R-PDCCH should be aligned with the transmission of data on the R-PDCH. The R-PDCCH consists of six information bits and an MSIB.

The R-PDCCH data should be encoded, repeated (the encoded sequence), modulated and orthogonally spread, quadrature spread, and filtered.

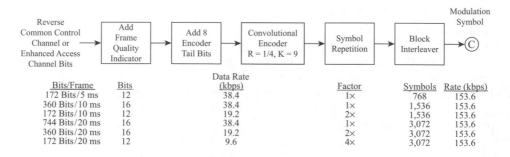

Bits/Frame	Bits	Data Rate (kbps)	Factor	Symbols	Rate (kbps)
172 Bits/5 ms	12	38.4	1×	768	153.6
360 Bits/10 ms	16	38.4	1×	1,536	153.6
172 Bits/10 ms	12	19.2	2×	1,536	153.6
744 Bits/20 ms	16	38.4	1×	3,072	153.6
360 Bits/20 ms	16	19.2	2×	3,072	153.6
172 Bits/20 ms	12	9.6	4×	3,072	153.6

Figure 8.5   Each data and R-CCCH structure for spreading rate 1

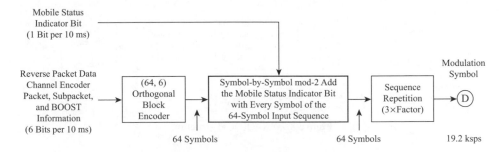

Figure 8.6   R-PDCCH structure

When the Physical Layer receives a PHY-RPDCCH.Request (SDU, BOOST, EP-SIZE, MSIB, SYS-Time) from the MAC Layer, the mobile station should store the arguments SDU, BOOST, EP-SIZE, MSIB, and SYS-TIME; set the information bits to SDU; set the Mobile Status Indicator Bit to MSIB; and transmit the information bits and the MSIB in a R-PDCCH frame using the appropriate row depending on the value of BOOST and EP-SIZE starting at SYS-TIME. The R-PDCCH structure is shown in Figure 8.6.

8.2.7 Reverse Request Channel (R-REQCH)

The R-REQCH is used by the MS to indicate its ability to the BS to transmit above the autonomous data rate on the R-PDCH. This channel is also used by the MS to update the BS with the amount of data in the buffer, the QoS requirement, and the available headroom.

The MS should transmit information on the R-REQCH at a fixed data rate of 3200 bps. For each frame (10 ms in duration), the MAC Layer instructs the Physical Layer whether to transmit to the R-REQCH. When the mobile station receives a PHY-RREQCH.Request primitive from the MAC Layer, the MS should transmit one R-REQCH frame.

A R-REQCH frame begins only when System Time is an integral multiple of 10 ms, and may be offset from the frames of the RTC operating with RCs 3 and 4. The RC is a set of Forward Traffic Channel and Reverse Traffic Channel transmission formats that are characterized by physical layer parameters such as data rates, modulation characteristics, and spreading rate.

The frame structure of the R-REQCH consists of 12 information bits, followed by a 12-bit CRC, and the 8-bit Encoder Tail (each set to "0"). The Frame Quality Indicator (CRC) is calculated on all bits within the frame, except the CRC itself and the Encoder Tail Bits. The R-REQCH uses a 12-bit CRC with the generator polynomial $g(x) = x^{12} + x^{11} + x^{10} + x^9 + x^8 + x^4 + x + 1$.

The R-REQCH data should be convolutionally encoded, repeated (each code symbol from the convolutional encoder), interleaved, modulated and orthogonally spread, quadrature spread, and baseband filtered.

When the Physical Layer receives a PHY-RREQCH.Request (SDU, SYS-TIME) from the MAC Layer, the MS should store arguments SDU and SYS-TIME, set the information bits to SDU (Service Data Unit), and transmit the information bits, a computed 12-bit CRC, and the 8-bit Encoder Tail on the R-REQCH frame starting at SYS-Time. Figure 8.7 depicts the R-REQCH structure.

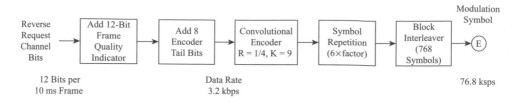

Figure 8.7   R-REQCH structure

8.2.8 Reverse Dedicated Control Channel (R-DCCH)

The R-DCCH is used for transmission of user and signaling information to the BS during a call. The RTC may contain up to one R-DCCH.

The MS transmits information on the R-DCCH at a fixed data rate of 9.6 or 14.4 Kbps using 20 ms frames, or 9.6 Kbps using 5 ms frames. The MS may also support flexible data rates. The R-DCCH frame should be 5 or 20 ms in duration. The R-DCCH frame structures for both non-flexible data rates and flexible data rates are shown in Tables 8.1 and 8.2, respectively.

As shown in Tables 8.1 and 8.2, the MS transmits information on the R-DCCH at a data rate of 9600 bps for RCs 3 and 5. If the MS supports flexible data rates, other fixed data rates from 1050 to 9600 bps using 20 ms can also be used for the R-DCCH in RCs 3 and 5. The MS transmits information on the R-DCCH at data rates of 14 400 bps for 20 ms and 9600 bps for 5 ms frames for RCs 4 and 6. The MS should support discontinuous transmission on the R-DCCH. The decision to enable or disable the R-DCCH should be made on a frame-by-frame basis (i.e. 5 or 20 ms).

The MS should support R-DCCH frames that are time offset by multiples of 1.25 ms. A zero-offset 5 ms R-DCCH should begin only when System Time is an integral multiple of 20 ms. A zero-offset 20 ms R-DCCH should begin only when System Time is an integral multiple of 5 ms.

The Frame Quality Indicator (CRC) should be computed on all bits within the frame, except CRC itself and the Encoder Tail Bits. The 20 ms R-DCCH should use a 12-bit CRC with the generator polynomial $g(x) = x^{12} + x^{11} + x^{10} + x^9 + x^8 + x^4 + x + 1$ for non-flexible

Table 8.1   R-DCCH frame structure summary for non-flexible data rates

Radio configuration	Frame length (ms)	Data rate (bps)	Number of bits per frame				
			Total	Reserved	Information	Frame Quality Indicator (CRC)	Encoder Tail
3 and 5	20	9600	192	0	172	12	8
4 and 6	20	14 400	288	1	267	12	8
3, 4, 5, and 6	5	9600	48	0	24	16	8

CRC: Frame Quality Indicator

Table 8.2 R-DCCH frame structure summary for flexible data rates

Radio configuration	Frame length (ms)	Data rate (bps)	Number of bits per frame			
			Total	Information	Frame Quality Indicator (CRC)	Encoder Tail
3 and 5	20	1250–9600	25–192	1–168	16	8
	20	1050–9550	21–191	1–171	12	8
4 and 6	20	1250–14 400	25–288	1–264	16	8
	20	1050–14 300, 14 400	21–286, 288	1–266, 268	12	8

CRC: Frame Quality Indicator

data rates. If flexible data rates are supported, either a 12-bit CRC with $g(x) = x^{12} + x^{11} + x^{10} + x^9 + x^8 + x^4 + x + 1$ or a 16-bit CRC with $g(x) = x^{16} + x^{15} + x^{14} + x^{11} + x^6 + x^5 + x^2 + x + 1$ should be used. The 5 ms R-DCCH should use a 16-bit CRC (Frame Quality Indicator) with $g(x) = x^{16} + x^{15} + x^{14} + x^{11} + x^6 + x^5 + x^2 + x + 1$.

The R-DCCH should process convolutional encoding, encoder output symbol repetition and puncturing, modulation symbols interleaving, R-DCCH data modulation and orthogonal spreading, quadrature spreading, and baseband filtered. These channel proceedings were fully covered in Chapter 2.

When the Physical Layer receives a PHY-DCCH.Request (SDU, FRAME-DURATION, NUM-BITS) from the MAC Layer, the MS should store the arguments SDU, FRAME-DURATION, and NUM-BITS; if SDU is not equal to NULL, set the information bits to SDU first and transmit NUM-BITS of SDU in a R-DCCH frame of duration FRAME-DURATION (5 or 20 ms). If a PHY-DCCH.Request primitive for a 5 ms frame is received coincident with a PHY-DCCH.Request primitive for a 20 ms frame or during transmission of a 20 ms frame, then the MS may pre-empt transmission of the 20 ms frame and transmit a 5 ms frame. Transmission of the 20 ms frame may start or resume after completion of the 5 ms frame. If transmission of the 20 ms frame is resumed after an interruption in transmission, then the relative power level of the R-DCCH modulation symbols should be equal to that of the modulation symbols sent prior to the pre-emption. Figure 8.8 shows the R-DCCH Structure for Radio Configuration 3.

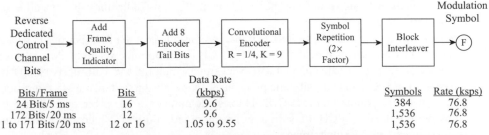

Note: If flexible data rates are supported, there can be 1 to 171 channel bits in a 20 ms frame and the encoded symbols will be repeated and then punctured to provide a 76.8 ksps modulation symbol rate.

Figure 8.8 R-DCCH structure for RC 3

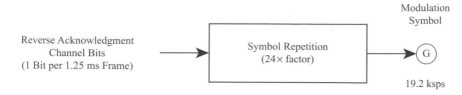

Figure 8.9   R-ACKCH structure

8.2.9 *Reverse Acknowledgment Channel (R-ACKCH)*

The R-ACKCH provides feedback for the Forward Packet Data Channel (F-PDCH). The MS transmits ACK or NAK responses to Forward Packet Data Control Channel (F-PDCCH) messages. When the R-ACKCH is transmitted, it should transmit an ACK-OR-NAK response (as specified by the MAC Layer) as a parameter in the PHY-RACKCH.Request primitive. ACK-OR-NAK should be set to ACK or NAK. If no PHY-RACKCH.Request primitive is received from the MAC Layer, the R-ACKCH should be gated off. A "0" bit will be transmitted for an ACK response and a "1" for a NAK response. A R-ACKCH frame should begin only when System Time is an integral multiple of 1.25 ms.

When the Physical Layer receives a PHY-RACKCH.Request (ACK-OR-NAK) from the MAC Layer, the MS should store the argument ACK-OR-NAK and transmit an ACK on the R-ACKCH if ACK-OR-NAK is equal to ACK; otherwise, it transmits a NAK on the R-ACKCH. Figure 8.9 illustrates the R-ACKCH structure.

8.2.10 *Reverse Channel Quality Indicator Channel (R-CQICH)*

The MS uses the R-CQICH to indicate the channel quality measurements of the member of the packet data channel active set from which the mobile station has selected to receive F-PDCH transmission. The BS uses this information to determine the transmission power levels and data transmission rates on the F-PDCH and the power level and duration of the F-PDCCH. It can also be used by the BS in the decision of when to schedule a particular mobile station on the F-PDCH.

When the F-PDCH is assigned to a mobile station, the MS sends the channel quality feedback information on the R-CQICH every 1.25 ms. When the MS is not indicating a cell switch, each transmission on the R-CQICH carries either a full channel quality indicator value or a different channel quality indicator value.

A distinct Walsh cover on the R-CQICH transmission indicates the member of the packet data channel active set selected by the MS for F-PDCH transmissions. When the MS selects a new member of the packet data channel active set from which the MS will receive F-PDCH transmissions, the MS invokes a switching procedure. To initiate the switch, the MS transmits a distinctive switching pattern on the R-CQICH for some number of 20 ms periods. During the switching period, the R-CQICH transmissions are modified to use Walsh cover of the target base station in certain 1.25 ms frames.

There are two modes of operation of the R-CQICH: full C/I feedback mode and differential C/I feedback mode. In the full C/I feedback mode, only full C/I reports are sent. In the differential C/I feedback mode, a pattern of full and differential C/I reports is sent. The BS can

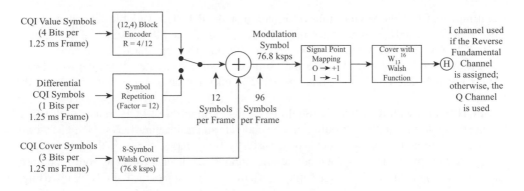

Figure 8.10 R-CQICH structure (Reproduced under written permission from Telecommunications Industry Association)

indicate that the MS is to alter the pattern of full and differential C/I reports to repeat the full C/I reports a total of two or four times. The R-CQICH of full and differential C/I reports is also modified by the current reverse link pilot gating rate.

The R-CQICH structure is shown in Figure 8.10. A R-CQICH frame should begin only when System Time is an integral multiple of 1.25 ms. The MAC Layer instructs the Physical Layer what to transmit on the R-CQICH for each frame. When the MS receives a PHY-RCQICH. Request primitive from the MAC Layer with CQI-VALUE set to a 4-bit value, the MS should map these 4 bits into 12 symbols using a (12, 4) block code consisting of the last 12 symbols of each of the 16-symbol Walsh functions. For example, a codeword for R-CQICH channel (12, 4) block code consists of Input (0010) and Output (0011 0011 0011).

When the MS receives a PHY-RCQICH.Request primitive from the MAC Layer with CQI-VALUE set to "UP" or "DOWN," the MS should map it into a Channel Quality Indicator Channel bit (a_0), and it should be repeated 12 times to form 12 symbols according to Figure 8.10. If the CQI-VALUE is "UP," the MS sets a_0 to "1." If the CQI-VALUE is "DOWN," the MS sets a_0 to "0."

An 8-ary Walsh function specified by the WALSH-COVER parameter of the PHY-RCQICH. Request primitive should be used to spread the R-CQICH transmission to indicate the BS identity.

The R-CQICH should be modulated and orthogonally spread, quadrature spread (R-CQICH is transmitted on *I*-channel if the Reverse Fundamental Channel (R-FCH) is assigned; otherwise, the R-CQICH should be transmitted on the *Q*-channel), and baseband filtered.

When the Physical Layer receives a PHY-RCQICH.Request (WALSH-COVER, CQI-VALUE, CQI-GAIN) from the MAC Layer, the MS should store the arguments WALSH-COVER, CQI-VALUE, and CQI-GAIN, and transmit the channel quality indicator value, CQI-VALUE, on the R-CQICH using the Walsh cover, WALSH-COVER, with the relative gain specified by CQI-GAIN.

8.2.11 Reverse Fundamental Channel (R-FCH)

The R-FCH is used for the transmission of user and signaling information to the BS during a call. The RTC may contain up to one R-FCH. When operating with RC 1, the MS transmits information on the R-FCH at variable data rates of 9600, 4800, 2400, and 1200 bps. When operating with RC 2, the MS transmits information on the R-FCH at variable data rates 14 400, 7200, 3600, and 1800 bps. When operating with RCs 3 and 5, the MS transmits information on the R-FCH at variable data rates of 9600, 4800, 2700, and 1500 bps during 20 ms frames or at 9600 bps during 5 ms frames. The MS may support flexible data rates. If so, the MS should support variable data rates corresponding to 1–171 information bits per 20 ms frame on the R-FCH.

R-FCH frames with RCs 1 and 2 should be 20 ms in duration. R-FCH frames with RCs 3 through 6 should be 5 or 20 ms in duration. The data rate and frame duration on a R-FCH within a radio configuration should be selected on a frame-by-frame basis. Although the data rate may vary on a frame-by-frame basis, the modulation symbol rate is kept constant by code repetition. An MS operating with RCs 3 through 6 may continue transmission of the R-FCH for up to three 5 ms frames in a 20 ms frame.

The MS supports R-FCHs frames that are time offset by multiples of 1.25 ms. The amount of the time offset is specified by FRAME-OFFSETs. A zero-offset 20 ms R-FCH frame should begin only when System Time is an integral multiple of 20 ms. A zero-offset 5 ms R-FCH frame should begin only when System Time is an integral multiple of 5 ms. An offset 20 ms R-FCH frame should begin $1.25 \times$ FRAME-OFFSETs ms later than the zero-offset 20 ms R-FCH frame. An offset 5 ms R-FCH frame should begin $1.25 \times$ [FRAME − OFFSETs mode 4] ms later than the zero-offset 5 ms R-FCH frame. The interleaver block for the R-FCH should be aligned with the R-FCH frame.

The 2400 and 1200 bps frames with RC 1 consist of the information bits followed by 8 Encoder Tail Bits. A 5 ms frame, all frames with RCs 3 and 5, and the 9600 and 4800 bps frames with RC 1 should consist of the information bits followed by a Frame Quality Indicator (CRC) and 8 Encoder Tail Bits. All 20 ms frames with RCs 2, 4, and 6 should consist of zero or one Reserved/Erasure Indicator Bits, followed by the information bits, Frame Quality Indicator (CRC), and 8 Encoder Tail Bits. The order of the bits in the R-FCH frame structure is shown in Figure 8.11. The Frame Quality Indicator supports two functions at the BS. The first function is to determine whether the frame is in error, while the second function is to assist in the determination of the data rate of the received frame. Encoder Tail Bits are a fixed sequence of 8 bits added to the end of a data block (a frame) to reset the convolutional encoder to a known state.

Each frame with RCs 2 through 6, and the 9600 and 4800 bps frames of RC 1, should include a Frame Quality Indicator (CRC), but no Frame Quality Indicator is used for 2400 and 1200 bps data rates of RC 1. The R-FCH frame structure with generator polynomial for non-flexible data rates is shown in Table 8.3.

The RTC preamble is transmitted on the R-PICH or R-FCH to aid the BS in acquiring the R-FCH transmissions.

Notation
R/E - Reserved/Erasure Indicator Bit
F - Frame Quality Indicator (CRC)
T - Encoder Tail Bits

Figure 8.11   R-FCH structure

Table 8.3 R-FCH frame structure with generator polynomial for non-flexible data rate

Radio configuration	Data rate (bps)	Frame Quality Indicator	Generator polynomial $g(x)$
3	9600	16	$x^{16} + x^{15} + x^{14} + x^{11} + x^6 + x^5 + x^2 + x + 1$
1	9600	12	$x^{12} + x^{11} + x^{10} + x^9 + x^8 + x^4 + x + 1$
2	1400	12	$x^{12} + x^{11} + x^{10} + x^9 + x^8 + x^4 + x + 1$
2	7200	10	$x^{10} + x^9 + x^8 + x^7 + x^6 + x^4 + x^3 + x + 1$
1	4800	8	$x^8 + x^7 + x^4 + x^3 + x + 1$
2	3600	8	$x^8 + x^7 + x^4 + x^3 + x + 1$
2	1800	6	$x^6 + x^2 + x + 1$
4	9600	16	$x^{16} + x^{15} + x^{14} + x^{11} + x^6 + x^5 + x^2 + x + 1$
6	7200	10	$x^{10} + x^9 + x^8 + x^7 + x^6 + x^4 + x^3 + x + 1$

The R-FCH should be convolutionally encoded. When generating R-FCH data, the encoder should be initialized to the all-zero state at the end of each 5 or 20 ms frame. The R-FCH code symbol should be repeated and puncturing. Within a puncturing pattern, a "0" means that the symbol should be deleted and "1" means that a symbol should be passed. For example, the 5-symbol puncturing pattern for RC 3 is 11110, meaning that the first, second, third, and fourth symbols are passed, while the fifth symbol is removed. The R-FCH should be interleaved prior to modulation and transmission.

When using RCs 1 and 2, the R-FCH data should be modulated, while using RCs 3 through 6, the R-FCH data should be modulated and orthogonally spread using the W_4^{16} Walsh function. W_4^{16} represents the fourth Walsh function of length 16.

The BS should perform the data burst randomizing function while transmitting on the R-FCH with RC 1 or 2. The transmission of the R-FCH with RCs 3, 4, 5, or 6 may be gated when no other RTC is assigned.

When operating in RC 1 or 2, the R-FCH should be spread by the long code for direct sequence spreading, quadrature spreading, and baseband filtering.

When the Physical Layer receives a PHY-FCH (SDU, FRAME-DURATION NUM-BITS) from the MAC LAYER, the MS should store the arguments SDU (Service Data Unit), FRAME-DURATION, and NUM-BITS. If SDU is not equal to NULL, set the information bits to SDU. If SDU is not equal to NULL, transmit NUM_BITS bits of SDU on the R-FCH frame of FRAME-DURATION 5 or 20 ms. If a PHY-FCH.Request primitive for a 20 ms frame or during transmission of a 20 ms frame, then the MS may pre-empt transmission of the 20 ms frame and transmit a 5 ms frame. Transmission of the 20 ms frame may start or resume after completion of the 5 ms frame.

8.2.12 Reverse Supplemental Channel (R-SCH)

The R-SCH is used for the transmission of higher-level data to the BS during a call. The R-SCH applies to RCs 3 through 6 only. The RTC contains up to two R-SCHs.

When transmitting on the R-SCH with a single assigned data rate in RC 3, the BS should submit information at fixed data rates of 307.2, 153.6, 76.8, 38.4, 19.2, 9.6, 4.8, 2.7, 2.4, 1.5, 1.35, or 1.2 Kbps.

When transmitting on the R-SCH with a single assigned data rate in RC 4, the MS should transmit information at fixed data rates of 230.4, 115.2, 57.6, 28.8, 14.4, 7.2, 3.6, or 1.8 Kbps.

When transmitting on the R-SCH with a single assigned data rate in RC 5, the MS should transmit information at fixed data rate of 614.4, 307.2, 153.6, 76.8, 38.4, 19.2, 9.6, 4.8, 2.7, 2.4, 1.5, 1.35, or 1.2 Kbps.

When transmitting on the R-SCH with a single assigned data rate in RC 6, the MS should transmit information at fixed data rate of 1036.8, 518.4, 460.8, 259.2, 230.4, 115.2, 57.6, 28.8, 14.4, 7.2, 3.6, or 1.8 Kbps.

When using variable-rate transmission on the R-SCH with multiple assigned data rates in RCs 3, 4, 5, and 6, the MS should transmit information at the maximal assigned data rate, or transmit information at the other assigned data rates with the same modulation symbol rate as that of the maximal assigned data rate. To achieve a higher modulation symbol rate, repetition or puncturing is applied to the specified data rate.

If the mobile station supports the R-SCH, the MS should support R-SCH frames that are 20, 40, or 80 ms in duration. The MS should support R-SCH frames that are time offset by multiples of 1.25 ms as specified by FRAME-OFFSETs.

All frames with RCs 3 and 5 and the frames with RCs 4 and 6 with data rates above 14 400 bps should consist of the information bits, followed by the Frame Quality Indicator (CRC) and 8 Encoder Tail Bits. All frames with RCs 4 and 6 with data rates equal to or less than 14 400 bps should consist of zero or one Reserved Bits, followed by the information bits, Frame Quality Indicator (CRC), and the 8-bit Encoder Tail.

Each frame includes a Frame Quality Indicator (CRC). The CRC should be calculated on all bits within the frame, except the CRC itself and the Reserved/Encoder Tail Bits. Table 8.4 shows the R-SCH frame structure with the generator polynomial.

Bit allocations in the R-SCH frame structure consist of zero or one Reserved Bits, followed by the information bits, Frame Quality Indicator (CRC), and the 8-bit Encoder Tail. For the convolutional encoder, the last 8-bit Reserved/Encoder Tail Bits should be set to "0." For the turbo encoder, the first two of the eight bits should each be set "0," and the turbo encoder will calculate and append the remaining six tail bits.

Table 8.4 R-SCH frame structure with generator polynomial

Radio configuration	Total bits in frame, L	Frame Quality Indicator	Generator polynomial $g(x)$
3 and 5	L > 192	16	$x^{16} + x^{15} + x^{14} + x^{11} + x^6 + x^5 + x^2 + x + 1$
4 and 6	L > 288	16	$x^{16} + x^{15} + x^{14} + x^{11} + x^6 + x^5 + x^2 + x + 1$
* 3 and 5	$97 \leq L \leq 192$	12	$x^{12} + x^{11} + x^{10} + x^9 + x^8 + x^4 + x + 1$
* 4 and 6	$145 \leq L \leq 288$	12	$x^{12} + x^{11} + x^{10} + x^9 + x^8 + x^4 + x + 1$
4 and 6	$73 \leq L \leq 144$	10	$x^{10} + x^9 + x^8 + x^7 + x^6 + x^4 + x^3 + x + 1$
3 and 5	$55 \leq L \leq 96$	8	$x^8 + x^7 + x^4 + x^3 + x + 1$
4 and 6	$37 \leq L \leq 72$	8	$x^8 + x^7 + x^4 + x^3 + x + 1$
3 and 5	$54 \leq L$	6	$x^6 + x^5 + x^2 + x + 1$
4 and 6	$36 \leq L$	6	$x^6 + x^5 + x^2 + x + 1$

*Frames in Radio Configurations 3 and 5 with 97–192 total bits and the Radio Configurations 4 and 6 with 145–288 total bits should use a 12-bit CRC with the generator polynomial $g(x) = x^{12} + x^{11} + x^{10} + x^9 + x^8 + x^4 + x + 1$.

The R-SCH should be convolutionally or turbo encoded. When generating R-SCH data, the encoder should be initialized to the all-zero state at the end of frame. The R-SCH code symbol of the encoder output should be processed as follows: symbol repetition, puncturing, interleaving, modulating and orthogonal spreading, quadrature spreading, and filtering.

8.2.13 Reverse Supplemental Code Channel (R-SCCH)

The R-SCCH applies to Radio Configurations 1 and 2 only. This channel is used for the transmission of higher-level data to the BS during a call. The RTC contains up to seven R-SCCHs. This channel frame should be 20 ms in duration.

When transmitting on R-SCCHs with RCs 1 and 2, the MS should transmit information at 96 000 and 14 400 bps, respectively. The MS should transmit R-SCCHs within 3/8 of a PN chip (305.1758 ns) of the R-FCH. The MS supports R-SCCH frames that are time offset by multiples of 1.25 ms.

The order of R-SCCH bit allocations with RC 1 frames consists of the information bits followed by a CRC and 8 Encoder Tail Bits, whereas the order with RC 2 frames consists of a Reversed Bit, followed by the information bits, a CRC, and 8 Encoder Tail Bits. Each frame with RCs 1 and 2 should include a 12-bit Frame Quality Indicator (CRC) with the generator polynomial $g(x) = x^{12} + x^{11} + x^{10} + x^9 + x^8 + x^4 + x + 1$.

The R-SCCH preamble is transmitted on the Reverse Supplemental Code Channel to aid the BS in acquiring the R-SCCH transmissions. The R-SCCH preamble consists of NUM-PREAMBLE-FRAMES frames of all-zeros that are transmitted with a 100% transmission duty cycle. The R-SCCH preamble should not include the CRC. For RC 1, each frame of the R-SCCH preamble should consist of 192 zeros that are transmitted at the 9600 bps rate. For RC 2, each frame of the R-SCCH preamble should consist of 288 zeros that are transmitted at the 14 400 bps rate. When discontinuous transmission is permitted on the R-SCCH, the BS may resume transmission following a break in the R-SCCH transmission.

The R-SCCH should be convolutionally encoded, code symbol repeated, modulated, spread by long code for direct sequence spreading, quadrature spread by the pilot PN sequences, and baseband filtered. The R-FCH and R-SCCH structure for RCs 1 and 2 are shown in Figure 8.12(a) and (b) respectively.

8.2.14 Reverse Packet Data Channel (R-PDCH)

The R-PDCH is used for the transmission of high-level data to the base stations by the mobile stations operating with Spreading Rate 1. The R-PDCH should transmit 174, 386, 770, 1538, 3074, 4610, 6146, 9218, 12 290, 15 362, or 18 434 information bits followed by the Frame Quality Indicator (12 or 16 bits) and Encoder Tail allowance (6 bits). These CRC and Encoder Tail bits should be appended to the information bits to form encoder packets. The encoder packets are encoded with a rate 1/5 turbo encoder and interleaved. The symbols from the interleaved sequence should then be selected for transmission as a subpacket. Encoder packets are transmitted as one, two, or three subpackets. Initially, the first subpacket is transmitted. Then, subsequent subpackets are transmitted if the transmitted subpacket is not the last subpacket, and the MS does not receive an acknowledgment from the BS. The symbols in a subpacket are formed by selecting a specific sequence of symbols from the interleaved turbo encoder output sequence. The resulting subpacket is a binary sequence of symbols for the modulator.

For the R-PDCH operating with RC 7, symbols from the subpacket symbol selection operation are mapped into a sequence of BPSK, QPSK, or 8-PSK symbols depending on the encoder packet size. For encoder packet sizes of 192, 408, 792, or 1560 bits, the modulated symbols should be spread with a 4-chip Walsh function. For an encoder packet size of 3096 bits, the modulated symbols should be spread with a 2-chip Walsh function. For encoder packet sizes of 4632, 6168, 9240, 12 312, 15 384, or 18 456 bits, the modulated symbols should be sequence demultiplexed and Walsh processed. Specifically, the encoder packet sizes of 4632, 6168, 9240, 12 312, and 15 384 bits should be QPSK modulated, while the encoder packet size of 18 456 bits should be 8-PSK modulated. The modulator outputs are demultiplexed into two sequences for covering by two different Walsh covers. The first one third of the modulated symbols of the subpacket should be demultiplexed to the output for covering by the W_2^4 Walsh function, and the

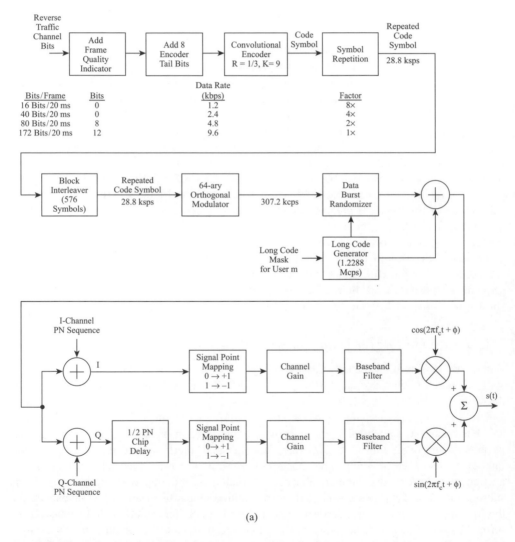

(a)

Figure 8.12　(a) R-FCH and R-SCCH structure for RC 1 (b) R-FCH and R-SCCH structure for RC 2 (Reproduced under written permission from Telecommunications Industry Association)

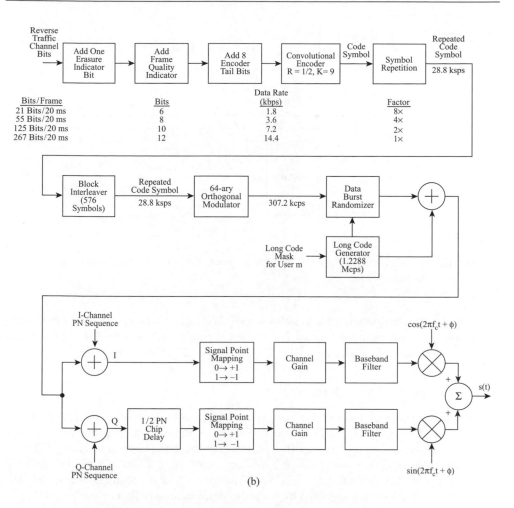

Figure 8.12 (*Continued*)

last two thirds of the modulated symbols of the subpacket should be demultiplexed to the output for covering by the W_1^2 Walsh function. After the Walsh covering, the symbols that have been covered with the W_1^2 Walsh function should be amplified by a power gain factor of two thirds and the symbols that have been covered by the W_2^4 Walsh function should be amplified by a power gain factor of one third. The resulting sequences are summed to obtain a single sequence. The Walsh functions for the R-PDCH are W_1^2, W_2^4, or both. Generally, the Walsh function W_n^N represents the nth Walsh function ($n = 0$ to $N-1$) of length N with the binary symbols of the Walsh function mapped to the symbols using the mapping "0" to $+1$ and "1" to -1.

Encoder packets with 192 bits use a 12-bit CRC with the generator polynomial $g(x) = x^{12} + x^{11} + x^{10} + x^9 + x^8 + x^4 + x + 1$. Encoder packets with 408, 792, 1560, 3096, 4632, 6168, 9240, 12312, 15384, or 18456 bits will use a 16-bit CRC with the generator polynomial $g(x) = x^{16} + x^{15} + x^{14} + x^{11} + x^6 + x^5 + x^2 + x + 1$. Encoder packets should be turbo coded with a code rate of 1/5 and include a 6-bit turbo encoder tail allowance. The turbo encoder output sequence should be interleaved and subpacket symbols are selected from the interleaver

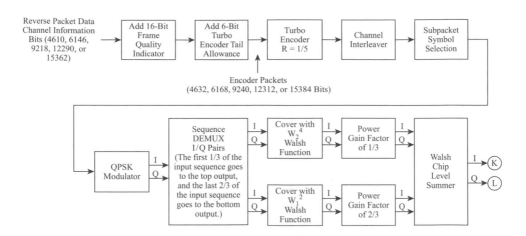

Figure 8.13   R-PDCH structure for RC 7 (Reproduced under written permission from Telecommunications Industry Association)

output sequence. The symbols from the subpacket symbols selection process should be modulated orthogonally spread, quadrature spread, and filtered. Figure 8.13 depicts the R-PDCH structure for RC 7 with encoder packet sizes of 4632, 6168, 9240, 12 312, and 15 384 bits.

8.3 Forward CDMA Channel

The Forward CDMA Channel transmitted by a BS consists of 20 code channels specified in Figure 8.14. The Forward Traffic Channel (FTC) is composed of five channels (i.e. the Forward Dedicated Control Channel (F-DCCH), Forward Fundamental Channel (F-FCH), Forward Supplemental Channel (F-SCH), Forward Supplemental Code Channel (F-SCCH), and Forward Packet Data Channel (F-PDCH)) sent to Mobile Stations (MSs). Signals transmitted to the FTC are specified by ten radio configurations (RCs).

Each code channel transmitted on the Forward CDMA Channel should be spread with a Walsh function or a quasi-orthogonal function at a fixed chip rate of 1.2288 Mcps to provide channelization among all code channels on a given Forward CDMA Channel. Thus, the assignment of code channels should be such that each code channel is orthogonal or quasi-orthogonal to all other code channels in use. The maximum length N_{max} of the associated Walsh functions for code channels, except for the Forward Auxiliary Pilot Channel (F-APICH) and Forward Auxiliary Transmit Diversity Pilot Channel (F-ATDPICH), transmitted on the Forward CDMA Channel is 256. One of N-ary ($N \leq N_{max}$) time-orthogonal Walsh functions is used. A code channel that is spread using Walsh function n from the N-ary orthogonal set ($0 \leq n \leq N - 1$) should be assigned to code channel W_n^N. The Walsh function spreading sequence should be spread with a period of ($N/1.2288$) μs.

The modulation parameters for some forward link channels on the Forward CDMA Channel operating in Spreading Rates (SRs) (3, 6, 7, 8, 9) and Radio Configurations (3, 4, 5) are classified by PN chip rate (Mcps), code rate (bits/code symbol), code symbol repetition, modulation symbol rate, Walsh length (PN chips), number of Walsh function repetitions per modulation symbol, and processing gain (PN chips/bit).

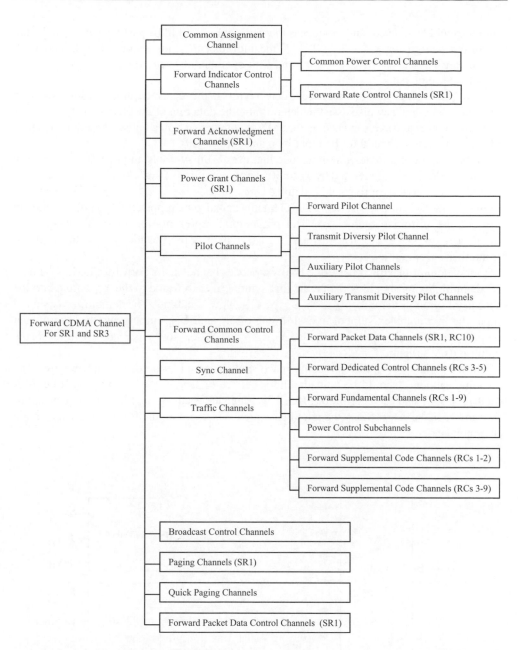

Figure 8.14 Forward CDMA Channel transmitted by a base station, stating the range of valid channels for each channel type (Reproduced under written permission from Telecommunications Industry Association)

The punctured codes used with convolutional code symbols, or turbo code symbols for all channels (except for the F-PDCCH and Forward Grant Channel (F-GCH), on the Forward CDMA Channel include the base code rate, puncturing ratio, and puncturing patterns that will be used for different radio configurations. Within a puncturing pattern, a "0" means that the symbol should be deleted, and "1" means that the symbol should be passed. For example, the puncturing pattern for RC 2 is "110101," meaning that the first, second, fourth, and sixth symbols are passed, whereas the third and fifth symbols of each consecutive group of six symbols are removed.

Frame Quality Indicator bits are used to detect errors in the received frames for some forward link channels. They may also assist in determining the data rate of the channel if the receiver performs blind rate detection such as the F-FCH supporting voice calls. The Frame Quality Indicator bits are appended to the input bits, and form a CRC.

The outer code is a Reed-Solomon code that uses 8-bit symbols and operates in a binary extension field $GF(2^8)$ generated by a primitive polynomial $p(x) = x^8 + x^4 + x^3 + x^2 + 1$. A primitive element α for this field is defined by $\alpha^8 + \alpha^4 + \alpha^3 + \alpha^2 + 1 = 0$. The elements of $GF(2^8)$ are $0, 1, \alpha, \alpha^2, \ldots, \alpha^{254}$. The elements can also be represented as polynomials in terms of the primitive element α. The polynomial representation of α^k is denoted by $a_7\alpha^7 + a_6\alpha^6 + \cdots + a_0 = \text{hex}(a_7, a_6, \ldots, a_0)$, which is the hexadecimal value of an 8-bit number $a_7, a_6, \ldots a_0$.

Symbol demultiplexing is performed on every code channel in the Forward CDMA Channel. The demultiplexing should generate the first symbol in each frame to the Y_{I1} output and the subsequent symbols to the $Y_{I2}, Y_{I3}, Y_{Q1}, Y_{Q2}, Y_{Q3}, Y_{I1}, \ldots$ outputs. The demultiplexer should output the first complex symbol in each frame of Y_{I1} and Y_{Q1} outputs, and the subsequent complex symbols to Y_{I2} and Y_{Q2}, Y_{I3} and Y_{Q3}, Y_{I1} and $Y_{Q1}, \ldots$ outputs. Figure 8.15 shows the demultiplexer structure for SR 3.

Following the orthogonal spreading, each code channel is spread in quadrature. The spreading sequence should be a quadrature sequence of length 2^{15} (i.e. 32 768 PN chips in length) for SR 1 and each carrier of SR 3. This sequence is called the pilot PN sequence. For SR 1 and each carrier of SR 3, the pilot PN sequence is based on the following characteristic polynomials:

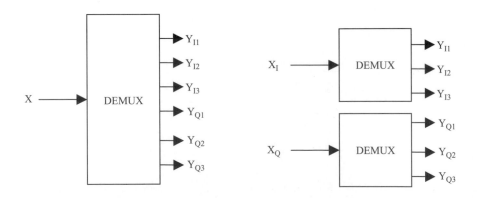

The DEMUX functions distribute input symbols sequentially
from the top to the bottom output paths.

Figure 8.15   Demultiplexer structure for Spreading Rate 3

$P_I(x) = x^{15} + x^{13} + x^9 + x^8 + x^7 + x^5 + 1$ for the in-phase (I) sequence and $P_Q(x) = x^{15} + x^{12} + x^{10} + x^6 + x^5 + x^4 + x^3 + 1$ for the quadrature-phase (Q) sequence

The reciprocal polynomials of $P_I(x)$ and $P_Q(x)$ of period 2^{15} are generated as follows:

$$i(x) = x^{15} P_I(x^{-1}) = x^{15} + x^{10} + x^8 + x^7 + x^6 + x^2 + 1$$

and

$$q(x) = x^{15} P_Q(x^{-1}) = x^{15} + x^{12} + x^{11} + x^{10} + x^9 + x^5 + x^4 + x^3 + 1$$

The maximum length of the LFSR sequences $i(n)$ and $q(n)$ based on the above reciprocal polynomials is $2^{15} - 1$ and can be generated by the following linear recursions:

$$i(n) = i(n-15) \oplus i(n-10) \oplus i(n-8) \oplus i(n-7) \oplus i(n-6) \oplus i(n-2)$$

and

$$q(n) = i(n-15) \oplus i(n-12) \oplus i(n-11) \oplus i(n-10) \oplus i(n-9) \oplus i(n-5) \oplus i(n-4) \oplus i(n-3)$$

were $i(n)$ and $q(n)$ are binary-valued ("0" and "1") and the additions are modulo-2. In order to obtain the I and Q pilot PN sequences of period 2^{15}, a "0" is inserted in $i(n)$ and $q(n)$ after 14 consecutive "0" outputs (this occurs only once in each period); therefore, the pilot PN sequences have one run of 15 consecutive "0" outputs instead of 14. The chip rate for SR 1 and each carrier of SR 3 should be 1.2288 Mcps. The pilot PN sequence period is 32 768/ 122 800 = 26.666... ms, and exactly 75 pilot PN sequence repetitions occur every 2 seconds. Each BS should use a time offset of the pilot PN sequence to identify a Forward CDMA Channel. Time offsets may be reused within a CDMA cellular system. Distinct pilot channels are identified by offset index 0 through 511 inclusive. This offset index specifies the offset time from the zero offset pilot PN sequence in multiples of 64 chips. The zero offset pilot PN sequence should be such that the start of the sequence should be output at the beginning of every second in time, referenced to the BS transmission time. For SR 1 and for each carrier of SR 3, the start of the zero offset pilot PN sequence for either the I or Q sequence should be defined as the state of the sequence for which the previous 15 outputs were "0"s. There are 512 unique values that are possible for the pilot PN sequence offset. The offset in chips for a given pilot PN sequence from the zero offset pilot PN sequence is equal to the index value multiplied by 64. As an example, if the pilot PN sequence offset index is 15, the pilot PN sequence offset will be $15 \times 64 = 960$ PN chips. The same pilot PN sequence offset should be used on all CDMA frequency assignment for a given BS.

Following the spreading operation, the I and Q impulses are applied to the inputs of the I and Q baseband filters.

8.3.1 Pilot Channels

The F-PICH, F-TDPICH, F-APICHs, and F-ATDPICHs are unmodulated spread spectrum signals used for synchronization by an MS operating within the coverage area of the BS.

Figure 8.16   Pilot channels for Spreading Rate 1

Each of these pilot channels should not be demultiplexed using a non-TD demultiplexer; that is, the TD demultiplexer is not allowed. But the F-PICH, F-APICH, and F-ATDPICH should be demultiplexed using the multiplexer shown in Figure 8.15.

8.3.1.1 Forward Pilot Channel (F-PICH)

The F-PICH is transmitted at all times by the BS on each active Forward CDMA Channel, unless the BS is classified as a hopping pilot beacon. If the F-PICH is transmitted by a hopping pilot beacon, then the timing requirements should apply as follows. The hopping pilot beacon is transmitted periodically. The transmission time of a hopping pilot beacon is defined by three parameters: NGHBR-TX-PERIOD (20 in units of 80 ms), NGHBR-TX-OFFSET (9 in units of 80 ms), and NGHBR-TX-DURATION (7 in units of 80 ms). NGHBR-TX-PERIOD is the period between pilot beacon transmissions. HGHBR-TX-OFFSET is the time offset of the pilot beacon transmission from the beginning of the transmission period. NGHBR-TX-DURATION is the duration of each pilot beacon transmission. The F-PICH should be spread by W_0^{64}, which is not used with a non-zero quasi-orthogonal function. Figure 8.16 shows pilot channels for SR 1.

8.3.1.2 Forward Transmit Diversity Pilot Channel (F-TDPICH)

When the F-TDPICH is transmitted, the BS should continue to use sufficient power on the F-PICH to ensure that a BS is able to acquire and estimate the Forward CDMA Channel without using energy from the F-TDPICH. If transmit diversity is supported on the Forward CDMA Channel, the F-TDPICH should be spread with W_{16}^{128}.

8.3.1.3 Forward Auxiliary Pilot Channel (F-APICH)

Zero or more Auxiliary Pilot Channels can be transmitted by the BS on an active Forward CDMA Channel. If transmit diversity is used on the Forward CDMA Channel associated with an F-APICH, the BS should transmit an Auxiliary Transmit Diversity Pilot.

Code multiplexed Auxiliary Pilots are generated by assigning a different Walsh function or different quasi-orthogonal function to each Auxiliary Pilot. The Walsh function length can be extended to increase the number of available Walsh functions or quasi-orthogonal functions. Every Walsh function W_i^m, where i is the index of the Walsh function and m is 256, may be used to generate N Walsh functions of order $N \times m$, where $N = 2^n$ (a non-negative integer power of 2). A Walsh function of order $N \times m$ can be constructed by concatenating N times W_i^m (i.e. $2^n \times W_i^{256}$), but concatenation of W_0^m is not allowed. Additionally, concatenation of W_{64}^{128} should not be allowed for SR 1 and concatenation of W_{64}^{256}, W_{128}^{256}, and W_{192}^{256} should not be allowed for SR 3. Walsh function time alignment should be such that the first Walsh chip begins at an even second time mark referenced by BS transmission time. The Walsh function spreading sequence will repeat with a period of $(N \times m)/1.2288 \,\mu s$.

The maximum length of the Walsh functions used for Walsh orthogonal or quasi-orthogonal spreading of an Auxiliary Pilot is 512. For SR 1, N equals 1, 2, or 4. For SR 3, N equals 1 or 2. For the case $N = 2$, the two possible Walsh functions of order $2 \times m$ are $W_i^m W_i^m = W_i^{2m}$ and $W_i^m \overline{W_i^m} = W_{i+1}^{2m}$, where the overbar denotes a polarity change and $i < m$. For the case $N = 4$, the four possible Walsh functions are $W_i^m W_i^m W_i^m W_i^m = W_i^{4m}$, $W_i^m \overline{W_i^m} W_i^m \overline{W_i^m} = W_{i+m}^{4m}$, $W_i^m W_i^m \overline{W_i^m} \overline{W_i^m} = W_{i+2m}^{4m}$, $W_i^m \overline{W_i^m} \overline{W_i^m} W_i^m = W_{i+3m}^{4m}$.

When the F-ATDPICH is transmitted, the BS should continue to use sufficient power on the F-APICH to ensure that a MS is able to acquire and estimate the Forward CDMA Channel without using energy from the F-ATDPICH.

8.3.1.4 Forward Auxiliary Transmit Diversity Pilot Channel (F-ATDPICH)

If transmit diversity is supported on the Forward CDMA Channel associated with an F-APICH, the F-ATDPICH should be spread with a Walsh function or a quasi-orthogonal function. The length of the Walsh function, the sign multiplier QOF mask, and the rotate enable Walsh function used to spread the F-ATDPICH shall be the same as the length of the Walsh function, the sign multiplier QOF mask, and the rotate enable Walsh function, respectively, that are used to spread the associated Auxiliary Pilot Channel.

8.3.2 Sync Channel (F-SYNCH)

The F-SYNCH is an encoded, interleaved, spread, and modulated spread spectrum signal that is used by MSs operating within the coverage of the BS to acquire initial time synchronization.

The bit rate for the Sync Channel is 1200 bps. A Sync Channel frame is 26.67 ms in duration. For a given BS, the I and Q channel pilot PN sequences for the Sync Channel use the same pilot PN sequence offset as for the F-PICH.

Once the MS achieves pilot PN sequence synchronization by acquiring the F-PICH, the synchronization for the F-SYNCH is immediately known. This is not only because the Sync Channel is spread with the same pilot PN sequence, but also the frame and interleaver timing on the F-SYNCH is aligned with the pilot PN sequence.

Figure 8.17 depicts the Sync Channel (F-SYNCH) structure for SR1. The F-SYNCH data is convolutionally encoded prior to transmission, code symbol repeating, modulation symbol interleaving, orthogonal spreading by Walsh functions, quadrature spreading using PN sequence, and baseband filtering.

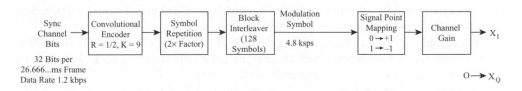

Figure 8.17   Sync channel for Spreading Rate 1

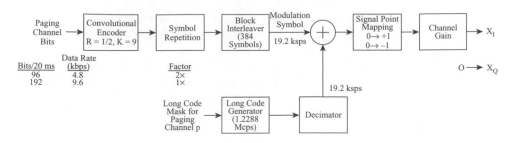

Figure 8.18   Paging channel for Spreading Rate 1

8.3.3 Paging Channel (F-PCH)

The base station uses the F-PCH to transmit system overhead information and MS-specific messages. The F-PCH applies to SR1 only. The F-PCH transmits information at a fixed data rate of 9600 or 4800 bps. An F-PCH will be divided into Paging Channel slots that are each 80 ms in duration. An F-PCH frame is 20 ms in duration. The first F-PCH frame should begin at the start of BS transmission time.

The F-PCH is an encoded, repeated, interleaved, spread, and modulated spread spectrum signal that is used by mobile stations operating within the coverage area of the BS. The F-PCH data should be convolutionally encoded, code-symbol repeated, modulation-symbol inter-leaved, orthogonally spread by W_i^{64} (where i denotes the F-PCH number), scrambled by long code mask, quadrature spread using the PN sequence, and filtered. Figure 8.18 illustrates the Paging Channel structure SR1).

When the Physical Layer receives a PHY-PCH.Request (SDU) from the MAC Layer, the BS should set the information bits to SDU and transmit a Paging Channel frame.

8.3.4 Quick Paging Channel (F-QPCH)

The F-QPCH is an uncoded, spread, and On-Off-Keying (OOK) modulated spread spectrum signal that is used by mobile stations within the coverage area of the base station. The BS uses the F-QPCH to inform MSs whether or not they should receive the F-CCCH or F-PCH starting in the next F-CCCH or F-PCH slot.

The F-QPCH transmits information at a fixed data rate of 4800 or 2400 bps. For a given base station, the I and Q channel pilot PN sequences for the F-QPCH should use the same pilot PN sequence offset as for the Pilot Channel. The F-QPCH slots should be aligned such that they begin 20 ms before the start of the zero-offset pilot PN sequence at every even second time mark.

The F-QPCH should be divided into F-QPCH slots that are each 80 ms in duration. F-QPCH slots should be divided into Paging Indicators, Configuration Change Indicators, and Broadcast Indicators. The indicator data rate is 9600 or 4800 bps.

The BS enables the Paging Indicators that are to receive the F-CCCH or F-PCH to start 20 ms following the end of the current F-QPCH slot. The BS enables two Paging Indicators in the F-QPCH slot for each mobile station that is to receive the next F-CCCH or F-PCH slot.

Configuration Change Indicators are only used on Quick Paging Channel 1. If the F-QPCH indicator data rate is 4800 bps (or 9600 bps), the least two indicators (or last four indicators) of

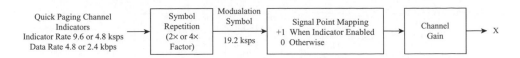

Figure 8.19   Quick paging channel for Spreading Rate 1

the first 40 ms of an F-QPCH slot and the last two indicators (or the last four indicators) of the F-QPCH slot are reserved as Configuration Change Indicators.

Broadcast Indicators are only used on Quick Paging Channel 1. If the F-QPCH indicator data rate is 4800 bps (or 9600 bps), the two indicators (or the four indicators) prior to the last two indicators (or the last four indicators) of the first 40 ms of an F-QPCH slot and the two indicators (or the four indicators) prior to the last two indicators (or the last four indicators) of the F-QPCH slot are reserved as Broadcast Indicators.

For SR 1, each Paging Indicator, Configuration Change Indicator, and Broadcast Indicator at the 9600 bps rate should be repeated one time (each indicator occurs two consecutive times) and each indicator at the 4800 bps rate should be repeated three times (each indicator occurs four consecutive times). For SR 3, each indicator at the 9600 bps rate should be repeated two times (each indicator occurs three consecutive times) and each indicator at the 4800 bps rate should be repeated five times (each indicator occurs six consecutive times). The F-QPCH structure for SR 1 is shown in Figure 8.19.

8.3.5 Broadcast Control Channel (F-BCCH)

The F-BCCH is an encoded convolutionally, interleaved, repeated, spread by a Walsh function, scrambled by long code mask, PN spread using the PN sequence, and filtered signal that is used by mobile stations operating within the coverage area of the base station.

The F-BCCH transmits information at a data rate of 19 200, 9600, or 4800 bps, which correspond to a slot duration of 40, 80, and 160 ms, respectively. The decision to enable or disable transmission should be made on an F-BCCH slot basis.

For a given BS, the I and Q channel pilot PN sequences for the F-BCCH should use the same pilot PN sequence offset as for the F-PICH. The first F-BCCH slot begins at the start of BS transmission time.

The F-BCCH is divided into F-BCCH slots that are 40, 80, or 160 ms in duration. For the 80 ms F-BCCH slot case, each F-BCCH slot consists of two 40 ms F-BCCH frames. For the 160 ms F-BCCH slot case, each F-BCCH slot consists of four 40 ms F-BCCH frames. The first F-BCCH frame of a Broadcast Control Channel slot consists of a sequence of encoded and interleaved symbols. The following F-BCCH frames of an F-BCCH slot should consist of the same sequence of encoded and interleaved symbols that were used on the first F-BCCH frame.

An F-BCCH frame consists of 768 bits that are composed of 744 information bits followed by 16 Frame Quality Indicator (CRC) bits and 8 Encoded Trial Bits. The F-BCCH structures for SR 1 with code rates $R = 1/4$ and $R = 1/2$ are illustrated in Figure 8.20(a) and (b).

8.3.6 Common Assignment Channel (F-CACH)

The F-CACH is especially designed to provide fast-response reverse link channel assignments to support transmission of random access packets on the reverse link. This channel controls the

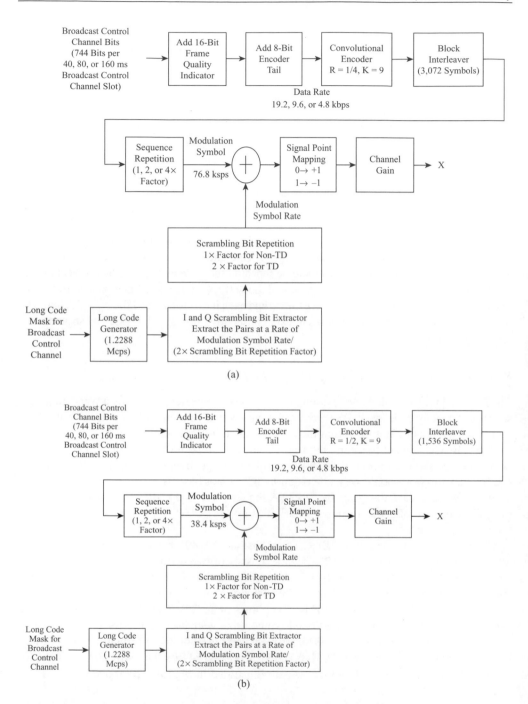

Figure 8.20 (a) F-BCCH structure for Spreading Rate 1 with R $= ^1/_4$ mode. (b) BCCH structure for Spreading Rate 1 with R $= ^1/_2$ mode (Reproduced under written permission from Telecommunications Industry Association)

R-CCCH and the associated common power control subchannel in the Reservation Access Mode. It also implements congestion control. The BS transmits information on the F-CACH at a fixed data rate of 9600 bps. The F-CACH frame length is 5 ms.

For a given base station, the I and Q channel pilot PN sequences for the F-CACH should use the same pilot PN sequence offset as for the F-PICH. The F-CACH block interleaver should always be aligned with the F-CACH frame. The base station supports discontinuous transmission in such a way that the decision to enable or disable the F-CACH should be made on a frame-by-frame (i.e. 5 ms basis).

F-CACH frames consist of 48 bits. These 48 bits are composed of 32 information bits followed by 8 Frame Quality Indicator (CRC) bits and 8 Encoder Tail Bits. The Frame Quality Indicator (CRC) can be calculated on all bits within the frame (48 bits), except the CRC itself and Encoder Tail Bits. The F-CACH uses an 8-bit CRC with the generator polynomial $g(x) = x^8 + x^7 + x^4 + x^3 + x + 1$. The Encoder Tail Bits are the last eight bits of each F-CACH frame. These eight bits should each be set to "0."

Figure 8.21 shows the F-CACH structure for SR 1 with $R = 1/4$ or $R = 1/2$ mode. The F-CACH is convolutionally encoded, interleaved, scrambled using long code mask, spread by a Walsh function, PN spread using PN sequence, and filtered.

When the Physical Layer receives a PHY-CACH.Request (SDU, CACH-ID, NUM-BITS) from the MAC Layer, the base station should store the arguments SDU, CACH-ID, and NUM-BITS, set the information bits to SDU, and transmit an F-CACH frame.

8.3.7 Forward Common Control Channel (F-CCCH)

The F-CCCH is transmitted to a mobile station at a variable data rate of 9600, 19 200, or 38 400 bps from frame to frame. An F-CCCH frame is 20, 10, or 5 ms in duration.

Figure 8.22 illustrates the F-CCCH structure for SR 1 with $R = 1/4$ or 1/2 mode. The F-CCCH is convolutionally encoded, block interleaved, scrambled with the long code mask, orthogonally spread by a Walsh function, PN spread with the PN sequence, and filtered. The MS uses this channel to transmit mobile station-specific messages.

The Frame Quality Indicator (CRC) can be calculated on all bits within the frame, except the CRC (12 or 16 bits) itself and Encoder Tail Bits (8 bits). The 20 ms F-CCCH uses a 12-bit CRC with the generator polynomial $g(x) = x^{12} + x^{11} + x^{10} + x^9 + x^8 + x^4 + x + 1$ for the 9600 bps frame and a 16-bit CRC with the generator polynomial $g(x) = x^{16} + x^{15} + x^{14} + x^{11} + x^6 + x^5 + x^2 + x + 1$ for the 38.4 and 19.2 Kbps frames. The 10 ms F-CCCH uses a 12-bit CRC with the generator polynomial $g(x) = x^{12} + x^{11} + x^{10} + x^9 + x^8 + x^4 + x + 1$ for the 19.2 Kbps frame and a 16-bit CRC with the generator polynomial $g(x) = x^{16} + x^{15} + x^{14} + x^{11} + x^6 + x^5 + x^2 + x + 1$ for the 38.4 Kbps frame. The 5 ms F-CCCH should use a 12-bit CRC with the generator polynomial $g(x) = x^{12} + x^{11} + x^{10} + x^9 + x^8 + x^4 + x + 1$.

The F-CCCH is divided into Forward Common Control Channel slots that are each 80 ms in duration. All frames consist of the information bits followed by a Frame Quality Indicator (CRC) and 8 Encoder Tail Bits that should each be set to "0."

When the Physical Layer receives a PHY-FCCCH.Request (SDU, FCCCH-ID, FRAME-DURATION, NUM-BITS) from the MAC Layer, the base station should store the arguments SDU, FCCCH-ID, FRAME-DURATION, and NUM-BITS, set the information bits to SDU, and transmit an F-CCCH frame of duration 5, 10, or 20 ms at a data rate that corresponds to 38.4, 19.2, or 38.4 Kbps (including 9.6 and 19.2 Kbps).

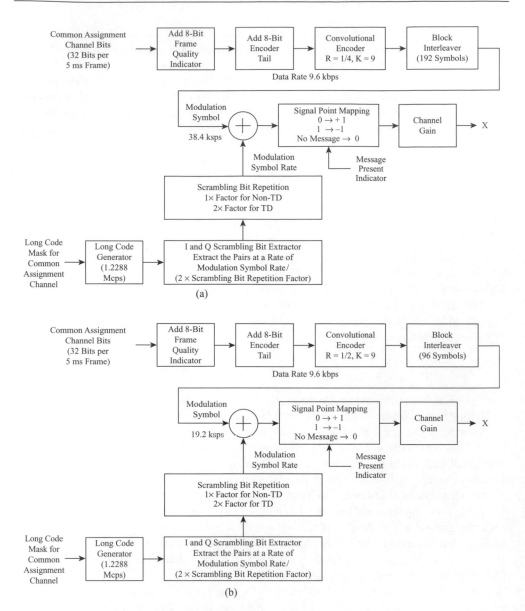

Figure 8.21 (a) F-CACH structure for Spreading Rate 1 with R = $\frac{1}{4}$ mode. (b) F-CACH structure for Spreading Rate 1 with R = $\frac{1}{2}$ mode (Reproduced under written permission from Telecommunications Industry Association)

8.3.8 Forward Indicator Control Channel (F-ICCH)

The F-ICCH for SR 1 consists of the F-CPCCH and the Forward Rate Control Channel (F-RCCH). The F-ICCH for SR 3 consists of only the F-CPCCH. The BS may support operation on one or more Forward Indicator Control Channels. The F-ICCH contains indicator

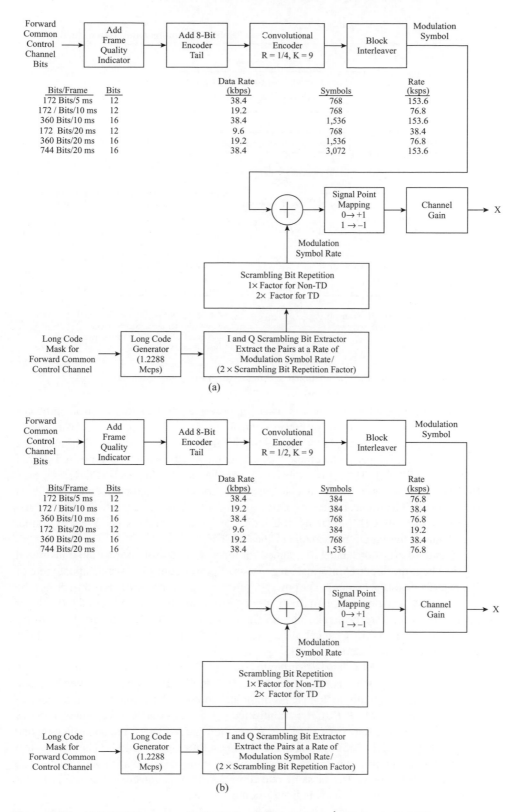

Figure 8.22 (a) F-CCCH structure for Spreading Rate 1 with R = $^1/_4$ Mode. (b) F-CCCH structure 3 for Spreading Rate 1 with R = $^1/_2$ mode

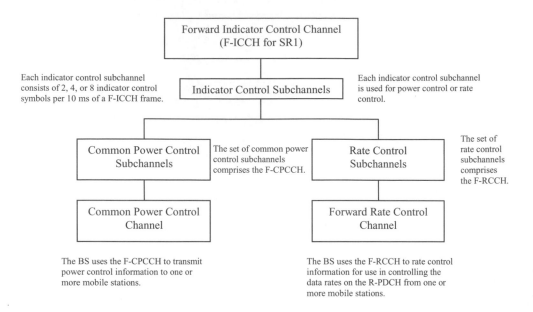

Figure 8.23   Functional schematic diagram of the F-ICCH

control subchannels. Each subchannel is used for power control or rate control. Indicator control subchannels that are used for power control are also called common power control subchannels, and indicator control subchannels that are used for rate control are also called rate control subchannels. The set of common power control subchannels within an F-ICCH comprises the F-CPCCH. The set of rate control subchannels within an F-ICCH comprises the F-RCCH. Each indicator control subchannel consists of 2, 4, or 8 indicator control symbols per 10 ms of an F-ICCH frame. Indicator control symbols are transmitted at 9.6 ksps on the I and the Q arms of the F-ICCH, producing 192 indicator control symbols per F-ICCH frame. Figure 8.23 shows the functional scheme of the F-ICCH.

One indicator control subchannel is assigned for the power control information transmitted to each MS. The power control information consists of power control bits at an update rate of 800, 400, or 200 updates per second. The power control information to a MS uses 8, 4, or 2 of the 192 indicator control symbols per 10 ms indicator control frame for update rates of 800, 400, or 200 updates per second, respectively. The BS may use the F-CPCCH to adjust the transmit power of the Reverse CDMA Channel, if they (all code channels within the Reverse CDMA Channel) are being transmitted by the MS.

The BS may use the F-RCCH to transmit rate control information for use in controlling the data rates on the R-PDCHs from one or more MSs. Each R-PDCH may be controlled by a rate control subchannel, or multiple R-PDCHs may be controlled by the same rate control subchannel. When a single R-PDCH is controlled by a rate control subchannel, the MS operates in the Dedicated Rate Control Mode. When multiple R-PDCHs are controlled by one rate control subchannel, the MSs operate in the Common Rate Control Mode. The rate control information on a rate control subchannel consists of rate control symbols that are updated at a rate of 100 updates per seconds. The same rate control symbol is transmitted on the 2, 4, or 8 indicator control symbols per 10 ms indicator control frame.

The F-ICCH forms indicator control subchannels by mapping input power control bits and rate control symbols into I-arm and Q-arm multiplexers and selecting the appropriate indicator control symbols of the multiplexer output sequences. The rate of the indicator control symbols at the outputs of the I-arm and Q-arm multiplexers should be 9.6 Ksps. The set of symbols into the multiplexers should be updated at rates of 800, 400, or 200 updates per second, resulting in update periods of 1.25, 2.5, or 5 ms, respectively. The update rate for the set of symbols into the multiplexers is called the F-ICCH update rate. The F-ICCH should have a 10 ms indicator control frame structure.

The channel structure for the F-ICCH is shown in Figure 8.24. The F-ICCH maps sequences of power control bits and rate control symbols to indicator control subchannels that are time multiplexed to form a sequence of indicator control symbols at a rate of 9.6 Ksps on the I and Q arms. There are $2N$ multiplexer input symbol positions numbered 0 through $2N - 1$. Multiplexer input symbol positions 0 through $N - 1$ are transmitted on the I arm, and multiplexer input symbol positions N through $2N - 1$ are transmitted on the Q arm. The multiplexer output sequences depend on a randomization parameter called *relative offset*. For SR 1, a decimation factor of 128 should be realized by outputting the first chip of each 128 chips output from the long code generator. For SR 3, a decimation factor of 384 ($=128 \times 3$) should be realized by outputting the first chip of each 384 chips output from the long code generator. The multiplexer output sequence on the I arm should be a sequence of N symbols from multiplexer input symbol positions ($i + N -$ relative offset) mod N, where $i = 0$ to

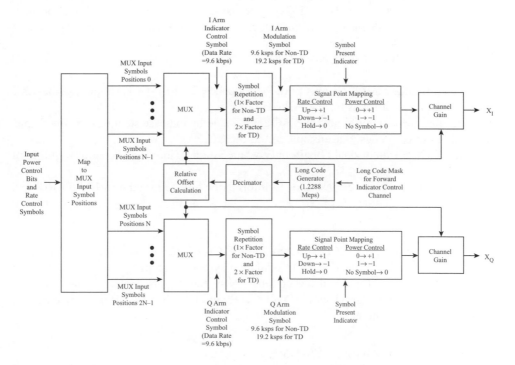

Figure 8.24   F-ICCH structure for Spreading Rate 1 (Reproduced under written permission from Telecommunications Industry Association)

$N - 1$, $i = 0$ corresponds to the first output symbol on the I arm and $i = N - 1$ corresponds to the last output symbol. The multiplexer output sequence on the Q arm should be a sequence of N symbols from multiplexer input symbol positions $N + ((i + N - \text{relative set}) \bmod N)$, where $i = 0$ to $N - 1$, $i = 0$ corresponds to the first output symbol on the Q arm and $i = N - 1$ corresponds to the last output symbol.

For SR 1, the indicator control symbols on the I and Q arms should not be repeated (i.e. each symbol is transmitted once) for the non-Transmit Diversity (TD) mode, and the I and Q arms should be repeated once (i.e. each symbol is transmitted twice) for the TD mode. When spread by appropriate Walsh function, the non-TD indicator control symbols may be time multiplexed with the TD indicator control symbols. For SR 3, the indicator control symbols on the I and Q arms should be repeated twice (i.e. each symbol is transmitted three times). The power control bits have one of two values where a "0" is mapped to a "$+1$" modulation symbol and a "1" is mapped to a "-1" modulation symbol. The rate control symbols will have one of three values where an "UP" is mapped to a "$+1$" modulation symbol, a "DOWN" is mapped to a "-1" modulation symbol, and a "HOLD" is mapped to a "0" modulation symbol.

The F-ICCH is spread by a Walsh function, PN spread using the PN sequence, randomized by the decimated output of the long code generator, and filtered.

8.3.9 Forward Grant Channel (F-GCH)

The base station uses the F-GCH to grant mobile stations operating with SR 1 permission to transmit on the R-PDCH. The F-GCH gives permission to the MS to transmit one or more encoder packets. The BS should transmit information on the F-GCH at a fixed data rate of 3200 bps and its duration is 10 ms.

The F-GCH has the following characteristic:

- For a given BS, the I and Q channel pilot PN sequence for the F-GCH should use the same pilot PN sequence offset as for the F-PICH.
- The decision to enable or disable the F-GCH is made by the base station frame-by-frame, supporting discontinuous transmission on the F-GCH.
- A 10-ms F-GCH frame should begin only when System Time is an integral multiple of 10 ms.
- The F-GCH frame structure consists of 14 information bits, followed by a 10-bit CRC with the generator polynomial $g(x) = x^{10} + x^9 + x^8 + x^7 + x^6 + x^4 + x^3 + x$ and an 8-bit (a "0"s) Encoded Tail Bit.

Figure 8.25 illustrates the F-GCH structure: The F-GCH data is convolutionally encoded, punctured (code symbols), interleaved (modulation symbols), scrambled (using the F-GCH long code mask), spread (by a Walsh function), PN spread (by PN sequence), and filtered.

When the Physical Layer receives a PHY-FGCH.Request (SDU, SYS-TIME) from the MAC Layer, the base station should

- Store the arguments SDU and SYS-TIME.
- Set the information bits to SDU.
- Transmit the F-GCH frame starting at SYS-TIME.

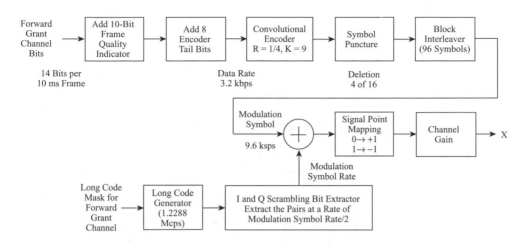

Figure 8.25   F-GCH structure

8.3.10 Forward Acknowledgment Channel (F-ACKCH)

The F-ACKCH provides feedback for the Reverse Packet Data Channel (R-PDCH) transmissions to mobile stations operating with SR 1 on the forward acknowledgment subchannels. The data rate on the feedback acknowledgment subchannel is 100 bps. The data rate on the F-ACKCH is 19.2 kbps and its frame is 10 ms in duration. A 10 ms F-ACKCH frame begins only when System Time is an integral multiple of 10 ms.

The forward acknowledgment subchannels are multiplexed in separate data streams on the *I* and *Q* arms of the F-ACKCH. For a given base station, the *I* and *Q* channel pilot PN sequences for the F-ACKCH should use the same pilot PN sequence offset as for the F-PICH.

The channel structure for the F-ACKCH is shown in Figure 8.26. The F-ACKCH should be spread by a Walsh function, sequence repetition (2 × Factor), PN spread using PN sequence, and filtered. For an F-ACKCH, there are 192 forward acknowledgment subchannels numbered from 0 through 191. Forward acknowledgment subchannels numbered 0 through 95 corresponds to the *I* arm and those numbered 96 through 191 corresponds to the *Q* arm.

When the forward acknowledgment subchannel is transmitted, an ACK-OR-NAK response should be transmitted as specified by the MAC Layer as a parameter in the PHY-FACKCH. Request primitive. On the other hand, if no PHY-FACKCH.Request primitive is received from the MAC Layer, the forward acknowledge subchannel should be gated off. A "+1" value should be transmitted for an ACK response and a "0" value should be transmitted for a NAK response. When a NAK is transmitted, the forward acknowledgment subchannel should be gated off.

As shown in Figure 8.26, every three forward acknowledgment subchannel is time multiplexed. Based on IQ multiplex Index (IQM-IDX), the 192 forward acknowledgment subchannels are multiplexed into *I* and *Q* arms, each consisting of 32 (=96/3) code groups identified by CDM-IDX. Each code group contains three forward acknowledgment subchannels that have the same IQM-IDX and CDM-IDX. The three forward acknowledgment subchannels in a group should be time multiplexed. The bit from the forward acknowledgment subchannel in the group with TDM-IDX equal to 0 is output first, 1 output second, and 2 output last.

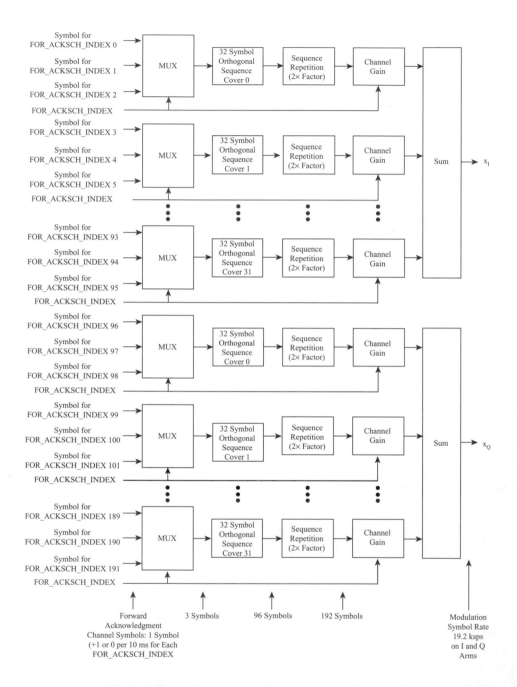

Figure 8.26 F-ACKCH structure (Reproduced under written permission from Telecommunications Industry Association)

The multiplex output symbols of the three forward acknowledgment subchannels in a group with the same sequence index should be spread by orthogonal covering sequence as shown below (first group only):

Sequence Index	Orthogonal Sequence of bits
0	00 00 00 00 00 00 00 00 00 00 00 00 00 00 00 00
1	00 10 01 00 01 01 11 11 01 10 01 11 00 00 11 01
2	01 01 00 10 00 10 11 11 10 11 00 11 10 00 01 10

In these sequences, a bit with a value of "0" means multiplying by "+1" and a bit with a value of "1" means multiplying by "−1." The sequence obtained after covering the F-ACKCH should be repeated once. The symbol after sequence repetition should be multiplied by the channel gain of the forward acknowledgment subchannel. The forward acknowledgment subchannels with IQM-IDX equal to 0 should be summed and assigned to the I arm, whereas the forward acknowledgment subchannels with IQM-IDX equal to 1 should be summed separately and assigned to the Q arm.

When the Physical Layer receives a PHY-FACKCH.Request (ACK-OR-NACK, SYS-TIME) from the MAC Layer, the base station should

• Store the arguments ACK-OR-NAK and SYS-TIME.
• If ACK-OR-NAK is equal to ACK, transmit an ACK on the F-ACKCH; otherwise, transmit a NAK on the F-ACKCH starting at SYS-TIME.

8.3.11 Forward Packet Data Control Channel (F-PDCCH)

The F-PDCCH is used by the BS for transmitting control information for the associated F-PDCH. A Forward CDMA Channel may contain up to two F-PDCCHs that are each identified by a channel identifier (PDCCH-ID). A channel identifier is set to "0" for the first F-PDCCH and to "1" for the second F-PDCCH. If the BS supports one F-PDCCH with RC 10, the BS should support one or two F-PDCCHs. If the BS supports two F-PDCCHs with RC 10, the BS should support two F-PDCCHs.

The BS transmits control information on the F-PDCCH at variable data rates of 29.6, 14.8, and 7.4 Kbps depending on the frame duration. The frame duration is NUM-SLOTS (1, 2, or 4) 1.25-ms slots. For a given base station, the I and Q pilot PN sequences for the F-PDCCH use the same pilot PN sequence offset as for the F-PICH.

The F-PDCCH frame to be transmitted consists of the scrambled Service Data Unit (SDU) [0...20] (13 bits), the 8-bit Frame Quality Indicator-covered SDU[0...7] (where SDU is a parameter passed by the MAC Layer in the PHY-FPDCCH.Request primitive), the 8-bit inner Frame Quality Indicator (CRC), and the 8-bit (all "0"s) Encoder Tail Bit. Note that the 13-bit scrambled SDU [8...20] is used to calculate the Outer Frame Quality Indicator, whereas the scrambled SDU[8...20] plus Frame Quality Indicator-covered SDU[0...7] (total 21 bits) is used to compute the Inner Frame Quality Indicator. SDU[8...20] is scrambled by bit-by-bit modulo-2 adding a 13-bit scrambler sequence that is equal to the 13 least significant bits of (SYS-TIME + NUM-SLOTS). The bit-by-bit modulo-2 additions will be such that SDU[8] is added to the most significant bit of the scrambler sequence and SDU[20] to the least significant bit of the scrambler sequence.

The 8-bit Frame Quality Indicator-covered SDU[0...7] (8 bits) should be generated by performing the modulo-2 addition of the SDU[0...7] passed by the MAC Layer with an Outer Frame Quality Indicator. The Outer Frame Quality Indicator is calculated on the scrambled SDU[8...20] with the generator polynomial $g(x) = x^8 + x^2 + x + 1$. The first input bit to the generator of the Output Frame Quality Indicator should be the scrambled SDU[8] and the last input bit will be the scrambled SDU[20]. The bit-by-bit modulo-2 addition of the SDU[0...7] with the Outer Frame Quality Indicator should be such that SDU[0] is added to the first output bit of the Outer Frame Quality Indicator and SDU[7] to the last output bit of the outer Frame Quality Indicator. The Frame Quality Indicator-covered SDU[0] will output followed by the Frame Quality Indicator-covered SDU[1], and so on.

The inner CRC is calculated on all bits within the frame, except the inner CRC itself and the Encoder Tail Bits. The F-PDCCH uses an 8-bit inner Frame Quality Indicator (CRC) with the generator polynomial $g(x) = x^8 + x^7 + x^4 + x^3 + x + 1$. The last eight bits of each F-PDCCH frame are called the Encoder Tail Bits.

The F-PDCCH is convolutionally encoded, code symbol repeated, punctured, block interleaved, spread by a Walsh function, PN spread and filtered.

The F-PDCCH structure is illustrated in Figure 8.27 where each functional block is as explained above.

When the Physical Layer receives a PHY-FPDCCH.Request (PDCCH-ID, SDU, NUM-SLOTS, SYS-TIME) from the MAC Layer, the base station should

- Store the arguments of these parameters.
- Scramble SDU[8...20].
- Use the scrambled SDU[8...20] and the SDU[0...7] to compute the 8-bit Frame Quality Indicator-covered SDU[0...7].
- Transmit the scrambled SDU[8...20], the computed Frame Quality Indicator-covered SDU[0...7], the computed 8-bit Inner Frame Quality Indicator, and 8 Encoder Tail Bits on the

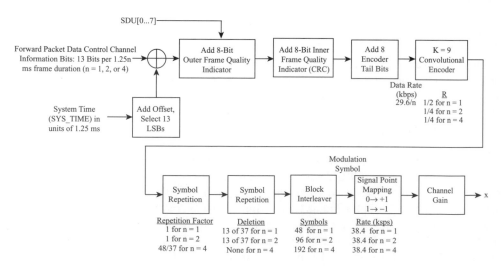

Figure 8.27 F-PDCCH structure (Reproduced under written permission from Telecommunications Industry Association)

F-PDCCH frame of duration NUM-SLOTS starting at SYS-TIME on the F-PDCCH with identified PDCCH-ID.

8.3.12 Forward Dedicated Control Channel (F-DCCH)

The F-DCCH is used for the transmission of user and signaling information to a specific mobile station during a call. Each Forward Traffic Channel may contain one F-DCCH. The frame length of an F-DCCH is 5 or 20 ms in duration.

For the 20 ms frame length, the base station transmits information on the F-DCCH at a fixed data rate of 9600 bps for RCs 3, 4, 6, and 7 and at a fixed data rate of 1400 bps for RCs 5, 8, and 9. For the frame length of 5 ms, the base station transmits information on the F-DCCH at a fixed data rate of 9600 bps for RCs 3, 4, 5, 6, 7, 8, and 9. On the other hand, for flexible data rates supported for 20 ms frames, the base station transmits information on the F-DCCH at a fixed data rate between 1050 and 14 400 bps in RCs 5, 8, and 9.

The base station supports F-DCCH frames that are time offset by multiples of 1.25 ms. A zero-offset 20 ms F-DCCH frame should begin only when System Time is an integral multiple of 20 ms. A zero offset 5 ms F-DCCH frame should begin only when System Time is an integral multiple of 5 ms. All F-DCCH frames that carry data consist of zero or one Reserved Bits and the information bits followed by a Frame Quality Indicator (CRC) and an 8-bit encoder tail. The Frame Quality Indicator should be calculated on all bits within the frame, except the CRC itself and the encoder tail bits. The 20 ms F-DCCH should use a 12-bit CRC with the generator polynomial $g(x) = x^{12} + x^{11} + x^{10} + x^9 + x^8 + x^4 + x + 1$. If flexible data rates are supported, a 16-bit CRC with the generator polynomial $g(x) = x^{16} + x^{15} + x^{14} + x^{11} + x^6 + x^5 + x^2 + x + 1$ may be used. The 5 ms F-DCCH should use a 16-bit CRC with the generator polynomial $g(x) = x^{16} + x^{15} + x^{14} + x^{11} + x^6 + x^5 + x^2 + x + 1$.

Figure 8.28 depicts the F-DCCH structure for RC 8. As seen from this figure, the F-DCCH should be able

- to add the Frame Quality Indicator and encoder tail bits to the information bits;
- to encode convolutionally;
- to repeat and then puncture the encoded symbols;
- to interleave the modulation symbols;
- to spread by a Walsh function or quasi-orthogonal function;
- to scramble the data by the public or private long code mask;
- to PN-spread and filter.

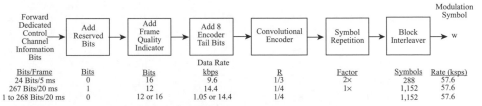

Bits/Frame	Bits	Bits	Data Rate kbps	R	Factor	Symbols	Rate (ksps)
24 Bits/5 ms	0	16	9.6	1/3	2×	288	57.6
267 Bits/20 ms	1	12	14.4	1/4	1×	1,152	57.6
1 to 268 Bits/20 ms	0	12 or 16	1.05 or 14.4	1/4		1,152	57.6

Note: If flexible data rates are supported, there can be 1 to 268 information bits in a 20 ms frame and the encoded symbols will be additionally repeated then punctured to provide a 57.6 ksps modulation symbol rate.

Figure 8.28   F-DCCH structure for RC 8

For a given base station, the I and Q channel pilot PN sequence for F-DCCH should use the same pilot PN sequence offset as for the F-PICH.

When the physical Layer receives a PHY-DCCH.Request (SDU, FRAME-DURATION, NUM-BITS) from the MAC Layer, the base station should perform the following transmission processing:

- Store the arguments SDU, FRAME-DURATION, and NUM-BITS.
- If SDU is not equal to NULL, set the information bits to SDU.
- If SDU is not equal to NULL, transmit NUM-BITS bits of SDU in an F-DCCH frame of duration FRAME DURATION (5 or 20 ms). If a PHY-DCCH.Request primitive for a 5 ms frame is received coincident with a PHY-DCCH.Request primitive for a 20 ms frame or during transmission of a 20 ms frame, then the base station may pre-empt transmission of the 20 ms frame and transmit a 5 ms frame. Transmission of the 20 ms frame may start or resume after completion of the 5 ms frame.

8.3.13 Forward Fundamental Channel (F-FCH)

The F-FCH is used for the transmission of user and signaling information to a specific mobile station during a call. Each Forward Traffic Channel may contain one F-FCH.

Table 8.5 summarizes the F-FCH bit allocations for non-flexible data rates.

Table 8.5 The F-FCH frame structure showing bit allocations for non-flexible data rates

Radio configuration	Data rate (bps)	Total	Number of bits per frame			
			Reserved/Flag	Information	Frame Quality Indicator	Encoder Tail
1	9600	192	0	172	12	8
	4800	96	0	80	8	8
	2400	48	0	40	0	8
	1200	24	0	16	0	8
2	14 400	288	1	267	12	8
	7200	144	1	125	10	8
	3600	72	1	55	8	8
	1800	36	1	21	6	8
3, 4, 6, and 7	9600 (5 ms)	48	0	24	16	8
	9600 (20 ms)	192	0	172	12	8
	4800	96	0	80	8	8
	2700	54	0	40	6	8
	1500	30	0	16	6	8
5, 8, and 9	9600	48	0	24	16	8
	14 400	288	1	267	12	8
	7200	144	1	125	10	8
	3600	72	1	55	8	8
	1800	36	1	21	6	8

When operating in RC 1, the BS should transmit information on the F-FCH at variable data rates of 9600, 4800, 2400, and 1200 bps. Number of bits per frame is 192, 96, 48, or 24, respectively. The 2400 and 1200 bps frames with RC 1 consist of information bits (40 and 16 bits each) followed by 8 Encoder Tail Bits with no CRC bits. The 9600 and 4800 bps frames with RC 1 consist of the information bits (172 and 80 bits each) followed by a Frame Quality Indicator (12 and 8 bits each) and 8 Encoder Tail Bits.

When operating in RC 2, the BS should transmit information on the F-FCH at variable data rates of 14 400, 7200, 3600, and 1800 bps. Number of bits per frame is 288, 144, 72 or 36, respectively. All frames with RC 2 consist of one Reserve/Flag bits followed by information bits (267, 125, 55, and 21 bits each), a Frame Quality Indicator (12, 10, 8, and 6 bits each), and 8 Encoder Tail Bits.

When operating in RCs 3, 4, 6, or 7, the BS should transmit information on the F-FCH at variable rates of 9600, 4800, 2700, and 1500 bps during 20 ms frames or at 9600 bps during 5 ms frames. All frames with RCs 3, 4, 6, and 7 consist of one Reserved/Flag bits followed by the information bits (24, 172, 80, 40, and 16 bits each), a Frame Quality Indicator (16, 12, 8, 6 and 6 bits each), and 8 Encoder Tail Bits.

When operating in RCs 5, 8, or 9, the BS should transmit information on the F-FCH at variable data rates of 14 400, 7200, 3600, and 1800 bps during 20 ms frames or at 9600 bps during 5 ms frames. All frames with RCs 5, 8, and 9 consist of zero or one Reserved/Flag bits followed by the information bits (24, 267, 125, 55, and 21 bits each), a Frame Quality Indicator (16, 12, 10, 8, and 6 bits), and 8 Encoder Tail Bits.

On the other hand, the base station may support flexible data rates. If flexible data rates are supported for 20 ms frames, the BS should support variable data rates corresponding to 1 through 171 information bits per 20 ms frame on the F-FCH. When operating in RCs 5, 8, or 9, the BS should support variable data rates of 850 through 14 400 bps corresponding to 1 through 258 information bits per 20 ms frame on the F-FCH. The minimum number of flexible data rates used in variable rate operation is not specified.

The data rate and frame duration on an F-FCH within a RC should be selected on a frame-by-frame basis. Although the data rate may vary on a frame-by-frame basis, the modulation symbol rate is kept constant by code repetition for data rates lower than the maximum. A base station operating with RCs 3 through 9 may discontinue transmission of the F-FCH for up to three 5 ms frames in a 20 ms frame.

The base station may support F-FCH frames that are time offset by multiples of 1.25 ms. A zero-offset 20 ms F-FCH frame should begin only when System Time is an integral multiple of 20 ms. A zero-offset 5 ms F-FCH frame should begin only when System Time is an integral multiple of 5 ms. The interleaver block for the F-FCH should be aligned with the F-FCH frame.

Figure 8.29 shows the F-FCH and F-SCH structure for RC 9.

The Reverse/Flag bit is used with RCs 2, 5, 8, and 9. The Reverse/Flag bit may be used on the F-FCH when one or more Forward Supplemental Code Channels (F-SCCHs) are in use; otherwise this bit is reserved and should be set to "0." The BS should set this bit to "1" if the BS will not transmit to the BS on the F-SCCHs in the second frame after the current frame.

Each frame with RCs 2 through 9, and the 9600 and 4800 bps frames of RC 1 should include a Frame Quality Indicator (CRC). No Frame Quality Indicator is used for the 2400 and 1200 bps data rates of RC 1. The 20 ms frames in RCs 5, 8, and 9 with more than 144 total bits

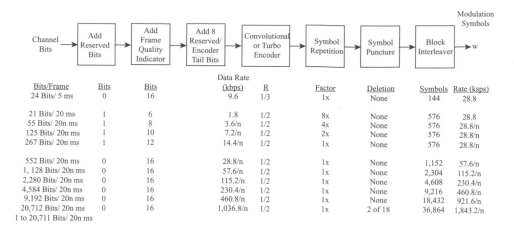

Bits/Frame	Bits	Bits	Data Rate (kbps)	R	Factor	Deletion	Symbols	Rate (ksps)
24 Bits/ 5 ms	0	16	9.6	1/3	1x	None	144	28.8
21 Bits/ 20 ms	1	6	1.8	1/2	8x	None	576	28.8
55 Bits/ 20n ms	1	8	3.6/n	1/2	4x	None	576	28.8/n
125 Bits/ 20n ms	1	10	7.2/n	1/2	2x	None	576	28.8/n
267 Bits/ 20n ms	1	12	14.4/n	1/2	1x	None	576	28.8/n
552 Bits/ 20n ms	0	16	28.8/n	1/2	1x	None	1,152	57.6/n
1,128 Bits/ 20n ms	0	16	57.6/n	1/2	1x	None	2,304	115.2/n
2,280 Bits/ 20n ms	0	16	115.2/n	1/2	1x	None	4,608	230.4/n
4,584 Bits/ 20n ms	0	16	230.4/n	1/2	1x	None	9,216	460.8/n
9,192 Bits/ 20n ms	0	16	460.8/n	1/2	1x	None	18,432	921.6/n
20,712 Bits/ 20n ms	0	16	1,036.8/n	1/2	1x	2 of 18	36,864	1,843.2/n
1 to 20,711 Bits/ 20n ms								

Figure 8.29   F-FCH and F-SCH structure for RC 9

should use a 12-bit CRC with the generator polynomial $g(x) = x^{12} + x^{11} + x^{10} + x^9 + x^8 + x^4 + x + 1$. A 16-bit CRC with the generator polynomial $g(x) = x^{16} + x^{15} + x^{14} + x^{11} + x^6 + x^5 + x^2 + x + 1$ may be used if flexible data rates are supported. The 20 ms frames in RCs 5, 8, and 9 with 73–144 total bits should use a 10-bit CRC with the generator polynomial $g(x) = x^{10} + x^9 + x^8 + x^7 + x^6 + x^4 + x^3 + 1$. The 20 ms frames in RCs 5, 8, and 9 with 37–72 total bits should use an 8-bit CRC. A 10-bit, 12-bit, and 16-bit CRCs may be used if flexible data rates are supported. The 20 ms frames in RCs 5, 8, and 9 with 36 or fewer total bits should use a 6-bit Frame Quality Indicator.

The F-FCH is used for the transmission of channel bits by the following procedure:

- Channel bits should be added Reserved Bits, Frame Quality Indicator, and 8 Reserved/ Encoder Tail Bits.
- Convolutional coding $(K = 9)$ is used for the F-FCH and turbo coding may be used for the F-SCH with 576 or more encoder input per frame. With convolutional coding, the Reserved/Encoder Tail Bits provide an encoder tail. With turbo coding, the first two of these bits are reserved bits that are encoded and the last six bits are replaced by an internally generated tail.
- The F-FCH code symbol should do repetition.
- Code symbols resulting from the symbol repetition should be punctured.
- The modulation symbols should be interleaved.
- The F-FCH should be spread by a Walsh function (RCs 1 through 9) or quasi-orthogonal function (RCs 3 through 9).
- The F-FCH data should be scrambled by using bits M_{41} through M_0 of the public long code mask; otherwise they should rely on the generation of the private long code mask.
- The F-FCH should be PN spread for quadrature spreading.
- The F-FCH should be filtered.

8.3.14 Forward Supplemental Channel (F-SCH)

The F-SCH is used for the transmission of high-level data to a specific MS during a call. Each Forward Traffic Channel contains up to two F-SCHs. This channel applies to RCs 3 through 9 only.

F-SCH frames are 20, 40, or 80 ms in duration. A BS may support discontinuous transmission of F-SCH frames. The F-SCH also supports outer coding in RC 5 at the data rate of 115 200 bps for 20 ms frames. The F-SCH outer coding buffer should be 1.28 s in duration. For a given BS, the *I* and *Q* channel pilot PN sequences for the F-SCH use the same pilot PN sequence offset as for the F-PICH.

The BS may support F-SCH frames that are time offset by multiples of 1.25 ms as specified by FRAME-OFFSET. The BS may also support 40 or 80 ms F-SCH frames that are time offset by multiples of 20 ms as specified by FOR-SCH-FRAME-OFFSET[i], where $i = 1$ and 2 for F-SCHs 1 and 2, respectively. A zero-offset 20 ms F-SCH frame should begin only when System Time is an integral multiple of 20 ms, whereas a zero-offset 40 ms F-SCH frame should begin only when System Time is an integral multiple of 40 ms. A zero-offset 80 ms F-SCH frame should begin only when System Time is an integral multiple of 80 ms.

Figure 8.29 illustrates the F-SCH frame structure for RC 9, which consists of zero or one Reserved bits and the information bits followed by a Frame Quality Indicator (CRC) and 8 Encoder Tail Bits. The last eight bits of each F-SCH frame are called the Reserved/Encoder Tail Bits. The data for F-SCHs should be convolutionally or turbo encoded. For the convolutional encoder, these 8 bits should be set to "0." For turbo encoder, the first two of the eight bits should be set to "0", and the turbo encoder will calculate and append the remaining six tail bits.

When outer coding is used on the F-SCH with RC 5, turbo coding should be used. The outer code is a Reed-Solomon block code that uses 8-bit symbols and operates in a Galois Field GF (2^8) generated by a primitive polynomial $p(x) = x^8 + x^4 + x^3 + x^2 + 1$. A primitive element α for this field is defined by $\alpha^8 + \alpha^4 + \alpha^3 + \alpha^2 + 1 = 0$. The elements of GF (2^8) are 0, 1, α, $\alpha^2, \ldots, \alpha^{254}$ (i.e. α^i, $0 \le i \le 254$). When outer coding is used, the information bits are stored in the outer coding buffer and then outer encoded. The outer coded symbols of the outer coding buffer are transmitted in 64 consecutive 20-ms F-SCH frames with RC 5 and data rate 115.2 Kbps format. The outer coding buffer is an array with 64 rows and 2280 columns (i.e. 145 920 cells). The outer coding buffers are divided into four outer coding sub-buffers, each with 16 rows and 2280 columns (i.e. 36 480 cells).

The F-SCH code symbols should be repeated, punctured, and interleaved. Each code channel transmitted on the Forward CDMA Channel should be spread with a Walsh function or a quasi-orthogonal function at a fixed chip rate of 1.2288 Mcps to provide channelization among all code channels on a given Forward CDMA Channel. The F-SCHs should be spread by a Walsh function or quasi-orthogonal function. The data for F-SCHs should be scrambled by using the long code mask. Following the orthogonal spreading, the F-SCH is PN spread in quadrature and filtered.

When transmitting on the F-SCH with a single assigned data rate in RC3 (as an example), the BS will transmit at a fixed rate of 153 600, 76 800, 38 400, 19 200, 9600, 4800, 2700, 2400, 1500, 1350, or 1200 bps. Similarly, when transmitting on the F-SCH with a single assigned rate in RCs 4, 5, 6, 7, 8, or 9, the BS will transmit at a respective fixed rate. If flexible data rates are supported, the BS should transmit information at a fixed rate corresponding to respective fixed total bits per frame in 1-bit increments.

As one example, Table 8.6 specifies the F-SCH bit allocations for non-flexible data rates.

Table 8.6   F-SCH frame structure summary for 20 ms frames for non-flexible data rates

Radio configuration	Data rate (bps)	Number of bits per frame				
		Total	Reserved	Information	Frame Quality Indicator	Reserved/Encoder Tail
3, 4, 6, and 7	614 400	12 288	0	12 264	16	8
	307 200	6144	0	6120	16	8
	153 600	3072	0	3048	16	8
	76 800	1536	0	1512	16	8
	38 400	768	0	744	16	8
	19 200	384	0	360	16	8
	9600	192	0	172	12	8
	4800	96	0	80	8	8
	2700	54	0	40	6	8
	1500	30	0	16	6	8
5, 8, and 9	1 036 800	20 736	0	20 712	16	8
	460 800	9216	0	9192	16	8
	230 400	4608	0	4584	16	8
	115 200	2304	0	2280	16	8
	57 600	1152	0	1128	16	8
	28 800	576	0	552	16	8
	14 400	288	1	267	12	8
	7200	144	1	125	10	8
	3600	72	1	55	8	8
	1800	36	1	21	6	8

Note: The 614 400 bps rate applies to Radio Configuration 7. The 307 200 bps rate applies to Radio Configurations 4, 6, and 7. The 1 036 800 bps rate applies to Radio Configuration 9. The 460 800 bps rate applies to Radio Configurations 8 and 9. The 115 200 bps rate in Radio Configuration 5 for 20 ms frames supports outer coding at code rates of 14/16, 13/16, 12/16, and 11/16.

8.3.15 Forward Supplemental Code Channel (F-SCCH)

The F-SCCH is used for the transmission of higher-level data to a specific mobile station during a call. The F-SCCH applies to RCs 1 and 2 only. Each Forward Traffic Channel contains up to seven F-SCCHs.

When transmitting on F-SCCHs with RC 1, the BS should transmit information at 9600 bps. When transmitting on F-SCCHs with RC 2, the BS should transmit information at 14 400 bps. All F-SCCH frames are 20 ms in duration. The BS may support F-SCCH frames that are time offset by multiples of 1.25 ms. The amount of the time offset is specified by FRAME-OFFSET. A zero-offset F-SCCH frame will begin only when System Time is an integral multiple of 20 ms. An offset F-SCCH frame should begin 1.25 × FRAME-OFFSET ms later than the zero-offset F-SCCH frame. The BS transmits frames on the F-SCCHs in time alignment with the F-FCH. The interleave block for the F-SCCH should be aligned with the F-SCCH frame. The BS transmits frames on the F-SCCHs in time alignment with the F-FCH frame. The interleave block for the F-SCCH should be aligned with the F-SCCH frame.

Total bits per frame are 192 bits for the 9600 bps data rate with RC 1 and 288 bits for the 14 400 bps data rate with RC 2. All frames should consist of zero or one Reserved bits and information bits (172 bits for the 9600 bps rate and 267 bits for the 14 400 bps rate) followed by a Frame Quality Indicator (a 12-bit CRC with RC 1 or 2) and the 8-bit Encoder Tail Bits. Each frame with RCs 1 and 2 should include a 12-bit Frame Quality Indicator with the generator polynomial $g(x) = x^{12} + x^{11} + x^{10} + x^9 + x^8 + x^4 + x + 1$. The last eight bits of each F-SCH frame are called the Reserved/Encoder Tail Bits that shall be set to "0."

The data for F-SCCHs should be convolutionally encoded. The encoder should be initialized to the all-zero state at the end of each 20 ms frame. For the F-SCH, the code output symbols from the convolutional encoder (i.e. the forward error correction encoder) should be repeated and code symbols resulting from the symbol repetition should be punctured. All repeated code symbols and subsequent puncturing on the F-SCCH should be interleaved prior to modulation and transmission. The data for F-SCCHs should be scrambled for spreading using the long code. This spreading operation involves modulo-2 addition of the 64-ary orthogonal modulation output stream and the long code. The long code is periodic with period $2^{42}-1$ chips and should satisfy the linear recursion specified by the following characteristic polynomial:

$$p(x) = x^{42} + x^{35} + x^{33} + x^{31} + x^{29} + x^{26} + x^{25} + x^{22} + x^{21} + x^{19} \\ + x^{18} + x^{17} + x^{16} + x^{10} + x^7 + x^6 + x^5 + x^3 + x^2 + x + 1$$

Each PN chip of the long code is generated by the modulo-2 inner product of a 42-bit mask and the 42-bit sequence generator. Punctured codes used with convolutional codes include the base code rate, puncturing ratio, and puncturing patterns that should be used for different RCs. Within a puncturing patterns, a "0" means that the symbol should be deleted and "1" denotes that a symbol should be passed. For example, consider a punctured code used with the convolutional code consisting of the base code rate = 1/2, puncturing ratio = 2 of 6, and puncturing pattern = 110101 for RC 2. This pattern "110101" means that the first, second, fourth, and sixth symbols are passed, whereas the third and fifth symbols of each consecutive group of six symbols are removed. All the modulation symbols after symbol repetition and subsequent puncturing should be block interleaved. When operating on the F-SCCHs with RCs 1 and 2, the data scrambling should be accomplished by performing the modulo-2 addition of the modulation symbol with the binary value of the long code PN chip that is valid at the start of the transmission period for that symbol. This PN sequence should be the equivalent of the long code operating at 1.2288 MHz clock rate. Only the first output of every 64 chips is used for the data scrambling. The F-SCCHs with RCs 1 and 2 should be spread by a Walsh function of length 64. Following the orthogonal spreading, each code channel is spread in quadrature. The spreading sequence should be a quadrature sequence of length 2^{15} (i.e. 32 768 PN chips in length) for SR 1. This sequence is called the pilot PN sequence. Following the spreading operation, the I and Q impulses are applied to the inputs of the I and Q baseband filters for the F-SCCH filtering.

Transmission processing for the F-SCCH is described below. When the Physical Layer receives a PHY-SCCH.Request (SDU, FRAME-DURATION, NUM-BITS) from the MAC Layer, the base station should perform the following:

- Store the arguments SDU, FRAME-DURATION, and NUM-BITS.
- If SDU is not equal to NULL, set the information bits to SDU.
- If SDU is not equal to NULL, transmit NUM-BITS bits of SDU in an F-SCCH frame.

8.3.16 Forward Packet Data Channel (F-PDCH)

The F-PDCH is used for the transmission of high-level data to the MS with SR 1. A Forward CDMA Channel may contain up to two F-PDCHs. Each F-PDCH transmits information to one specific MS at a time.

The F-PDCH transmits 194, 386, 770, 1538, 2306, 3074, or 3842 information bits. 16 CRC bits and 6 turbo encoder tail allowance bits should be added to the information bits to form encoder packets. The encoder packets are encoded with a rate-1/5 turbo encoder, interleaved, and scrambled. Then, symbols from the scrambled sequence should be selected for transmission as a *subpacket*. The selected symbols may or may not include all of the scrambled output symbols with some symbols repeated one or more times. The selected subpacket symbols should be modulated into QPSK, 8-PSK, or 16-QAM symbols and demultiplexed into one to 28 32-chip Walsh channels used for that F-PDCH. Each of these Walsh channels is spread with a different 32-chip Walsh function. Thus, the spread symbols on the Walsh channels should be summed to obtain a single sequence of I/Q-symbols. Then, the F-PDCH should be PN spread and filtered. Figure 8.30 illustrates the F-PDCH structure for RC 10, explaining the information packet flow through the channel as described above.

So far we have described the overall information structure of all code channels within the Reverse CDMA Channel (14 different channels) and Forward CDMA Channel (16 different channels) to be used for service interfaces required by the CDMA2000 1x EV-DV system.

8.4 CDMA2000 Entities and Service Interfaces

This section provides the brief description of all component entities and interfaces exchanged between entities within all layers in the CDMA2000 family of standards. cdma One IS-95A/B has a layered structured providing voice and packet data (up to 60 Kbps) services. CDMA2000 provides protocols and services that correspond to the bottom two layers of the ISO/OSI Reference Model specified by the ITU for IMT-2000 systems. The Physical Layer and the Link Layer in IMT-2000 Systems correspond to OSI Layers 1 and 2. Layer 2 is further subdivided into the LAC sublayer and the MAC sublayer. Applications and upper layer protocols, corresponding to OSI Layer 3 through 7, use the services provided by CDMA2000: Link Access Control (LAC) services. Those services may include signaling services, voice services, packet data applications, and circuit data applications.

The CDMA2000 1x EV-DV structural model from the mobile station perspective is introduced below.

8.4.1 CDMA2000 1x EV-DV Service Interface Structure (Mobile Station)

Figure 8.31 depicts the CDMA2000 1x EV-DV structure, which shows interconnection between entities and services based on the mobile station.

MAC (Medium Access Control) is the entity that controls the access to and from Upper Layer Signaling, Data Services, and Voice Services to Physical Layer resources. LAC is the entity that

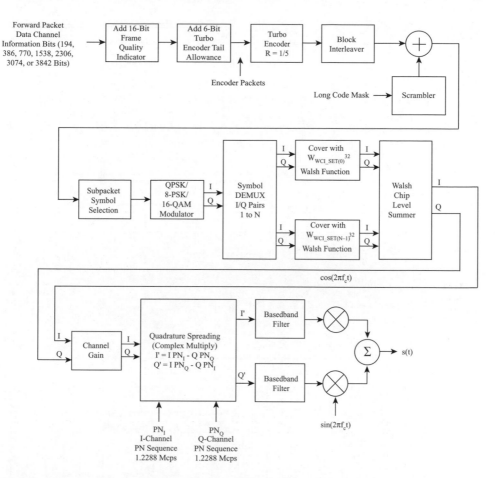

Figure 8.30 F-PDCH structure for RC 10 (Reproduced under written permission from Telecommunications Industry Association)

provides assured mode or unassured mode delivery of information across the air interface between the MS and the BS.

The multiplex sublayer in MAC has both a transmitting and receiving function. The multiplex sublayer transmitting function combines the information from Upper Layer Signaling, Data Services, and Voice Services. It forms Physical Layer Service Data Units (PHY-SDUs) and Packet Data Channel Control Function Service Data Units (PDCHCF-SDUs) for transmission. The multiplex sublayer receiving function separates the information contained in Physical Layer (PHY) and PDCHCF SDUs, and directs the information to the correct entity of Upper Layer Signaling, Data Services, or Voice Service.

8.4.1.1 Multiplex Sublayer Operations

The transmission function, under QoS control, solicits information bits from signaling and connected services. Information bits are exchanged between signaling and the multiplex

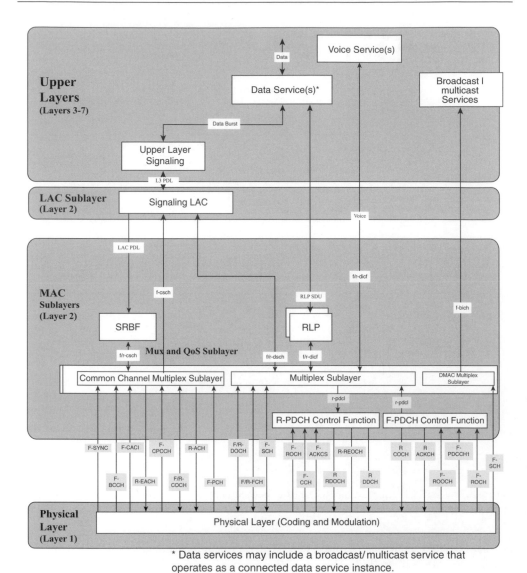

Figure 8.31 CDMA2000 architecture (mobile station) (Reproduced under written permission from Telecommunications Industry Association)

sublayer using the primitives: the multiplex sublayer sends a MAC-Availability. Indication primitive to the Signaling LAC entity to request information bits from signaling. The multiplex sublayer processes information that is received in a MAC-DATA.Request primitive from the Signaling LAC entity. The multiplex sublayer sends a MAC-Data.Indication primitive to the Signaling LAC entity to indicate that data for signaling LAC entity has been received. The multiplex sublayer converts the information bits received from the Signaling LAC entity into a data block. Information bits are exchanged between a connected service or a logical channel and the multiplex sublayer in a unit called a data block. The multiplex sublayer multiplexes one or more data blocks into MuxPDU and combines one or more MuxPDUs into a Physical Layer SDU for transmission by the Physical Layer. Physical Layer SDUs are exchanged between the multiplex sublayer and the Physical Layer using the service interface operations. The multiplex sublayer operates in time synchronization with the Physical Layer. The multiplex sublayer delivers a Physical Layer SDU to the Physical Layer using a physical-channel specific service interface set of primitives. The Physical Layer delivers a Physical Layer SDU to the multiplex sublayer using a physical channel specific *Receive Indication* service interface operation.

8.4.1.2 Packet Data Channel (PDCH) Control Function

When the PDCH is assigned, the Multiplex Sublayer combines one or more MuxPDUs into a PDCH Control Function SDU for transmission by the F-PDCH Control Function (F-PDCHCF) or the R-PDCH Control Function (R-PDCHCF). PDCHCF SDUs are exchanged between the multiplex sublayer and the F-PDCHCF using the service interface procedures as shown in the following. The F-PDCHCF delivers a PDCHCF SDU to the multiplex sublayer using an FPDCH-Data.Indication primitive (sdu, frame-duration, num-bits, frame-quality) with the following arguments:

- *sdu:* The F-PDCH Control Channel sets to the PDCHCF sdu if the F-PDCHCF received a forward packet data channel encoder packet.
- *frame-duration:* The F-PDCHCF sets to NULL.
- *num-bit:* The F-PDCHCF sets to the number of bits in the received SDU.
- *frame-quality:* The F-PDCHCF sets to sufficient.

The base station F-PDCHCF sends an FPDCH-Availability.Indication primitive to the base station multiplex sublayer to indicate that the base station multiplex sublayer is to deliver an SDU to the base station F-PDCHCF to be delivered to the mobile station over the f-bdch. The base station multiplex sublayer sends an FPDCH-Data.Request primitive to the base station F-PDCHCF to indicate that the base station F-PDCHCF is to deliver an SDU to the mobile station over the f-pdch.

Consider the multiplex sublayer transmitting function. If an FCH physical channel has been assigned, the multiplex sublayer should assemble a Physical Layer FCH SDU every 20 ms (Mode A). The multiplex sublayer should combine available data blocks from signaling and all connected services, according to the related priority of each, to form a MuxPDU. The multiplex sublayer may determine the relative priority between traffic supplied by Signaling and other Services (data, voice, and broadcast/multicast) using the information provided by the parameters in the MAC-SDUReady.Request primitive.

If an FCH physical channel has been assigned, the Multiplex Sublayer should assemble a Physical Layer FCH SDU at the beginning of every 20 ms interval. The multiplex sublayer should not assemble a 20 ms FCH SDU at any time other than the beginning of 20 ms interval (Mode B). The multiplex sublayer may assemble a 5 ms FCH SDU at the beginning of any 5 ms interval. Note that Mode A is the multiplex sublayer operational mode in which the RC is less than or equal to 2, whereas Mode B is the case when the RC is greater than 2.

If a DCCH physical channel has been assigned, the Multiplex Sublayer should assemble a Physical Layer DCCH SDU at the beginning of every 20 ms interval. The Multiplex Sublayer should not assemble a 20 ms DCCH SDU at any time other than the beginning of a 20 ms interval. The Multiplex Sublayer may assemble a 5 ms DCCH SDU at the beginning of any 5 ms interval (Mode B).

For each SCCH physical channel that has been assigned (Mode A), the Multiplex Sublayer should assemble a Physical Layer SCCH SDU every 20 ms. The Multiplex Sublayer should form a MuxPDU using a single data block from a connected service, requesting the data block from each connected service in order of the relative priority of each connected service. If all connected services supply Blank data blocks, the Multiplex Sublayer should create a Null MuxPDU. If a connected service supplies a non-Blank data block and if the SCCH is used with Rate Set 1 or 2, the Multiplex Sublayer should use the supplied data block to form a MuxPDU Type 1 or 2, respectively. For each assigned SCH physical channel, the Physical Layer delivers a Physical Layer SCH SDU every 20 ms, every 40 ms, or every 80 ms according to the SCH frame length configured. Each physical layer SCH SDU contains one or more MuxPDUs.

Since the Multiplex Sublayer function is very important, as described above, its operating functions will be reiterated here once again: The Multiplex Sublayer has both a transmitting and a receiving function. The Multiplex Sublayer transmitting function combines information from various sources (i.e. Upper Layer Signaling, Data service, and Voice Service) and forms Physical Layer SDUs and PDCHCF SDUs for transmission. The Multiplex Sublayer receiving function separates the information contained in Physical Layer and PDCHCF SDUs, and directs the information to the correct entity (i.e. Upper Layer Signaling, Data Service, and Voice Service). The Physical Layer delivers a Physical Layer SDU to the Multiplex Sublayer using a physical channel specific *Receive Indication* service interface operation. The MS's multiplex sublayer should categorize each MuxPDU in each received Physical Layer SDU and should supply the category when it delivers a data block from the SDU to the logical channel (logical connection between peer entities).

8.4.1.3 Common Channel Multiplex Sublayer and SRBP

This section will provide an informative overview of the SRBP (Signaling Radio Burst Protocol) entity and of channel procedures associated with the Common Channel Multiplex Sublayer. SRBP is an entity that provides connectionless protocol for signaling messages.

The Forward Sync Channel (F-SYNC) is used to provide time and frame synchronization to the MS. The Forward Common Control Channel (F-CCCH) is used to send control information to MSs that have not been assigned to a Traffic Channel. The Forward Paging Channel (F-PCH) is used to send control information to MSs that have not been assigned to a Traffic Channel. If the Common Channel Multiplex Sublayer receives a PHY-(PCH or SYNC).Indication (sdu) from the Physical Layer, the Common Channel Multiplex Sublayer entity should send a MAC-Data.Indication primitive with channel-id (a unique channel identifier for the physical channel), channel-type (F-(PCH or SYNC) frame), data (sdu), size (size of data in bits), and system-time (System Time associated with the Physical Layer frame which carries the F-(PCH or SYNC) SDU). The Broadcast Control channel is used to send control information to mobile stations that have not been assigned a Traffic Channel.

The entire process of sending one Layer 2 (LAC and MAC Sublayers) encapsulated PDU and receiving an acknowledgment for the PDU is called an access attempt. One access attempt consists of one or more access sub-attempts. Each transmission in the access sub-attempt is called an access probe. Each access probe consists of an R-ACH preamble and an R-ACH message capsule. Within an access sub-attempt, access probes are grouped into access probe sequences. The R-ACH used for each access probe sequence is chosen pseudorandomly from among all the R-ACHs associated with the current F-PCH. If there is only one R-ACH associated with the current F-PCH, all access probes within an access probe sequence are transmitted on the same R-ACH. If there are more than one R-ACH associated with the current F-PCH, access probes within an access probe sequence may be transmitted on different R-ACHs associated with the current F-PCH. There are two different access modes that can be used when transmitting messages using the Enhanced Access Procedures: Basic Access mode and Reservation Access mode. The SRBP entity uses an algorithm to select an access mode prior to each Layer 2 encapsulated PDU transmission, based on the length of the Layer 2 encapsulated PDU and configuration parameters from the base station.

8.4.1.4 F-PDCH Control Function Operation

If the MS supports the Forward Packet Data Channel (F-PDCH), then the mobile station must support a F-PDCHCF entity. The F-PDCHCF entity terminates all of the physical channels associated with the operation of the F-PDCH (i.e. the F-PDCCHs, the R-ACKCH, the R-CQICH, and the F-PDCH). The F-PDCHCF provides an automatic retransmission (ARQ) protocol that ensures the delivery of encoder packets from the BS to the MS by retransmitting portions of the turbo-coded packets based on feedback from the MS to indicate successful (ACK) or unsuccessful (NAK) reception and decoding of the encoder packet. The F-PDCHCF uses two additional techniques to enhance the performance of F-PDCH transmission, providing four independent ARQ channels (to permit the BS to have up to four outstanding encoder packets) and code division multiplexing (to permit the BS to optionally transmit encoder packets to two different mobile stations).

The physical channels that are terminated in the F-PDCHCF are used in the following manner:

- The F-PDCH physical layer channel carries subpackets of turbo-encoded encoder packets from the base station. The F-PDCH is time division multiplexed between two mobile stations concurrently.
- The F-PDCCH0 and F-PDCCH1 physical layer channels carry control information to direct the mobile stations that are assigned to the F-PDCH. Whenever the MS is assigned to the F-PDCH, the MS must monitor the two F-PDCCH channels.
- The R-ACKCH carries acknowledgements of successful transmissions (ACKs) and negative acknowledgements for unsuccessful transmissions (NAKs) from the MS to the BS following the reception and attempted decoding of the subpackets transmitted on the F-PDCH.
- The R-CQICH carries feedback information from the MS about the received signal quality (Channel Quality Indicator) for the F-PDCH. This information can be used by the BS to control transmission power to the MS, determine data rate, trigger forward packet rate channel handoff, and determine scheduling for the F-PDCH.

Thus, if the MS supports the F-PDCH, the MS should support an F-PDCHCF entity. The F-PDCHCF provides the F-PDCH operating procedures and performs all functions to follow the timing of F-PDCH operations as controlled by transmission from the BS on the F-PDCH, the F-PDCCHs, and the R-ACKCH.

8.4.1.5 R-PDCH Control Functions Operation

If the MS supports the R-PDCH, then the MS must support a Reverse Packet Data Channel Control Function (R-PDCHCF) entity. The R-PDCHCF entity terminates all the physical channels associated with the operation of the R-PDCH (i.e. the R-PDCCH, the R-REQCH, the F-ACKCH, the F-GCH, and the F-RCCH). The R-PDCHCF provides a synchronous automatic retransmission (ARQ) protocol that ensures the delivery of encoder packets from the MS to the BS by retransmitting portions of the turbo-coded encoder packets based on the feedback from the BS to indicate ACK or NAK reception and decoding of the encoder packet. The R-PDCHCF uses four independent ARQ channels to permit the MS up to four outstanding (i.e. unacknowledged by the BS) encoder packets at any given time.

The physical channels that are terminated in the R-PDCHCF are used as follows:

- The R-PDCH physical channel carries subpackets of the turbo-encoded encoder packet from the MS.
- The R-PDCCH physical channel carries transmission format of the MS. It is transmitted whenever the MS is transmitting a subpacket. The R-PDCCH is time aligned with the R-PDCH transmission. The R-PDCCH conveys the encoder packet size, subpacket ID, boost indicator, and mobile status indicator bit.
- The R-REQCH physical channel carries the information about power headroom at the MS, amount of data in the MS buffer, and the associated service reference identifier. One type of event that causes the R-REQCH transmission is power headroom, that is, when the difference between the MS maximum transmit power and its current transmit power set points reaches a threshold established by Upper Layer Signaling, or when the MS has data in its buffer and has not transmitted any requests for a period of time established by Upper Layer Signaling. The other two types of events that cause R-REQCH transmissions are buffer status (MS implementation-dependent reports of buffer size) and watermark crossing (when the amount of data in the MS transmit buffer crosses certain threshold established by Upper Layer Signaling).
- The F-ACKCH physical channel carries acknowledgments of ACKs and NAKs from the BS to the MS following the reception and attempted decoding of the subpackets transmitted on the R-PDCH.
- The F-GCH physical channels carry information to the MS on the encoder packet size it is to use.
- The F-RCCH physical channel carries an indication to the MS about whether to increase its authorized traffic to pilot ratio on the R-PDCH (i.e. UP), and decrease its authorized traffic to pilot ratio (i.e. DOWN), or keep its authorized traffic to pilot ratio unchanged (i.e. HOLD).

If the BS supports R-PDCH, the BS should support a R-PDCHCF entity. The R-PDCHCF (Reverse Packet Data Channel Control Function) provides the R-PDCH operating procedures and performs all functions to follow the timing of R-PDCH operations as controlled by transmissions on the R-PDCH, the R-PDCCH, the R-REQCH, the F-ACKCH, the F-GCH, and the F-RCCH.

8.4.1.6 Broadcast/Multicast Medium Access Control (BMAC)

The Physical Layer delivers a Physical Layer broadcast F-SCH Outer Code SDU to the BMAC multiplex Sublayer using a PHY-OuterCode. Indication primitive. If a broadcast F-SCH has been assigned and outer block coding is not used, the BMAC Multiplex Sublayer may transmit a broadcast F-SCH SDU at the System Time boundary (20, 40, or 80 ms) corresponding to the configured broadcast F-SCH frame length. The forward broadcast traffic channel (f-btch) is a point-to-point *logical channel* that carries Broadcast/Multicast data over the `f-btch`.

If outer block coding is not used on the supplemental channel, the BMAC multiplex sublayer interfaces with the F-SCH (Forward Supplemental Channel) using the PHY-SCH. Request and PHY-SCH. Indication primitives; otherwise, the BMAC multiplexer sublayer interfaces with the Physical Layer outer block coding entity using the PHY-SCHOuterCode. Request and PHY-SCHOuterCode.Indication primitives. To deliver a Physical Layer broadcast F-SCH SDU to the Physical Layer, the BMAC multiplex sublayer should send a PHY-SCH.Request to the Physical Layer with the primitive arguments: sdu, frame-duration, and num-bits. The Physical Layer delivers a Physical Layer broadcast F-SCH SDU to the BMAC multiplex sublayer using PHY-SCH.Indication primitive arguments of sdu, frame duration, num-bits, and frame-quality.

To deliver a Physical Layer broadcast F-SCH Outer Code SDU to the Physical Layer, the BMAC multiplex sublayer should send a PHY-SCHOuterCode. Request primitive to the Physical Layer with the primitive arguments of sdu, num-bits, and sys-time. The Physical Layer delivers a Physical Layer broadcast F-SCH Outer Code SDU to the multiplex sublayer using a PHY-SCHOuterCode.Indication primitive with its arguments: sdu$[i-1]$ carried in the ith Physical Layer frame, num-bits and num-frames, frame-quality$[i-1]$ (frame quality of the ith received Physical Layer frame), and sys-time.

CDMA2000 1x EV-DV structure (Base Station) will have exactly the same structure as that of (Mobile Station) except that all the unilateral arrow heads should be pointed in the opposite direction.

9

Advanced Encryption Standard and Elliptic Curve Cryptosystems

In the late 1960s, IBM initiated a Lucifer research project for computer cryptography. The Lucifer project ended in 1971 and IBM embarked on another effort to develop a commercial encryption scheme called Data Encryption Standard (DES), which was a refined version of Lucifer. In 1973, the National Bureau of Standards (NBS), now the National Institute of Standards and Technology (NIST), issued a public request for proposals for a US national cipher standard. IBM submitted the research results of the DES project as a possible candidate. The NBS requested the National Security Agency (NSA) to evaluate the feasibility relating to the algorithm's security and to determine its suitability as a federal standard. In November 1976, the DES was adopted as a Federal Standard and authorized for use on all unclassified US government communications. The official description of DES was published in FIPS PUB 46 on 15 January 1977. The DES algorithm was the best proposed standard and was adopted in 1977 even though there were many criticisms of its key length and the design criteria on the internal structure of the S-box. Nevertheless, DES became a basic security algorithm for organizations worldwide. In fact, DES survived years of intense cryptanalysis remarkably well and was a worldwide standard for over 20 years.

In 1993, The NIST solicited a review to assess the continued adequacy of DES to protect computer data, because there was no alternative to DES. After reviewing many comments and technical inputs, NIST recommended that the useful lifetime of DES would end in the late 1990s. In 2001, the Advanced Encryption Standard (AES), developed by Daemen and Rijmen in 1999 and known as the Rijndael algorithm, became a FIPS-approved advanced symmetric cipher algorithm. AES is a strong advanced algorithm in lieu of DES. AES is expected to provide the data security algorithm for network communications and access control systems. AES deals with the algorithm specifications such as the key expansion routine, encryption by cipher, and decryption by inverse cipher.

The concept of the Elliptic Curve Cryptosystem (ECC) was introduced by Koblitz and Miller. The elliptic curve discrete algorithm problem appears to be substantially more difficult and somewhat harder than the existing discrete logarithm problem. Implementations can explore this difference to provide both increased speed and decreased key size for a given level of security. Assuming an equivalent level of security, ECC uses smaller parameters than the conventional discrete logarithm systems. Elliptic Curves (ECs) have been well studied by the mathematicians for many years, and in the latter half of the twentieth century this research yielded some very significant results.

All practical public-key systems, such as Diffie-Hellman, RSA, ElGamal, DSA, and many other public-key methods, have exploited the properties of cryptographic techniques of encryption/decryption, authentication, and signature over the large finite fields. For those systems, the security depends directly on the relative difficulty of performing group operations such as exponentiation versus discrete logarithm. Computation of exponentiation is much easier than its inverse operation, that is, discrete logarithm. In the commonly used groups, discrete log is hard to compute when the modulus is very large, making large exponentiation expensive.

All commercial public-key cryptosystems rely on the difficulty of a discrete log problem. As discrete log gets easier, long bit-lengths are required to keep the algorithms safe. Discrete logs in ordinary prime number fields Z_p are much easier to solve than in EC fields. Thus, the discrete log problem for ordinary fields has been getting steadily easier due to successive refinement in the Number Field Sieve (NFS) techniques. In contrast, EC discrete log techniques have not seen significant improvement in the past 20 years. This difference accounts for today's reduced key-size requirement for ECs. Cracking RSA has never been proven to be as hard as prime factoring, while factoring has never been proven to be as hard as discrete log. The only way for future breakthroughs is to resort to the best mathematicians for help.

In this chapter, we first review the concept of an elliptic curve and then discuss the comparative study of encryption/decryption and digital signature problems against existing public-key algorithms over the finite fields. ECs over the finite prime field Z_p or the finite Galois field $GF(2^m)$ are particularly interesting because they have the potential to provide faster public-key cryptosystems with smaller key sizes.

9.1 Advanced Encryption Standard (AES)

The AES Rijndael algorithm, is a symmetric block cipher that can process data block of 128 bits long using cryptographic keys of 128, 192, and 256 bits, respectively.

9.1.1 Notational Conventions

- The cipher key for the Rijndael algorithm is a sequence of 128, 192, or 256 bits such that the index attached to a bit falls in between the range $0 \leq i \leq 128$, $0 \leq i \leq 192$, or $0 \leq i \leq 256$, respectively.
- All byte values of the AES Rijndael algorithm are presented by a vector notation (b_7, b_6, b_5, b_4, b_3, b_2, b_1, b_0), which corresponds to a polynomial representation as:

$$b_7x^7 + b_6x^6 + b_5x^5 + b_4x^4 + b_3x^3 + b_2x^2 + b_1x + b_0 = \sum_{i=0}^{7} b_ix^i$$

For example, $(01001011) \rightarrow x^6 + x^3 + x + 1$.

- In the case of an additional bit b_8 to the left of an 8-bit byte, it will appear immediately to the left of the left bracket such as $1(00101110) = 1(2e)$.
- Arrays of bytes, $a_0, a_1, a_2, \ldots, a_{15}$, are defined from the 128-bit input sequence, $ip_0, ip_1, ip_2, \ldots, ip_{126}, ip_{127}$, as follows:

$$
\begin{aligned}
a_0 &= (ip_0, ip_1, \ldots, ip_7) \\
a_1 &= (ip_8, ip_9, \ldots, ip_{15}) \\
&\vdots \\
a_{15} &= (ip_{120}, ip_{121}, \ldots, ip_{127})
\end{aligned}
$$

where ip_k, denotes $input_k$ for $k = 0, 1, 2, \ldots, 127$.

In general, the pattern extended to longer sequence such as 192-bit and 256-bit keys is expressed as:

$$a_n = (ip_{8n}, ip_{8n+1}, \ldots, ip_{8n+7}), \quad 16 \leq n \leq 31$$

- The AES algorithm's operations are internally performed on a two-dimensional array of bytes called the *State*. The state consists of four rows of bytes. The state array is $S_{r,\,c}$ with its row number r, $0 \leq r < 4$, and its column number c, $0 \leq c < Nb$, where Nb bytes are the block length divided by 32.

The input-byte array $(in_0, in_1, \ldots, in_{15})$ at the cipher is copied into the state array according to the following scheme:

$$S_{r,c} = in(r + 4c) \quad \text{for} \quad 0 \leq r < 4 \text{ and } 0 \leq c < Nb.$$

and at the inverse cipher, the state is copied into the output array as follows:

$$out(r + 4c) = S_{r,c} \quad \text{for} \quad 0 \leq r < 4 \text{ and } 0 \leq c < Nb.$$

An individual byte of the state is referred to as either $S_{r,c}$ or $S(r,c)$.

The cipher and inverse cipher operations are conducted on the state array as illustrated in Figure 9.1.

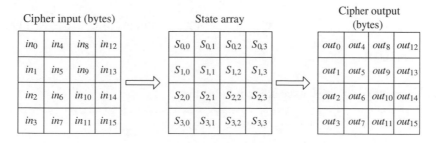

Figure 9.1 State array input and output

For example, if $r = 0$ and $c = 3$, then $in(0 + 12) = in(12) = S_{0,3}$:

$$if \ r = 3 \ and \ c = 2, \ then \ in(3+8) = in(11) = S_{3,2}.$$

The four bytes in each column of the state form a 32-bit word, where the row number r provides an index for the four bytes within each word, and the column number c provides an index representing the column in this array.

9.1.2 Mathematical Operations

Finite field elements (all bytes in the AES algorithm) can be added and multiplied. The basic mathematical operations are introduced below:

- *Addition:* The addition of two elements in a finite field is achieved by XORing the coefficients for the corresponding powers in the polynomials for two elements. For example,

$$
\begin{aligned}
&(x^5 + x^3 + x^2 + 1) + (x^7 + x^5 + x + 1) = x^7 + x^3 + x^2 + x \quad \text{(polynomial)} \\
&(00101101) \oplus (10100011) = (10001110) \quad \text{(binary)} \\
&(2d) \oplus (a3) = (8e) \quad \text{(hexadecimal)}
\end{aligned}
$$

- *Multiplication:* The polynomial multiplication in $GF(2^8)$ corresponds to the multiplication of polynomials modulo $m(x)$. This modulo $m(x)$ is an irreducible (or primitive) polynomial of degree 8. For the AES algorithm:

$$m(x) = x^8 + x^4 + x^3 + x + 1$$

Example 9.1. Prove $(73) \bullet (a5) = (e3)$:

$$
\begin{aligned}
&(01110011) \bullet (10100101) \\
&= (x^6 + x^5 + x^4 + x + 1) \bullet (x^7 + x^5 + x^2 + 1) \\
&= x^{13} + x^{12} + x^{10} + x^9 + x^6 + x^4 + x^3 + x^2 + x + 1
\end{aligned}
$$

The modular reduction by $m(x)$ results in

$$
\begin{aligned}
&(x^{13} + x^{12} + x^{10} + x^9 + x^6 + x^4 + x^3 + x^2 + x + 1) \mathrm{mod}(x^8 + x^4 + x^3 + x + 1) \\
&= x^7 + x^6 + x^5 + x + 1 \\
&= (11100011) = (e3)
\end{aligned}
$$

□

Since the multiplication is associative, it holds that

$$a(x)(b(x) + c(x)) = a(x)b(x) + a(x)c(x)$$

The element $(01) = (00000001)$ is called the multiplicative identity.

For any polynomial $b(x)$ of degree less than 8, the multiplicative inverse of $b(x)$, denoted by $b^{-1}(x)$, can be found by using the extended Euclidean algorithm such that

$$b(x)a(x) + m(x)c(x) = 1$$

from which $b(x)a(x) \bmod m(x) \equiv 1$. Thus, the multiplicative inverse of $b(x)$ becomes

$$b^{-1}(x) = a(x) \bmod m(x)$$

The set of 256 possible byte values has the structure of the finite field $GF(2^8)$ by means of XOR used as addition and multiplication.

Multiplication by x: Let the binary polynomial be $b(x) = \sum_{i=0}^{7} b_i x^i$. Multiplying $b(x)$ by x results in $xb(x) = \sum_{i=0}^{7} b_i x^{i+1}$, but it can be reduced by modulo $m(x)$. If $b_7 = 1$, the reduction is achieved by shifting one place to the left first and then XORing (1b). It follows that multiplication by x (i.e., $(00000010)_{(2)} = (02)_{(16)}$) can be implemented at the byte level with a left shift and bitwise XOR with (1b) if necessary. This operation on bytes is denoted by *xtime*(). Multiplication by higher powers of x can be implemented by repeated application of *xtime*().

Example 9.2. Compute $(57) \bullet (13) = (fe)$:

$$
\begin{aligned}
(57) &= (01010111) \\
(57) \bullet (02) &= \text{xtime}(57) = (10101110) = (ae) \\
(57) \bullet (04) &= \text{xtime}(ae) = (01011100) \oplus (00011011) \\
&= (01000111) = (47) \\
(57) \bullet (08) &= \text{xtime}(47) = (10001110) = (8e) \\
(57) \bullet (10) &= \text{xtime}(8e) = (00011100) \oplus (00011011) = (07)
\end{aligned}
$$

Thus, it follows that

$$
\begin{aligned}
(57) \bullet (13) &= (57) \bullet \{(01) \oplus (02) \oplus (10)\} \\
&= (57) \oplus (57) \bullet (02) \oplus (57) \bullet (10) \\
&= (57) \oplus (ae) \oplus (07) \\
&= (01010111) \oplus (10101110) \oplus (00000111) \\
&= (11111110) = (fe) \qquad \qquad \square
\end{aligned}
$$

Polynomials with finite field elements in $GF(2^8)$: A polynomial $a(x)$ with byte-coefficient in $GF(2^8)$ can be expressed in word form as:

$$a(x) = a_3 x^3 + a_2 x^2 + a_1 x + a_0 \Leftrightarrow a = (a_0, a_1, a_2, a_3)$$

To illustrate the addition and multiplication operations, let

$$b(x) = b_3 x^3 + b_2 x^2 + b_1 x + b_0 \Leftrightarrow b = (b_0, b_1, b_2, b_3)$$

be a second polynomial. Addition is performed by adding the finite field coefficients of x such that

$$a(x) + b(x) = (a_3 \oplus b_3)x^3 + (a_2 \oplus b_2)x^2 + (a_1 \oplus b_1)x + (a_0 \oplus b_0)$$

This addition corresponds to an XOR operation between the corresponding bytes in each of the words.

Multiplication is achieved as shown below:

The polynomial product $c(x) = a(x) \bullet b(x)$ is expanded and like powers are collected to give

$$
\begin{aligned}
c(x) &= a(x) \bullet b(x) = c_6 x^6 + c_5 x^5 + c_4 x^4 + c_3 x^3 + c_2 x^2 + c_1 x + c_0 \\
&= (c_6, c_5, c_4, c_3, c_2, c_1, c_0)
\end{aligned}
$$

where

$$
\begin{array}{ll}
c_0 = a_0 b_0 & c_4 = a_3 b_1 \oplus a_2 b_2 \oplus a_1 b_3 \\
c_1 = a_1 b_0 \oplus a_0 b_1 & c_5 = a_3 b_2 \oplus a_2 b_3 \\
c_2 = a_2 b_0 \oplus a_1 b_1 \oplus a_0 b_2 & c_6 = a_3 b_3 \\
c_3 = a_3 b_0 \oplus a_2 b_1 \oplus a_1 b_2 \oplus a_0 b_3 &
\end{array}
$$

The next step is to reduce $c(x) \bmod (x^4 + 1)$ for the AES algorithm, so that $x^i \bmod (x^4 + 1) = x^{i \bmod 4}$.

The modular product, $a(x) \otimes b(x)$, of two four-term polynomials $a(x)$ and $b(x)$ is given by

$$d(x) = a(x) \otimes b(x) = d_3 x^3 + d_2 x^2 + d_1 x^1 + d_0$$

where

$$
\begin{aligned}
d_0 &= a_0 b_0 \oplus a_3 b_1 \oplus a_2 b_2 \oplus a_1 b_3 \\
d_1 &= a_1 b_0 \oplus a_0 b_1 \oplus a_3 b_2 \oplus a_2 b_3 \\
d_2 &= a_2 b_0 \oplus a_1 b_1 \oplus a_0 b_2 \oplus a_3 b_3 \\
d_3 &= a_3 b_0 \oplus a_2 b_1 \oplus a_1 b_2 \oplus a_0 b_3
\end{aligned}
$$

Thus, $d(x)$ in the matrix form is written as:

$$
\begin{bmatrix} d_0 \\ d_1 \\ d_2 \\ d_3 \end{bmatrix}
\begin{bmatrix} a_0 & a_3 & a_2 & a_1 \\ a_1 & a_0 & a_3 & a_2 \\ a_2 & a_1 & a_0 & a_3 \\ a_3 & a_2 & a_1 & a_0 \end{bmatrix}
=
\begin{bmatrix} b_0 \\ b_1 \\ b_2 \\ b_3 \end{bmatrix}
$$

The AES algorithm also defines the inverse polynomials as:

$$
\begin{aligned}
a(x) &= (03)x^3 + (01)x^2 + (01)x + (02) \\
a^{-1}(x) &= (0b)x^3 + (0d)x^2 + (09)x + (0e)
\end{aligned}
$$

9.1.3 AES Algorithm Specification

The AES algorithm defines that:

- Nb denotes the number of 32-bit words with respect to the 128-bit block of the input, output, or state:

$$128 = Nb \times 32 \text{ from which } Nb = 4$$

- Nk represents the number of 32-bit words with respect to the cipher-key length of 128, 192, or 256 bits:

$$128 = Nk \times 32, Nk = 4$$
$$192 = Nk \times 32, Nk = 6$$
$$256 = Nk \times 32, Nk = 8$$

- The number of rounds is 10, 12, and 14, respectively.

9.1.4 Key Expansion

The AES algorithm takes the cipher key K and performs a key expansion routine to generate a key schedule. The key expansion generates a total of $Nb(Nr + 1)$ words: an initial set of Nb words for $Nr = 0$, and $2Nb$ for $Nr = 1$, $3Nb$ for $Nr = 2, \ldots, 11Nb$ for $Nr = 10$. Thus, the resulting key schedule consists of a linear array of 4-byte words $[w_i]$, $0 \leq i < Nb(Nr + 1)$.

RotWord() takes a four-byte input word $[a_0, a_1, a_2, a_3]$ and performs a cyclic permutation like $[a_1, a_2, a_3, a_0]$.

SubWord() takes a four-byte input word and applies the S-box (Figure 9.2) to each of the four bytes to produce an output word.

Rcon[i] represents the round constant word array and contains the values given by $[x^{i-1}, \{00\},\{00\},\{00\}]$ where i starts at 1.

		0	1	2	3	4	5	6	7	8	9	a	b	c	d	e	f
	0	63	7c	77	7b	f2	6b	6f	c5	30	01	67	2b	fe	d7	ab	76
	1	ca	82	c9	7d	fa	59	47	f0	ad	d4	a2	af	9c	a4	72	c0
	2	b7	fd	93	26	36	3f	f7	cc	34	a5	e5	f1	71	d8	31	15
	3	04	c7	23	c3	18	96	05	9a	07	12	80	e2	eb	27	b2	75
	4	09	83	2c	1a	1b	6e	5a	a0	52	3b	d6	b3	29	e3	2f	84
	5	53	d1	00	ed	20	fc	b1	5b	6a	cb	be	39	4a	4c	58	cf
	6	d0	ef	aa	fb	43	4d	33	85	45	f9	02	7f	50	3c	9f	a8
x	7	51	a3	40	8f	92	9d	38	f5	bc	b6	da	21	10	ff	f3	d2
	8	cd	0c	13	ec	5f	97	44	17	c4	a7	7e	3d	64	5d	19	73
	9	60	81	4f	dc	22	2a	90	88	46	ee	b8	14	de	5e	0b	db
	a	e0	32	3a	0a	49	06	24	5c	c2	d3	ac	62	91	95	e4	79
	b	e7	c8	37	6d	8d	d5	4e	a9	6c	56	f4	ea	65	7a	ae	08
	c	ba	78	25	2e	1c	a6	b4	c6	e8	dd	74	1f	4b	bd	8b	8a
	d	70	3e	b5	66	48	03	f6	0e	61	35	57	b9	86	c1	1d	9e
	e	e1	f8	98	11	69	d9	8e	94	9b	1e	87	e9	ce	55	28	df
	f	8c	a1	89	0d	bf	e6	42	68	41	99	2d	0f	b0	54	bb	16

Figure 9.2 AES S-box

Example 9.3. Compute the round constant words Rcon[i]:

$$\text{Rcon}[i] = [x^{i-1}, \{00\}, \{00\}, \{00\}]$$
$$\text{Rcon}[1] = [x^0, \{00\}, \{00\}, \{00\}] = [\{01\}, \{00\}, \{00\}, \{00\}] = 01000000$$
$$\text{Rcon}[2] = [x^1, \{00\}, \{00\}, \{00\}] = 02000000$$
$$\text{Rcon}[3] = [x^2, \{00\}, \{00\}, \{00\}] = 04000000$$
$$\text{Rcon}[4] = [x3, \{00\}, \{00\}, \{00\}] = 08000000$$
$$\text{Rcon}[5] = [x4, \{00\}, \{00\}, \{00\}] = 10000000$$
$$\text{Rcon}[6] = [x5, \{00\}, \{00\}, \{00\}] = 20000000$$
$$\text{Rcon}[7] = [x6, \{00\}, \{00\}, \{00\}] = 40000000$$
$$\text{Rcon}[8] = [x7, \{00\}, \{00\}, \{00\}] = 80000000$$
$$\text{Rcon}[9] = [x8, \{00\}, \{00\}, \{00\}] = [x7 \bullet x, \{00\}, \{00\}, \{00\}] = 1b000000$$

$$x^7 \bullet x = \text{xtime}(x^7) = \text{xtime}(80) = \{\text{leftshift}(80)\} \oplus \{1b\} = 1b$$

$$\text{Rcon}[10] = [x^9, \{00\}, \{00\}, \{00\}] = [x^8 \bullet x, \{00\}, \{00\}, \{00\}] = 36000000$$
$$\text{Rcon}[11] = [x^{10}, \{00\}, \{00\}, \{00\}] = [x^9 \bullet x, \{00\}, \{00\}, \{00\}] = 6c000000$$
$$\text{Rcon}[12] = [x^{11}, \{00\}, \{00\}, \{00\}] = [x^{10} \bullet x, \{00\}, \{00\}, \{00\}] = d8000000$$
$$\text{Rcon}[13] = [x^{12}, \{00\}, \{00\}, \{00\}] = [x^{11} \bullet x, \{00\}, \{00\}, \{00\}] = ab000000$$

$$x^{11} \bullet x = \text{xtime}(x^{11}) = \text{xtime}(d8) = \{\text{leftshift}(d^8)\} \oplus \{1b\} = ab \qquad \square$$

Rcon[i] is a useful component for the round constant word array in order to compute the key expansion routine.

The input key expansion into the key schedule proceeds as shown in Figure 9.3.

```
KeyExpansion(byte key[4×Nk], word w[Nb×(Nr + 1)], Nk)

begin

    i=0

    while (i < Nk)
        w[i] = word[key[4×i],key[4×i+1],key[4×i+2],key[4×i+3]]
        i = i + 1
    end while

    i = Nk
    while (i < Nb×(Nr + 1))
        word temp = w[i - 1]
        if (i mod Nk = 0)
            temp = SubWord(RotWord(temp)) xor Rcon[i / Nk]
        else if (Nk = 8 and i mod Nk = 4)
            temp = SubWord(temp)
        end if
        w[i] = w[i - Nk] xor temp
        i = i + 1
    end while

end
```

Figure 9.3   Pseudo code for key expansion

Example 9.4. Suppose the cipher key K is given as

$$K = 10\ 21\ 32\ 43\ 54\ 65\ 76\ 87\ 98\ a9\ ba\ cb\ dc\ ed\ fe\ 0f$$

The first four words of K for $Nk = 4$ results in

$$w[0] = 10213243, w[1] = 54657687, w[2] = 98a9bacb, w[3] = dcedfe0f$$

Computation of $w[4]$ for $i = 4$ is as follows:

$$\text{Temp} = w[3] = dcedfe0f$$

A cyclic permutation of $w[3]$ by one byte produces

$$\text{RotWord}(w[3]) = edfe0fdc$$

Taking each byte of RotWord($w[3]$) at a time and applying to the S-box yields

$$\text{SubWord}(edfe0fdc) = 55bb7686$$

Compute a round constant Rcon[i/Nk]:

$$\text{Rcon}[4/4] = \text{Rcon}[1] = 01000000$$
$$\text{XORing SubWord() with Rcon}[1] \text{ yields}$$
$$\text{SubWord}() \oplus \text{Rcon}[1] = 54bb7686$$
$$w[i-Nk] = w[0] = 10213243$$

Finally, $w[4]$ is computed as:

$$w[4] = 54bb7686 \oplus 10213243 = 449a44c5$$

Continuing in this fashion, the remaining $w[i]$, $4 \leq i \leq 43$, can be computed as shown in Table 9.1. $\qquad\qquad\square$

9.1.5 AES Cipher

The 128-bit cipher input is fed in a column-by-column manner, comprising each column with a four-byte word. In other words, the input is copied to the state array as shown in Table 9.2.

The cipher is described in the pseudo code in Figure 9.4.

The individual transformations for the pseudo code computation consist of SubBytes(), ShiftRows(), MixColumns(), and AddRoundKey(). These transformations play a role for processing the state and are briefly described below:

- *SubBytes() Transformation:* The SubBytes() transformation is a nonlinear byte substitution that operates independently on each byte of the state using an S-box (see Figures 9.2 and 9.5).

Table 9.1   AES key expansion

i	Temp	After RotWord	After SubWord	Rcon[i/Nk]	After XOR with Rcon	$w[i] = $ Temp $\oplus$ $w[i - Nk]$
4	dcedfe0f	edfe0fdc	55bb7686	01000000	54bb7686	449a44c5
5	449a44c5					10ff3242
6	10ff3242					88568889
7	88568889					54bb7686
8	54bb7686	bb768654	ea384420	02000000	e8384420	aca200e5
9	aca200e5					bc5d32a7
10	bc5d32a7					340bba2e
11	340bba2e					60b0cca8
12	60b0cca8	b0cca860	e74bc2d0	04000000	e34bc2d0	4fe9c235
13	4fe9c235					f3b4f092
14	f3b4f092					c7bf4abc
15	c7bf4abc					a70f8614
16	a70f8614	0f8614a7	7644fa5c	08000000	7e44fa5c	31ad3869
17	31ad3869					c219c8fb
18	c219c8fb					05a68247
19	05a68247					a2a90453
20	a2a90453	a90453a2	d3f2ed3a	10000000	c3f2ed3a	f25fd553
21	f25fd553					30461da8
22	30461da8					35e09fef
23	35e09fef					97499bbc
24	97499bbc	499bbc97	3b146588	20000000	1b146588	e94bb0db
25	e94bb0db					d90dad73
26	d90dad73					eced329c
27	eced329c					7ba4a920
28	7ba4a920	a4a9207b	49d3b721	40000000	09d3b721	e09807fa
29	e09807fa					3995aa89
30	3995aa89					d5789815
31	d5789815					aedc3135
32	aedc3135	dc3135ae	86c796e4	80000000	06c796e4	e65f911e
33	e65f911e					dfca3b97
34	dfca3b97					0ab2a382
35	0ab2a382					a46e92b7
36	a46e92b7	6e92b7a4	9f4fa949	1b000000	844fa949	62103857
37	62103857					bdda03c0
38	bdda03c0					b768a042
39	b768a042					130632f5
40	130632f5	0632f513	6f23e67d	36000000	5923e67d	3b33de2a
41	3b33de2a					86e9ddea
42	86e9ddea					31817da8
43	31817da8					22874f5d

Table 9.2 A 16-byte cipher input array

Row no (r)	Mapping of input block into column-by-column array			
0	a_0	a_4	a_8	a_{12}
1	a_1	a_5	a_9	a_{13}
2	a_2	a_6	a_{10}	a_{14}
3	a_3	a_7	a_{11}	a_{15}

For example, if $s_{2,1} = \{8f\}$, then the substitution value is determined by the intersection of the row with index 8 and the column with index f in Figure 9.2. The resulting $s'_{2,1}$ would be a value of $\{73\}$.

- *ShiftRows() Transformation:* In the ShiftRows(), the first row (Row 0) is not shifted and the rest of the rows proceed as follows:

$$s'_{r,c} = s'_r, (c + \text{shift}(r, Nb))\bmod Nb, \quad \text{for} \quad 0 < r < 4 \text{ and } 0 \le c < Nb$$

where the shift value $shift(r, Nb) = shift(r, 4)$ depends on the row number r as follows:

$$\text{shift}(1,4) = 1; \quad \text{shift}(2,4) = 2; \quad \text{shift}(3,4) = 3;$$

This has the effect of cyclic shifting the leftmost bytes around into the rightmost positions over different numbers of bytes in a given row.

- *MixColumns() Transformation:* The MixColumns() transformation operates on the state column-by-column, treating each column as a four-term polynomial over $GF(2^8)$ and multiplied modulo $x^4 + 1$ with a fixed polynomial $a(x)$ as:

$$s'(x) = a(x) \otimes s(x)$$

```
Cipher(byte in[4 * Nb], byte out[4 * Nb], word w[Nb * (Nr + 1)])

begin
      byte    state[4, Nb]

      state = in

      AddRoundKey(state, w)

      for round = 1 step 1 to Nr – 1
          SubBytes(state)
          ShiftRows(state)
          MixColumns(state)
          AddRoundKey(state, w + round * Nb)
      end for

      SubBytes(state)
      ShiftRows(state)
      AddRoundKey(state, w + Nr * Nb)

      out = state
end
```

Figure 9.4 Pseudo code for the cipher

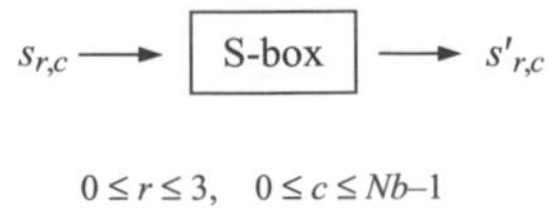

$$s_{r,c} \longrightarrow \boxed{\text{S-box}} \longrightarrow s'_{r,c}$$

$$0 \leq r \leq 3, \quad 0 \leq c \leq Nb{-}1$$

Figure 9.5   SubBytes() transformation by the S-box

where $a(x) = \{03\}x^3 + \{01\}x^2 + \{01\}x + \{02\}$, and $s(x)$ is the input polynomial and $s'(x)$ is the corresponding polynomial after the MixColumns() transformation.

The matrix multiplication of $s'(x)$ is

$$
\begin{bmatrix} s'_{0,c} \\ s'_{1,c} \\ s'_{2,c} \\ s'_{3,c} \end{bmatrix}
=
\begin{bmatrix}
02 & 03 & 01 & 01 \\
01 & 02 & 03 & 01 \\
01 & 01 & 02 & 03 \\
03 & 01 & 01 & 02
\end{bmatrix}
\begin{bmatrix} s_{0,c} \\ s_{1,c} \\ s_{2,c} \\ s_{3,c} \end{bmatrix}
\quad \text{for} \quad 0 \leq c < Nb
$$

The four bytes in a column after the matrix multiplication are

$$s'_{0,c} = (\{02\} \bullet s_{0,c}) \oplus (\{03\} \bullet s_{1,c}) \oplus s_{2,c} \oplus s_{3,c}$$
$$s'_{1,c} = s_{0,c} \oplus (\{02\} \bullet s_{1,c}) \oplus (\{03\} \bullet s_{2,c}) \oplus s_{3,c}$$
$$s'_{2,c} = s_{0,c} \oplus s_{1,c} \oplus (\{02\} \bullet s_{2,c}) \oplus (\{03\} \bullet s_{3,c})$$
$$s'_{3,c} = (\{03\} \bullet s_{0,c}) \oplus s_{1,c} \oplus s_{2,c} \oplus (\{02\} \bullet s_{3,c})$$

- *AddRoundKey() Transformation:* In AddRoundKey() transformation, a Round Key is added to the state by a simple bitwise XOR operation. Each Round Key consists of Nb words from the key schedule. These Nb words are each added into the columns of the state such that

$$[s'_{0,c}, s'_{1,c}, s'_{2,c}, s'_{3,c}] = [s_{0,c}, s_{1,c}, s_{2,c}, s_{3,c}] \oplus [w_{\text{round} * Nb + c}] \quad \text{for} \quad 0 £ c < Nb$$

where $[w_i]$ are the key schedule words, and *round* is a value in the range $0 \leq round \leq Nr$. The initial Round key addition occurs when $round = 0$, prior to the first application of the round function. The application of the AddRoundKey() transformation to the Nr rounds of the Cipher occurs when $1 \leq round \leq Nr$.

Example 9.5.   Assume that the input block and a cipher key (whose length is 16 bytes each) are given as

Plaintext $= 37$ 0f d1 2a 01 33 45 cf d7 01 1d 86 17 fa 6d 08
Cipherkey $= 10$ 21 32 43 54 65 76 87 98 a9 ba cb dc ed fe 0f

Using the algorithm for the pseudo code computation described in Figure 9.4, the intermediate values in the state array are given in Table 9.3. The round key values $w[i]$ are taken from Example 9.4.

Table 9.3 Cipher encryption for ciphertext computation

r	Start of round				After SubByte				After ShiftRows				After MixColumns				After XOR with w[]			
0	37	01	d7	17													27	55	4f	cb
	0f	33	01	fa													2e	56	a8	17
	d1	45	1d	6d													e3	33	a7	93
	2a	cf	86	08													69	48	4d	07
1	27	55	4f	cb	cc	fc	84	1f	cc	fc	84	1f	d2	9b	5b	4d	96	8b	d3	19
	2e	56	a8	17	31	b1	c2	f0	b1	c2	f0	31	94	e5	1e	c0	0e	1a	48	7b
	e3	33	a7	93	11	c3	5c	dc	5c	dc	11	c3	91	8d	a0	8d	d5	bf	28	fb
	69	48	4d	07	f9	52	e3	c5	c5	f9	52	e3	33	e8	d2	0e	f6	aa	5b	88
2	96	8b	d3	19	90	3d	66	d4	90	3d	66	d4	36	c1	00	64	9a	7d	34	04
	0e	1a	48	7b	ab	a2	52	21	a2	52	21	ab	57	ca	8d	b8	f5	97	86	08
	d5	bf	28	fb	03	08	34	0f	34	0f	03	08	0d	b7	ae	24	0d	85	14	e8
	f6	aa	5b	88	42	ac	39	c4	c4	42	ac	39	ae	9e	cb	b6	4b	39	e5	1e
3	9a	7d	34	04	b8	ff	18	f2	b8	ff	18	f2	60	01	a5	80	2f	f2	62	27
	f5	97	86	08	e6	88	44	30	88	44	30	e6	d4	72	08	5e	3d	c6	b7	51
	0d	85	14	e8	d7	97	fa	9b	fa	9b	d7	97	49	58	ab	51	8b	a8	e1	d7
	4b	39	e5	1e	b3	12	d9	72	72	b3	12	d9	45	b8	eb	d5	70	2a	57	c1
4	2f	f2	62	27	15	89	aa	cc	15	89	aa	cc	6d	b6	ff	73	5c	74	fa	d1
	3d	c6	b7	51	27	b4	a9	d1	b4	a9	d1	27	0d	83	b1	84	a0	9a	17	2d
	8b	a8	e1	d7	3d	c2	f8	0e	f8	0e	3d	c2	c2	cf	35	99	fa	07	b7	9d
	70	2a	57	c1	51	e5	5b	78	78	51	e5	5b	83	85	d8	1c	ea	7e	9f	4f
5	5c	74	fa	d1	4a	92	2d	3e	4a	92	2d	3e	6a	ed	f7	59	98	dd	c2	ce
	a0	9a	17	2d	e0	b8	f0	d8	b8	f0	d8	e0	45	0c	02	6a	1a	4a	e2	23
	fa	07	b7	9d	2d	c5	a9	5e	a9	5e	2d	c5	2c	4c	a1	39	f9	51	3e	a2
	ea	7e	9f	4f	87	f3	db	84	84	87	f3	db	dc	16	7f	ca	8f	be	90	76
6	98	dd	c2	ce	46	c1	25	8b	46	c1	25	8b	67	63	17	41	8e	ba	fb	3a
	1a	4a	e2	23	a2	d6	98	26	d6	98	26	a2	04	d7	77	dc	4f	da	9a	78
	f9	51	3e	a2	99	d1	b2	3a	b2	3a	99	d1	a7	b8	c3	30	17	15	f1	99
	8f	be	90	76	73	ae	60	38	38	73	ae	60	de	1c	97	35	05	6f	0b	15
7	8e	ba	fb	3a	19	f4	0f	80	19	f4	0f	80	33	a5	99	fe	d3	9c	4c	50
	4f	da	9a	78	84	57	b8	bc	57	b8	bc	84	16	dd	cf	53	8e	48	b7	8f
	17	15	f1	99	f0	59	a1	ee	a1	ee	f0	59	fc	36	ab	cb	fb	9c	33	fa
	05	6f	0b	15	6b	a8	2b	59	59	6b	a8	2b	6f	87	16	10	95	0e	03	25
8	d3	9c	4c	50	66	de	29	53	66	de	29	53	c6	40	63	28	20	9f	69	8c
	8e	48	b7	8f	19	52	a9	73	52	a9	73	19	a3	ca	75	63	fc	00	c7	0d
	fb	9c	33	fa	0f	de	c3	2d	c3	2d	0f	de	e8	53	a2	60	79	68	01	f2
	95	0e	03	25	2a	ab	7b	3f	3f	2a	ab	7b	45	a9	4a	c4	5b	3e	c8	73
9	20	9f	69	8c	b7	db	f9	64	b7	db	f9	64	23	4c	8f	ae	41	f1	38	bd
	fc	00	c7	0d	b0	63	c6	d7	63	c6	d7	b0	7a	f5	3f	38	6a	2f	57	3e
	79	68	01	f2	b6	45	7c	89	7c	89	b6	45	a6	5f	94	7d	9e	5c	34	4f
	5b	3e	c8	73	39	b2	e8	8f	8f	39	b2	e8	d8	4b	0e	92	8f	8b	4c	67
10	41	f1	38	bd	83	a1	07	7a	83	a1	07	7a					**b8**	**27**	**36**	**58**
	6a	2f	57	3e	02	15	5b	b2	15	5b	b2	02					**26**	**b2**	**33**	**85**
	9e	5c	34	4f	0b	4a	18	84	18	84	0b	4a					**c6**	**59**	**76**	**05**
	8f	8b	4c	67	73	3d	29	85	85	73	3d	29					**af**	**99**	**95**	**74**

□

9.1.6 AES Inverse Cipher

The cipher transformation can be implemented in reverse order to produce an inverse cipher for the AES algorithm. The individual transformations used in the inverse cipher are InvShiftRows(), InvSubBytes(), InvMixColumns(), and AddRoundKey(). These inverse transformations process the state as follows:

- *InvShiftRows() Transformation:* InvShiftRows() is the inverse of the ShiftRows() transformation. The first row (Row 0) is not shifted. The bytes in the last three rows (Row 1, Row 2, Row 3) are cyclically shifted over different numbers of bytes as follows:

    $shift(r, Nb)$: shift values, where r is a row number and $Nb = 4$.
    $shift(1, 4) = 1$, $shift(2, 4) = 2$, $shift(3, 4) = 3$, respectively.

    Specifically, the InvShiftRows() transformation proceeds as

    $$s'_{r(c + \text{shift}(r,Nb))\text{mod}Nb,} = s_{r,c}, \quad \text{for} \quad 0 < r < 4 \text{ and } 0 \le c < Nb$$

- *InvSubBytes() Transformation:* InvSubBytes() is the inverse of the byte substitution transformation, in which the inverse S-box is applied to each byte of the state. The inverse S-box used for the InvSubBytes() transformation is presented in Figure 9.6.
- *InvMixColumns() Transformation:* InvMixColumns() is the inverse of the MixColumns() transformation. This transformation operates on the state column-by-column, treating each column as a four-term polynomial. The columns are considered as polynomials over $GF(2^8)$ and multiplied modulo $x^4 + 1$ with a fixed polynomial $a^{-1}(x)$.

If the inverse state $s'(x)$ is written as a matrix multiplication, then it follows that

$$s'(x) = a^{-1}(x) \otimes s(x)$$

where $a^{-1}(x) = \{0b\}x^3 + \{0d\}x^2 + \{09\}x + \{0e\}$.

		y															
		0	1	2	3	4	5	6	7	8	9	a	b	c	d	e	f
x	0	52	09	6a	d5	30	36	a5	38	bf	40	a3	9e	81	f3	d7	fb
	1	7c	e3	39	82	9b	2f	ff	87	34	8e	43	44	c4	de	e9	cb
	2	54	7b	94	32	a6	c2	23	3d	ee	4c	95	0b	42	fa	c3	4e
	3	08	2e	a1	66	28	d9	24	b2	76	5b	a2	49	6d	8b	d1	25
	4	72	f8	f6	64	86	68	98	16	d4	a4	5c	cc	5d	65	b6	92
	5	6c	70	48	50	fd	ed	b9	da	5e	15	46	57	a7	8d	9d	84
	6	90	d8	ab	00	8c	bc	d3	0a	f7	e4	58	05	b8	b3	45	06
	7	d0	2c	1e	8f	ca	3f	0f	02	c1	af	bd	03	01	13	8a	6b
	8	3a	91	11	41	4f	67	dc	ea	97	f2	cf	ce	f0	b4	e6	73
	9	96	ac	74	22	e7	ad	35	85	e2	f9	37	e8	1c	75	df	6e
	a	47	f1	1a	71	1d	29	c5	89	6f	b7	62	0e	aa	18	be	1b
	b	fc	56	3e	4b	c6	d2	79	20	9a	db	c0	fe	78	cd	5a	f4
	c	1f	dd	a8	33	88	07	c7	31	b1	12	10	59	27	80	ec	5f
	d	60	51	7f	a9	19	b5	4a	0d	2d	e5	7a	9f	93	c9	9c	ef
	e	a0	e0	3b	4d	ae	2a	f5	b0	c8	eb	bb	3c	83	53	99	61
	f	17	2b	04	7e	ba	77	d6	26	e1	69	14	63	55	21	0c	7d

Figure 9.6 AES algorithm inverse S-box

The matrix multiplication can be expressed as

$$
\begin{bmatrix} s'_{0,c} \\ s'_{1,c} \\ s'_{2,c} \\ s'_{3,c} \end{bmatrix} = \begin{bmatrix} 0e & 0b & 0d & 09 \\ 09 & 0e & 0b & 0d \\ 0d & 09 & 0e & 0b \\ 0b & 0d & 09 & 0e \end{bmatrix} \begin{bmatrix} s_{0,c} \\ s_{1,c} \\ s_{2,c} \\ s_{3,c} \end{bmatrix} \quad \text{for} \quad 0 \le c < Nb
$$

This multiplication will result in the four bytes in a column as follows:

$$
s'_{0,c} = (\{0e\} \bullet s_{0,c}) \oplus (\{0b\} \bullet s_{1,c}) \oplus (\{0d\} \bullet s_{2,c}) \oplus (\{09\} \bullet s_{3,c})
$$
$$
s'_{1,c} = (\{09\} \bullet s_{0,c}) \oplus (\{0e\} \bullet s_{1,c}) \oplus (\{0b\} \bullet s_{2,c}) \oplus (\{0d\} \bullet s_{3,c})
$$
$$
s'_{2,c} = (\{0d\} \bullet s_{0,c}) \oplus (\{09\} \bullet s_{1,c}) \oplus (\{0e\} \bullet s_{2,c}) \oplus (\{0b\} \bullet s_{3,c})
$$
$$
s'_{3,c} = (\{0b\} \bullet s_{0,c}) \oplus (\{0d\} \bullet s_{1,c}) \oplus (\{09\} \bullet s_{2,c}) \oplus (\{0e\} \bullet s_{3,c})
$$

- *Inverse of AddRoundKey() Transformation:* AddRoundKey() is its own inverse because it involves only application of the XOR operation.

For decrypting the ciphertext, the inverse cipher is described in the pseudo code shown in Figure 9.7.

Example 9.6. The input to the inverse cipher is the cipher encryption values obtained from Example 9.5.

Ciphertext = b8 26 c6 af 27 b2 59 99 36 33 76 95 58 85 05 74
Cipherkey = 10 21 32 43 54 65 76 87 98 a9 ba cb dc ed fe 0f

The Round Key values are the same as used in Example 9.5. Table 9.4 shows the state array values in inverse cipher operation for recovering the plaintext. □

```
EqInvCipher(byte in[4 * Nb], byte out[4 * Nb], word dw[Nb * (Nr + 1)])
begin
      byte    state[4,Nb]

      state = in

      AddRoundKey(state, dw + Nr * Nb)

      for round = Nr - 1 step -1 to 1
          InvSubBytes(state)
          InvShiftRows(state)
          InvMixColumns(state)
          AddRoundKey(state, dw + round * Nb)
      end for

      InvSubBytes(state)
      InvShiftRows(state)
      AddRoundKey(state, dw)

      out = state
end
```

Figure 9.7   Pseudo code for the inverse cipher

Table 9.4 Inverse ciphering process for plaintext recovery

r	Start of round				After InvShiftRows				After InvSubBytes				After XOR with w[]				After InvMixColumns			
0	b8	27	36	58									83	a1	07	7a				
	26	b2	33	85									15	5b	b2	02				
	c6	59	76	05									18	84	0b	4a				
	af	99	95	74									85	73	3d	29				
1	83	a1	07	7a	83	a1	07	7a	41	f1	38	bd	23	4c	8f	ae	b7	db	f9	64
	15	5b	b2	02	02	15	5b	b2	6a	2f	57	3e	7a	f5	3f	38	63	c6	d7	b0
	18	84	0b	4a	0b	4a	18	84	9e	5c	34	4f	a6	5f	94	7d	7c	89	b6	45
	85	73	3d	29	73	3d	29	85	8f	8b	4c	67	d8	4b	0e	92	8f	39	b2	e8
2	b7	db	f9	64	b7	db	f9	64	20	9f	69	8c	c6	40	63	28	66	de	29	53
	63	c6	d7	b0	b0	63	c6	d7	fc	00	c7	0d	a3	ca	75	63	52	a9	73	19
	7c	89	b6	45	b6	45	7c	89	79	68	01	f2	e8	53	a2	60	c3	2d	0f	de
	8f	39	b2	e8	39	b2	e8	8f	5b	3e	c8	73	45	a9	4a	c4	3f	2a	ab	7b
3	66	de	29	53	66	de	29	53	d3	9c	4c	50	33	a5	99	fe	19	f4	0f	80
	52	a9	73	19	19	52	a9	73	8e	48	b7	8f	16	dd	cf	53	57	b8	bc	84
	c3	2d	0f	de	0f	de	c3	2d	fb	9c	33	fa	fc	36	ab	cb	a1	ee	f0	59
	3f	2a	ab	7b	2a	ab	7b	3f	95	0e	03	25	6f	87	16	10	59	6b	a8	2b
4	19	f4	0f	80	19	f4	0f	80	8e	ba	fb	3a	67	63	17	41	46	c1	25	8b
	57	b8	bc	84	84	57	b8	bc	4f	da	9a	78	04	d7	77	dc	d6	98	26	a2
	a1	ee	f0	59	f0	59	a1	ee	17	15	f1	99	a7	b8	c3	30	b2	3a	99	d1
	59	6b	a8	2b	6b	a8	2b	59	05	6f	0b	15	de	1c	97	35	38	73	ae	60
5	46	c1	25	8b	46	c1	25	8b	98	dd	c2	ce	6a	ed	f7	59	4a	92	2d	3e
	d6	98	26	a2	a2	d6	98	26	1a	4a	e2	23	45	0c	02	6a	b8	f0	d8	e0
	b2	3a	99	d1	99	d1	b2	3a	f9	51	3e	a2	2c	4c	a1	39	a9	5e	2d	c5
	38	73	ae	60	73	ae	60	38	8f	be	90	76	dc	16	7f	ca	84	87	f3	db
6	4a	92	2d	3e	4a	92	2d	3e	5c	74	fa	d1	6d	b6	ff	73	15	89	aa	cc
	b8	f0	d8	e0	e0	b8	f0	d8	a0	9a	17	2d	0d	83	b1	84	b4	a9	d1	27
	a9	5e	2d	c5	2d	c5	a9	5e	fa	07	b7	9d	c2	cf	35	99	f8	0e	3d	c2
	84	87	f3	db	87	f3	db	84	ea	7e	9f	4f	83	85	d8	1c	78	51	e5	5b
7	15	89	aa	cc	15	89	aa	cc	2f	f2	62	27	60	01	a5	80	b8	ff	18	f2
	b4	a9	d1	27	27	b4	a9	d1	3d	c6	b7	51	d4	72	08	5e	88	44	30	e6
	f8	0e	3d	c2	3d	c2	f8	0e	8b	a8	e1	d7	49	58	ab	51	fa	9b	d7	97
	78	51	e5	5b	51	e5	5b	78	70	2a	57	c1	45	b8	eb	d5	72	b3	12	d9
8	b8	ff	18	f2	b8	ff	18	f2	9a	7d	34	04	36	c1	00	64	90	3d	66	d4
	88	44	30	e6	e6	88	44	30	f5	97	86	08	57	ca	8d	b8	a2	52	21	ab
	fa	9b	d7	97	d7	97	fa	9b	0d	85	14	e8	0d	b7	ae	24	34	0f	03	08
	72	b3	12	d9	b3	12	d9	72	4b	39	e5	1e	ae	9e	cb	b6	c4	42	ac	39
9	90	3d	66	d4	90	3d	66	d4	96	8b	d3	19	d2	9b	5b	4d	cc	fc	84	1f
	a2	52	21	ab	ab	a2	52	21	0e	1a	48	7b	94	e5	1e	c0	b1	c2	f0	31
	34	0f	03	08	03	08	34	0f	d5	bf	28	fb	91	8d	a0	8d	5c	dc	11	c3
	c4	42	ac	39	42	ac	39	c4	f6	aa	5b	88	33	e8	d2	0e	c5	f9	52	e3
10	cc	fc	84	1f	cc	fc	84	1f	27	55	4f	cb	**37**	**01**	**d7**	**17**				
	b1	c2	f0	31	31	b1	c2	f0	2e	56	a8	17	**0f**	**33**	**01**	**fa**				
	5c	dc	11	c3	11	c3	5c	dc	e3	33	a7	93	**d1**	**45**	**1d**	**6d**				
	c5	f9	52	e3	f9	52	e3	c5	69	48	4d	07	**2a**	**cf**	**86**	**08**				

9.2 Elliptic Curve Cryptosystem (ECC)

The elliptic curve discrete logarithm problem appears to be substantially harder than the existing discrete logarithm problem. Considered with equal levels of security, ECC uses smaller parameters than the conventional discrete logarithm systems.

In this section we start by presenting the concept of an elliptic curve and then discuss its applications based on existing public-key algorithms. Finally, we take a look at cryptographic algorithms with elliptic curves for encryption and digital signatures over the prime or finite Galois fields.

9.2.1 Elliptic Curves

Elliptic curve groups have been studied for many years, and those over the prime field Z_p or the finite Galois field $GF(2^m)$ are particularly interesting because they provide a way of constructing cryptographic algorithms. ECs have the potential to provide faster public-key cryptosystems with smaller key sizes.

Let $EC(a, b, p)$ denote the elliptic curve group (mod p) over a finite field. There exist two formulas for EC (a, b, p) as follows:

1. $y^2 \equiv x^3 + ax + b \pmod{p}$ over Z_p
   where $a, b \in Z_p$ such that $4a^3 + 27b^2 \pmod{p} \neq 0$.
2. $y^2 + xy \equiv x^3 + ax^2 + b \pmod{p}$ over $GF(2^m)$ with the primitive polynomial $p(x)$ where $b \neq 0$ and $a, b, x, y \in$ primitive roots of $GF(2^m)$.

9.2.2 Elliptic Curves Over Prime Field Z_p

Figure 9.8 shows the elliptic curve $y^2 \equiv x^3 + ax + b$ defined over Z_p where $a, b \in Z_p$. Z_p is called a prime field if and only if $p > 3$ is an odd prime. An elliptic curve $EC(a, b, p)$ can be made into an abelian group with an addition or doubling operation on all points on EC, including the point at infinity O under the condition of $4a^3 + 27b \neq 0 \pmod{p}$. If two distinct points $P(x_1, y_1)$ and $Q(x_2, y_2)$ are on an elliptic curve, the third point R is defined as $P + Q = R$ (x_3, y_3) (see Figure 9.8(a)). The rule for adding two points P and Q on the elliptic curve is defined as follows:

- Draw a straight line through two points P and Q, and then find the intersection point $-R$ on the elliptic curve. $-R$ denotes the point $(x_3, -y_3)$.
- Draw a vertical line through point $-R$ that meets the elliptic curve at the third point R and the point of infinity O. These two points have the same x coordinate, but a negative y coordinate, as shown in Figure 9.8(a).
- When doubling two points, P and Q, into a single point $P = Q$ on the elliptic curve (EC), draw the tangential line to the EC at point P and find the third intersection point $-R$, and then draw a vertical line through point $-R$ and find point R as shown in Figure 9.8(b). Thus, $P + Q = 2P = R$.

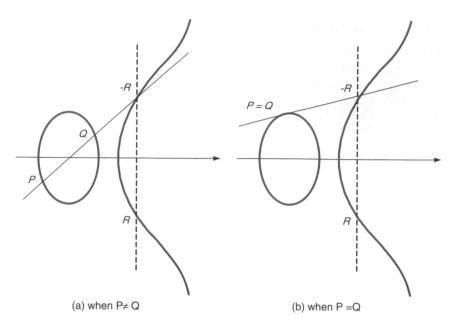

(a) when P≠ Q   (b) when P =Q

Figure 9.8   Elliptic curve

Multiplication of a point P on an elliptic curve by a positive integer k is defined as the sum of k copies of P. All arithmetic operations are written additively for a point $P \in \mathrm{EC}\ (a,\ b,\ p)$ as follows:

$$2P = P + P, \quad 3P = P + P + P = 2P + P, \quad 4P = 3P + P, \quad \text{and so on.}$$

If the points on an elliptic curve $y^2 \equiv x^3 + ax + b \pmod{p}$ over Z_p are denoted by the points $P(x_1,\ y_1)$, $Q(x_2,\ y_2)$, and $R(x_3,\ y_3) = P + Q$, our analyses will be result in the following:

1. Consider the case where $P \neq Q$.

   When the linear curve $y = \alpha x + c$ passes through the points P and Q, α and c are written as $\alpha = \frac{y_2 - y_1}{x_2 - x_1}$ and $c = y_1 - \alpha x_1$, respectively. If the point $(x, y) = (x, \alpha x + c)$ on $\overline{PQ}$ meets the condition to be on the EC, it should be satisfied by $(\alpha x + c)^2 = x^3 + ax + b$ or $x^3 - \alpha^2 x^2$ $(a - 2\alpha c)x + b - c^2 = 0$. An algebraic equation having three roots is expressed by $(x - x_1)$ $(x - x_2)(x - x_3) = 0$ or $x^3 - (x_1 + x_2 + x_3)x^2 + (x_1 x_2 + x_2 x_3 + x_3 x_1)x - x_1 x_2 x_3 = 0$. The coefficients of x^2 in both cubic equations yield $\alpha^2 = x_1 + x_2 + x_3$, from which we obtain $x_3 = \alpha^2 - x_1 - x_2$. From the linear curve $y = \alpha x + c$, we can obtain $-y_3 = \alpha x_3 + c$. Thus, $y_3 = \alpha(x_1 - x_3) - y_1$. Hence, the third point $R(x_3, y_3)$ is the result of $P + Q$.

2. Consider the case where $P = Q$.

   This is the case where two points are doubling. Draw the tangential line at P and compute the slope as follows:

Using $y^2 = x^3 + ax + b$, we have

$$2y\left(\frac{dy}{dx}\right) = 3x^2 + a$$

or $\quad \left(\frac{dy}{dx}\right) = \frac{3x^2 + a}{2y} = \beta$

Thus, $x^3 = \beta^2 - 2x_1$ and $y_3 = \beta(x_1 - x_3) - y_1$.

Figure 9.8(b) shows a geometric description of the doubling of an EC point $2P = R(x_3, y_3)$.

3. When $P = -Q$, it is obvious that $P + Q = O$.

Computation for the addition or doubling of two points on EC is shown below:

Summary of computations for $P(x_1, y_1) + Q(x_2, y_2) = R(x_3, y_3)$ on EC $y^2 = x^3 + ax + b$ over Z_p under $4a^3 + 27b^2 \neq 0 \pmod{p}$

$P \neq Q$	$P = Q$
$\alpha = \dfrac{y_2 - y_1}{x_2 - x_1}$	$\beta = \dfrac{3x_1^2 + a}{2y_1}$
$x_3 = \alpha^2 - x_1 - x_2$	$x_3 = \beta^2 - 2x_1$
$y_3 = \alpha(x_1 - x_3) - y_1$	$y_3 = \beta(x_1 - x_3) - y_1$

Example 9.7. Let $p = 17$. Choose $a = 1$ and $b = 5$ such that the elliptic curve over Z_{17} becomes $y^2 \equiv x^3 + x + 5 \pmod{17}$.

$$4a^3 + 27b^2 = 4 + 675 = 679 \equiv 16 \pmod{17}$$

Hence the given equation is indeed an elliptic curve.

1. Let $P = (3, 1)$ and $Q = (8, 10)$ be two points on the EC. Then $P + Q = R(x_3, y_3)$ is computed as follows:

$$P + Q = (3, 1) + (8, 10)$$

$$x_3 = \left(\frac{y_2 - y_1}{x_2 - x_1}\right)^2 - x_1 - x_2$$

$$= \left(\frac{9}{5}\right)^2 - 3 - 8$$

Since $9 \times 5^{-1} \pmod{17} = 9 \times 7 \pmod{17} = 12$, it gives

$$x^3 = (12^2 - 3 - 8)(\bmod\ 17) \equiv 14$$

$$y_3 = \left(\frac{9}{5}\right) \times (3 - 14) - 1 = 12 \times (-11) - 1 = -133\ (\bmod\ 17) \equiv 3$$

Hence $P + Q = R(14, 3)$.

2. Let $P = (3, 1)$. Then $2P = P + P = (x_3, y_3)$ is computed as follows:

$$2P = (3, 1) + (3, 1)$$

$$x_3 = \left(\frac{3x_1{}^2 + a}{2y_1}\right)^2 - 2x_1$$

$$= \left(\frac{27 + 1}{2}\right)^2 - 6$$

$$= 14^2 - 6 = 196 - 6 = 190\ (\bmod\ 17) \equiv 3$$

and

$$y_3 = \left(\frac{3x_1{}^2 + a}{2y_1}\right)(x_1 - x_3) - y_1$$

$$= 14(3 - 3) - 1 = -1\ (\bmod\ 17) \equiv 16$$

Hence $2P = (3, 16)$.   □

Quadratic Residue

If p is an odd prime, $0 < z < p$, and $\gcd(z, p) = 1$, then z is called a quadratic residue modulo p if and only if $y^2 \equiv z \pmod{p}$ has a solution for some y; otherwise z is called a quadratic nonresidue. For example, the quadratic residues modulo 13 are determined as follows:

$$Z_{13}^* = \{1, 2, 3, \ldots, 12\} \quad \text{(Least Residue Set)}$$

The squares of the integers in Z_{13}^* for modulo 13 is computed as:

$$\{1^2, 2^2, 3^2, \ldots, 11^2, 12^2\} \pmod{13} = \{1, 3, 4, 9, 10, 12\}$$

Hence the quadratic nonresidues modulo 13 are $\{2, 5, 6, 7, 8, 11\}$. Now you can see that the least residue set $Z_{13}^* = \{1, 2, 3, \ldots, 12\}$ is equally divided into quadratic residues and nonresidues. In general, there are precisely $(p - 1)/2$ quadratic residues and $(p - 1)/2$ quadratic nonresidues of p.

Euler's Criterion

Let p be an odd prime and $\gcd(z, p) = 1$. Using the Fermat's theorem $z^{p-1} \equiv 1 \pmod{p}$, or $z^{p-1} - 1 \equiv 0 \pmod{p}$, it gives $(z^{(p-1)/2} - 1)(z^{(p-1)/2} + 1) \equiv 0 \pmod{p}$ from which z is a quadratic residue of p if $z^{(p-1)/2} \equiv 1 \pmod{p}$; and a quadratic nonresidue of p if and only if $z^{(p-1)/2} \equiv -1 \pmod{p}$.

Legendre Symbol $\left(\frac{z}{p}\right)$

If $p > 2$ is a prime, $0 < z < p$, and $\gcd(z, p) = 1$, the Legendre symbol $\left(\frac{z}{p}\right)$ is a characteristic function of the set of quadratic residues modulo p as follows:

$$\left(\frac{z}{p}\right) = \begin{cases} 1 & \text{if } z \text{ is a quadratic residue of } p \\ -1 & \text{if } z \text{ is a quadratic nonresidue of } p \end{cases}$$

Example 9.8. Let $p = 17$, $a = 6$, and $b = 5$. Then the elliptic curve is defined as $y^2 \equiv x^3 + 6x + 5$ over Z_{17}. Note that $4a^3 + 27b^2 = 1539 \pmod{17} \equiv 9$, so the given EC is indeed an elliptic curve. The points in $EC(Z_{17})$ are $\{O\} \cup \{(2, 5), (2, 12), \ldots, (16, 10)\}$ as shown in Table 9.5. Let us first determine the points on EC. Compute $y^2 = x^3 + 6x + 5 \pmod{17}$ for each possible $x \in Z_{17}$. It is indeed required to check whether or not $z \equiv x^3 + 6x + 5 \pmod{17}$ is a quadratic residue for a given value of x. If z is a quadratic residue, then y can be computed by solving $y^2 \equiv z \pmod{17}$.

Table 9.5   Quadratic residues and points on EC $y^2 = x^3 + 6x + 5 = z$ over Z_{17}

x	$z \pmod{17}$	Quadratic residue $z^{(p-1)/2} \equiv 1$ or $\left(\frac{z}{p}\right) = 1$	Point (x, y) on EC
0	5	−1	
1	12	−1	
2	8	1	(2, 5) (2, 12)
3	16	1	(3, 4) (3, 13)
4	8	1	(4, 5) (4, 12)
5	7	−1	
6	2	1	(6, 6) (6, 11)
7	16	1	(7, 4) (7, 13)
8	4	1	(8, 2) (8, 15)
9	6	−1	
10	11	−1	
11	8	1	(11, 2) (11, 15)
12	3	−1	
13	2	1	(13, 6) (13, 11)
14	11	−1	
15	2	1	(15, 6) (15, 11)
16	15	1	(16, 7) (16, 10)

For $x = 0$, it gives $z = 5$. Hence $Z^{(p-2)/2}(\bmod Z) \equiv 5^8 \ (\bmod 17) \equiv 16 \ (\bmod\ 17) \equiv -1$
    (quadratic nonresidue)
For $x = 1$, then $z = 12$. Hence $12^8 \ (\bmod\ 17) \equiv 16(\bmod 17) \equiv -1$ (quadratic nonresidue)
For $x = 2$, then $z = 25$. Hence $25^8 \ (\bmod\ 17) \equiv 1$ (quadratic residue)
Then, solving $y^2 \equiv 25 \ (\bmod\ 17)$, we obtain its roots $y = 5$ and $y = 12$.
Two points on the elliptic curve are found as (x, y): $(2, 5)$ and $(2, 12)$.

$$\text{Check}: 5^2 \ (\bmod\ 17) = 25 \ (\bmod\ 17) \equiv 8 \text{ and } 12^2 \ (\bmod\ 17) = 144 \ (\bmod\ 17) \equiv 8.$$

Hence, $y = 5$ and $y = 12$ are checked as two solutions.
Continuing in this way, the quadratic residues and all remaining points on EC can be
computed as shown in Table 9.5.   □

Let EC be an elliptic curve over Z_p. Hasse states that the number of points on an elliptic curve,
including the point at infinity 0, is $\#EC(Z_p) = p + 1 - t$ where $|t| \leq 2\sqrt{p}$. $\#EC(Z_p)$ is called the
order of EC and t is called the trace of EC.

Example 9.9. Let EC be the elliptic curve $y^2 \equiv x^3 + x + 6$ over Z_{11}. All points on EC can be
determined as:

$$EC(Z_{11}) = \{(2,4),(2,7),(3,5),(3,6),(5,2),(5,9),\ (7,2),$$
$$(7,9),(8,3),(8,8),(10,2),(10,9)\} \cup \{O\}$$

Any point other than the point at infinity can be a generator G of EC. If we pick $G = (8, 3)$ as
the generator, the multiples of G can be computed as follows:
    When $P = Q$, $2G = (8, 3) + (8, 3)$. Using $x_3 = \beta^2 - 2x_1$ and $y_3 = \beta(x_1 - x_3) - y_1$ where
$\beta = \frac{3x_1^2 + a}{2y_1}(\bmod\ p)$, $2G(x_3, y_3)$ is computed as follows:

$$\text{Since } \beta = \frac{3 \times 8^2 + 1}{2 \times 3}(\bmod\ 11) \equiv 1, x_3 = 1^2 - 16 \ (\bmod\ 11) \equiv 7 \text{ and } y_3 = 1(8-7) - 3 \ (\bmod\ 11) \equiv 9$$

We have $2G = (7, 9)$.
    For $3G = 2G + G = (7, 9) + (8, 3)$ may be expressed as $P = 2G$ and $Q = G$. Since $P \neq Q$, we
use $x_3 = \alpha^2 - x_1 - x_2$ and $y_3 = \alpha(x_1 - x_3) - y_1$ where $\alpha = \frac{y_2 - y_1}{x_2 - x_1}$. Compute α first as:
$\alpha = \frac{3-9}{8-7}(\bmod\ 11) \equiv 5$.
Thus, $x_3 = 5^2 - 7 - 8 \ (\bmod\ 11) \equiv 10$ and $y_3 = 5(7 - 10) - 9 \ (\bmod\ 11) \equiv 9$.
Hence $3G = (10, 9)$.
For $4G = 3G + G = (10, 9) + (8, 3)$:
$\alpha = 3$, $x_3 \equiv -9 \ (\bmod\ 11) \equiv 2$, $y_3 \equiv 15 \ (\bmod\ 11) \equiv 4$
$4G = (2, 4)$.
For $5G = 4G + G = (2, 4) + (8, 3)$:
$\alpha = 9$, $x_3 \equiv 71 \ (\bmod\ 11) \equiv 5$, $y_3 \equiv -31 \ (\bmod\ 11) \equiv 2$
$5G = (5, 2)$.

For $6G = 5G + G = (5, 2) + (8, 3)$:
$\alpha = 4$, $x_3 \equiv (16 - 5 - 8)$ (mod 11) $\equiv 3$, $y_3 \equiv (4(5 - 3) - 2)$ (mod 11) $\equiv 6$
$6G = (3, 6)$.
Continuing computation in this way, the remaining multiples can be evaluated as shown in Table 9.6.

Note that α denotes β for the doubling of an EC point, $G = (8, 3)$. The generator $G = (8, 3)$ is called a primitive element that generates the multiples.   □

Example 9.10. Consider the elliptic curve $y^2 \equiv x^3 + x + 6$ over Z_{11}. Let $z \equiv x^3 + x + 6$ (mod 11) and check whether or not z is a quadratic residue of $p = 11$ when $z^5 \equiv 1$ (mod 11). If z is a quadratic residue, then y is computed solving $y^2 \equiv z$ (mod 11). When $x = 0$, $z = 6$ is not a quadratic residue because z^5 (mod p) $\equiv 6^5$ (mod 11) $\equiv -1$. When $x = 1$, $z = 8$ is a quadratic nonresidue because z^5 (mod p) $\equiv 8^5$ (mod 11) $\equiv -1$. When $x = 2$, $z = 16$ is a quadratic residue because z^5 (mod p) $\equiv 16^5$ (mod 11) $\equiv 1$. Solving $y^2 \equiv 16$ (mod 11), $y = 4$ and $y = 7$ are two roots of this quadratic congruence. Hence, two points on this elliptic curve are found as (x, y): $(2, 4)$ and $(2, 7)$.

When $x = 3$, $z = 36$ and z^5 (mod p) $\equiv 36^5$ (mod 11) $\equiv 1$. Hence, $z = 36$ is a quadratic residue of $p = 11$. $y^2 \equiv 36$ (mod 11) has two roots of $y = 5$ and 6. Consequently the two coordinates of (x, y) on the elliptic curve can be found as $(3, 5)$ and $(3, 6)$. When $x = 4$, $z \equiv 74$ (mod 11) $\equiv 8$, but there are no solutions of $y^2 \equiv 8$ (mod 11) due to the fact that $z = 8$ is not a quadratic residue because z^5 (mod 11) $\equiv -1$. When $x = 5$, $z \equiv 136$ (mod 11) $\equiv 4$. Since z^5 (mod 11) $\equiv 4^5$ (mod 11) $\equiv 1$, $z = 4$ is a quadratic residue of $p = 11$. Solve then $y^2 \equiv 4$ (mod 11). The roots of this quadratic congruence are $y = 2$ and $y = 9$, respectively. Thus two points on this elliptic curve are found as (x, y): $(5, 2)$ and $(5, 9)$.

Beyond $x = 6$, the search for all points on the elliptic curve is left to the reader as an exercise.   □

Table 9.6   Multiples of generator $G = (8, 3)$

α	xG ($2 \leq x \leq 12$)
$1(\beta^*)$	$2G = (7, 9)$
5	$3G = (10, 9)$
3	$4G = (2, 4)$
9	$5G = (5, 2)$
4	$6G = (3, 6)$
6	$7G = (3, 5)$
4	$8G = (5, 9)$
9	$9G = (2, 7)$
3	$10G = (10, 2)$
5	$11G = (7, 2)$
1	$12G = (8, 8)$

9.2.3 Elliptic Curve Over Finite Galois Field GF(2^m)

An elliptic curve over $GF(2^m)$ is defined by the following equation:

$$y^2 + xy = x^3 + ax^2 + b$$

where a, $b \in GF(2^m)$ and $b \neq 0$. The set of EC points at $GF(2^m)$ consists of all points (x, y), x, $y \in GF(2^m)$ that satisfy the above defining equation, together with the point at infinite O.

9.2.3.1 Addition and Doubling

Adding of points on an EC over $GF(2^m)$ will give a third EC point. The set of EC points forms a group with O (point at infinity) serving as its identity. The algebraic formula for the sum of two points and the doubling point are defined as follows:

1. If $P \in EC(GF(2^m))$, then $P + (-P) = O$, where $P = (x, y)$ and $-P = (x, x + y)$ are indeed the points on the EC.
2. If P and Q (but $P \neq Q$) are the points on the $EC(GF(2^m))$, then

$$P + Q = P(x_1, y_1) + Q(x_2, y_2) = R(x_3, y_3), \text{ where}$$

$$x_3 = \lambda^2 + \lambda + x_1 + x_2 + a$$

and $y_3 = \lambda(x_1 + x_3) + x_3 + y_1$ where $\lambda = \dfrac{y_1 + y_2}{x_1 + x_2}$

3. If P is a point on the EC $(GF(2^m))$, but $(P \neq -P)$, then the point of doubling is $2P = R(x_3, y_3)$, where

$$x_3 = x_1^2 + \frac{b}{x_1^2} \quad \text{and} \quad y_3 = x_1^2 + \left(x_1 + \frac{y_1}{x_1}\right)x_3 + x_3$$

Example 9.11. Consider $GF(2^4)$ whose primitive polynomial is $p(x) = x^4 + x + 1$ of degree 4. If α is a root of $p(x)$, then the field elements of $GF(2^4)$ generated by $p(x)$ are shown in Table 9.7. Since $p(\alpha) = \alpha^4 + \alpha + 1 = 0$, that is, $\alpha^4 = \alpha + 1$, the field elements of $GF(2^4)$ are expressed by 4-tuple vectors such as $1 = (1000)$, $\alpha = (0100)$, $\alpha^2 = (0010), \ldots, \alpha^{14} = (1001)$.

Choosing $a = \alpha^4$ and $b = 1$, the EC equation over $GF(2^4)$ becomes

$$y^2 + xy = x^3 + a^4 x^2 + 1$$

Table 9.7 Field elements of $GF(2^4)$ using $\alpha^4 = \alpha + 1$

$\alpha^i, 0 \leq i \leq 14$	Polynomial expression	Vector from			
α^0	1	1	0	0	0
α^1	α	0	1	0	0
α^2	α^2	0	0	1	0
α^3	α^3	0	0	0	1
α^4	$1 + \alpha$	1	1	0	0
α^5	$\alpha + \alpha^2$	0	1	1	0
α^6	$\alpha^2 + \alpha^3$	0	0	1	1
α^7	$1 + \alpha + \alpha^3$	1	1	0	1
α^8	$1 + \alpha^2$	1	0	1	0
α^9	$\alpha + \alpha^3$	0	1	0	1
α^{10}	$1 + \alpha + \alpha^2$	1	1	1	0
α^{11}	$\alpha + \alpha^2 + \alpha^3$	0	1	1	1
α^{12}	$1 + \alpha + \alpha^2 + \alpha^3$	1	1	1	1
α^{13}	$1 + \alpha^2 + \alpha^3$	1	0	1	1
α^{14}	$1 + \alpha^3$	1	0	0	1

Check whether one element (α^3, α^8) satisfies the EC equation over $GF(2^4)$:

$$(\alpha^8)^2 + (\alpha^3)(\alpha^8) = (\alpha^3)^3 + \alpha^4(\alpha^3)^2 + 1$$
$$\alpha^{16} + \alpha^{11} = \alpha^9 + \alpha^{10} + 1$$
$$(0100) + (0111) = (0101) + (1110) + (1000)$$
$$(0011) = (0011)$$

Thus, the points on the $EC(GF(2^4))$ are O (point at infinity) and the following 15 elements:

$$
\begin{array}{lllll}
(0,1) & (1, \alpha^6) & (1, \alpha^{13}) & (\alpha^3, \alpha^8), & (\alpha^3, \alpha^{13}) \\
(\alpha^5, \alpha^3) & (\alpha^5, \alpha^{11}) & (\alpha^6, \alpha^8) & (\alpha^6, \alpha^{14}) & (\alpha^9, \alpha^{10}) \\
(\alpha^9, \alpha^{13}) & (\alpha^{10}, \alpha) & (\alpha^{10}, \alpha^8) & (\alpha^{12}, 0), & (\alpha^{12}, \alpha^{12})
\end{array}
$$

$\square$

Example 9.12. Consider the elliptic curve $y^2 + xy = x^3 + \alpha^4 x^2 + 1$ over $GF(2^4)$ used in Example 9.11. Then the point addition $P(\alpha^6, \alpha^8) + Q(\alpha^3, \alpha^{13}) = R(x_3, y_3)$ is computed as follows:

Since $\lambda = \dfrac{\alpha^8 + \alpha^{13}}{\alpha^6 + \alpha^3} = \alpha$, we have

$$x_3 = \lambda^2 + \lambda + x_1 + x_2 + a$$
$$= \alpha^2 + \alpha + \alpha^6 + \alpha^3 + \alpha^4 = 1$$

and $y_3 = \lambda(x_1 + x_3) + x_3 + y_1$
$$= \alpha(\alpha^6 + 1) + 1 + \alpha^8$$
$$= \alpha(\alpha^{13}) + \alpha^2 = \alpha^{13}$$

Hence $P + Q = R(1, \alpha^{13})$.

Next, the point doubling problem of $2P = P + P = R(x_3,\ y_3)$ is evaluated as shown below:

$$x_3 = x_1^2 + \frac{b}{x_1^2} = \alpha^{12} + \frac{1}{\alpha^{12}} = \alpha^{12} + \alpha^3 = \alpha^{10}$$

(Use that the inverse of α^i is $\alpha^{-i} = \alpha^{-i+15 \ (\text{mod } 15)}$.

$$\text{and } y_3 = x_1^2 + \left(x_1 + \frac{y_1}{x_1}\right)x_3 + x_3$$

$$= \alpha^{12} + \left(\alpha^6 + \frac{\alpha^8}{\alpha^6}\right)\alpha^{10} + \alpha^{10}$$

$$= \alpha^{12} + \alpha^{13} + \alpha^{10} = (1010) = \alpha^8$$

Hence $2P = R(x_3,\ y_3) = (\alpha^{10},\ \alpha^8)$.   □

9.3 Elliptic Curve Cryptosystem versus Public-Key Cryptosystems

All practical public-key systems, such as Diffie-Hellman, RSA, ElGamal, DSA, and many other public-key algorithms, can be implemented with elliptic curves over large finite fields. Some examples of the elliptic curve and traditional public-key protocols are introduced in this section.

9.3.1 Diffie–Hellman Key Exchange

The security of the Diffie-Hellman (D-H) key exchange algorithm is based on the difficulty of calculating discrete logarithms in a finite field, as compared with the easy computation of exponentiation in the same field. This algorithm was extended to elliptic curves, providing both increased speed and decreased key size for a given level of security.

9.3.1.1 D-H Key Exchange Algorithm

x and y: Two large integers chosen by User A and B, respectively
g: a primitive element of the prime integer p

User A	User B
$X \equiv g^x(\text{mod } p) \rightleftharpoons Y \equiv g^y \ (\text{mod } p)$	
Let Z be the negotiated secret key.	
$Z \equiv Y^x \ (\text{mod } p)$	$Z \equiv X^y \ (\text{mod } p)$
$\equiv g^{yx} \ (\text{mod } p)$	$\equiv g^{xy} \ (\text{mod } p)$

Example 9.13. Choose $g = 2$ of $p = 11$.

User A	User B
$x = 5$	$y = 7$
$X \equiv g^x \pmod{p}$	$Y \equiv g^y \pmod{p}$
$\equiv 2^5 \ (\mathrm{mod} \ 11 \equiv 10)$	$\equiv 2^7 (\mathrm{mod} \ 11) \equiv 7$
$Z \equiv Y^x \pmod{p}$	$Z \equiv X^y \pmod{p}$
$\equiv 7^5 (\mathrm{mod} \ 11) \equiv 10$	$\equiv 10^7 (\mathrm{mod} \ 11) \equiv 10$

Thus, the negotiated common key is $Z = 10$. ☐

9.3.2 Elliptic Curve Diffie–Hellman Key Exchange

In this section, the elliptic curve D-H key exchange algorithm over Z_p and $\mathrm{GF}(2^m)$ will be considered.

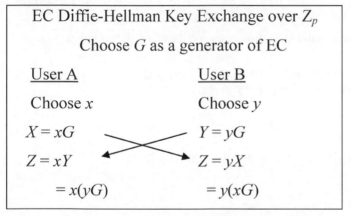

EC Diffie-Hellman Key Exchange over Z_p

Choose G as a generator of EC

User A — Choose x — $X = xG$ — $Z = xY = x(yG)$

User B — Choose y — $Y = yG$ — $Z = yX = y(xG)$

Example 9.14. Choose a generator $G = (2, 7)$ for EC $y^2 + xy \equiv x^3 + x + 6 \pmod{11}$. Scalar multiplication of a point on EC is simply the repeated addition of a point with itself. When letting $x = 2$ and $y = 3$, $X = 2G$ and $Y = 3G$ are computed as follows:

$$X = 2G = G + G = (2, 7) + (2, 7)$$

Since

$$P = Q, \ \beta \equiv \frac{3x_1^2 + a}{2y_1} \pmod{p} \equiv \frac{3 \times 2^2 + 1}{2 \times 7} \pmod{11} \equiv \frac{2}{3} \pmod{11} \equiv 8$$

$$x_3 \equiv \beta^2 - 2x_1 \ (\mathrm{mod} \ p) \equiv 64 - 4 \ (\mathrm{mod} \ 11) \equiv 5$$

$$y_3 \equiv \beta(x_1 - x_3) - y_1 \pmod{p} \equiv 8(2 - 5) - 7 \pmod{11} \equiv -31 \pmod{11} \equiv 2$$

Thus

$$2G = (5,2)$$
$$Y = 3G = 2G + G = (5,2) + (2,7)$$

For

$$P \neq Q, \ \alpha \equiv \frac{y_2 - y_1}{x_2 - x_1} (\text{mod} \, p) \equiv \frac{7-2}{2-5} (\text{mod} \, 11) \equiv -5 \times 4 (\text{mod} \, 11) \equiv 2$$

$$x_3 \equiv \alpha^2 - x_1 - x_2 \ (\text{mod} \, p) \equiv (4-5-2)(\text{mod} \, 11) \equiv 8$$

$$y_3 \equiv \alpha(x_1 - x_3) - y_1 (\text{mod} \, p) \equiv (2(5-8)-2)(\text{mod} \, 11) \equiv -8(\text{mod} \, 11) \equiv 3$$

Hence

$$3G = (8,3)$$

Continuing in this way, the remaining multiples are evaluated as shown below:

$$G = (2,7), 2G = (5,2), 3G = (8,3), 4G = (10,2), 5G = (3,6), 6G = (7,9), 7G = (7,2), 8G$$
$$= (3,5), 9G = (10,9), 10G = (8,8), 11G = (5,9), 12G = (2,4).$$

User A Computation

$$Z = x(yG) = 2(3G) = 2(8,3) = (8,3) + (8,3)$$

Since

$$P = Q, \ \beta \equiv \frac{3x_1^2 + a}{2y_1} (\text{mod} \, p) \equiv \frac{3 \times 64 + 1}{2 \times 3} (\text{mod} \, 11) \equiv \frac{193}{6} (\text{mod} \, 11) \equiv 1$$

$$x_3 \equiv \beta^2 - 2x_1 \ (\text{mod} \, p) \equiv -15(\text{mod} \, 11) \equiv 7$$

$$y_3 \equiv \beta(x_1 - x_3) - y_1 \ (\text{mod} \, p) \equiv (8-7)-3 \ (\text{mod} \, 11) \equiv -2 \ (\text{mod} \, 11) \equiv 9$$

Thus

$$Z = (x_3, y_3) = (7,9)$$

User B Computation

$$Z = y(2G) = 3(2G) = (2G + 2G) + 2G = P + Q, \text{ where } 2G = (5,2)$$

For $P = (2G + 2G)$,

$$\beta \equiv \frac{3x_1^2 + a}{2y_1} \pmod{p} \equiv \frac{76}{4} \pmod{11} \equiv 19 \pmod{11} \equiv 8$$

$$x_3 \equiv \beta^2 - 2x_1 \pmod{p} \equiv (64-10) \pmod{11} \equiv 10$$
$$y_3 \equiv \beta(x_1-x_3) - y_1 \pmod{p} \equiv (8(5-10)-2) \pmod{11} \equiv 2$$

Thus

$$Z = (10,2) + (5,2)$$

Since $P \neq Q$,

$$\alpha \equiv \frac{y_2 - y_1}{x_2 - x_1} \pmod{p} \equiv \frac{2-2}{5-10} \pmod{11} \equiv 0$$

$$x_3 \equiv \alpha^2 - x_1 - x_2 \pmod{p} \equiv -15 \pmod{11} \equiv 7$$
$$y_3 \equiv \alpha(x_1 - x_3) - y_1 \pmod{p} \equiv -2 \pmod{11} \equiv 9$$
$$Z = (x_3, y_3) = (7,9)$$

Thus, the common session key between A and B is computed as $Z = (7, 9)$ is negotiated.

$\square$

EC Diffie-Hellman Key Exchange over GF(2^m)

EC Diffie-Hellman Key Exchange over GF(2^m)
Select the base point (generator) G

User A	User B
Choose x_A from $1 \leq x_A \leq n$	Choose x_B from $1 \leq x_B \leq n$
Compute $X = x_A G$	Compute $Y = x_B G$
A sends X to B	B Sends Y to A
Compute the common keys	
$K = x_A Y = x_A(x_B G)$	$K = x_B X = x_B(x_A G)$

Example 9.15. Consider an EC equation $y^2 + xy = x^3 + \alpha^4 x^2 + 1$ for $a = \alpha^4$ and $b = 1$. Let the base point (generator) be $G = (\alpha^6, \alpha^8)$ over GF(2^4)

User A Computation
Choose $x_A = 2$ from $1 \leq x_A \leq 15$.
   Compute $X = x_A G = 2G = (\alpha^6, \alpha^8) + (\alpha^6, \alpha^8)$:

$$x_3 = (\alpha^6)^2 + \frac{1}{(\alpha^6)^2} = \alpha^{12} + \alpha^3 = \alpha^{10}$$

$$y_3 = (\alpha^6)^2 + \left(\alpha^6 + \frac{\alpha^8}{\alpha^6}\right)\alpha^{10} + \alpha^{10} = \alpha^8$$

Thus, we have

$$X = (\alpha^{10}, \alpha^8)$$

A sends $X = (\alpha^{10}, \alpha^8)$ to B

User B Computation
Choose $x_B = 3$ from $1 \le x_B \le 15$.
   Compute

$$\begin{aligned}
Y &= x_B G = 3(\alpha^6, \alpha^8)\\
&= 2(\alpha^6, \alpha^8) + (\alpha^6, \alpha^8)\\
&= X + (\alpha^6, \alpha^8) = (\alpha^{10}, \alpha^8) + (\alpha^6, \alpha^8)
\end{aligned}$$

$$\lambda = \frac{\alpha^8 + \alpha^8}{\alpha^{10} + \alpha^6} = 0$$

$$x_3 = \alpha^{10} + \alpha^6 + \alpha^4 = \alpha^3$$

$$y_3 = \alpha^2 + \alpha^8 = \alpha^{13}$$

Thus, we have

$$Y = (\alpha^3, \alpha^{13})$$

B sends $Y = (\alpha^3, \alpha^{13})$ to A.
Now, the common key K is computed as follows:
   User A computes $K = x_A Y$:

$$\begin{aligned}
K &= 2(\alpha^3, \alpha^{13}) = (\alpha^3, \alpha^{13}) + (\alpha^3, \alpha^{13})\\
x_3 &= (\alpha^3)^2 + \frac{1}{(\alpha^3)^2} = \alpha^6 + \alpha^9 = \alpha^5
\end{aligned}$$

$$\begin{aligned}
y_3 &= (\alpha^3)^2 + \left(\alpha^3 + \frac{\alpha^{13}}{\alpha^3}\right)\alpha^5 + \alpha^5\\
&= \alpha^6 + \alpha^8 + \alpha^{15} + \alpha^5 = \alpha^{11}\\
K &= (\alpha^5, \alpha^{11})
\end{aligned}$$

User B computes the common key $K = x_B X$:

$$K = 3(\alpha^{10}, \alpha^8) = (\alpha^{10}, \alpha^8) + (\alpha^{10}, \alpha^8) + (\alpha^{10}, \alpha^8)$$

$$x_3 = (\alpha^{10})^2 + \frac{1}{(\alpha^{10})^2} = \alpha^5 + \alpha^{10} = 1$$

$$y_3 = (\alpha^{10})^2 + \left(\alpha^{10} + \frac{\alpha^8}{\alpha^{10}}\right) + 1 = \alpha^5 + \alpha^{10} + \alpha^{13} + 1 = \alpha^{13}$$

$$K = (1, \alpha^{13}) + (\alpha^{10}, \alpha^8)$$

$$\lambda = \frac{\alpha^{13} + \alpha^8}{1 + \alpha^{10}} = \alpha^{13}$$

$$x_3 = \alpha^{26} + \alpha^{13} + 1 + \alpha^{10} + \alpha^4 = \alpha^5$$

$$y_3 = \alpha^{13}(1 + \alpha^5) + \alpha^5 + \alpha^{13} = \alpha^3 + \alpha^5 = \alpha^{11}$$

$K = (\alpha^5, \alpha^{11})$ as expected.

Thus, the secret common key (i.e. session key) has been exchanged between A and B.

□

9.3.3 RSA Signature Algorithm

The RSA algorithm (1977) was invented by Rivest, Shamir, and Adleman for message encryption or digital signature. The security of RSA depends on the problem of prime factorization of large numbers. Even though extensive cryptanalysis has progressed on RSA's security for so many years, it is still popular and maintains a confidence level. To keep this security level, more than 150-digit values for n are required. The RSA does not beat the DES as regards speed, however, because the software of DES is about 100 times faster than RSA.

Given the public key e and the modulus n, the private key d for decryption has to be found by factoring n. Choose two large prime numbers, p and q, and compute n which is the product of two primes, $n = p \times q$. Choose the encryption key (public key) such that e and $\phi(n)$ are coprime, that is, $\gcd(e, \phi(n)) = 1$, in which $\phi(n) = (p - 1) \times (q - 1)$ is called Euler's totient function.

RSA Signature Algorithm

---

p and q: Two large primes (secret)

$n = p \times q$: A composite integer (public)

$\phi(n)$: lcm $(p - 1, q - 1)$ (a positive number)

key-pair generation:

   e: User B's public key, relatively prime to $\phi(n)$

   d: User B's private key, coprime to $\phi(n)$

   $d \equiv e^{-1} \pmod{\phi(n)}$ from $ed \equiv 1 \pmod{\phi(n)}$

Encryption: (encrypt the message m with B's public key)

   $c \equiv m^e \pmod{n}$ (signature, i.e., encrypted ciphertext)

Decryption: (decrypt the ciphertext c with B's private key)

   $m \equiv c^d \pmod{n}$

Example 9.16. Given $p = 11$ and $q = 17$

$$n = p \times q = 11 \times 17 = 187$$

$$\phi(n) = \mathrm{lcm}(p-1, q-1) = \mathrm{lcm}(10, 16) = 80$$

Key Generation

Take User B's public key $e = 27$. User B can then compute its private key as follows:

$$27d \equiv 1 \ (\text{mod} \ 80)$$

$$d \equiv 81/27 = 3 \ (\text{User B's private key})$$

Encryption

Suppose User A chooses the message as $m = 55$.

Encrypt m with B's public key e:

$$c \equiv m^e \ (\text{mod} \ n)$$

$$\equiv 55^{27} \ (\text{mod} \ 187) = 132$$

This is the signature to be sent to User B.

Decryption

Upon received the ciphertext c, User B decrypts c with its private key d as follows:

$$m \equiv c^d (\text{mod} \ n)$$

$$\equiv 132^3 \ (\text{mod} \ 187) = 55$$

Thus, the message m is decrypted.   □

Note that User A can encrypt the message m with B's private key d.

9.3.4 Elliptic Curve RSA Signature Algorithm

The primary interest of this algorithm is an elliptic curve over Z_p with p elements for some prime p. An elliptic curve over Z_p with parameters a, b with x, $y \in Z_p$ is $y^2 \equiv x^3 + ax + b$, together with a special element O (point at infinity).

Let $\text{EC}(Z_p)$ or $\text{E}(a, b, p)$ denote the elliptic curve $y^2 \equiv x^3 + ax + b$ over the prime field Z_p. Hasse suggested that the order of $\text{EC}(Z_p)$ can be computed from the rule of $p + 1 - 2\sqrt{p} < \#E(Z_p) < p + 1 + 2\sqrt{p}$. Schoof proposed another algorithm for computing $\#E(Z_p)$, but it is quite impractical for large p. Although these two algorithms can guide us towards computing the range of $\#E(Z_p)$, they may not give us a satisfactory solution. The EC RSA algorithm seems insoluble from one perspective, but a solution can be found using the following Lemma:

If p is an odd prime satisfying $p \equiv 2 \ (\text{mod} \ 3)$, then $\#E(0, b, p) = p + 1$.

EC RSA Signature Algorithm

EC curve: $y^2 \equiv x^3 + b$ for $a = 0$

p and q: Two large primes satisfying $p \equiv q \equiv 2 \ (\text{mod} \ 3)$

n: $p \times q$

N_n: lcm $(\#E(0, b, p), \#E(0, b, q)) \equiv \text{lcm} \ (p + 1, q + 1)$

Key Generation:
  e: Public key such that $\gcd(e, N_n) = 1$ (coprime)
  d: Private key calculated from $ed \equiv 1 \pmod{N_n}$
Encryption: (User A)
  $M(m_x, m_y)$: This message should be a point on the elliptic curve $E(0, b, n)$ and $m_x, m_y \in Z_p$. b can be calculated from $b = y^2 - x^3$
  $C \equiv eM$ over $E(0, b, n)$ and send it to User B
Decryption: (User B)
  User B decrypts $C(c_x, c_y)$ with
  $M \equiv d.C$ over $E(0, b, n)$

In general, the RSA public-key algorithm for encryption, as well as for digital signatures, is almost impossible to apply for transferring from the PK RSA algorithm to the EC RSA algorithm, except for a special case illustrated in the following example.

Example 9.17. Consider the elliptic curve $y^2 \equiv x^3 + 1 \pmod{11}$ for $a = 0$ and $b = 1$. Choose $p = 11$ and $q = 5$ to be satisfied by $11 \equiv 5 \equiv 2 \pmod 3$.

$$n = p \times q = 11 \times 5 = 55$$

$$N_n = N_{55} = \mathrm{lcm}(p + 1, q + 1) = \mathrm{lcm}(12, 6) = 12$$

Choose $e = 5$ (encryption key) such that $\gcd(5, 12) = 1$.
d (private key) can be calculated from $d \equiv 5^{-1} \pmod{12} = 5$.
Choose a message $M(2, 3)$ in $E(0, b, p) = E(0, 1, 11)$: $y^2 \equiv x^3 + 1 \pmod{11}$.

Encryption

$$C \equiv eM \equiv 5(2,3) = (2,3) + (2,3) + (2,3) + (2,3) + (2,3)$$

For $C_1 = (2,3) + (2,3)$, we compute $\beta = 2, x_3 = 0$, and $y_3 = 1$, so $C_1 = (0,1)$

For $C_2 = C_1 + (2,3) = (0,1) + (2,3)$, we evaluate $\alpha = 1, x_3 = 54$, and $y_3 = 0$,

hence $C_2 = (54, 0)$
For $C_3 = C_2 + (2,3) = (54,0) + (2,3), \alpha = 1, x_3 = 0$, and $y_3 = 54$, hence $C_3 = (0,54)$
For $C_4 = C_3 + (2,3) = (54,0) + (2,3) = C$,
$$\alpha = 2, x_3 = 2, \text{ and } y3 = 52, \text{ finally } C = (2,52)$$

Decryption

$$M \equiv dC \equiv 5(2,52) = (2,52) + (2,52) + (2,52) + (2,52) + (2,52)$$

Let $M_1 = (2,52) + (2,52), \beta = 53, x_3 = 0$, and $y_3 = 54$,

$$M_1 = (0, 54)$$
$$M_2 = M_1 + (2, 52) = (0, 54) + (2, 52), \alpha = 54, x_3 = 54, \text{ and } y_3 = 0,$$
$$M_2 = (54, 0)$$
$$M_3 = M_2 + (2, 52) = (54, 0) + (2, 52), \alpha = 54, x_3 = 0, \text{ and } y_3 = 1,$$
$$M_3 = (0, 1)$$

Finally, $M = M_4 = (0,1) + (2,52)$, $\alpha = 53$, $x_3 = 2$, and $y_3 = 3$.

We compute $M = (2, 3)$.

Thus, the message M is recovered by decryption of the signature C and authentication is proved. $\square$

9.3.5 ElGamal Public-Key Encryption

ElGamal proposed a public-key cryptosystem in 1985. The ElGamal algorithm can be used for both encryption and digital signatures and it gets its security from the discrete logarithm, which is intractable. To form a cryptosystem using elliptic curves, it is relatively easy to compute $P = kG$ given k and G, but it is somewhat harder to determine k given P and G.

To encrypt the message m with the conventional ElGamal scheme, first choose a random number k that is relatively prime to $p - 1$, and let g be a primitive root of Z_p^*. The values of p, g and y are public, and x is secret. Then compute the pair (r, s) of ciphertext.

$$r \equiv g^k \pmod{p}$$
$$s \equiv (y^k \bmod p)(m (\bmod p - 1))$$

To decrypt r and s, compute $s/r^x \pmod{p}$, which reproduces the message m.

ElGamal Public-key Encryption

Choose p (a prime), $g < p$, and $x < p$ (a private key)

Public key: (p, g, y) where

$\quad y \equiv g^x \pmod{p}$

Encryption:

$\quad r \equiv g^k \pmod{p}$

$\quad s \equiv (y^k \bmod p) (m \ (\bmod p - 1)) \equiv my^k \pmod{p}$

where k is a random number, relatively prime to $p - 1$, and m is the message

Decryption:

$\quad m = s/r^x \pmod{p}$, $0 < m \leq p - 1$

Example 9.18. Choose $p = 11$, $g = 6$, $x = 3$ and $m = 7$.

Compute

$$y \equiv g^x \pmod{p} \equiv 6^3 \pmod{11} \equiv 7$$
$$\text{Public key}: \quad (y, g, p) = (7, 6, 11)$$
$$\text{Private key}: \quad x = 3 < p$$

To encrypt the message $m = 7$, first choose $k = 7$ and then compute:

$$r \equiv g^k (\bmod p) \equiv 6^7 (\bmod 11) \equiv 8$$
$$s \equiv (y^k \bmod p)(m(\bmod p - 1)), m < p - 1$$
$$\equiv 7^7 \times 7 \ (\bmod 11) \equiv 9$$

To decrypt the message m, first compute:

$$r^x (\bmod p) \equiv 8^3 (\bmod 11) \equiv 6 \text{ and take the ratio :}$$
$$m \equiv \frac{s}{r^x} (\bmod p) \equiv \frac{9}{6} (\bmod 11) \equiv 3 \times 2^{-1} (\bmod 11) \equiv 18 (\bmod 11) \equiv 7$$

Thus, the message $m = 7$ is completely recovered. □

9.3.6 Elliptic Curve ElGamal Encryption

In this section, the elliptic curve ElGamal encryption algorithm over Z_p and $GF(2^m)$ is considered.

EC ElGamal Encryption Over Z_p

Public Key: (Y, G, p)
  $Y = xG$
Private Key: $x < p$
k: a random number, relatively prime to $p - 1$
Encryption:
  $R = kG$
  $S = kY + M$
Decryption:
  $M = S - xR$

Example 9.19. Choose a generator $G = (2, 7)$ that is a base point on EC $y^2 = x^3 + x + 6$ over Z_{11} and picks User B's private key $x = 7$.
  Compute first xG.

$$2G = (2, 7) + (2, 7) \text{ for } P = Q$$
$$\beta = 8, x_3 = 5 \text{ and } y_3 = 2$$

Then, $2G = (5, 2)$
$$3G = 2G + G = (5, 2) + (2, 7) \text{ for } P \neq Q$$
$$\alpha = 2, x_3 = 8 \text{ and } y_3 = 3$$

Then, $3G = (8, 3)$

Repeating the same processes yields the following results:

$$G = (2,7), 2G = (5,2), 3G = (8,3), 4G = (10,2), 5G = (3,6), 6G = (7,9), 7G = (7,2),$$

$$8G = (3,5), 9G = (10,9), 10G = (8,8), 11G = (5,9), 12G = (2,4)$$

Using the above listing, the public key Y can be calculated as follows:

$$Y = xG = 7G = 7(2,7) = (7,2)$$

Public key : $(G, Y = 7G, p)$

Encryption
User A chooses a random number $k = 3$ $(1 \leq k \leq 12)$ and wants to send the plaintext $M = (10,9)$ to User B by Encryption.

$$R = kG = 3(2,7) = (8,3),$$

and $\quad S = kY + M = 3(7,2) + (10,9) = (3,5) + (10,9) = (10,2)$

User A then sends $R = (8, 3)$ and $S = (10, 2)$ to User B.

Decryption
User B decrypts the ciphertext to recover the message using the following formula.

$$
\begin{aligned}
M &= S - xR \\
&= (10,2) - 7(8,3) = (10,2) - (3,5) \\
&= (10,2) + (3,6) = (10,9)
\end{aligned}
$$

Thus, the message (10, 9) is successfully recovered.   □

EC ElGamal Encryption Over GF(2^m)

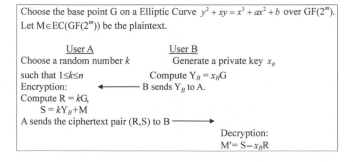

Example 9.20. Consider $EC(GF(2^4)) = y^2 + xy = x^3 + a^4x + 1$ for $a = \alpha^4$ and $b = 1$. Select a generator $G = (\alpha^3, \alpha^8)$ and choose $x_B = 2$.

Compute $Y_B = x_B G = 2(\alpha^3, \alpha^8) = (\alpha^3, \alpha^8) + (\alpha^3, \alpha^8)$

$$x_3 = (\alpha^3)^2 + \frac{1}{(\alpha^3)^2} = \alpha^6 + \alpha^{-6} = \alpha + \alpha^2 = \alpha^5$$

$$y_3 = \alpha^6 + (\alpha^3 + \alpha^5)\alpha^5 + \alpha^5 = \alpha^3$$
$$Y_B = (a^5, a^3)$$

Encryption

Choose $k = 3$ and the message $M = (\alpha^6, \alpha^8)$.
Compute $R = kG = 3(\alpha^3, \alpha^8) = (\alpha^5, \alpha^3) + (\alpha^3, \alpha^8)$.

$$\lambda = \frac{\alpha^3 + \alpha^8}{\alpha^5 + \alpha^3} = \frac{\alpha^{13}}{\alpha^{11}} = \alpha^2$$

$$x_3 = \alpha^4 + \alpha^2 + \alpha^5 + \alpha^3 + \alpha^4 = \alpha^2 + \alpha^5 + \alpha^3 = \alpha + \alpha^3 = \alpha^9$$
$$y_3 = \alpha^2(\alpha^5 + \alpha^4) + \alpha^9 + \alpha^3 = 1 + \alpha + \alpha^2 = \alpha^{10}$$
$$R = (\alpha^9, \alpha^{10})$$

Compute

$$S = kY_B + M$$
$$= 3(a^5, a^3) + (a^6, a^8)$$
$$= (\alpha^{10}, \alpha) + (\alpha^6, \alpha^8)$$

$$\lambda = \frac{\alpha + \alpha^8}{\alpha^{10} + \alpha^6} = \frac{\alpha^{10}}{\alpha^7} = a^3$$

$$x_3 = \alpha^6 + \alpha^3 + \alpha^{10} + \alpha^6 + \alpha^4 = \alpha^6$$
$$y_3 = \alpha^3(\alpha^{10} + \alpha^6) + \alpha^6 + \alpha^4 = 1 + \alpha^3 = \alpha^{14}$$
$$S = (\alpha^6, \alpha^{14})$$

User A sends the ciphertext (R,S) to User B.

Decryption

$$M' = S - x_B R$$
$$= (\alpha^6, \alpha^{14}) - 2(\alpha^9, \alpha^{10})$$
$$= (\alpha^6, \alpha^{14}) - (\alpha^{10}, \alpha)$$

Since $-P(x, y) = P(x, x + y)$, we have

$$M' = (\alpha^6, \alpha^{14}) + (\alpha^{10}, \alpha^{10} + \alpha)$$
$$= (\alpha^6, \alpha^{14}) + (\alpha^{10}, \alpha^8)$$

$$\lambda = \frac{\alpha^{14} - \alpha^8}{\alpha^6 + \alpha^{10}} = \frac{\alpha^6}{\alpha^7} = \alpha^{-1} = \alpha^{14}$$

$$x_3 = \alpha^{28} + \alpha^{14} + \alpha^6 + \alpha^{10} + \alpha^4 = \alpha^2 + \alpha^3 = \alpha^6$$
$$y_3 = \alpha^{14}(\alpha^6 + \alpha^6) + \alpha^6 + \alpha^{14} = 1 + \alpha^2 = \alpha^8$$
$$M' = (\alpha^6, \alpha^8)$$

Thus, since $M = M' = (\alpha^6, \alpha^8)$, the EC ElGamal encryption scheme works well. □

A small example is presented below to help the reader understand EC cryptosystem analysis. The following example illustrates dealing with a practical program.

Example 9.21. A practical problem on an EC ElGamal (WAP Forum Elliptic Curve, Assigned No. 8):

- Field Size: 112
- EC Curve: $y^2 = x^3 + ax + b$ over Z_p
- Prime: $p =$ ffff ffffffff ffffffff fffffde7
- $a = 0$, $b = 3$: $y^2 = x^3 + 3$ over Z_p
- Generating point: $G = (1, 2)$
- Order of G: 010000 00000000 01ecea55 1ad837e9
- Cofactor: $K = 1$

With these given parameters, the EC ElGamal encryption problem is solved as follows.
User A wants to send the message M to User B.
Start by choosing a key pair.
User B chooses the private key $d = 0x$ 1f2.

$$d = 0x1f2$$
$$Q = dG$$
$$= \begin{pmatrix} 0x\ 00004819 & 5c3cafd6 & b7ca5871 & 3448426f, \\ 0x\ 00002cc6 & 2a922993 & 62fefd26 & c3fdb0a4 \end{pmatrix}$$

Encryption
User A chooses the random number $k = 0x$ 10.

$$\text{Message } M = \begin{pmatrix} 0x\ 000061fa & 8fd6ac43 & 10f7a280 & f14ba5a9, \\ 0x\ 000035fc & 15c1afc0 & 8364f4c5 & ce4ee589 \end{pmatrix}$$

$$R = kG$$
$$= \begin{pmatrix} 0x\ 00004a4f & 12e271f3 & b7c309a9 & 0c7175ec, \\ 0x\ 0000cecd & eba7aade & 64ceb131 & 8d8bb35c \end{pmatrix}$$

$$S = kQ + M$$
$$= \begin{pmatrix} 0x\ 000046d0 & 9b75142f & 51f59f70 & e7eeef15, \\ 0x\ 00004ede & 205d35a6 & 57ea5b1d & 3cdb03f0 \end{pmatrix}$$

Decryption
User B receives two points R and S.

$$M = S{-}dR$$
$$= \begin{array}{llll} (0x\ 000061\text{fa} & 8\text{fd6ac43} & 10\text{f7a280} & \text{f14ba5a9}, \\ 0x\ 000035\text{fc} & 15\text{c1afc0} & 8364\text{f4}c5 & \text{ce4ee589}) \end{array}$$

Our analysis is complete.   ☐

Note that the WAP Forum used the private key d instead of x and the public key Q instead of Y.

9.3.7 Schnorr's Authentication Algorithm

In 1990, Schnorr introduced his authentication and signature schemes based on discrete logarithms. First, choose two primes, p and q, such that q $(1 < q < p - 1)$ is a prime factor of $p - 1$. To generate a public key, choose $a \neq 1$ such that $a \equiv h^{(p-1)/q}$ (mod p), that is, $a^q \equiv h^{p-1}$ (mod p). If h is relatively prime to p, by Fermat's theorem it can then be written as $h^{p-1} \equiv 1$ (mod p). As a result, we have $a^q \equiv 1$ (mod p), $1 < a < p - 1$. All these numbers, p, q, and a, can be freely published and shared to a group of users.

To generate a key pair, choose a random number $s < q$ that is used as the private key. After that, compute $\gamma \equiv a^{-s}$ (mod p), which is the public key.

Now, User A picks a random number $r < q$ and computes $x \equiv a^r$ (mod p). User B picks a random number t and sends it to User A, where $t \in (0, 1, 2, \ldots, 2^v - 1)$ indicates the security level. Schnorr recommends the value of $t = 72$ to ensure sufficient security. User A computes $y \equiv r + st$ (mod q) and sends it to the User B. Thus, User B tests verification of authenticity such that $x \equiv a^y \gamma^t$ (mod p). Schnorr's authentication scheme is illustrated below.

Schnorr's Authentication Algorithm

Preprocessing:
Choose two primes, p and q, such that q is a prime factor of $p{-}1$.
Choose a such that $a^q \equiv 1$ (mod p).
Key generation:
Choose a random number $s < q$ (private key)
Compute $\gamma \equiv a^{-s}$ (mod p) (public key)
User A   User B
Choose a random number $r < q$.
Compute $x \equiv a^r$ (mod p).   Pick a random number t such that
$0 < t < 2^v{-}1$.
←   Send t to the user A.
Compute $y \equiv r + st$ (mod q)
Send y to the user B   →   Verify that $x \equiv a^y \gamma^t$ (mod p)

Example 9.22. Choose two primes $p = 23$ and $q = 11$ such that $q = 11$ is a prime factor of $p - 1 = 22$. Choose $a = 3$ satisfying $a^q \equiv 1 \pmod{p}$, that is, $3^{11} \equiv 1 \pmod{23}$. Choose $s = 8 < q$ as the private key and compute the public key such that $\gamma \equiv a^{-s} \pmod{p} \equiv 3^{-8} \pmod{23}$. Compute the multiplicative inverse of $a = 3$: $aa^{-1} \equiv 1 \pmod{p}$, $3a^{-1} \equiv 1 \pmod{23}$ from which $a^{-1} = 8$. Thus, $\gamma \equiv 8^8 \pmod{23} \equiv 4$.

The sender picks $r = 5 < q$ and computes

$$x \equiv a^r \pmod{p}$$
$$\equiv 3^5 \pmod{23} \equiv 13$$

The receiver sends $t = 15$ to the sender and the sender computes

$$y \equiv r + st \pmod{q}$$
$$\equiv (5 + 8 \times 15) \pmod{11}$$
$$\equiv 125 \pmod{11} \equiv 4$$

To verify $x \equiv a^y \cdot \gamma^t \pmod{p} \equiv 13$, we compute

$$x \equiv (3^4)(4^{15}) \pmod{23}$$
$$\equiv 12 \times 3 \pmod{23} \equiv 13$$

Since $a^r \pmod{p} \equiv a^y \gamma^t \pmod{p} \equiv 13$, the authentication is accepted. □

9.3.8 EC Schnorr's Authentication Protocol

EC Schnorr's Authentication Algorithm

```
Preprocessing:
    Choose two primes, p and q.
    Choose a base point A.
Key generation:
    Choose a random number s < q (private key)
    Compute Γ = s(–A)   (public key)
User A                   User B
Choose a random number r <q.
Compute X ≡ rA              Pick a random number t such that
                                0< t< 2^v –1.
                   ←   Send t to the user A.
Compute y ≡ r +st (mod q)
Send y to the user B       →
                                Verify that X = yA + tΓ
```

Example 9.23. Choose $p = 23$, $q = 7$, $A = (2, 4)$, $s = 2$, $r = 3$, $t = 4$.

$$EC : y^2 = x^3 + x + 6 \text{ over } Z_p$$

- Compute $\Gamma = s(-A)$ (public key)
  $$= 2(2, -4) = (2, -4) + (2, -4)$$
  $\beta = 7, x_3 = 22, y_3 = 2$, hence $\Gamma = (22, 2)$

- Compute $X = rA = 3(2, 4) = (2, 4) + (2, 4) + (2, 4)$
  For $X_1 \quad = (2, 4) + (2, 4), \beta = 16, x_3 = 22, y_3 = 21$,
  hence $X_1 \quad = (22, 21)$
  For $X \quad = X_2 = (22, 21) + (2, 4), \alpha = 2, x_3 = 3, y_3 = 17$,
  hence $X \quad = (3, 17)$

using $t = 4, y = r + st = 3 + 2 \times 4 = 11 \pmod 7 = 4$
$X' = yA + t\Gamma = 4(2, 4) + 4(22, 2) = X'_1 + X'_2$
$X'_1 = 2(2, 4) = (3, 6)$
$X'_2 = 4(22, 2) = (2, 19)$
   Therefore $X' = X'_1 + X'_2 = (3, 6) + (2, 19) = (3, 17)$
   Since $X = X' = (3, 17)$, authentication is proved. $\quad\quad\quad\square$

9.3.9 Public-Key Digital Signature Algorithm

DSA Signature Scheme

Key pair generation:
  p: a prime number between 512 and 1024 bits long
  q: a prime factor of $p - 1$, 160 bits long
  $g \equiv \lambda^{(p-1)/q} \pmod p > 1$, and $\lambda < p - 1$
  $(p, q$ and $g)$: public parameters
  $x < q$: the private key, 160 bits long
  $y \equiv g^x \pmod p$: the public key, 160 bits long
Signing process (sender):
  $k < q$: a random number
  $r \equiv (g^k \bmod p) \pmod q$
  $s \equiv k^{-1} (h + xr) \pmod q$
where $h = H(m)$ is a one-way hash function of the message m
  (r, s): signature
Verifying signature (receiver):
  $w \equiv s^{-1} \pmod q$
  $u_1 \equiv h \times w \pmod q$
  $u_2 \equiv r \times w \pmod q$
  $v \equiv (g^{u_1} y^{u_2} \pmod p)) \pmod q$
If $v = r$, then the signature is verified

Example 9.24.

- Choose $p = 23$ and $q = 11$ (q is a prime factor of $p - 1$).
- Choose $\lambda = 16 < p - 1$ such that $g \equiv 16^2 \pmod{23} \equiv 3$.
- Choose the private key $x = 7 < q$.
- The public key $y \equiv g^x \pmod{p} \equiv 3^7 \pmod{23} \equiv 2$.

Sender (Signing)

- Choose $k = 5$.
- $r \equiv (g^k \bmod p) \pmod{q}$.
- $\equiv (3^5 \bmod 23) \pmod{11} \equiv 13 \pmod{11} \equiv 2$.
- Assume $h = H(m) = 10$.
- $s \equiv k^{-1} (h + xr) \pmod{q}$.
- $\equiv 5^{-1} (10 + 7 \times 2) \pmod{11} \equiv (9 \times 24) \pmod{11} \equiv 7$.
- $(5k^{-1} \equiv 1 \pmod{11}, k^{-1} = 9)$.

Receiver (Verifying)

- $w \equiv s^{-1} \pmod{q}$.
- $\equiv 7^{-1} \pmod{11} \equiv 8$.
- $u_1 \equiv h \times w \pmod{q}$.
- $\equiv (10 \times 8) \pmod{11} \equiv 3$.
- $u_2 \equiv r \times w \pmod{q}$.
- $\equiv (2 \times 8) \pmod{11} \equiv 5$.
- $v \equiv (g^{u_1} y^{u_2} \pmod{p}) \pmod{q}$.
- $\equiv ((3^3 \times 2^5) \bmod 23) \pmod{11}$.
- $\equiv ((864 \pmod{23})) \pmod{11} \equiv 2$.
- Since $v = r = 2$, the signature is verified. $\square$

9.3.10 Elliptic Curve Digital Signature Algorithm

The Elliptic Curve Digital Signature Algorithm (ECDSA) was first proposed by Scott Vanstone in 1992 and was accepted in 1999 as an ANSI standard and in 2000 as IEEE and NIST standards. ECDSA is the elliptic curve analog of DSA specified in DSS. Elliptic Curve Cryptosystems (ECCs) are viewed as elliptic curve analogs to the conventional discrete logarithm cryptosystems in which the subgroup of Z_p^* is replaced by the group of points on an elliptic curve over a finite field. The security of elliptic curve cryptosystems is based on the computational intractability of the elliptic curve discrete logarithm problem. The ECDSA signature and verification algorithms are presented in this section.

The procedures for generating and verifying signatures using ECDSA are described below.

9.3.10.1 Domain Parameters

The domain parameters for ECDSA consist of a proper elliptic curve, EC, defined over a prime field Z_p of elements p, or a extension field $GF(2^m)$ of characteristic 2 and a base point $G \in EC(Z_p)$. The order of the underline finite field Z_p or $GF(2^m)$ is p or 2^m. A set of EC domain parameters is comprised of

$$D = (q, \mathrm{FR}, a, b, G, n, \lambda)$$

where

- q: a field size either p or 2^m;
- FR: field representation used for elements of Z_p or $\mathrm{GF}(2^m)$;
- $a, b \in Z_p$ or $\mathrm{GF}(2^m)$: Two field elements that define an elliptic curve, EC,

$$y^2 = x^3 + ax + b \quad \text{over } Z_p, p > 3$$
$$y^2 + xy = x^3 + ax^2 + b \text{ over } \mathrm{GF}(2^m), p = 2^m$$

- G: the base point, $G \in \mathrm{EC}\ (Z_p$ or $\mathrm{GF}(2^m))$;
- n: The order of the point G, with $n > 2^{160}$ (ANSI X.9.62) and $n > 4\sqrt{q}$;
- λ: the cofactor is defined to be $\lambda = \#\mathrm{EC}(Z_p$ or $\mathrm{GF}(2^m))/n$.

9.3.10.2 Generation and Verification of a Random Elliptic Curve

The method for generating an elliptic curve verifiably at random is presented herein to help the reader gain some assurance regarding the possible future discovery of new and rare classes of weak elliptic curves.

Case Z_p

- Input: a field size p (an odd prime);
- Output: a bit string E of length $g \geq 160$ bits and field elements $a, b \in Z_p$ that define an elliptic curve EC: $y^2 = x^3 + ax + b$ over Z_p.

Algorithm

1. Choose an arbitrary bit string E of length $g \geq 160$ bits.
2. Compute the hash code $h = \mathrm{SHA}\text{-}1(E)$ and let c_0 be the bit stream of length v bits obtained by taking the v rightmost bits of h, where $v = t - 160 \times s$.
3. $t = \lceil \log_2 b \rceil$ and $s = \lfloor (t-1)/160 \rfloor$.
4. W_0 is the v-bit stream taken by setting the leftmost bit of c_0 to zero.
5. The integer z whose binary expansion is the g-bit stream E.
6. For i from 1 to s do:
7. Let s_i be the g-bit string of the integer $(z + i) \bmod 2^g$.
8. Compute $W_i = \mathrm{SHA}\text{-}1(s_i)$.
9. W is the bit string obtained by concatenation like $W = W_0 \| W_1 \| \ldots \| W_s$.
10. r is the integer whose binary expansion is W.
11. If $r = 0$ or $4r + 27 \equiv 0 \pmod{p}$, then go to step 1.
12. Choose $a \neq 0$, $b \neq 0 \in Z_p$ such that $rb^2 \equiv a^3 \pmod{p}$. If this condition is met, then accept, otherwise reject.
13. Output (E, a, b).

14. If the bit string is $W' = W_0 \| W_1 \| \ldots \| W_s$ and r' is the integer whose binary expansion is given by W', then the condition for acceptance is $r'b^2 \equiv a^3 \pmod{p}$, and otherwise reject.

Case GF(2^m)

For the case of GF(2^m), $s = \lfloor (m-1)/160 \rfloor$ and $v = m - 160 \times s$ are used.

- Input: a field size 2^m.
- Output: a bit string E of length $g \geq 160$ bits and field elements $a, b \in$ GF(2^m) that define an elliptic curve EC: $y^2 + xy = x^3 + ax^2 + b$ over GF(2^m).

Algorithm

1. Choose an arbitrary bit string E of length $g \geq 160$ bits.
2. Compute the hash code $h = $ SHA-1(E) and let b_0 be the bit string of length v bits obtained by taking the v rightmost bits of h.
3. Let z be the integer whose binary expansion is the g-bit stream E.
4. For i from 1 to s do:
5. Let s_i be the g-bit string of the integer $(z + i)$ mod 2^g.
6. Compute $b_i = $ SHA-1(s_i).
7. Let b be the field element obtained by concatenation as $b = b_0 \| b_1 \| \ldots \| b_s$.
8. If $b = 0$, then go to step 1.
9. Let a be an arbitrary element of GF(2^m).
10. Output (E, a, b).
11. Let b' be the field element such that $b' = b_0 \| b_1 \| \ldots \| b_s$.
12. If $b' = b$, then accept, and otherwise reject.

9.3.10.3 Key Pair Generation

An ECDSA key pair is associated with a particular set of EC domain parameters D $= (q, $ FR, $a,$ $b, G, n, \lambda)$ that must be valid prior to key generation.

   User A selects a random integer k for $1 \leq k \leq n - 1$ and computes $Q = kG$ where Q is A's public key and k is A's private key.

- Choose that $Q \neq 0$.
- Check whether a public key $Q = (x_Q, y_Q)$ is properly represented by the elements of Z_p over $(0, p - 1)$ and m-bit string over GF(2^m) of 2^m.
- Check that Q lies on the elliptic curve defined by a and b.
- Check $nQ = 0$.
- If any check fails, then Q is invalid; otherwise Q is valid.

Elliptic Curve DSA Signature Scheme over Z$_p$

User A	User B
Key pair (d, Q)	Verification of (r, s) from User A
$\quad d$: Private key	Compute:

User A	User B
$Q = dG$: Public key	$w \equiv s^{-1} \pmod{n}$
Select k (a random integer)	$u_1 \equiv hw \pmod{n}$
Compute:	$u_2 \equiv rw \pmod{n}$
$R = kG = (x_1, y_1)$	$X = u_1 G + u_2 Q = (x_2, y_2)$
Set $r = x_1$	Set $v = x_2$
$s \equiv k^{-1}(h + dr) \pmod{n}$	Accept the signature if $r = v$
where $h = \text{SHA}-1(m)$	
Send (r, s) to User B	

Example 9.25. Consider the elliptic curve $y^2 = x^3 + x + 6$ over Z_{11}. $n = 13$ is the order of the curve. Choose the key pair (d, Q) in which $d = 2$ (A's private key) and $Q = dG = (7, 9)$ (A's public key). Pick $k = 5$ (a random integer) and $G = (8, 3)$ (a generator).

User A

- $R = kG = (5, 2)$, then $r = 5$.
- Assume $h = \text{SHA-1}(m) = 8$.
- $k^{-1} = 8$, $s \equiv k^{-1}(h + dr) \pmod{13} \equiv 8(8 + 2 \times 5) \pmod{13} \equiv 1$.
- $(r, s) = (5,1) \rightarrow$ User B.

User B

- $w \equiv s^{-1} \pmod{n} \equiv 1^{-1} \pmod{13} \equiv 1$.
- $u_1 \equiv hw \pmod{n} \equiv 8 \times 1 \pmod{13} \equiv 8$.
- $u_2 \equiv rw \pmod{n} \equiv 5 \times 1 \pmod{13} \equiv 5$.
- $X = (x_2, y_2) = u_1 G + u_2 Q$.
- $= 8(8, 3) + 5(7, 9)$.
- $= (5, 9) + (10, 2) = (5, 2)$.
- $v = x_2 = 5$, $r = 5$.

Since $v = r = 5$, the signature verification is accepted. □

Example 9.26

Elliptic Curve DSA Signature Scheme over GF(2^m)
Select the base point $G = (\alpha^{12}, 0)$. Let $X \in EC(GF(2^4))$ be the plaintext. Choose a prime $n = 17 > 15$.

Signature Generation at User A
Choose key pair (d, Q), where $d = 2$ is the private key and Q is the public key, $Q = (\alpha^9, \alpha^{10})$.
    User A chooses a random integer $k = 3$ between $1 \leq k \leq 15$. User A computes $kQ = (x_1, y_1) = 3(\alpha^9, \alpha^{10}) = (\alpha^{10}, \alpha)$.

User A converts $x_1 = \alpha^{10}$ into an integer by means of the following integer conversion mapping:

$$
\begin{array}{cc}
r & x_1 \\
1 & \leftrightarrow \alpha^0 \\
2 & \leftrightarrow \alpha^1 \\
\vdots & \\
15 & \leftrightarrow \alpha^{14}
\end{array}
$$

For example, when α^{10} is converted into an integer, it would be 11, that is, $r = 11$.

Suppose the message digest is $h = 8$:

$$
\begin{aligned}
\text{User A calculates } s &= k^{-1}(h + dr)(\bmod n) \\
&= 6(8 + 2 \times 11)(\bmod 17) = 10
\end{aligned}
$$

User A's signature for the message is $(r,s) = (11,10)$.
User A sends the signature $(r,s) = (11,10)$ to User B.

Signature Verification at User B
User B computes:

$$
\begin{aligned}
w &= s^{-1}(\bmod n) = 10^{-1}(\bmod 17) = 12 \\
u_1 &= hw(\bmod n) = 8 \times 12(\bmod 17) = 11 \\
u_2 &= rw(\bmod n) = 11 \times 12(\bmod 17) = 13
\end{aligned}
$$

$$
\begin{aligned}
\text{Finally, User B computes } X &= u_1 G + u_2 Q \\
&= 11(\alpha^{12}, 0) + 13(\alpha^9, \alpha^{10}) \\
&= (\alpha^{10}, \alpha^8)
\end{aligned}
$$

We can again convert α^{10} into an integer $v = 11$. Since $v = r = 11$, the EC DSA signature is accepted. $\square$

10

Hash Function, Message Authentication Code, and Data Expansion Function

The digital signature technologies are widely used in electronic applications where sensitive information is being transmitted and in security services such as user authentication, message integrity, and non-repudiation by being digitally signed. One authentication problem in the CDMA System is how to confirm the identity of the mobile station by exchanging information between a mobile station and a base station. When a mobile station attempts to register by sending a register request message on the access channel, the authentication procedure should be performed.

Signing the message digest rather than the message often improves the efficiency of the process because the message digest is much smaller than the message. Several algorithms are introduced in order to compute message digests by employing several hash functions. The hash functions dealt with in this chapter are MD5 (1992), SHA-1 (1995), and HMAC (1996). At the end of this chapter the technique of data expansion is also discussed.

10.1 MD5 Message-Digest Algorithm

The MD5 message-digest algorithm was development by Ronald Rivest at MIT in 1992. This algorithm takes an input message of arbitrary length and produces a 128-bit hash value of the message. The input message is processed in 512-bit blocks, which can be divided into 16 32-bit subblocks. The message digest is a set of four 32-bit blocks, which concatenate to form a single 128-bit hash code. MD5 (1992) is an improved version of MD4, but is slightly slower than MD4 (1990).

The following steps are performed to compute the message digest of the input message.

10.1.1 Append Padding Bits

The message is padded so that its length (in bits) is congruent to 448 modulo 512. That is, the padded message is just 64 bits short of being a multiple of 512. This padding is formed by appending a single "1" bit to the end of message, and then "0" bits are appended, as many as

---

needed, such that the length (in bits) of the padded message becomes congruent to 448 ($= 512 - 64$), modulo 512.

10.1.2 Append Length

A 64-bit representation of the original message length is appended to the result of the previous step. If the original length is greater than 2^{64}, then only the low-order 64 bits of the length are used for appending two 32-bit words.

The resulting message has a length that is an exact multiple of 512 bits. Equivalently, this message has a length that is an exact multiple of 16 (32-bit) words. Let M[0...$N-1$] denote the word of the resulting message, with N an integer multiple of 16.

10.1.3 Initialize MD Buffer

A four-word buffer represents four 32-bit registers (A, B, C, D). This 128-bit buffer is used to compute the message digest. These registers are initialized to the following values in hexadecimal (low-order bytes first):

$$
\begin{aligned}
A &= 01\,23\,45\,67 \\
B &= 89\,ab\,cd\,ef \\
C &= fe\,dc\,ba\,98 \\
D &= 76\,54\,32\,10
\end{aligned}
$$

These four variables are copied into different variables: A as AA, B as BB, C as CC, and D as DD.

10.1.4 Define Four Auxiliary Functions (F, G, H, I)

F, G, H, and I are four basic MD5 functions. Each of these four nonlinear functions takes three 32-bit words as input and produces one 32-bit word as output. They are, one for each round, expressed as:

$$
\begin{aligned}
F(X,Y,Z) &= (X \cdot Y) + (\bar{X} \cdot Z) \\
G(X,Y,Z) &= (X \cdot Z) + (Y \cdot \bar{Z}) \\
H(X,Y,Z) &= X \oplus Y \oplus Z \\
I(X,Y,Z) &= Y \oplus (X + \bar{Z})
\end{aligned}
$$

where $X \cdot Y$ denotes the bitwise AND of X and Y; $X + Y$ denotes the bitwise OR of X and Y; $\bar{X}$ denotes the bitwise complement of X, that is, NOT(X); and $X \oplus Y$ denotes the bitwise XOR of X and Y.

These four auxiliary functions are designed in such a way that if the bits of X, Y, and Z are independent and unbiased, then at each bit position the function F acts as a conditional: if X then Y else Z. The functions G, H, and I are similar to the function F, in that they act in "bitwise parallel" to their product from the bits of X, Y, and Z. Notice that the function H is the bitwise XOR function of its inputs.

The truth table for computation of four nonlinear functions (F, G, H, I) is shown in Table 10.1.

Table 10.1   Truth table of four nonlinear functions

XYZ	FGHI
000	0001
001	1010
010	0110
011	1001
100	0011
101	0101
110	1100
111	1110

10.1.5 FF, GG, HH, and II Transformations for Rounds 1, 2, 3, and 4

If M[k], $0 \le k \le 15$, denotes the kth subblock of the message, and $\lll$s represents a left shift s bits, the four operations are defined as follows:

FF(a, b, c, d, M[k], s, i): a = b + ((a + F(b, c, d) + M[k] + T[i]) $\lll$ s)
GG(a, b, c, d, M[k], s, i): a = b + ((a + G(b, c, d) + M[k] + T[i]) $\lll$ s)
HH(a, b, c, d, M[k], s, i): a = b + ((a + H(b, c, d) + M[k] + T[i]) $\lll$ s)
II(a, b, c, d, M[k], s, i): a = b + ((a + I(b, c, d) + M[k] + T[i]) $\lll$ s)

Computation uses a 64-element table T[i], $i = 1, 2, \ldots, 64$, which is constructed from the sine function. T[i] denotes the ith element of the table, which is equal to the integer part of 4 294 967 296 times abs (sin(i)), where i is in radians:

$$T[i] = \text{integer part of} [2^{32} * |\sin(i)|]$$

where $0 \le |\sin(i)| \le 1$ and $0 \le 2^{32} * |\sin(i)| \le 2^{32}$.

Computation of T[i] for $1 \le i \le 64$ is tabulated as shown in Table 10.2.

Table 10.2   Computation of T[i] for $1 \le i \le 64$

T[1] = d76aa478	T[17] = f61e2562	T[33] = fffa3942	T[49] = f4292244
T[2] = e8c7b756	T[18] = c040b340	T[34] = 8771f681	T[50] = 432aff97
T[3] = 242070db	T[19] = 265e5a51	T[35] = 69d96122	T[51] = ab9423a7
T[4] = c1bdceee	T[20] = e9b6c7aa	T[36] = fde5380c	T[52] = fc93a039
T[5] = f57c0faf	T[21] = d62f105d	T[37] = a4beea44	T[53] = 655b59c3
T[6] = 4787c62a	T[22] = 02441453	T[38] = 4bdecfa9	T[54] = 8f0ccc92
T[7] = a8304613	T[23] = d8a1e681	T[39] = f6bb4b60	T[55] = ffeff47d
T[8] = fd469501	T[24] = e7d3fbc8	T[40] = bebfbc70	T[56] = 85845dd1
T[9] = 698098d8	T[25] = 21e1cde6	T[41] = 289b7ec6	T[57] = 6fa87e4f
T[10] = 8b44f7af	T[26] = c33707d6	T[42] = eaa127fa	T[58] = fe2ce6e0
T[11] = ffff5bb1	T[27] = f4d50d87	T[43] = d4ef3085	T[59] = a3014314
T[12] = 895cd7be	T[28] = 455a14ed	T[44] = 04881d05	T[60] = 4e0811a1
T[13] = 6b901122	T[29] = a9e3e905	T[45] = d9d4d039	T[61] = f7537e82
T[14] = fd987193	T[30] = fcefa3f8	T[46] = e6db99e5	T[62] = bd3af235
T[15] = a679438e	T[31] = 676f02d9	T[47] = 1fa27cf8	T[63] = 2ad7d2bb
T[16] = 49b40821	T[32] = 8d2a4c8a	T[48] = c4ac5665	T[64] = eb86d391

10.1.6 Computation of Four Rounds (64 Steps)

Each round consists of 16 operations. Each operation performs a nonlinear function on three of A, B, C, and D.

We now show FF, GG, HH, and II transformations for rounds 1, 2, 3, and 4 in what follows.

Round 1

Let FF[a, b, c, d, M[k], s, i] denote the operation

```
a = b + ((a + F(b, c, d) + M[k] + T[i]) <<< s).
```

Then the following 16 operations are computed:

```
FF[a, b, c, d, M[0], 7, 1], FF[d, a, b, c, M[1], 12, 2], FF[c, d, a, b, M
[2], 17, 3],
FF[b, c, d, a, M[3], 22, 4], FF[a, b, c, d, M[4], 7, 5], FF[d, a, b, c, M
[5], 12, 6],
FF[c, d, a, b, M[6], 17, 7], FF[b, c, d, a, M[7], 22, 8], FF[a, b, c, d, M
[8], 7, 9],
FF[d, a, b, c, M[9], 12, 10], FF[c, d, a, b, M[10], 17, 11], FF[b, c, d,
a, M[11], 22, 12],
FF[a, b, c, d, M[12], 7, 13], FF[d, a, b, c, M[13], 12, 14], FF[c, d, a,
b, M[14], 17, 15],
FF[b, c, d, a, M[15], 22, 16]
```

The basic MD5 operation for FF transformations of round 1 is plotted as shown in Figure 10.1. GG, HH, and II transformations for rounds 2, 3, and 4 are similarly sketched.

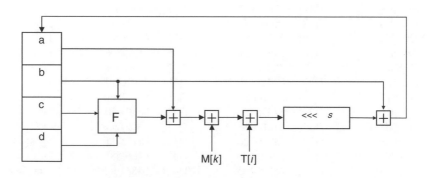

$$a = b + ((a + F(b,c,d) + M[k] + T[i]) <<< s)$$

Figure 10.1   Basic MD5 operation

Round 2

Let GG[a, b, c, d, M[*k*], s, *i*] denote the operation

```
a = b + ( (a + G(b, c, d) + M[k] + T[i] ) <<< s) .
```

Then the following 16 operations are computed:

```
GG[a, b, c, d, M[1], 5, 17], GG[d, a, b, c, M[6], 9, 18], GG[c, d, a, b, M
[11], 14, 19],
GG[b, c, d, a, M[0], 20, 20], GG[a, b, c, d, M[5], 5, 21], GG[d, a, b, c, M
[10], 9, 22],
GG[c, d, a, b, M[15], 14, 23], GG[b, c, d, a, M[4], 20, 24], GG[a, b, c,
d, M[9], 5, 25],
GG[d, a, b, c, M[14], 9, 26], GG[c, d, a, b, M[3], 14, 27], GG[b, c, d, a,
M[8], 20, 28],
GG[a, b, c, d, M[13], 5, 29], GG[d, a, b, c, M[2], 9, 30], GG[c, d, a, b, M
[7], 14, 31],
GG[b, c, d, a, M[12], 20, 32],
```

Round 3

Let HH[a, b, c, d, M[*k*], s, *i*] denote the operation

```
a = b + ( (a + H(b, c, d) + M[k] + T[i] ) <<< s)
```

Then the following 16 operations are computed:

```
HH[a, b, c, d, M[5], 4, 33], HH[d, a, b, c, M[8], 11, 34], HH[c, d, a, b, M
[11], 16, 35],
HH[b, c, d, a, M[14], 23, 36], HH[a, b, c, d, M[1], 4, 37], HH[d, a, b, c,
M[4], 11, 38],
HH[c, d, a, b, M[7], 16, 39], HH[b, c, d, a, M[10], 23, 40], HH[a, b, c,
d, M[13], 4, 41],
HH[d, a, b, c, M[0], 11, 42], HH[c, d, a, b, M[3], 16, 43], HH[b, c, d, a,
M[6], 23, 44],
HH[a, b, c, d, M[9], 4, 45], HH[d, a, b, c, M[12], 11, 46], HH[c, d, a, b,
M[15], 16, 47],
HH[b, c, d, a, M[2], 23, 48],
```

Round 4

Let II[a, b, c, d, M[*k*], s, *i*] denote the operation

```
a = b + ( (a + I(b, c, d) + M[k] + T[i] ) <<< s)
```

Then the following 16 operations are computed:

```
II[a, b, c, d, M[0], 6, 49], II[d, a, b, c, M[7], 10, 50], II[c, d, a, b, M
[14], 15, 51],
II[b, c, d, a, M[5], 21, 52], II[a, b, c, d, M[12], 6, 53], II[d, a, b, c,
M[3], 10, 54],
II[c, d, a, b, M[10], 15, 55], II[b, c, d, a, M[1], 21, 56], II[a, b, c,
d, M[8], 6, 57],
II[d, a, b, c, M[15], 10, 58], II[c, d, a, b, M[6], 15, 59], II[b, c, d,
a, M[13], 21, 60],
II[a, b, c, d, M[4], 6, 61], II[d, a, b, c, M[11], 10, 62], II[c, d, a, b,
M[2], 15, 63],
II[b, c, d, a, M[9], 21, 64],
```

After all the above steps, A, B, C, and D are added to their respective increments AA, BB, CC, DD, as follows:

$$A = A + AA, B = B + BB$$
$$C = C + CC, D = D + DD$$

and the algorithm continues with the resulting block of data. The final output is the concatenation of A, B, C, and D.

Example 10.1. Suppose the given message is

```
M = a2015613 ac67d14f 7ba011e6 33c71025 28ef51
```

Let us calculate the hash value of M using MD5 algorithm.

In the MD5 algorithm the input message is processed in 512-bit blocks that can be divided into 16 32-bit subblocks and a 128-bit hash value is computed. Each subblock is manipulated in little endian form. The message digest is a set of four 32-bit blocks, which concatenate to form a single 128-bit hash code.

The given message is of 152 bits = 0x 38 bits. So, 360 more bits are padded to make it 512 bits. This padding is formed by appending a single "1" bit to the end of the message, and then "0" bits are appended as needed such that the length (in bits) of the padded message becomes to 448. Then a 64-bit representation of the original message length is appended to make the length of the resulting message an exact multiple of 512 bits.

Thus after appending 360 bits accordingly, the 16 32-bit subblocks (in little endian form) are as follows:

```
M[0]  :135601a2        M[1]  :4fd167ac
M[2]  :e611a07b        M[3]  :2510c733
M[4]  :8051ef28        M[5]  :00000000
M[6]  :00000000        M[7]  :00000000
M[8]  :00000000        M[9]  :00000000
M[10]:00000000         M[11]:00000000
M[12]:00000000         M[13]:00000000
M[14]:00000098         M[15]:00000000
```

The initial buffer contents are given as follows:

a = 67 45 23 01
b = ef cd ab 89
c = 98 ba dc fe
d = 10 32 54 76

MD5 consists of 64 operations, grouped in four rounds of 16 operations. The nonlinear functions F, G, H, and I are used in the first, second, third, and fourth round respectively. These functions are summarized in the truth table given in Table 10.1.

One MD5 operation can be expressed as

a = b + ((a + f(b,c,d) + M[k] + T[i]) <<< s)

where, f stands for four nonlinear functions (F, G, H, and I), M[k] denotes a 32-bit block of the message input, T[i] denotes a 32-bit constant, different for each operation, and s denotes the number of bit-positions to be shifted circularly left. The value of T[i] is given in Table 10.2.   □

Example 10.2. Suppose the original message is the bit string

01100001   01100010   01100011

This message has length I = 24. After "1" is appended, we have 01100001 01100010 011000111. The number of bits of this bit string is 25 because I = 24. Therefore, we should append 423 "0"s and the 2-word representation of 24, that is, 00000000 00000018 (in hexs) for forming the final padded message as follows:

61626380	00000000	00000000	00000000
00000000	00000000	00000000	00000000
00000000	00000000	00000000	00000000
00000000	00000000	00000000	00000018

This final padded message consisting of one block contains 16 words = $16 \times 8 \times 4 = 512$ bits for $n = 1$ in this case.   □

Round 1 Computation for FF[a,b,c,d,M[k],s,i]:
a = b + ((a + F(b,c,d) + M[k] + T[i]) <<< s)
  = b + U <<< s
  = b + U'
where, 0 ≤ k ≤ 15,
1 ≤ i ≤ 16,
U = ((a + F(b,c,d) + M[k] + T[i]), and U' = U <<< s
  = the 32-bit value obtained by circularly shifting U left by s bit positions

1. First-word block process:

a=67452301, b=efcdab89,
c=98badcfe, d=10325476,
M[0]=135601a2, s=7, T[1]=d76aa478

Using Table 10.1, F(b,c,d) is computed as shown below:

```
b=efcdab89=1110 1111 1100 1101 1010 1011 1000 1001
c=98badcfe=1001 1000 1011 1010 1101 1100 1111 1110
d=10325476=0001 0000 0011 0010 0101 0100 0111 0110
```

```
F(b,c,d) = 1001 1000 1011 1010 1101 1100 1111 1110
```

```
So F(b,c,d) = 0x 98badcfe
Now compute U'=(a+F(b,c,d)+M[0]+T[1])<<<s, s = 7
a = 67452301
F(b,c,d) = 98badcfe
M[0] = 135601a2
T[1] = d76aa478
    U = eac0a619
Since, U' = U <<< s
= eac0a619 <<< 7
= 1110 1010 1100 0000 1010 0110 0001 1001 <<< 7
= 0110 0000 0101 0011 0000 1100 1111 0101
= 60530cf5
It gives, a = b + U'
= efcdab89 + 60530cf5
= 5020b87e
```

Thus, FF[a,b,c,d,M[0],7,1] of NO.1 operation can be computed as:

```
a =5020b87e, b=efcdab89, c=98badcfe, d=10325476
```

2. Second-word block process:

```
a=5020b87e, b=efcdab89,
c=98badcfe, d=10325476,
M[1]=4fd167ac, s=12, T[2]=e8c7b756
```

Using Table 10.1, F(a,b,c) is computed as shown below:

```
a=5020b87e=0101 0000 0010 0000 1011 1000 0111 1110
b=efcdab89=1110 1111 1100 1101 1010 1011 1000 1001
c=98badcfe=1001 1000 1011 1010 1101 1100 1111 1110
 F(a,b,c) =1100 1000 1001 1010 1110 1100 1000 1000
```

```
So F(a,b,c) = c89aec88
Now compute U'=(b+F(a,b,c)+M[1]+T[2])<<<s, s = 12
d = 10325476
F(a,b,c) = c89aec88
M[1] = 4fd167ac
T[2] = e8c7b756
    U = 11666000
```

Since, U' = U <<< s
= 11666000 <<< 12
= 00010001011001100110000000000000 <<< 12
= 0110 0110 0000 0000 0000 0001 0001 0110
= 66000116
We have, d = a + U'
= 5020b87e + 66000116
= b620b994

Thus the outcome of FF[2] is

a =5020b87e, b=efcdab89, c=98badcfe, d= b620b994

All FF transformations for round 1 are similarly computed and consist of the following results from the 16 operations:

```
                a        b        c        d
FF[1]  : 5020b87e efcdab89 98badcfe 10325476
FF[2]  : 5020b87e efcdab89 98badcfe b620b994
FF[3]  : 5020b87e efcdab89 0704b349 b620b994
FF[4]  : 5020b87e bf0fe287 0704b349 b620b994
FF[5]  : 48c915ed bf0fe287 0704b349 b620b994
FF[6]  : 48c915ed bf0fe287 0704b349 aaed46b8
FF[7]  : 48c915ed bf0fe287 e704e0b9 aaed46b8
FF[8]  : 48c915ed 481ea9ac e704e0b9 aaed46b8
FF[9]  : dfe96876 481ea9ac e704e0b9 aaed46b8
FF[10]: dfe96876 481ea9ac e704e0b9 ce5ab259
FF[11]: dfe96876 481ea9ac 9b181cf9 ce5ab259
FF[12]: dfe96876 0d803a35 9b181cf9 ce5ab259
FF[13]: 778942c0 0d803a35 9b181cf9 ce5ab259
FF[14]: 778942c0 0d803a35 9b181cf9 abab9858
FF[15]: 778942c0 0d803a35 b2b46a8d abab9858
FF[16]: 778942c0 5a4421df b2b46a8d abab9858
```

Round 2 Computation for GG[a,b,c,d,M[k],s,i]:
a = b + ((a + G(b,c,d) + M[k] + T[i]) <<< s)
= b + V <<< s
= b + V'
where $0 \le k \le 15$,
$17 \le i \le 32$,
V =((a + G(b,c,d) + M[k] + T[i]), and V' = V <<< s

1. First-word block process:
a=778942c0, b=5a4421df,
c=b2b46a8d, d=abab9858,
M[1]=4fd167ac, s=5, T[17]=f61e2562

Using Table 10.1, G(b,c,d) is computed as shown below:

```
b=5a4421df=0101 1010 0100 0100 0010 0001 1101 1111
c=b2b46a8d=1011 0010 1011 0100 0110 1010 1000 1101
d=abab9858=1010 1011 1010 1011 1001 1000 0101 1000
  G(b,c,d)=0001 1010 0001 0100 0110 0010 1101 1101
                    1 a       1      4      6      2 d     d
So G(b,c,d)= 1a1462dd
```

```
Now compute V'=(a+G(b,c,d)+M[1]+T[17])<<<s, s = 5
a = 778942c0
G(b,c,d)= 1a1462dd
M[1]= 4fd167ac
T[17]= f61e2562
   V = d78d32ab
Since, V' = V <<< s
= d78d32ab <<< 5
= 1101 0111 1000 1101 0011 0010 1010 1011 <<< 5
= 1111 0001 1010 0110 0101 0101 0111 1010
= f1a6557a
It gives, a = b + V'
= 5a4421df + f1a6557a
= 4bea7759
```

```
Hence, GG[a,b,c,d,M[1],5,17] of NO.1 operation can be computed as:
a=4bea7759 b=5a4421df c=b2b46a8d d=abab9858
```

Through the 16 operations, the GG transformation for round 2 can be accomplished as shown below:

```
                 a          b          c          d
GG[1]  : 4bea7759 5a4421df b2b46a8d abab9858
GG[2]  : 4bea7759 5a4421df b2b46a8d e5485ec6
GG[3]  : 4bea7759 5a4421df 2c316585 e5485ec6
GG[4]  : 4bea7759 b7519c9a 2c316585 e5485ec6
GG[5]  : a8aa43d3 b7519c9a 2c316585 e5485ec6
GG[6]  : a8aa43d3 b7519c9a 2c316585 8343ad18
GG[7]  : a8aa43d3 b7519c9a 4a1b9147 8343ad18
GG[8]  : a8aa43d3 976e3d7a 4a1b9147 8343ad18
GG[9]  : 5438208c 976e3d7a 4a1b9147 8343ad18
GG[10] : 5438208c 976e3d7a 4a1b9147 41fba4c3
GG[11] : 5438208c 976e3d7a 24ad8e21 41fba4c3
GG[12] : 5438208c 8f5bb95e 24ad8e21 41fba4c3
GG[13] : fed237c2 8f5bb95e 24ad8e21 41fba4c3
GG[14] : fed237c2 8f5bb95e 24ad8e21 9d13616b
GG[15] : fed237c2 8f5bb95e 8b4483d6 9d13616b
GG[16] : fed237c2 287f004a 8b4483d6 9d13616b
```

Round 3 Computation for HH[a,b,c,d,M[k],s,i]:

a = b + ((a + H(b,c,d) + M[k] + T[i]) <<< s)

= b + W <<< s

= b + W'

where, 0 ≤k≤15,

33≤i≤48,

W = ((a + H(b,c,d) + M[k] + T[i]), and W' = W <<< s

1. First-word block process:

a=fed237c2, b=287f004a,

c=8b4483d6, d=9d13616b,

M[5]=00000000, s=4, T[33]=fffa3942

Using Table 10.1, H(b,c,d) is computed as shown below:

b=287f004a=0010 1000 0111 1111 0000 0000 0100 1010
c=8b4483d6=1000 1011 0100 0100 1000 0011 1101 0110
d=9d13616b=1001 1101 0001 0011 0110 0001 0110 1011
$\overline{\text{H(b,c,d)=0011 1110 0010 1000 1110 0010 1111 0111}}$

So H(b,c,d)= 0x 3e28e2f7

Now compute W' = (a+H(b,c,d)+M[1]+T[33]) <<<s, s = 4

a = fed237c2

H(b,c,d)= 3e28e2f7

M[5]= 00000000

T[33]= fffa3942

    W = 3cf553fb

Since, W' = W <<< s

= 3cf553fb <<< 4

= 0011 1100 1111 0101 0101 0011 1111 1011 <<< 4

= 1100 1111 0101 0101 0011 1111 1011 0011

= cf553fb3

it gives, a = b + W'

= 287f004a + cf553fb3

= f7d43ffd

Thus, HH[a,b,c,d,M[5],4,33] of NO.1 operation is obtained as:

a=f7d43ffd, b=287f004a, c=8b4483d6, d=9d13616b

Through 16 operations, HH transformation for round 3 can be computed as shown below:

	a	b	c	d
HH[1] :	f7d43ffd	287f004a	8b4483d6	9d13616b
HH[2] :	f7d43ffd	287f004a	8b4483d6	a076abc8
HH[3] :	f7d43ffd	287f004a	19ee2487	a076abc8
HH[4] :	f7d43ffd	ea287cfb	19ee2487	a076abc8

```
HH[5]  : eb80d70f ea287cfb 19ee2487 a076abc8
HH[6]  : eb80d70f ea287cfb 19ee2487 5b513b36
HH[7]  : eb80d70f ea287cfb 5bfaa6d9 5b513b36
HH[8]  : eb80d70f 81c4b09b 5bfaa6d9 5b513b36
HH[9]  : da7ce534 81c4b09b 5bfaa6d9 5b513b36
HH[10]: da7ce534 81c4b09b 5bfaa6d9 353f2800
HH[11]: da7ce534 81c4b09b 517fec82 353f2800
HH[12]: da7ce534 fca230f9 517fec82 353f2800
HH[13]: cfeccf7d fca230f9 517fec82 353f2800
HH[14]: cfeccf7d fca230f9 517fec82 2e942b6f
HH[15]: cfeccf7d fca230f9 6cf9ba6c 2e942b6f
HH[16]: cfeccf7d 98942b2f 6cf9ba6c 2e942b6f
```

Round 4 Computation for II[a,b,c,d,M[k],s,i]:

$a = b + ((a + I(b,c,d) + M[k] + T[i]) <<< s)$

$= b + X <<< s$

$= b + X'$

where, $0 \leq k \leq 15$,

$49 \leq I.64$,

$X = ((a + I(b,c,d) + M[k] + T[i]))$, and $X' = X <<< s$

1. First-word block process:

a=cfeccf7d, b=98942b2f,

c=6cf9ba6c, d=2e942b6f,

M[0]=135601a2, s=6, T[49]=f4292244

Using Table 10.1, I(b,c,d) is computed as shown below:

```
b=98942b2f=1001 1000 1001 0100 0010 1011 0010 1111
c=6cf9ba6c=0110 1100 1111 1001 1011 1010 0110 1100
d=2e942b6f=0010 1110 1001 0100 0010 1011 0110 1111
I(b,c,d) =1011 0101 0000 0110 0100 0101 1101 0011
```

```
Thus I(b,c,d) = 0x b50645d3
Now compute X' = (a+I(b,c,d)+M[0]+T[49]) <<<s, s = 6
a = cfeccf7d
I(b,c,d) = b50645d3
M[0] = 135601a2
T[49] = f4292244
    X = 8c723936
```

```
Since, X' = X <<< s
= 8c723936 <<< 6
= 1000 1100 0111 0010 0011 1001 0011 0110 <<< 6
= 0001 1100 1000 1110 0100 1101 1010 0011
= 1c8e4da3
```

it gives, a = b + X'
= 98942b2f + 1c8e4da3
= b52278d2
Thus, II[a,b,c,d,M[0],6,49] of NO.1 operation is obtained as:
a= b52278d2 b=98942b2f c=6cf9ba6c d=2e942b6f

The results from 16 operations for II transformation are listed below:

```
                a         b         c         d
II[1]  : b52278d2 98942b2f 6cf9ba6c 2e942b6f
II[2]  : b52278d2 98942b2f 6cf9ba6c 7b2a8357
II[3]  : b52278d2 98942b2f 4202f4c2 7b2a8357
II[4]  : b52278d2 461bd86a 4202f4c2 7b2a8357
II[5]  : 1cd287d1 461bd86a 4202f4c2 7b2a8357
II[6]  : 1cd287d1 461bd86a 4202f4c2 ce8d547d
II[7]  : 1cd287d1 461bd86a 3b42e716 ce8d547d
II[8]  : 1cd287d1 608d119f 3b42e716 ce8d547d
II[9]  : 2ea5bc14 608d119f 3b42e716 ce8d547d
II[10]: 2ea5bc14 608d119f 3b42e716 da48b97f
II[11]: 2ea5bc14 608d119f 9113a190 da48b97f
II[12]: 2ea5bc14 49aad69f 9113a190 da48b97f
II[13]: 07930a07 49aad69f 9113a190 da48b97f
II[14]: 07930a07 49aad69f 9113a190 6c659cfe
II[15]: 07930a07 49aad69f 52c56aef 6c659cfe
II[16]: 07930a07 f8fee230 52c56aef 6c659cfe
```

Hence, after 64 operations, the content of the buffers are:

a = 07930a07   c = 52c56aef
b = f8fee230   d = 6c659cfe

The final values of a,b,c, and d are computed by adding the initial buffer values accordingly as shown below:

a = 07930a07 + 67452301 = 6ed82d08
b = f8fee230 + efcdab89 = e8cc8db9
c = 52c56aef + 98badcfe = eb8047ed
d = 6c659cfe + 10325476 = 7c97f174

The 128-bit hash value is finally computed by concatenating the content of the four buffers as shown below:

6ed82d08 e8cc8db9 eb8047ed 7c97f174 (little endian format)
or
082dd86e b98dcce8 ed4780eb 74f1977c (big endian format)

In CDMA cellular mobile communications, a shared secret data (SSD) is a 128-bit pattern stored in semi-permanent memory in the mobile station. SSD is partitioned into two 64-bit distinct subsets, SSD-A and SSD-B. SSD-A is used to support the authentication process, whereas SSD-B is used to support voice privacy and message confidentiality.

SSD data subsets are generated from the message digest as follows:

```
SSD-A: 6ed82d08e8cc8db9,
SSD-B: eb8047ed7c97f174
```

10.2 Secure Hash Algorithm (SHA-1)

The Secure Hash Algorithm (SHA) was developed by the National Institute of Standards and Technology (NIST) for use with the Digital Signature Algorithm (DSA) and published as a Federal Information Processing Standards (FIPS PUB 180) in 1993. The Secure Hash Standard (SHS) specifies a SHA-1 for computing a hash value of a message or a data file. When a message of any length of less than 2^{64} bits is input, the SHA-1 produces a 160-bit output called a message digest (or a hash code). The message digest can then be input to the DSA which generates or verifies the signature for the message. Signing the message digest rather than the message often improves the efficiency of the process because the message digest is usually much smaller in size than the message.

The SHA-1 (FIPS 180-1, 1995) is a technical revision of SHA (FIPS 180, 1993). The SHA-1 is secure because it is computationally infeasible to find a message that corresponds to a given message digest, or to find two different messages that produce the same message digest. Any change to a message in transit will result in a different message digest, and the signature will fail to verify. The SHA-1 is based on the MD4 message digest algorithm and its design is closely modeled after that algorithm.

10.2.1 Message Padding

The message padding is provided to make a final padded message that is a multiple of 512 bits. The SHA-1 sequentially processes blocks of 512 bits when computing the hash value (or message digest) of a message or data file that is provided as input. Padding is exactly the same as in MD5. The following specifies how this padding is performed. As a summary, first append a "1" followed by as many "0"s as necessary to make it 64 bits short of a multiple of 512 bits, and finally a 64-bit integer is appended to the end of the zero-appended message to produce a final padded message of length $n \times 512$ bits. The 64-bit integer "I" represents the length of the original message. Now, the padded message is processed by the SHA-1 as $n \times 512$ bit blocks.

10.2.2 Initialize 160-Bit Buffer

The 160-bit buffer consists of five 32-bit registers (A, B, C, D, E). Before processing any blocks, these registers are initialized to the following hexadecimal values:

$H_0 = 67\ 45\ 23\ 01$
$H_1 = ef\ cd\ ab\ 89$

H_2 = 98 ba dc fe
H_3 = 10 32 54 76
H_4 = c3 d2 e1 f0

Note that the first four values are the same as those used in MD5. The only difference is the use of a different rule for expressing the values, that is, high-order octets first for SHA and low-order octets first for MD5.

10.2.3 Functions Used

A sequence of logical functions $f_0, f_1, \ldots, f_{79}$ is used in the SHA-1. Each function f_t, $0 \le t \le 79$, operates on three 32-bit words B, C, D and produces a 32-bit word as output. Each operation performs a nonlinear operation of three of A, B, C, and D, and then does shifting and adding similar to MD5. The set of SHA primitive functions, f_t (B, C, D), is defined as follows:

$$f_t(B,C,D) = (B \cdot C) + (\bar{B} \cdot D), 0 \le t \le 19$$
$$f_t(B,C,D) = B \oplus C \oplus D, 20 \le t \le 39$$
$$f_t(B,C,D) = (B \cdot C) + (B \cdot D) + (C \cdot D), 40 \le t \le 59$$
$$f_t(B,C,D) = B \oplus C \oplus D, 60 \le t \le 79$$

where B·C = bitwise logical "AND" of B and C; $B \oplus C$ = bitwise logical XOR of B and C; $\bar{B}$ = bitwise logical "complement" of B; $\boxplus$ = addition modulo 2^{32}.

As you can see, only three different functions are used. For $0 \le t \le 19$, the function f_t acts as a conditional: if B then C else D. For $20 \le t \le 39$ and $60 \le t \le 79$, the function f_t is true if two or three of the arguments are true.

Table 10.3 is a truth table of these functions.

10.2.4 Constants Used

Four distinct constants are used in the SHA-1. In hexadecimal, these values are given by

$$K_t = 5a827999, \quad 0 \le t \le 19$$
$$K_t = 6ed9eba1, \quad 20 \le t \le 39$$
$$K_t = 8fbbcdc, \quad 40 \le t \le 59$$
$$K_t = ca62c1d6, \quad 60 \le t \le 79$$

Table 10.3 Truth table of four nonlinear functions for SHA-1

B	C	D	$f_{0, 1, \ldots, 19}$	$f_{20, 21, \ldots, 39}$	$f_{40, 41, \ldots, 59}$	$f_{60, 61, \ldots, 79}$
0	0	0	0	0	0	0
0	0	1	1	1	0	1
0	1	0	0	1	0	1
0	1	1	1	0	1	0
1	0	0	0	1	0	1
1	0	1	0	0	1	0
1	1	0	1	0	1	0
1	1	1	1	1	1	1

10.2.5 Computing the Message Digest

The message digest is computed using the final padded message. To generate the message digest, the 16-word blocks (M_0 to M_{15}) are processed in order. The processing of each M_i involves 80 steps. That is, the message block is transformed from 16 32-bit words (M_0 to M_{15}) to 80 32-bit words (W_0 to W_{79}) using the following algorithm:

. Divide M_i into 16 words W_0, W_1, ..., W_{15}, where W_0 is the leftmost word. For t = 0 to 15, $W_t = M_t$]]>
. For t = 16 to 79, $W_t = S^1 (W_{t-16} \oplus W_{t-14} \oplus W_{t-8} \oplus W_{t-3})$
. Let A = H_0, B = H_1, C = H_2, D = H_3, E = H_4
. For t = 0 to 79 do
    TEMP = S^5(A) + F_t (B, C, D) + E + W_t + K_t ;
    E = D ; D = C ; C = S^{30}(B) ; B = A ; A = TEMP

where A, B, C, D, E are five words of the buffer; t is a round number, $0 \le t \le 79$; S^i is a circular left shift by i bits; W_t is a 32-bit word derived from the current 512-bit input block; K_t is an additive constant; and $\boxplus$ is addition modulo 2^{32}

After all N 512-bit blocks have been processed, the output from the Nth stage is the 160-bit message digest, represented by the five words H_0, H_1, H_2, H_3, H_4.

The SHA-1 operation looking at the logic in each of 80 rounds of one 512-bit block is shown in Figure 10.2.

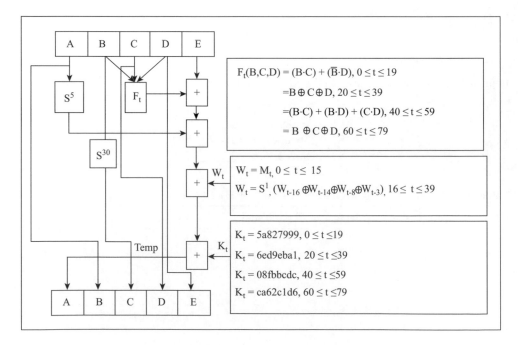

Figure 10.2 SHA-1 operation

Example 10.3. Show how to derive the 32-bit words $W_t, 0 \le t \le 79$, from the 512-bit message.

t	W_t
0	$W_0 = M_0$
1	$W_1 = M_1$
15	$W_{15} = M_{15}$
16	$W_{16} = S^1 (W_0 \oplus W_2 \oplus W_8 \oplus W_{13})$
17	$W_{17} = S^1 (W_1 \oplus W_3 \oplus W_9 \oplus W_{14})$
30	$W_{30} = S^1 (W_{14} \oplus W_{16} \oplus W_{22} \oplus W_{27})$
31	$W_{31} = S^1 (W_{15} \oplus W_{17} \oplus W_{23} \oplus W_{28})$
59	$W_{59} = S^1 (W_{43} \oplus W_{45} \oplus W_{51} \oplus W_{56})$
60	$W_{60} = S^1 (W_{44} \oplus W_{46} \oplus W_{52} \oplus W_{57})$
78	$W_{78} = S^1 (W_{62} \oplus W_{64} \oplus W_{70} \oplus W_{75})$
79	$W_{79} = S^1 (W_{63} \oplus W_{65} \oplus W_{71} \oplus W_{76})$

$\square$

Example 10.4. Let the original message be 1a7fd53b4c. Then, the final padded message consists of the following 16 words:

```
1a7fd53b  4c800000  00000000  00000000
00000000  00000000  00000000  00000000
00000000  00000000  00000000  00000000
00000000  00000000  00000000  00000028
```

The initial hex values of $\{H_i\}$ are
$H_0 = 67452301$
$H_1 = efcdab89$
$H_2 = 98badcfe$
$H_3 = 10325476$
$H_4 = c392e1f0.$

The hex values of A, B, C, D, and E after pass t $(0 \le t \le 79)$ are computed as follows:

	Register output				
t	A	B	C	D	E
0	ba346dee	67452301	7bf36ae2	98badcfe	10325476
1	f9be8ae4	ba346dee	59d148c0	7bf36ae2	98badcfe
2	84e1fdf6	f9be8ae4	ae8d1b7b	59d148c0	7bf36ae2
3	1b82edab	84e1fdf6	3e6fa2b9	ae8d1b7b	59d148c0
4	531f1a75	1b82edab	a1387f7d	3e6fa2b9	ae8d1b7b
5	926052f7	531f1a75	c6e0bb6a	a1387f7d	3e6fa2b9

	Register output				
t	A	B	C	D	E
6	c71cfaac	926052f7	54c7c69d	c6e0bb6a	a1387f7d
7	341b3a4b	c71cfaac	e49814bd	54c7c69d	c6e0bb6a
8	79a59326	341b3a4b	31c73eab	e49814bd	54c7c69d
9	d47fe3c4	79a59326	cd06ce92	31c73eab	e49814bd
10	185db57b	d47fe3c4	9e6964c9	cd06ce92	31c73eab
11	3569d479	185db57b	351ff8f1	9e6964c9	cd06ce92
12	6b01c842	3569d479	c6176d5e	351ff8f1	9e6964c9
13	5d3c5387	6b01c842	4d5a751e	c6176d5e	351ff8f1
14	04434893	5d3c5387	9ac07210	4d5a751e	c6176d5e
15	c1456f97	04434893	d74f14e1	9ac07210	4d5a751e
16	a44dbea6	c1456f97	c110d224	d74f14e1	9ac07210
17	ef0512e1	a44dbea6	f0515be5	c110d224	d74f14e1
18	f3c545ab	ef0512e1	a9136fa9	f0515be5	c110d224
19	b78ca1cc	f3c545ab	7bc144b8	a9136fa9	f0515be5
20	a3d6efd7	b78ca1cc	fcf1516a	7bc144b8	a9136fa9
21	c3880afc	a3d6efd7	2de32873	fcf1516a	7bc144b8
22	a25fd097	c3880afc	e8f5bbf5	2de32873	fcf1516a
23	2263e9cb	a25fd097	30e202bf	e8f5bbf5	2de32873
24	cd820d01	2263e9cb	e897f425	30e202bf	e8f5bbf5
25	9824bad0	cd820d01	c898fa72	e897f425	30e202bf
26	59e04bcd	9824bad0	73608340	c898fa72	e897f425
27	b7581fd3	59e04bcd	26092eb4	73608340	c898fa72
28	7efb6e25	b7581fd3	567812f3	26092eb4	73608340
29	18d1583d	7efb6e25	edd607f4	567812f3	26092eb4
30	42659f77	18d1583d	5fbedb89	edd607f4	567812f3
31	22b4bfef	42659f77	4634560f	5fbedb89	edd607f4
32	a9390191	22b4bfef	d09967dd	4634560f	5fbedb89
33	ffd2919f	a9390191	c8ad2ffb	d09967dd	4634560f
34	a0585c33	ffd2919f	6a4e4064	c8ad2ffb	d09967dd
35	8fae2fc9	a0585c33	fff4a467	6a4e4064	c8ad2ffb
36	5337d670	8fae2fc9	e816170c	fff4a467	6a4e4064
37	7044d0fe	5337d670	63eb8bf2	e816170c	fff4a467
38	78304e61	7044d0fe	14cdf59c	63eb8bf2	e816170c
39	2c5ca6b0	78304e61	9c11343f	14cdf59c	63eb8bf2
40	f304b895	2c5ca6b0	5e0c1398	9c11343f	14cdf59c
41	e89d0d8b	f304b895	b1729ac	5e0c1398	9c11343f
42	79f30210	e89d0d8b	7cc12e25	b1729ac	5e0c1398
43	f37223c6	79f30210	fa274362	7cc12e25	0b1729ac
44	f53bdd27	f37223c6	1e7cc084	fa274362	7cc12e25
45	b1cf753c	f53bdd27	bcdc88f1	1e7cc084	fa274362
46	d9030e9b	b1cf753c	fd4ef749	bcdc88f1	1e7cc084
47	9bf173ff	d9030e9b	2c73dd4f	fd4ef749	bcdc88f1

48	bae46f3c	9bf173ff	f640c3a6	2c73dd4f	fd4ef749
49	e8be1481	bae46f3c	e6fc5cff	f640c3a6	2c73dd4f
50	4a0bb5b8	e8be1481	2eb91bcf	e6fc5cff	f640c3a6
51	6d99dcd5	4a0bb5b8	7a2f8520	2eb91bcf	e6fc5cff
52	5e0e5623	6d99dcd5	1282ed6e	7a2f8520	2eb91bcf
53	422c7e52	5e0e5623	5b667735	1282ed6e	7a2f8520
54	e6ca43ae	422c7e52	d7839588	5b667735	1282ed6e
55	835bd439	e6ca43ae	908b1f94	d7839588	5b667735
56	32a7862d	835bd439	b9b290eb	908b1f94	d7839588
57	250ada00	32a7862d	60d6f50e	b9b290eb	908b1f94
58	a46d627b	250ada00	4ca9e18b	60d6f50e	b9b290eb
59	0588823a	a46d627b	942b680	4ca9e18b	60d6f50e
60	2d9bba2e	588823a	e91b589e	0942b680	4ca9e18b
61	8d8fb303	2d9bba2e	8162208e	e91b589e	0942b680
62	860d6a4f	8d8fb303	8b66ee8b	8162208e	e91b589e
63	14b64733	860d6a4f	e363ecc0	8b66ee8b	8162208e
64	7f486fbe	14b34733	e1835a93	e363ecc0	8b66ee8b
65	7d3d3745	7f486fbe	c52cd1cc	e1835a93	e363ecc0
66	d17b4506	7d3d3745	9fd21bef	c52cd1cc	e1835a93
67	2e4967ee	d17b4506	5f4f4dd1	9fd21bef	c52cd1cc
68	cc1e45de	2e4967ee	b45ed141	5f4f4dd1	9fd21bef
69	b3f80c20	cc1e45de	8b9259fb	b45ed141	5f4f4dd1
70	f124837a	b3f80c20	b3079177	8b9259fb	b45ed141
71	56ed70b1	f124837a	2cfe0308	b3079177	8b9259fb
72	d8b0d990	56ed70b1	bc4920de	2cfe0308	b3079177
73	1d849b17	d8b0d990	55bb5c2c	bc4920de	2cfe0308
74	84257988	1d849b17	362c3664	55bb5c2c	bc4920de
75	9eec3055	84257988	c76126c5	362c3664	55bb5c2c
76	6240e72c	9eec3055	21095e62	c76126c5	362c3664
77	8243ecda	6240e72c	67bb0c15	21095e62	c76126c5
78	a8342af0	8243ecda	189039cb	67bb0c15	21095e62
79	e1426096	a8342af0	a090fb36	189039cb	67bb0c15

After all 512-bit blocks have been processed, the output represented by the five words, H_0, H_1, H_2, H_3, H_4 will comply with the 160-bit message digest as shown below:

H_0: 48878397
H_1: 9801d679
H_2: 394bd834
H_3: 28c28e41
H_4: 2b8dee05

The 160-bit message digest is then data concatenation of $\{H_i\}$:

$$H_0\|H_1\|H_2\|H_3\|H_4 = 488783979801d679394bd83428c28e412b8dee05 \qquad \square$$

As discussed previously, the digitized document or message of any length can create a 160-bit message digest, which is produced by applying the SHA-1 algorithm.

Any change to a digitized message in transit results in a different message digest. In fact, changing a single bit of the data modifies at least half of the resulting digest bits. Furthermore, it is computationally infeasible to find two meaningful messages that have the same 160-bit digest. On the other hand, given a 160-bit message digest, it is also impossible to find a meaningful message with that digest.

10.3 Hashed Message Authentication Codes (HMAC)

The keyed Hashing Message Authentication Code (HMAC) is a key-dependent one-way hash function that provides both data integrity and data origin authentication for files sent between two users. HMACs have the same properties as the one-way hash functions discussed previously in this chapter, but they also include a secret key. HMACs can be used to authenticate data or files between two users (data authentication). They can also be used by a single user to determine whether or not his/her files have been altered (data integrity).

An HMAC mechanism can be used with any iterative hash functions in combination with a secret key. MD5 and SHA-1 are examples of such hash functions. Message authentication codes are used between two parties (for example, client and server) that share a secret key in order to validate information transmitted between them. The MAC mechanism should allow for easy replacement of the embedded hash function in case that faster or more secure hash functions are found or required. That is, if it is desired to replace a given hash function in an HMAC implementation, all that is required is simply to remove the existing hash function module and replace it with the new more secure module.

10.3.1 HMAC Structure

The HMAC is a secret key authentication algorithm that provides both data integrity and data origin authentication for packets sent between two parties. The HMAC definition requires a cryptographic hash function H and a secret key K. H denotes a hash function where the message is hashed by iterating a basic compressing function on data blocks. Let b denote the block length of $b = 64$ bytes $= 512$ bits for all hash functions such as MD5 and SHA-1. h denotes the length of hash values, that is, $h = 16$ bytes $= 128$ bits for MD5 and $h = 20$ bytes $= 160$ bits for SHA-1. The secret key K can be of any length up to $b = 512$ bits.

10.3.2 HMAC Computation Using RFC Method

To compute HMAC over the message, the HMAC equation is expressed as follows:

$$\text{HMAC} = \text{H}[(K \oplus \text{opad}) || \text{H}[(K \oplus \text{ipad}) || M]]$$

where $ipad = 00110110(0x\,36)$ repeated 64 times (512 bits); $opad = 01011100$ $(0 \times 5c)$ repeated 64 times (512 bits); $ipad$ is inner padding and $opad$ is outer padding.

We explain the HMAC equation in words below:

1. Append zeros to the end of K to create a 64-byte string K' (i.e. if $K = 160$ bits in length and $b = 512$ bits, then K should append 352 zero bits or 44 zero bytes to generate K').
2. XOR (bitwise exclusive-OR) K' with $ipad$ to produce the b-bit block Ω_i,

where

$$K' = K||(0 \times 00\ldots00) \quad \text{(b bits)}$$
$$\Omega_i = K' \oplus \text{ipad} \quad \text{(b bits)}$$

3. Concatenate M' to the b-bit string Ω_i obtained from step 2.
4. Apply H to the stream $\Omega_i||M'$ generated in step 3.
5. XOR (bitwise exclusive-OR) K' with *opad* to produce the b-bit string Ω_0, as shown in step 2, $\Omega_0 = K' \oplus opad$ (b bits).
6. Concatenate the padded hash h' from step 4 to the b-bit string Ω_0 resulting from step 5.
7. Apply H to the stream generated in step 6 and obtain the resulting output of HMAC(M).

Figure 10.3 illustrates the overall operation of HMAC, explaining the steps listed above.

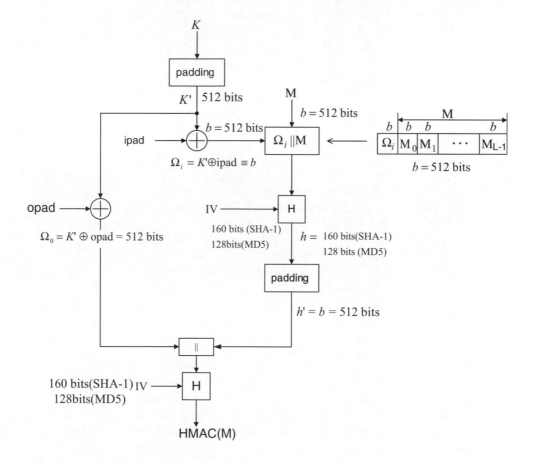

Figure 10.3 Overall operation of HMAC computation using either MD5 or SHA-1 (message length computation is based on M)

Example 10.5. Consider the HMAC computation by using a hash function SHA-1. Assume that the message (M), the key (K), and the initialization vector (IV) are given as follows:

```
M : 0x 1a7fd53b4c
K : 0x 31fa7062c45113e32679fd1353b71264
IV : A = 0x67452301, B = 0xefcdab89, C = 0x98badcfe,
     D = 0x10325476, E = 0xc3d2e1f0
```

Referring to Figure 10.3, the HMAC-SHA-1 calculation is carried out with the following steps:

```
K' = K||(0x00...00) (512 bits)
   = 31fa7062      c45113e3      2679fd13      53b71264
     00000000      00000000      00000000      00000000
     00000000      00000000      00000000      00000000
     00000000      00000000      00000000      00000000
Ωi = K ⊕ ipad = K' ⊕(0x3636...36)
   = 07cc4654      f26725d5      104fcb25      65812452
     36363636      36363636      36363636      36363636
     36363636      36363636      36363636      36363636
     36363636      36363636      36363636      36363636
M' = 1a7fd53b      4c800000      00000000      00000000
     00000000      00000000      00000000      00000000
     00000000      00000000      00000000      00000000
     00000000      00000000      00000000      00000028
Ωi || M' :
     07cc4654      f26725d5      104fcb25      65812452
     36363636      36363636      36363636      36363636
     36363636      36363636      36363636      36363636
     36363636      36363636      36363636      36363636
     1a7fd53b      4c800000      00000000      00000000
     00000000      00000000      00000000      00000000
     00000000      00000000      00000000      00000000
     00000000      00000000      00000000      00000028
h    = H[(Ωi || M'), IV)]
     = 9691eb0c    d263a12f    ab7e0e2f    e60ced5f    546c857a
Ωo = K'⊕opad = K' ⊕(0x5c5c...5c)
   = 6da62c3e      980d4fbf      7a25a1    f 0feb4e38
     5c5c5c5c      5c5c5c5c      5c5c5c5c      5c5c5c5c
     5c5c5c5c      5c5c5c5c      5c5c5c5c      5c5c5c5c
     5c5c5c5c      5c5c5c5c      5c5c5c5c      5c5c5c5c
h' = 9691eb0c      d263a12f      ab7e0e2f      e60ced5f
     546c857a      80000000      00000000      00000000
     00000000      00000000      00000000      00000000
     00000000      00000000      00000000      000000a0
```

$\Omega_o \, || \, h'$:

6da62c3e	980d4fbf	7a25a14f	0feb4e38
5c5c5c5c	5c5c5c5c	5c5c5c5c	5c5c5c5c
5c5c5c5c	5c5c5c5c	5c5c5c5c	5c5c5c5c
5c5c5c5c	5c5c5c5c	5c5c5c5c	5c5c5c5c
9691eb0c	d263a12f	ab7e0e2f	e60ced5f
546c857a	80000000	00000000	00000000
00000000	00000000	00000000	00000000
00000000	00000000	00000000	000000a0

HMAC $[\Omega_o \, || \, h']$ = Outer SHA-1

= c19e1236	ae346195	16594259	4c5202b3	4a85c5e	□

Example 10.6. Here's the HMAC-MD5 computation using the RFC method:

```
Data : 0x  2143f501  f014a713  c1059e23  7123fd68
Key  : 0x  31fa7062  c45113e3  2679fd13  53b71264
```

	A	B	C	D
IV	67452301	efcdab89	98badcfe	10325476
$H[(K \oplus \text{ipad})||M]$	4f556d1d	62d021b7	6db31022	00219556
$H[(K \oplus \text{opad})||H[(K \oplus \text{ipad})||M]]$	b1c3841c	73b63dff	1a22d4bd	f468e7b4

HMAC-MD5 = 0x b1c3841c 73b63dff 1a22d4bd f468e7b4   □

10.3.3 HMAC Computation (Alternative Method)

The alternative operation for computation of HMAC-MD5 or HMAC-SHA-1 is based on the following expression:

```
HMAC = H[H[M, (IV)ᵢ], (IV)ₒ]
(IV)ᵢ= f[(K' ⊕ ipad), IV]
(IV)ₒ = f[(K' ⊕ opad), IV]
   K' = K||(0x00...0) (512 bits)
       ipad = 00110110(0x36) repeated 64 times (512 bits)
       opad = 01011100 (0x5c) repeated 64 times (512 bits)
ipad: inner padding, opad: outer padding
```

We explain the steps of the operation procedure in words as follows:

1. Append zeros to K to create a b-bit string K', where $b = 512$ bits.
2. XOR K' (padding with zero) with *ipad* to produce the b-bit block.
3. Apply the compression function $f(K' \oplus ipad, IV)$ to produce $(IV)_I = 160$ bits for SHA-1.
4. Compute the hash code h with $(IV)_i$ and M_i.

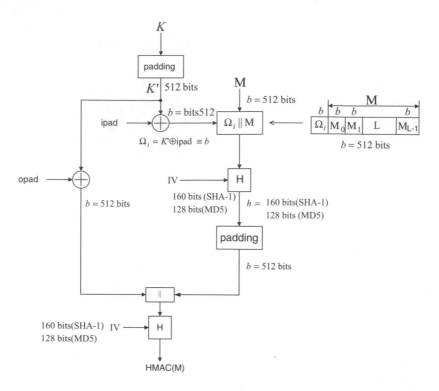

Figure 10.4   Alternative operation of HMAC computation using MD5 or SHA-1 (message length computation is based on M only.)

5. Raise the hash value computed from step 4 to a b-bit string.
6. XOR K' (padded with zeros) with *opad* to produce the b-bit block.
7. Apply the compression function $f(K' \oplus opad, \text{IV})$ to produce $(\text{IV})_o = 160$ bits for SHA-1.
8. Compute the HMAC with $(\text{IV})_o$ and the raised hash value resulting from step 5.

Figure 10.4 shows the alternative scheme based on the above steps.

Example 10.7. Consider the HMAC computation by the alternative method. Assume that the message (M), the key (K), and the initialization vector (IV) are given as follows:

```
 M : 0x 1a7fd53b4c
 K : 0x 31fa7062c45113e32679fd1353b71264
IV : A = 0x67452301, B = 0xefcdab89, C = 0x98badcfe,
     D = 0x10325476, E = 0xc3d2e1f0
```

Referring to Figure 10.4, the HMAC-SHA-1 calculation is carried out with the following steps:

```
K' = K||(0x00...00) (512 bits)
   = 31fa7062     c45113e3     2679fd13     53b71264
     00000000     00000000     00000000     00000000
     00000000     00000000     00000000     00000000
     00000000     00000000     00000000     00000000
```

Ω_i = K' $\oplus$ ipad = K' $\oplus$ (0x363636)

```
   = 07cc4654     f26725d5     104fcb25     65812452
     36363636     36363636     36363636     36363636
     36363636     36363636     36363636     36363636
     36363636     36363636     36363636     36363636
```

$(IV)_i$ = f(Ω_i, IV)

```
   = c6edf676     ef938cee     84dd1b00     5b3b8996     cb172ad4
M' = 1a7fd53b     4c800000     00000000     00000000
     00000000     00000000     00000000     00000000
     00000000     00000000     00000000     00000000
     00000000     00000000     00000000     00000028
```

```
 h = H(M', (IV)_i) = Inner SHA-1
   = 613f6cbd     b336740e     8af4b185     367b1773     d260afce
```

Ω_o = K' $\oplus$ opad = K' $\oplus$ (0x5c5c...5c)

```
   = 6da62c3e     980d4fbf     7a25a14f     0feb4e38
     5c5c5c5c     5c5c5c5c     5c5c5c5c     5c5c5c5c
     5c5c5c5c     5c5c5c5c      c5c5c5c     5c5c5c5c
     5c5c5c5c     5c5c5c5c     5c5c5c5c     5c5c5c5c
```

$(IV)_o$ = f(Ω_o, IV)

```
   = a46e7eba   64c80ca4   c42317b3   dd2b4f1e   81c21ab0
Outer SHA-1
   = af625840     ed120ccd     ba408de3     b259a95b     d4d98eda  □
```

Example 10.8. Here's the HMAC-MD5 computation using the alternative method:

```
Data : 0x 2143f501   f014a713   c1059e23   7123fd68
Key  : 0x 31fa7062   c45113e3   2679fd13   53b71264
             A          B          C          D
IV      67452301   efcdab89   98badcfe   10325476
f[(K⊕ipad),IV]=(IV)_i 13fbaf34  034879ab  35e73505  526a8d28
H[M, (IV)_i]     90c6d9b0   0f281bc8   94d04b33   7f0f4265
f[(K⊕opad),IV]=(IV)_o 5f8647d7  fa8e9afa  bffa4989  3cd471d1
H[H[M, (IV)_i], (IV)_o] 2c47cd5b  68830268  7d255059  45c7bef0
HMAC-MD5=0x    2c47cd5b     68830268     7d255059     45c7bef0     □
```

The HMAC is a cryptographic checksum with the highest degree of security against attacks. HMACs are used to exchange information between two parties, where both have knowledge of the secret key. A digital signature does not require any secret key to be verified.

10.4 Data Expansion Function

The Pseudorandom Function (PRF) can be used to expand data into blocks for the purposes of key generation or validation. The PRF takes relatively small values such as a secret, a seed, and identifying label as input and generates an output of arbitrary longer blocks of data.

The data expansion function, `P_hash(secret, data)`, uses a single hash function to expand a secret and seed into an arbitrary quantity of output. The data expansion function is defined as follows:

```
P_hash (secret, seed) =     HMAC_hash (secret, A(1)‖ seed)‖
                            HMAC_hash (secret, A(2)‖ seed)‖
                            HMAC_hash (secret, A(3)‖ seed)‖
```

where A() is defined as A(0) = seed; A(i) = `HMAC_hash` (secret, A(i − 1)); and ‖ indicates concatenation.

Applying A(i), i = 0, 1, 2, . . . , to `P_hash`, the resulted sketch can be depicted as shown in Figure 10.5. As you can see, `P_hash` is iterated as many times as necessary to produce the required quantity of data. Thus the data expansion function makes use of the HMAC algorithm with either MD5 or SHA-1 as the underlying hash function. As an example, consider SHA-1 whose value is 20 bytes (160 bits). If P_SHA-1 is used to create 64 bytes (512 bits) of data, it will have to be iterated four times up to A(4), creating $20 \times 4 = 80$ bytes (640 bits) of output data. Hence, the last 16 bytes (128 bits) of the final iteration A(4) must be discarded, leaving $(80 - 16) = 64$ bytes of output data. On the other hand, MD5 produces 16 bytes (128 bits). In order to generate a 80-byte output, P_MD5 should be exactly iterated through A(5), whereas P_SHA-1 will only iterate through A(4) as described above. In fact, alignment to a shared 64-byte output will be required to discard the last 16 bytes from both P_SHA-1 and P_MD5.

The PRF for the Transport Layer Security (TLS) protocol is created by splitting the secret into two halves (S1 and S2) and using one half to generate data with P_MD5 and the other half to generate data with P_SHA-1. These two results are then XORed to produce the output. S1 is taken from the first half of the secret and S2 from the second half. Their length is respectively created by rounding up the length of the overall secret divided by two. Thus, if the original secret is an *odd* number of bytes long, the last bytes of S1 will be the same as the first byte of S2.

```
L_S = length in bytes of secret
L_S1 = L_S2 = ceil (L_S/2)
```

The PRF is then defined as the result of mixing the two pseudorandom streams by XORing them together. PRF is defined as

```
PRF(secret, label, seed) = P_MD5(S1, label‖ seed) ⊕ P_SHA-1(S2,
label‖ seed)
```

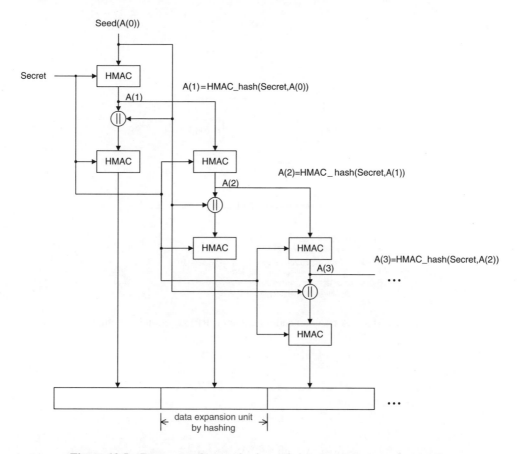

Figure 10.5 Data expansion mechanism using P_hash(secret, seed)

The label is an ASCII string. Figure 10.6 illustrates the PRF generation scheme to expand secrets into blocks of data.

Example 10.9. Refer to Figure 10.6. Suppose the following parameters are given as below:

```
seed = 0x 80 af 12 5c 7e 36 f3 21
label = rocky mountains
= 0x 72 6f 63 6b 79 20 6d 6f 75 6e 74 61 69 6e 73
secret = 0x 35 79 af 12 c4
Then label ‖ seed = 0x 72 6f 63 6b 79 20 6d 6f 75 6e 74 61 69 6e 73
          ‖ 0x 80 af 12 5c 7e 36 f3 21
= A(0)
S1 = 0x 35 79 af for P-MD5, S2 = 0x af 12 c4 for P-SHA-1
Data expansion by P-MD5
A(1) = HMAC_MD5(S1, A(0))
=0x d0 de 36 53 79 78 04 a0 21 b8 6f f8 29 60 d5 f7
HMAC_MD5(S1, A(1) ‖ A(0))
```

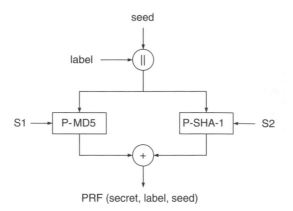

S1 : First half of the secret
S2 : Second half of the secret

P-MD5   : Data expansion function to expand a secret
           S1 and (seed||secret) using MD5
P-SHA-1 : Data expansion function to expand a secret
           S2 and (seed||secret) using SHA-1

Figure 10.6   A pseudorandom function (PRF) generation scheme

```
= 32 fd b3 70 eb 36 11 70 a4 3b 50 a9 fb ea 2a ec
A(2) = HMAC_MD5(S1, A(1))
= 8c ce 5b 50 02 af 75 91 e7 20 cd 86 d9 3e 67 9d
HMAC_MD5(S1, A(2) || A(0))
= 1f a8 4c af 5d e1 20 01 ea b0 38 6a a5 76 f9 8e
A(3) = HMAC_MD5(S1, A(2))
= 45 48 5d 00 4e 64 07 45 eb 2c 18 60 7c e6 fa 1f
HMAC_MD5(S1, A(3) || A(0))
= f0 23 29 d9 5e 89 4b 70 cc 45 f8 aa 1f 58 8e 55
A(4) = HMAC_MD5(S1, A(3))
= 87 39 c6 d3 7a b f8 e3 29 79 3a ae 63 24 6a ff
HMAC_MD5(S1, A(4) || A(0))
= 2e 0c 27 26 d0 b4 78 85 09 a2 69 1c 1b 1b d7 8d
A(5) = HMAC_MD5(S1, A(4))
= 3a 2c aa d8 b3 ec 2e 5d 40 1c 39 bd 3e 48 1a d9
HMAC_MD5(S1, A(5) || A(0))
= 92 f2 63 5d 88 3a dd bf 8d ec e1 cf 0c 5c 8f 4c
where S1 = 0x 35 79 af = first half of the secret, and A(0) = label || seed.
Thus, P-MD5 =
    32  fd  b3  70  eb  36  11  70  a4  3b  50  a9  fb  ea  2a  ec
    1f  a8  4c  af  5d  e1  20  01  ea  b0  38  6a  a5  76  f9  8e
    f0  23  29  d9  5e  89  4b  70  cc  45  f8  aa  1f  58  8e  55
    2e  0c  27  26  d0  b4  78  85  09  a2  69  1c  1b  1b  d7  8d
    92  f2  63  5d  88  3a  dd  bf  8d  ec  e1  cf  0c  5c  8f  4c
(80 bytes)
```

Data expansion by P-SHA-1
A(1) = HMAC_SHA1(S2, A(0))
= aa ea 46 1b a6 ad 43 34 51 f8 c6 ef 70 dd f4 60 ca b9 40 2f
HMAC_SHA1(S2, A(2) ‖ A(0))
= d0 8a d5 07 e0 b8 30 78 70 d9 c8 bb dd ba f5 a3 d0 77 49 e8
A(2) = HMAC_SHA1(S2, A(1))
= 33 fd 23 41 01 ce 06 f8 c0 2b b3 e6 54 21 1c f4 6c 88 ab da
HMAC_SHA1(S2, A(3) ‖ A(0))
= 64 b5 cc 3f 79 31 5b 5d e6 e4 4f eb 98 a8 bf 3f 97 13 38 e1
A(3) = HMAC_SHA1(S2, A(2))
= 86 1f a3 a5 37 58 41 71 f1 9f a5 f3 48 2e 5d 84 7c a8 b6 52
HMAC_SHA1(S2, A(4) ‖ A(0))
= 03 26 11 02 ce 69 74 4a 21 f4 76 55 13 af 77 80 2d fb 2f 36
A(4) = HMAC_SHA1(S2, A(3))
= 9c 4d 01 3a 8c 48 54 42 68 07 4d f1 f0 a9 78 c3 6f ab d8 b4
HMAC_SHA1(S2, A(5) ‖ A(0))
= 48 56 04 b5 b4 5f 9b d8 c7 2f 28 f6 9e 1d 8a c4 72 9a b9 32
where S2 = 0x af 12 c4 = second half of the secret, and A(0) = label ‖
seed.
Thus, P_SHA1=

d0	8a	d5	07	e0	b8	30	78	70	d9	c8	bb	dd	ba	f5	a3
d0	77	49	e8	64	b5	cc	3f	79	31	5b	5d	e6	e4	4f	eb
98	a8	bf	3f	97	13	38	e1	03	26	11	02	ce	69	74	4a
21	f4	76	55	13	af	77	80	2d	fb	2f	36	48	56	04	b5
b4	5f	9b	d8	c7	2f	28	f6	9e	1d	8a	c4	72	9a	b9	32

(80 bytes)
Finally, P-MD5 $\oplus$ P-SHA-1 =

e2	77	66	77	0b	8e	21	08	d4	e2	98	12	26	50	df	4f
cf	df	05	47	39	54	ec	3e	93	81	63	37	43	92	b6	65
68	8b	96	e6	c9	9a	73	91	cf	63	e9	a8	d1	31	fa	1f
0f	f8	51	73	c3	1b	0f	05	24	59	46	2a	53	4d	d3	38
26	ad	f8	85	4f	15	f5	49	13	f1	6b	0b	7e	c6	36	7e

(80 bytes)   □

This is the end of Example 10.9.

Bibliography

1 3GPP TR 25.848 (V4.0.0). *Physical layer aspects of UTRA High Speed Downlink Packet Access.*
2 3GPP TR 25.855 (V5.0.0). *High Speed Downlink Packet Access (HSDPA): Overall UTRAN description.*
3 3GPP TR 25.869 (V1.2.0). *Tx diversity solutions for multiple antennas.*
4 3GPP TR 25.876 (V7.0.0) *Multiple Input Multiple Output in UTRA (Release 7).*
5 3GPP TR 25.892 (V6.0.0). *Feasibility Study for Orthogonal Frequency Division Multiplexing (OFDM) for UTRAN Enhancement.*
6 3GPP TR 25.896 (V6.0.0). *Feasibility Study for enhanced uplink for UTRA FDD.*
7 3GPP TR 25.922. *Radio resource management strategies.*
8 3GPP TR 25.944. *Channel coding and multiplexing examples.*
9 3GPP TR 25.990. *Vocabulary.*
10 3GPP TR 25.996 (V6.0.0). *Spatial channel model for Multiple Input Multiple Output (MIMO) simulations.*
11 3GPP TR 31.900 V5.1.0. *SIM/USIM internal and external interworking aspect (Release 5).*
12 3GPP TR 33.901 V4.0.0. *3G Security: Criteria for cryptographic algorithm design process (Release 4).*
13 3GPP TR 55.919 V6.1.0. *Specification of the A5/3 encryption algorithms for GSM and ECSD, and the GEA3 encryption algorithm for GPRS; Document 4: Design and evaluation report (Release 6).*
14 3GPP TS 23.101. *General UMTS architecture.*
15 3GPP TS 23.110. *UMTS access stratum services and functions.*
16 3GPP TS 25.101. *UE Radio transmission and reception (FDD).*
17 3GPP TS 25.104. *BTS radio transmission and reception (FDD).*
18 3GPP TS 25.105. *BTS radio transmission and reception (TDD).*
19 3GPP TS 25.201. *Physical layer – General description.*
20 3GPP TS 25.211. *Physical channels and mapping of transport channels onto physical channels (FDD).*
21 3GPP TS 25.212. *Multiplexing and channel coding (FDD).*
22 3GPP TS 25.213. *Spreading and modulation (FDD).*
23 3GPP TS 25.214. *Physical layer procedures (FDD).*
24 3GPP TS 25.221. *Physical channels and mapping of transport channels onto physical channels (TDD).*
25 3GPP TS 25.222. *Multiplexing and channel coding (TDD).*
26 3GPP TS 25.223. *Spreading and modulation.*
27 3GPP TS 25.224. *Physical layer procedures (TDD).*

28 3GPP TS 25.225. *Physical layer – Measurements (TDD)*.
29 3GPP TS 25.301. *Radio interface protocol architecture*.
30 3GPP TS 25.302. *Services provided by the physical layer*.
31 3GPP TS 25.306. *UE radio access capabilities*.
32 3GPP TS 25.308 (V5.2.0). *UTRA High Speed Downlink Packet Access (HSDPA): Overall description; Stage 2*.
33 3GPP TS 25.309. *FDD enhanced uplink. Overall description*.
34 3GPP TS 25.321. *Medium Access Control (MAC) protocol specification*.
35 3GPP TS 25.331. *Radio Resource Control (RRC)*.
36 3GPP TS 25.401. *UTRAN overall description*.
37 3GPP TS 25.402. *Synchronization in UTRAN, Stage 2*.
38 3GPP TS 25.414. *UTRAN Iu interface data transport and transport signaling*.
39 3GPP TS 25.420. *UTRAN Iur interface general aspects and principles*.
40 3GPP TS 25.421. *UTRAN Iur interface. Layer 1*.
41 3GPP TS 25.422. *UTRAN Iur interface. Signaling transport*.
42 3GPP TS 25.423. *UTRAN Iur interface. RNSAP Signaling*.
43 3GPP TS 25.424. *UTRAN Iur interface. Data transport & transport signaling*.
44 3GPP TS 25.425. *UTRAN Iur interface. User plane protocols for common transport channel data streams*.
45 3GPP TS 25.426. *UTRAN Iur & Iub Interface. Data transport & transport signaling for DCH data streams*.
46 3GPP TS 25.427. *UTRAN Iub/Iur interface. User plane protocol for DCH data streams*.
47 3GPP TS 25.430. *UTRAN Iub Interface. General aspects and principles*.
48 3GPP TS 25.431. *UTRAN Iub interface. Layer 1*.
49 3GPP TS 25.432. *UTRAN Iub interface. Signaling transport*.
50 3GPP TS 25.433. *UTRAN Iub Interface. NBAP signaling*.
51 3GPP TS 25.434. *UTRAN Iub interface. Data transport & transport signaling for common transport channel data streams*.
52 3GPP TS 25.435. *UTRAN Iub interface. User plane protocols for common transport channel data streams*.
53 3GPP TS 25.442. *UTRAN implementation specific O&M transport*.
54 3GPP TS 25.950 (Release 4, Version 4). *UTRA High Speed Downlink Packet Access*.
55 3GPP TS 33.102 V7.1.0. *3G Security architecture (Release 5)*.
56 3GPP TS 33.105 V4.1.0. *3G Security: Cryptographic algorithm requirements (Release 4)*.
57 3GPP TS 35.202 (V7.0.0, Release 7). *Universal Mobile Telecommunications System (UMTS): Specification of the 3GPP confidentiality and integrity algorithms; Document 2. Kasumi specification*.
58 3GPP TS 55.205 V6.0.0. *GSM authentication and key generation function A3 and A8 (Release 6)*.
59 3GPP TS 55.216 V6.1.0. *Specification of the A5/3 encryption algorithm for GSM and ECSD, and the GEA3 encryption algorithm for GPRS; Document 1: A5/3 and GEA3 specification (Release 6)*.
60 3GPP TS 55.217 (V6.1.0). *KGCORE*.
61 3GPP2 C.S0001-D v2.0, *Introduction to cdma2000 standards for spread spectrum systems*, September 2005.
62 3GPP2 C.S0002-D v2.0, *Physical layer standard for cdma2000 spread spectrum systems*, September 2005.
63 3GPP2 C.S0003-D v2.0, *Medium Access Control (MAC) standard for cdma2000 spread spectrum systems, September 2005*.
64 3GPP2 C.S0004-D v2.0, *Signaling Link Access Control (LAC) standard for cdma2000 spread spectrum systems, September 2005*.

65 3GPP2 C.S0005-D v2.0, *Upper layer (Layer 3) signaling standard for 1 cdma2000 spread spectrum systems, September 2005.*

66 3GPP2 C.S0007-0 v2.0, *Direct spread specification for spread spectrum systems on ANSI-41 (DS-41) – Upper layers interface.*

67 3GPP2 C.S0009-0 v1.0, *Speech service option standard for wideband spread spectrum systems, December 1999.*

68 3GPP2 C.S0012-0 v1.0, *Recommended minimum performance standard for digital cellular wideband spread spectrum speech service option 1, December 1999.*

69 3GPP2 C.S0015-A0, *Short Message Service for spread spectrum systems, December 1999.*

70 3GPP2 C.S0016-AC v1.0, *Over-the-air service provisioning of mobile stations in spread spectrum systems, November 2004.*

71 3GPP2 C.S0017-0 v5.0, *Data service options for spread spectrum systems, February 2003.*

72 3GPP2 C.S0020-0-1A v1.0, *High rate speech service option 17 for wideband spread spectrum communication systems, May 2004.*

73 3GPP2 C.S0024-A v2.0, *cdma2000 high rate packet data air interface specification, August 2005.*

74 3GPP2 C.S0072-0 v1.0, *Mobile Equipment Identifier (MEID) support for cdma2000 spread spectrum systems, August 2005.*

75 Andrews, J.G., Ghosh, A. and Muhamed, R., *Fundamentals of WiMAX: Understanding Broadband Wireless Networking*, Englewood Cliffs, NJ, Prentice-Hall, 2007.

76 ANSI/TIA/EIA-95-B-99, *Mobile station–base station compatibility standard for wideband spread spectrum cellular systems, February, 1999.*

77 Bluetooth SIG, specification of the Bluetooth System (Vol. 1, Version 1.1), February 2001.

78 Cheng, P. and Glenn, R. Test Cases for HMAC-MD5 and HMAC-SHA-1, RFC 2202, September 1997.

79 Cimini, L. Jr.,"Analysis and simulation of a digital mobile channel using orthogonal frequency division multiplexing". *IEEE Transactions on Communications*, **33**(7), 1985, 665–675.

80 Cingular Wireless, Orange, 3, Telecom Italia, T-Mobile, Vodafone Group, Reference Scenario for the Selection of the UTRA MIMO Scheme, R1-051626, R1 reflector, December 2005.

81 Daemen, J. and Rijmen, V.,"AES Proposal. Rijndael, AES Algorithm Submission", September 1999.

82 Dannan, A. and Kaiser, S.,"Transmit/receive – Antenna diversity techniques for OFDM systems". *European Transaction on Telecommunications*, **13**(5), 2002, 531–538.

83 Douillard, C. and Berrou, C.,"Turbo codes with rate – m/(m+1) constituent convolutional codes", *IEEE Transactions on Communications*, **53**(10), 2005, 1630–1638.

84 EIA/IS-19-B, *Recommended minimum standards for 800-MHz cellular subscriber units*, May 1988.

85 Ericsson, Qualcomm Europe, Transmit Diversity Operation in MIMO Mode, R1-071167, TS25.211 CR 238.

86 FIPS Publication ZZZ, *Announcing the Advanced Encryption Standard (AES)*, US DoC/NIST, 2001.

87 GSM 02.60. *Digital cellular telecommunications system (Phase 2 +): General Packet Radio Service (GPRS); Service Description; State 1.*

88 GSM 03.60. *Digital cellular telecommunications system (Phase 2 +): General Packet Radio Service (GPRS); Service Description; State 2.*

89 GSM 04.64. *Digital cellular telecommunications system (Phase 2 +): General Packet Radio Service (GPRS); Mobile Station-Serving GPRS Support Node (MS_SGSN) Logical Link Control (LLC) layer specification.*

90 IEEE Standard P802.16 Rev2/DL, Part 16. *Air interface for fixed and mobile broadband wireless access system*, October 2007.

91 Johnson, D., Menezes, A. and Vanstone, S., *The Elliptic Curve Digital Signature Algorithm*, Berlin and Heidelberg, Springer-Verlag, 2001, pp. 36–63.

92 Koblitz, N., "Elliptic Curves Cryptosystems". *Mathematics of Computing*, **48** (177), 1987, 203–209.

93 Koblitz, N., Constructing Elliptic Curves Cryptosystems in Characteristic 2, *Advances in Cryptology–Crypt'91*, Berlin and Heidelberg, Springer-Verlag, 1991, 156–167.

94 Krawczyk, H., Bellare, M. and Canetti, R. (February 1997) HMAC. Keyed-Hashing for Message Authentication, RFC 2104.

95 Madson, C. and Glenn, R. (November 1998) The Use of HMAC-MD5-96 within ESP and AH, RFC 2403.

96 Madson, C. and Glenn, R. (November 1998) The Use of HMAC-SHA-1-96 within ESP and AH, RFC 2404.

97 Menezes, A.J. and Vanstone, S.A. 1993. Elliptic curve cryptosystems and their implementation. *Journal of Cryptology*, **6** (4), 209–224.

98 Motorola, (May 19–22, 2003) *A comparison of ideal relative data throughput of MIMO and Release 5, R1-030556*, Marne la Vallee, France.

99 Philips, Ericsson, Qualcomm, Nokia, Siemens, Motorola, *Coding of HS-SCCH to support FDD MIMO*, R1-071165, TS25.212 CR 241r4.

100 Qualcomm Europe, *Introduction of MIMO*, R1-071215, TS25.201 CR 032.

101 Qualcomm Europe, Philips, Ericsson, Nokia, *Definition of MIMO operation on HS-PDSCH, preferred precoding and CQI reporting procedures, modified CQI tables*, R1-071229, TS25.214 CR 430r10.

102 Qualcomm Europe, Nokia, *Introduction of MIMO in NBAP*, R3-070362, TS25.433 CR 1342.

103 R2A010002. *Support of standalone carrier for DSCH*, Nortel.

104 R2A010010. *HSDPA radio interface protocol architecture*, Ericsson, Motorola.

105 R2A010015. *HSDPA signaling requirements*, Motorola.

106 R2A010016. *Dual-channel stop-and-wait HARQ*, Motorola.

107 R2A010017. *Fast cell selection and handovers in HSDPA*, Motorola.

108 R2A010021. *ARQ technique for HSDPA*, Lucent.

109 RFC 2104 HMAC. Keyed-Hashing for Message Authentication.

110 Sin, Susan, COMP128, http://www.comsec.uwaterloo.ca/~ssjsin/.

111 TCR-TR 030. Security Techniques Advisory Group (STAG): A Guide to Specifying Requirements for Cryptographic Algorithms.

112 TIA/EIA-95-B, *Mobile station–base station compatibility standard for dual-mode spread spectrum cellular systems*, 1996.

113 TSB51, Cellular Radio-Telecommunications Intersystem Operations. Authentication, Signaling Message Encryption and Voice Privacy, May 1993.

114 TSB64, IS-41-B Support for Dual-Mode Wideband Spread Spectrum Mobile Stations, January 1994.

115 TSGR1#12(00) 0556, Feasibility Study of Advanced techniques for High Speed Downlink Packet Access, Motorola.

116 WiMAX Forum, "A Comparative Analysis of Mobile WiMAX Deployment Alternatives in the Access Network", white paper, May 2007.

117 WiMAX Forum, "Network Architecture – Stage 3: Detailed Protocols and Procedures", Release 1.1.0, July 2007.

Index

1xEV-DV, 120, 124, 127
2.5G GPRS, 95
2.5G technology, 95
3G Americas and 3G.IP, 121
3G CDMA2000 1x networks, 122
3G FOMA service, 125
3G FOMA technology, 95
3G mobile radio technologies, 122, 124
3G partnership project (3GPP), 119–121, 123,
    124, 126
3G technologies, 119, 120, 125
3G WCDMA networks, 120
3GPP f8 confidentiality algorithm
4G environment, 126
8-ary Walsh function, 259, 260, 261, 265
8-PSK, 332
10 ms F-ACKCH frame, 321
16-QAM symbols, 332

AAL5 pre-defined virtual connections
    (PVCs), 142
access attempt, 254, 336
access channel, 28, 29, 58, 60–64, 199, 200, 217,
    222, 234, 235, 239–243, 251, 252, 256, 257,
    291, 292, 294
access channel baseband filtering, 64
access channel direct sequence spreading, 64
access channel frame(s), 60, 291
access channel long code masks, 242
access channel MAC layer capsule, 243
access channel MAC layer packet, 235, 239
access channel MAC protocol, 199, 200,
    239, 240
access channel message capsule, 60, 61
access channel modulation, 64

access channel physical layer packet, 251, 256
access channel preamble, 60, 291
access channel probe, 240
access channel quadrature spreading, 64
access channel transmission, 291
access cycle duration, 240
access cycle number, 242
access grant channel (AGCH), 9
access link control application part
    (ALCAP), 138, 165
access probe, 254, 291, 337
access probe sequences, 240, 241, 337
access sub-attempts, 336
access-terminal-directed messages, 283
access terminal identifier (ATI), 198
ACK channel, 254, 255, 256, 260, 265, 266,
    267, 273
ACK channel gain, 254, 266
ACK detection threshold, 174
activate command, 254, 266
active state, 220
adaptive modulation and coding (AMC) 167, 184
    scheme, 184
    principle, 184
adaptive multirate code (AMR), 140
Address Management, 198, 199, 208–213
Address Management Protocol (AMP), 210, 213
    setup state, 209, 218
address mapping, 100
address translation, 100
AddRoundKey(), 349, 351, 354, 355
AddRoundKey() transformation, 352, 355
admission control, 162
advanced encryption standard (AES), 128,
    341, 342